Wind Energy – The Facts

A guide to the technology, economics and future of wind power

European Wind Energy Association (EWEA)

earthscan
from Routledge

First published by Earthscan in the UK and USA in 2009

For a full list of publications please contact:
Earthscan
2 Park Square, Milton Park, Abingdon, Oxfordshire OX14 4RN
711 Third Avenue, New York, NY 10017

First issued in paperback 2015

Earthscan is an imprint of the Taylor & Francis Group, an informa business

ISBN 13: 978-1-138-88126-6 (pbk)
ISBN 13: 978-1-8440-7710-6 (hbk)

Typeset by Techset Composition Ltd
Index compiled by Indexing Specialists (UK) Ltd
Cover design by Susanne Harris

A catalogue record for this book is available from the British Library

Library of Congress Cataloging-in-Publication Data

Wind energy – the facts : a guide to the technology, economics and future of wind power / European Wind Energy Association.
    p. cm.
  Includes bibliographical references and index.
  ISBN 978-1-84407-710-6 (hardback)
  1. Wind power industry–European Union countries. 2. Wind power–European Union countries. I. European Wind Energy Association.
HD9502.5.W553W57 2009
  333.9'2–dc22
                        2008046551

TABLE OF CONTENTS

LIST OF ACRONYMS AND ABBREVIATIONS

AA	Appropriate assessment		FACT	Flexible AC transmission system device
AAUs	Assigned amount units		FEE	French Wind Energy Association
AC	Alternating current		FIT	Feed-in tariff
ACER	Agency for the Coordination of Energy Regulators in Europe		FRT	Fault ride through
AEE	Spanish Wind Energy Association		GHG	Greenhouse gas
AGC	Automatic generation control		GIS	Green Investment Scheme
APPA	Association of Renewable Energy Producers		GO	Guarantee of origin
			GW	Gigawatt
			GWh	Gigawatt hour
BDIG	Brushless doubly fed induction generator		HV	High voltage
BOP	Balance of plant		HVAC	High voltage AC
			HVDC	High voltage DC
CAES	Compressed-air energy storage			
CCGT	Combined-cycle gas turbine		IC	Impact category
CDM	Clean Development Mechanism		IGBT	Insulated gate bipolar transistor
CDM	Coastal discontinuity model		ISET	Institut fur Solare Energietechnik (Kassel, Germany)
CERs	Certified emissions reductions			
CFD	Computational fluid dynamics		ISO	Independent system operator
CHP	Combined heat and power			
			kV	Kilovolt
DEA	Danish Energy Agency			
DFIG	Doubly fed induction generator		LCA	Life-cycle analysis
DGS	Distributed generation systems		LCI	Life-cycle inventory
DSCR	Debt service cover ratio		LFN	Low frequency noise
DSM	Demand-side management			
DSO	Distribution system operator		MCP	Measure correlate predict
DTI	Department for Trade and Industry (UK)		MNRE	Ministry of New and Renewable Energy (India)
			MV	Medium voltage
ECCP	European Climate Change Programme		MVAR	Megavolt ampere reactive
			MW	Megawatt
EIA	Environmental impact assessments		MWh	Megawatt hour
EIA	Environmental impact investigation			
ENTSO-E	European Network of Transmission System Operators for Electricity		NTC	Net transfer capacity
			NWP	Numerical weather prediction
ERUs	Emission reduction units			
ES	Environmental statement		ORCU	Offshore Renewables Consents Unit (UK)
EWIS	European Wind Integration Study		OTC	Over the counter

PAC	Pumped (hydro) accumulation storage		TCG	Tradeable Green Certificate
PCC	Point of common coupling		TEN-E	Trans-European Energy Networks
PMG	Permanent magnet generator		TSO	Transmission system operator
PPA	Power purchase agreement		TSPs	Total suspended particles
PTC	Production Tax Credit (US)		TW	Terawatt
PWEA	Polish Wind Energy Association		TWh	Terawatt hour
			UCTE	Union for the Coordination of Transmission of Electricity
QUELRO	Quantified emission reduction or limitation obligation			
			VAWT	Vertical-axis wind turbine
RES-E	Electricity from renewable energy sources		VPS	Virtual power station
REZ	Renewable Energy Zones (UK)		VSC	Voltage source converter
RMSE	Root mean square error		VSL	Value of statistical life
ROC	Renewables Obligation Certificate (UK)			
RPM	Regulating power market		WEPP	Wind energy power plant
RPS	Renewable portfolio standard		WFCM	Wind farm cluster management
RTD	Research and technological development		WFDT	Wind farm design tool
			WHS	Wind home system
SACs	Special Areas of Conservation (EU)		WRIG	Wound rotor induction generator
SCADA	Supervisory control and data acquisition			
SCI	Site of Community Importance (EU)		UNFCCC	United Nations Framework Convention on Climate Change
SEA	Strategic environmental assessment			
SERCs	State Electricity Regulatory Commissions (India)		YOLL	Years of life lost
SHS	Solar home system			
SPAs	Special Protection Areas (EU)		ZTV	Zone of theoretical visibility
SPV	Special purpose vehicle		ZVI	Zone of visual influence
SWT	Small wind turbine			
SVC	Static Var Compensator			

ACKNOWLEDGEMENTS

The European Wind Energy Association would very much like to thank all the authors that contributed to this new edition of *Wind Energy – The Facts*.

Part I

Paul Gardner, Andrew Garrad, Lars Falbe Hansen, Peter Jamieson, Colin Morgan, Fatma Murray and Andrew Tindal of Garrad Hassan and Partners, UK (www.garradhassan.com); José Ignacio Cruz and Luis Arribas of CIEMAT, Spain (www.ciemat.es); Nicholas Fichaux of the European Wind Energy Association (EWEA) (www.ewea.org).

Part II

Frans Van Hulle of EWEA and Paul Gardner of Garrad Hassan and Partners.

Part III

Poul Erik Morthorst of Risø DTU National Laboratory, Technical University of Denmark (www.risoe.dk); Hans Auer of the Energy Economics Group, University of Vienna; Andrew Garrad of Garrad Hassan and Partners; Isabel Blanco of UAH, Spain (www.uah.es).

Part IV

Angelika Pullen of the Global Wind Energy Council (GWEC) (www.gwec.net); Keith Hays of Emerging Energy Research (www.emerging-energy.com); Gesine Knolle of EWEA.

Part V

Carmen Lago, Ana Prades, Yolanda Lechón and Christian Oltra of CIEMAT, Spain (www.ciemat.es); Angelika Pullen of GWEC; Hans Auer of the Energy Economics Group, University of Vienna.

Part VI

Arthouros Zervos of the National Technical University of Athens, Greece (www.ntua.gr); Christian Kjaer of EWEA.

Co-ordinated by

Zoé Wildiers, Gesine Knolle and Dorina Iuga, EWEA

Edited by

Christian Kjaer, Bruce Douglas, Raffaella Bianchin and Elke Zander, EWEA

Language Editors

Rachel Davies, Sarah Clifford and Chris Rose.

For invaluable data and insightful additions throughout the text, we would like to thank the organisations listed below and many other EWEA members who have given their support and input.

Association of Renewable Energy Producers
 (APPA, Spain)
Austrian Wind Energy Association
British Wind Energy Association
Bulgarian Wind Energy Association
Cyprus Institute of Energy
Czech Society for Wind Energy
Danish Wind Industry Association
ECN, The Netherlands
Ente Per Le Nuove Tecnologie, l'Energia e
 l'Ambiente-Centro Ricerche (ENEA, Italy)
Estonian Wind Power Association
Finnish Wind Energy Association
French Agency for Environment and Energy
 Management (ADEME, France)
German Wind Energy Association
GE Wind Energy, Germany

Hellenic Wind Energy Association

Horvath Engineering, Hungary

Instituto de Engenharia Mecanica e Gestao Industrial
  (INEGI, Portugal)

Irish Wind Energy Association

KEMA Power Generation and Sustainables,
  The Netherlands

Latvian Wind Energy Association

Romanian Wind Energy Association

Slovakian Wind Energy Association

Swedish Defence Research Agency

Vestas Wind Systems, Denmark

VIS VENTI Association for Supporting Wind Energy,
  Poland

Finally, the European Wind Energy Association would like to thank the European Commission's Directorate General for Transport and Energy (DG TREN) for the valuable support and input it has given to this project (No EIE/07/230/SI2.466850).

The information in *Wind Energy – The Facts* does not necessarily reflect the formal position of the European Wind Energy Association or the European Commission.

FOREWORD

by

Hans van Steen

Head of Unit, Regulatory Policy and Promotion of Renewable Energy,

Directorate General for Energy and Transport (DG TREN), European Commission

It is a pleasure for me to introduce the new edition of *Wind Energy – The Facts*, produced by the European Wind Energy Association (EWEA) and supported by the European Commission in the framework of the Intelligent Energy Europe programme.

At the European Union (EU) level, the challenges related to energy and climate change are at the very top of the political agenda and, more than ever before, wind energy is proving its potential as an important part of the solution. Through its concrete contribution to clean and secure power generation, wind ensures that an increasing amount of electricity is produced without using fossil fuels, without using precious fresh water for cooling purposes, and without emitting greenhouse gases or harmful air pollutants.

The deployment of wind energy continues to be a success story in the EU. More and more Member States have joined the initial front-runners, and wind continues to be one of the fastest growing forms of electricity generation in Europe. Wind turbine technology continues to improve, larger and more efficient technologies are being deployed, and offshore applications are coming on-stream.

The European Commission is convinced that there is a large untapped potential of renewable energy in Europe. The agreed target of a 20 per cent share of renewable energy in the EU's energy mix in 2020 is thus at the same time both ambitious and achievable.

But meeting the target will not become a reality without strong commitments at all levels, including national governments and the renewables industry itself. Large-scale integration of wind into electricity grids and markets poses significant challenges to the sector – challenges that will require researchers, transmission system operators, energy companies, regulators, policymakers and other stakeholders to work closely together and constructively consider appropriate solutions.

This publication provides an excellent, highly readable and comprehensive overview of the diverse issues of relevance to wind power. Given the growing importance of wind in the European energy sector, it is a useful reference document, not just for the sector itself, but also more widely for policy- and decision-makers.

EWEA FOREWORD

by
Arthouros Zervos
President, European Wind Energy Association

It has been five years since I wrote the foreword to the 2004 edition of *Wind Energy – The Facts*, and in that short time there have been major, positive changes in the European wind power industry. For the most part, the environmental, regulatory, technological, financial and political questions surrounding the industry in 2004 have now been satisfactorily answered.

Cumulative installed wind capacity is perhaps the most relevant proof of this amazing success story. By the end of 2003, the EU-15 had installed more than 28,000 megawatts (MW) of wind turbine capacity. By the end of 2007, the enlarged EU-27 had in excess of 56,000 MW of capacity.

These 56,000 MW met 3.7 per cent of total EU electricity demand, provided power equivalent to the needs of 30 million average European households and avoided 91 million tonnes of carbon dioxide emissions. In addition, there were billions of euros saved on imported fuel costs in 2007, while more than 11 billion was invested in installing wind turbines in Europe.

A deepening political concern has also emerged over the past five years with regard to climate change and energy. Politicians are seeking viable and powerful solutions to the challenges of escalating oil prices, depleting fossil fuel reserves, dependence on foreign energy supplies and the potential ravages of global warming. Now, more than ever, our elected leaders are looking for solutions to these complex and critically important issues.

As a result, the EU has set a binding target of 20 per cent of its energy supply to come from wind and other renewable resources by 2020. To meet this target, more than one-third of European electrical demand will need to come from renewables, and wind power is expected to deliver 12 to 14 per cent (180 GW) of the total demand. Thus wind energy will play a leading role in providing a steady supply of indigenous, green power.

We feel it is the right time to update *Wind Energy – The Facts*, in order to address further changes in this rapidly expanding industry, both in Europe and globally. Both market and turbine sizes have grown immeasurably since 2003, prompting an entirely new set of considerations. With all this additional power being created, issues such as grid access, new and strengthened transmission lines, and system operations have to be dealt with in a fair, efficient and transparent way. The relatively new offshore wind industry has astonishingly rich potential, but needs to be helped through its infancy, and the supply chain bottlenecks created by this rapid growth still need to be overcome.

I hope this latest edition of *Wind Energy – The Facts* helps to provide a pathway to a truly sustainable future. I remain confident that the wind power industry will overcome the challenges it faces and achieve even greater success.

Arthouros Zervos
President, EWEA

EXECUTIVE SUMMARY

EXECUTIVE SUMMARY

Since the last edition of *Wind Energy – The Facts* was published in February 2004, the wind energy sector has grown at an astonishing rate and is now high on the political agenda. With the looming energy crisis, calls are increasing for an immediate and concrete solution to the many energy and climate challenges the world is currently facing – and wind energy offers just this.

In order to facilitate informed choices and policy decisions, a clear and profound understanding of the wind power sector is required. This volume aims to contribute to this knowledge dissemination by providing detailed information on the wind power sector. *Wind Energy – The Facts* provides a comprehensive overview of the essential issues concerning wind power today: wind resource estimation, technology, wind farm design, offshore wind, research and development, grid integration, economics, industry and markets, environmental benefits, and scenarios and targets.

Since 2004, wind energy deployment has risen dramatically. Global installed capacity increased from 40,000 megawatts (MW) at the end of 2003 to 94,000 MW at the end of 2007, at an average annual growth rate of nearly 25 per cent. Europe is the undisputed global leader in wind energy technology. Sixty per cent of the world's capacity was installed in Europe by the end of 2007, and European companies had a global market share of 66 per cent in 2007. Penetration levels in the electricity sector have reached 21 per cent in Denmark and about 7 and 12 per cent in Germany and Spain respectively. Achievements at the regional level are even more impressive: the north German state of Schleswig-Holstein, for example, has over 2500 MW of installed wind capacity, enough to meet 36 per cent of the region's total electricity demand, while in Navarra, Spain, some 70 per cent of consumption is met by wind power.

A huge step forward was taken in March 2007, when EU Heads of State adopted a binding target of 20 per cent of energy to come from renewables by 2020. And in January 2008, the European Commission released a renewables legislation draft, proposing a stable and flexible EU framework, which should ensure a massive expansion of wind energy in Europe. If such positive policy support continues, EWEA projects that wind power will achieve an installed capacity of 80,000 MW in the EU-27 by 2010. This would represent

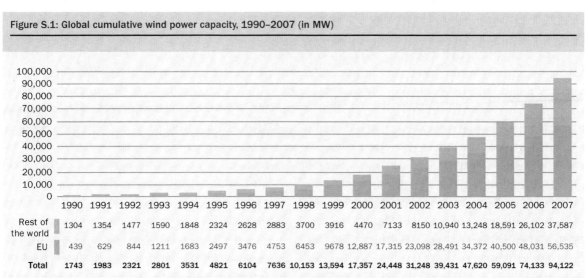

Figure S.1: Global cumulative wind power capacity, 1990–2007 (in MW)

	1990	1991	1992	1993	1994	1995	1996	1997	1998	1999	2000	2001	2002	2003	2004	2005	2006	2007
Rest of the world	1304	1354	1477	1590	1848	2324	2628	2883	3700	3916	4470	7133	8150	10,940	13,248	18,591	26,102	37,587
EU	439	629	844	1211	1683	2497	3476	4753	6453	9678	12,887	17,315	23,098	28,491	34,372	40,500	48,031	56,535
Total	1743	1983	2321	2801	3531	4821	6104	7636	10,153	13,594	17,357	24,448	31,248	39,431	47,620	59,091	74,133	94,122

Source: GWEC/EWEA (2008)

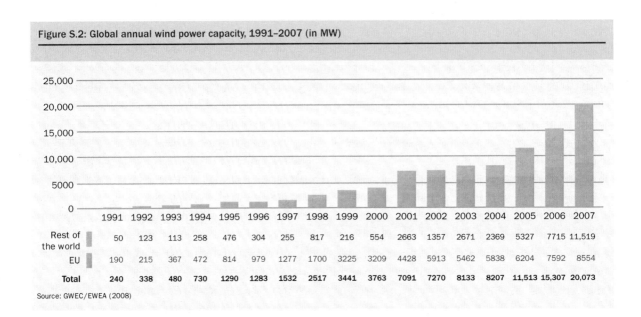

Figure S.2: Global annual wind power capacity, 1991–2007 (in MW)

	1991	1992	1993	1994	1995	1996	1997	1998	1999	2000	2001	2002	2003	2004	2005	2006	2007
Rest of the world	50	123	113	258	476	304	255	817	216	554	2663	1357	2671	2369	5327	7715	11,519
EU	190	215	367	472	814	979	1277	1700	3225	3209	4428	5913	5462	5838	6204	7592	8554
Total	**240**	**338**	**480**	**730**	**1290**	**1283**	**1532**	**2517**	**3441**	**3763**	**7091**	**7270**	**8133**	**8207**	**11,513**	**15,307**	**20,073**

Source: GWEC/EWEA (2008)

an overall contribution to electricity supply of 5 per cent. By 2020, this figure is expected to increase to 12–14 per cent, with wind power providing energy equal to the demand of 107 million average European households.

Part I: Technology

Part I covers all aspects of the technology of the wind industry, which has made rapid advances in all areas. Much has been learned, but there is much still to be discovered, both in fundamental meteorology, aerodynamics and materials science and in highly applied areas such as maintenance strategies, wind farm design and electricity network planning. And there are still untried concepts for turbine design which may be worth serious consideration. This part describes the fundamentals of wind technology, the current status and possible future trends.

WIND RESOURCE ESTIMATION

The methods for carrying out wind resource estimation are well established. Chapter I.1 describes wind resource estimation for large areas, which is carried out in order to establish both the available resource in a region and the best areas within that region. It also covers wind resource and energy production estimation for specific sites. The accuracy of the energy production estimate is of crucial importance both to the owner of the project and to organisations financing it, and the chapter explains the many factors which can affect energy production.

Forecasting is also covered, in Chapter I.2, as this is now an important part of the wind industry. Depending on the structure of the electricity market, the owner of the project or the purchaser of the energy may be able to reap significant financial benefits from the accurate forecasting of wind production. Operators of electricity systems with high wind penetration also need forecasts in order to optimise the operation of their systems.

WIND TURBINE TECHNOLOGY

The rapid technical advances are most apparent in wind turbine technology. Chapter I.3 shows how turbine size, power and complexity have developed

Figure S.3: European Wind Atlas, onshore

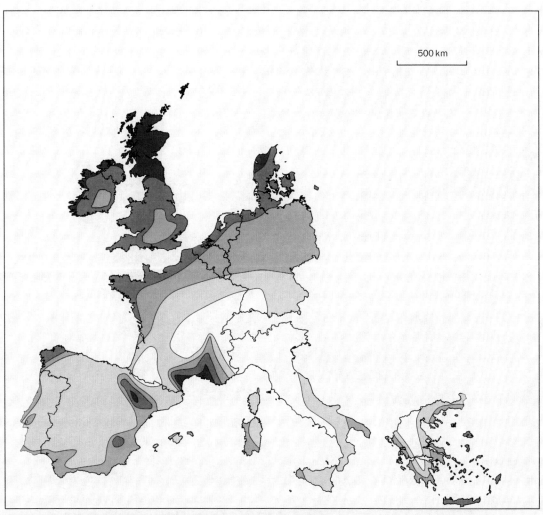

Wind resources at 50 metres above ground level for five different topographic conditions										
	Sheltered terrain		Open terrain		At a sea coast		Open sea		Hills and ridges	
	m/s	W/m^2	m/s	W/m^2	m/s	W/m	m/s	W/m^2	m/s	W/m^2
	>6.0	>250	>7.5	>500	>8.5	>700	>9.0	>800	>11.5	>1800
	5.0–6.0	150–250	6.5–7.5	300–500	7.0–8.5	400–700	8.0–9.0	600–800	10.0–11.5	1200–1800
	4.5–5.0	100–150	5.5–6.5	200–300	6.0–7.0	250–400	7.0–8.0	400–600	8.5–10.0	700–1200
	3.5–4.5	50–100	4.5–5.5	100–200	5.0–6.0	150–250	5.5–7.0	200–400	7.0–8.5	400–700
	<3.5	<50	<4.5	<100	<5.0	<150	<5.5	<200	<7.0	<400

Source: Risø DTU

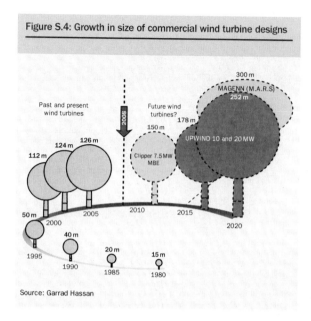

Figure S.4: Growth in size of commercial wind turbine designs

Source: Garrad Hassan

extremely fast, best evidenced by the increase in commercial turbine size by a factor of around 100 in 20 years (Figure S.4). Wind turbines may appear to be simple machines, but there are some fundamental requirements which make this branch of engineering unlike any other:

- The machine has to operate as a power station, unattended, and provide more to the electricity network than simply energy.
- The wind is variable on timescales from seconds to years, which introduces uncertainty into everything from mechanical loads to energy production.
- The technology has to compete on cost of energy against other renewables and against conventional generation.

Chapter I.3 thus discusses the evolution of wind turbine design and explains why three-bladed, upwind, variable-speed, pitch-regulated turbines currently predominate. The principal design drivers are now grid compatibility, cost of energy (which includes reliability), acoustic emissions, visual appearance and suitability for site conditions.

However, there are still many unresolved technical issues. For example, large turbines currently in production include:

- concepts with large-diameter, slow-speed generators;
- concepts with high-speed generators and gearboxes; and
- intermediate arrangements with medium-speed generators and reduced stages of gearing.

Similarly, it is perhaps surprising that the optimum size of a wind turbine for 'standard' onshore wind farms is not yet obvious. The chapter explains some of these technical issues, and concludes by reviewing some radical alternative concepts.

WIND FARM DESIGN

Chapter I.4 describes how wind turbines are grouped into wind farms, the factors affecting siting and how they are built. Wind farm design is a critical area for cost reduction and public acceptability, both onshore and offshore, especially with some now being bigger than large conventional electricity-generating plants.

The arrangement of wind turbines within the wind farm clearly affects not only the energy production, but also the visual appearance and the noise influence on neighbours. This chapter explains how the layout can be optimised to take account of such constraints, using software designed specifically for the wind industry.

The chapter also discusses the important issues in 'balance of plant' design, including civil and electrical works. As the wind industry gains experience in constructing projects in different conditions, the costs and other important issues are becoming clearly understood, and the risks should be no greater than other civil engineering or power station projects of similar size.

OFFSHORE WIND POWER

Chapter I.5 covers offshore wind, and in particular extends the discussion of onshore issues in

Chapters I.2–I.4 to the offshore case. Although this market is currently substantially smaller than the onshore one, it is now a fundamental part of several nations' energy policies, and expectations are high. The offshore wind market is characterised by projects which are significantly larger and more risky than most onshore projects, and it appears likely that different organisations will develop and construct these projects. Special vessels and techniques for erecting turbines have been developed, and the means of access to offshore turbines has emerged as a major issue influencing cost, availability and safety.

The turbine technology too is different for offshore projects: there are strong reasons why individual turbine size is significantly larger, and turbines of 5 MW and more are being aimed at this market. More subtle differences in technology are also emerging, due to the different environment and increased requirements for reliability.

There is perhaps greater probability of truly innovative designs emerging for the offshore market than for the onshore market, and the chapter concludes by reviewing innovative concepts such as floating turbines.

SMALL WIND TURBINES

At the other end of the scale, Chapter I.6 describes the small and very small wind turbines that are emerging to meet several distinct needs. As well as the traditional areas of rural electrification and providing power to isolated homes, boats and telecommunications facilities, the prospects for significant demand for 'micro-generation' in urban areas is prompting technical developments in small wind turbine design, which could result in significant improvements in the economics. Furthermore, increasing fuel costs are encouraging developments in the technically demanding field of high-penetration wind-diesel systems. This wide range of markets, each with its own characteristics, means that the small wind turbine field shows much greater variety than that of conventional large wind turbines. There is great potential for growth in many of these markets.

RESEARCH AND DEVELOPMENT

Chapter I.7 describes research and development (R&D) efforts in wind technology. A common misunderstanding is to consider wind energy as a mature technology, which could lead to a reduced R&D effort. In addition, the European 20 per cent target for the promotion of energy production from renewable sources poses new challenges. In its recently published Strategic Research Agenda, the European Technology Platform for Wind Energy, TPWind (www.windplatform.eu), proposes an ambitious vision for Europe. In this vision, 300 GW of wind energy capacity is implemented by 2030, representing up to 28 per cent of EU electricity consumption. Moreover, the TPWind vision includes a sub-objective on offshore wind energy, which should represented some 10 per cent of EU electricity consumption by 2030.

An intermediate step is the implementation of 40 GW by 2020, compared to the 1 GW installed today.

R&D is needed to ensure the efficient implementation of the TPWind vision for wind energy so that its targets can be achieved, and TPWind has established R&D priorities in order to implement its 2030 vision for the wind energy sector. Four thematic areas have been identified:

1. wind conditions;
2. wind turbine technology;
3. wind energy integration;
4. offshore deployment and operation.

In order to implement the TPWind 2030 vision and enable the large-scale deployment of wind energy, the support of a stable and well-defined market, policy and regulatory environment is essential. The Market Deployment Strategy includes, amongst other aims, cost reduction and the effective integration of wind into the natural environment.

One main concern is the R&D funding effort. Indeed the total current R&D effort for wind energy in the EU is insufficient to reach European objectives regarding the share of renewable sources of energy in the energy mix and satisfy the Lisbon objectives for growth and jobs.

The most critical component is the European contribution. The Strategic Energy Technology Plan (SET-Plan) proposes a series of instruments to solve this situation, such as the European Industrial Initiatives, which include the European Wind Initiative.

Part II: Grid Integration

Wind power varies over time, mainly under the influence of meteorological fluctuations, on timescales ranging from seconds to years. Understanding these variations and their predictability is of primary importance for the integration and optimal utilisation of wind power in the power system. These issues are discussed in Chapters II.1 and II.2. Electric power systems are inherently variable, in terms of both demand and supply. However, they are designed to cope effectively with these variations through their configuration, control systems and interconnection.

In order to reduce variability, wind plant output should be aggregated to the greatest extent possible. As well as reducing fluctuations, geographically aggregating wind farm output results in an increased amount of firm wind power capacity in the system. Predictability is key to managing wind power's variability. The larger the area, the better the overall prediction of aggregated wind power, with a beneficial effect on the amount of balancing reserves required, especially when gate closure times in the power market take into account the possible accuracy levels of wind power forecasting.

In addition to the advantage of reducing the fluctuations, the effect of geographically aggregating wind farm output is an increased amount of firm wind power capacity in the system.

The large-scale integration of wind energy is seen in the context that wind will provide a substantial share of future European electricity demand. While wind energy covered around 4 per cent of electricity demand in 2008, EWEA targets for 2020 and 2030 are for penetration levels of 12–14 per cent and 21–28 per cent respectively, depending on future electricity demand.

DESIGN AND OPERATION OF POWER SYSTEMS

The established control methods and system reserves available for dealing with variable demand and supply

Figure S.5: Example of smoothing effect by geographical dispersion

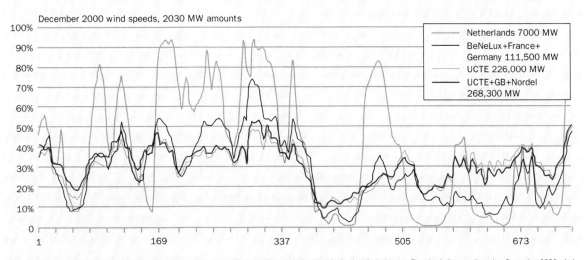

December 2000 wind speeds, 2030 MW amounts

Legend:
— Netherlands 7000 MW
— BeNeLux+France+ Germany 111,500 MW
— UCTE 226,000 MW
— UCTE+GB+Nordel 268,300 MW

Note: The figure compares the hourly output of wind power capacity in four situations, calculated with simulated wind power. The simulations are based on December 2000 wind speeds and wind power capacity estimated for 2030.

Source: www.trade-wind.eu

are more than adequate for dealing with the additional variability at wind energy penetration levels up to around 20 per cent, though the exact level depends on the nature of the specific system. The estimate for extra reserve requirements is around 2–4 per cent of the installed wind power capacity at 10 per cent wind energy penetration, depending on power system flexibility, short-term forecast quality and the gate closure times in the power markets. At higher penetration levels, changes to systems and their method of operation may be required in order to accommodate the further integration of wind energy, and this issue is covered in Chapter II.3. In order to reduce integration efforts and costs, power system design needs to be more flexible. This can be achieved by a combination of flexible generating units, storage systems, flexibility on the demand side, availability of interconnection capacity and more flexible rules in the power market.

Table S.1 gives a detailed overview and categorisation of the power system effects of wind power.

A graphical overview of the various impacts of wind power in the power system is given in Figure S.6, which clearly shows both the local and system-wide impacts and the short- and long-term impacts for the various affected aspects of the power system, including grid infrastructure, system reserves and system adequacy.

GRID INFRASTRUCTURE UPGRADE

Wind energy, as a distributed and variable-output generation source, requires infrastructure investments and the implementation of new technology and grid-management concepts; these are presented in Chapter II.4. The large-scale integration of wind power requires a substantial increase in transmission capacity and other upgrade measures, both within and between the European Member States. Significant improvements can be achieved by network optimisation and other 'soft' measures, but the construction of new lines will also be necessary. At the same time, adequate and

Table S.1: Power system impacts of wind power, causing integration costs

	Effect or impacted element	Area	Timescale	Wind power contribution
Short-term effects	Voltage management	Local/regional	Seconds/ minutes	Wind farms can provide (dynamic) voltage support (design-dependent).
	Production efficiency of thermal and hydro	System	1–24 hours	Impact depends on how the system is operated and on the use of short-term forecasting.
	Transmission and distribution efficiency	System or local	1–24 hours	Depending on penetration level, wind farms may create additional investment costs or benefits. Spatially distributed wind energy can reduce network losses.
	Regulating reserves	System	Several minutes to hours	Wind power can partially contribute to primary and secondary control.
	Discarded (wind) energy	System	Hours	Wind power may exceed the amount the system can absorb at very high penetrations.
Long-term effects	System reliability (generation and transmission adequacy)	System	Years	Wind power can contribute (capacity credit) to power system adequacy.

Source: EWEA

fair procedures need to be developed to provide grid access to wind power even where grid capacity is limited. A transnational offshore grid would not only provide access to the huge offshore resource, but would also improve the cross-border power exchange between countries and alleviate congestion on existing interconnectors. Improving the European networks requires the coordination in network planning to be strengthened at the European level and greater cooperation between all parties involved, especially transmission system operators (TSOs). At the distribution level, more active network management is required. Enhancing the grid's suitability for increased transnational and regional electricity transport is in the interest of both the wind industry and the internal electricity market.

Figures S.7–S.9 show three examples of offshore grid configurations in the North Sea.

GRID CONNECTION REQUIREMENTS

Specific technical requirements within grid codes in terms of tolerance, control of active and reactive power, protective devices, and power quality are changing as penetration increases and wind power assumes additional power plant capabilities, such as active control and the provision of grid support services (Chapter II.5). There may also be a move towards markets for control services, rather than mandatory requirements. In principle, this would make economic sense, as the generator best able to provide the service would be contracted.

Figure S.6: System impacts of wind power

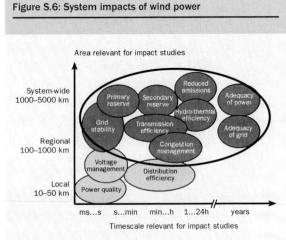

Note: Issues which are within the scope of Task 25 are circled in black.

Sources: IEA Wind Task 25; Holttinen et al. (2007)

Figure S.7: Vision of high voltage 'super grid' to transmit wind power through Europe

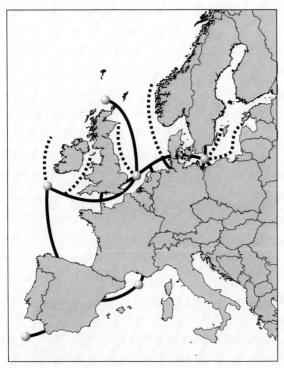

Source: Dowling and Hurley (2004)

Figure S.8: Offshore grid proposal by Statnett

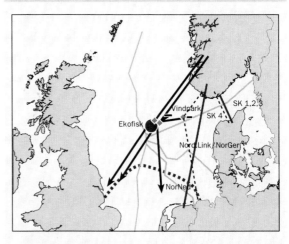

Source: Statnett (2008)

As wind energy penetration increases, there is a greater need to develop a harmonised set of grid code requirements, which would require a concerted effort by the wind power industry and system operators.

WIND POWER'S CONTRIBUTION TO SYSTEM ADEQUACY

For low wind energy penetration levels, the relative capacity credit of wind power (that is 'firm' capacity as a fraction of total installed wind power capacity) is close to the average production (load factor) during the period under consideration – usually the time of highest demand. For north European countries, this is typically 25 to 30 per cent onshore and up to 50 per cent offshore.

With increasing penetration levels of wind energy in the system, its relative capacity credit reduces. However, this does not mean, as Chapter II.6 shows, that less conventional capacity can be replaced, but rather that adding a new wind plant to a system with high wind power penetration levels will substitute less than the first wind plants in the system.

MARKET DESIGN

In the interest of economical wind power integration, changes in market rules throughout Europe are required, so that markets operate faster and on shorter gate closure times (typically three hours ahead or less). This will minimise forecasting uncertainty and last-minute balancing needs. Further substantial economic benefits are expected from the geographical enlarge-ment of market and balancing areas, and from appropri-ate market rules in cross-border power exchange.

ECONOMICS OF WIND POWER INTEGRATION

The introduction of significant amounts of wind energy into the power system brings with it a series of economic

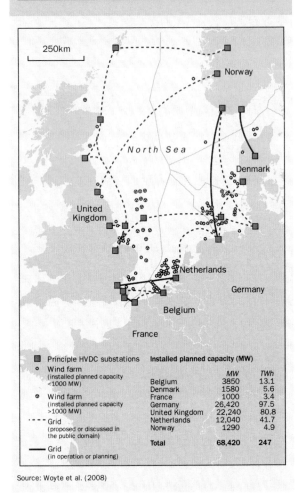

Figure S.9: Offshore grid examined in the Greenpeace study

Installed planned capacity (MW)		
	MW	TWh
Belgium	3850	13.1
Denmark	1580	5.6
France	1000	3.4
Germany	26,420	97.5
United Kingdom	22,240	80.8
Netherlands	12,040	41.7
Norway	1290	4.9
Total	**68,420**	**247**

■ Principle HVDC substations
○ Wind farm (installed planned capacity <1000 MW)
◉ Wind farm (installed planned capacity >1000 MW)
- - - Grid (proposed or discussed in the public domain)
—— Grid (in operation or planning)

Source: Woyte et al. (2008)

Figure S.10: Estimates for the increase in balancing and operating costs due to wind power

Legend:
- Nordic 2004
- Finland 2004
- UK
- Ireland
- Greennet Germany
- Greennet Denmark
- Greennet Finland
- Greennet Norway
- Greennet Sweden

Source: Holttinen et al. (2007)

impacts, both positive and negative. Two main factors determine wind energy integration costs: balancing needs and grid infrastructure (Chapter II.7). The additional balancing cost in a power system arises from the inherent variable nature of wind power, requiring changes for other generators to deal with unpredicted deviations between supply and demand. Evidence from national studies shows that these additional costs represent only a small fraction of the generation costs of wind energy and the overall balancing costs of the power system.

Figure S.10 illustrates the costs from several studies as a function of wind power penetration. Balancing costs increase on a linear basis with wind power penetration, but the absolute values are moderate and always less than €4/MWh at the 20 per cent level (and more often below €2/MWh).

Network upgrade costs arise from the need to connect wind plants and from the extra capacity required to carry the increased power flows in the transmission and distribution networks. Networks also need to be adapted to improve voltage management, and additional interconnection capacity between countries is required to optimally capture the benefits of the continental nature of the wind resource. Any infrastructure improvement fulfilling these needs would provide multiple benefits to the system, and so its cost should not be allocated only to wind power generation.

The cost of modifying power systems with significant amounts of wind energy increases in a quasi linear fashion with wind energy penetration. Identifying an 'economic optimum' is not easy, as costs are accompanied by benefits. Benefits include significant reductions in fossil fuel consumption and cost reductions due to decreased energy dependency, and they are already visible as lower prices in the power exchange markets where large amounts of wind power are offered. From the studies carried out so far, when extrapolating the results to high penetration levels, it is clear that the integration of more than 20 per cent of wind power into the EU power system would be economically beneficial.

Experience and studies provide positive evidence on the feasibility and solutions for integrating the expected wind power capacity in Europe for 2020, 2030 and beyond. Today, the immediate questions mainly relate to the most economic way of dealing with the issues of power system design and operation, electrical network upgrade, connection rules, and electricity market design.

One of the challenges is the creation of appropriate market rules, including incentives to enable power generation and transmission to develop towards being able to accommodate variable output and decentralised generation, notably by becoming more flexible and providing more interconnection capacity. Studies are required at the European level to provide a technical and scientific basis for grid upgrade and market organisation.

Part III: The Economics of Wind Power

Wind power is developing rapidly at both European and global levels. Over the past 15 years, the global installed capacity of wind power has increased from around 2.5 GW in 1992 to more than 94 GW at the end of 2007 – an average annual growth of more than 25 per cent. Due to ongoing improvements in turbine efficiency and higher fuel prices, wind power is increasing in economic competitiveness against conventional power production. And at sites with high wind speeds

on land, wind power is considered to be fully commercial today.

ONSHORE WIND POWER

Capital costs of onshore wind energy projects, covered in Chapter III.1, are dominated by the cost of the wind turbine. The total investment cost of an average turbine installed in Europe is around €1.23 million/MW, including all additional costs for foundations, electrical installation and consultancy (2006 prices). The main costs are divided as follows (approximate levels): turbine 76 per cent, grid connection 9 per cent and foundations 7 per cent. Other cost components, such as control systems and land, account for a minor share of the total costs. The total cost per kW of installed wind power capacity differs significantly between countries, from around €1000/kW to €1350/kW.

In recent years, three major trends have dominated the development of grid-connected wind turbines:
1. turbines have become larger and taller;
2. the efficiency of turbine production has increased steadily; and
3. in general, the investment costs per kW have decreased, although there has been a deviation from this trend in the last three to four years.

In 2007, turbines of the MW-class (above 1 MW) represented a market share of more than 95 per cent,

Table S.2: Cost structure of a typical 2 MW wind turbine installed in Europe (2006-€)

	Investment (€1000/MW)	Share (%)
Turbine (ex-works)	928	75.6
Foundations	80	6.5
Electric installation	18	1.5
Grid connection	109	8.9
Control systems	4	0.3
Consultancy	15	1.2
Land	48	3.9
Financial costs	15	1.2
Road	11	0.9
Total	1227	100

Note: Calculated by the author based on selected data for European wind turbine installations.

Source: Risø DTU

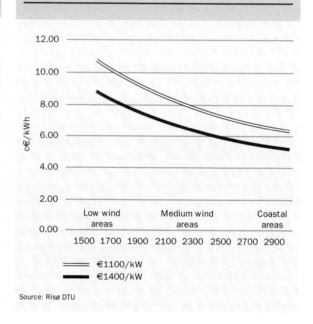

Figure S.11: Calculated costs per kWh of wind generated power as a function of the wind regime at the chosen site (number of full load hours)

Source: Risø DTU

leaving less than 5 per cent for the smaller machines. Within the MW-segment, turbines with capacities of 2.5 MW or above are becoming increasingly important, even for onshore siting. The wind regime at the chosen site, the turbine hub height and the efficiency of production determine the turbine's power production. So just increasing the height of turbines has resulted in higher power production. Similarly, the methods for measuring and evaluating the wind speed at a given site have improved significantly in recent years, and thus improved the siting and economics of new turbines.

Electricity production efficiency, owing to better equipment design, has also improved dramatically. From the late 1980s until 2004, overall investments per unit of swept rotor area decreased by more than 2 per cent per annum. However, in 2006, total investment costs rose by approximately 20 per cent compared to 2004, mainly due to a marked increase in global demand for wind turbines, combined with rising commodity prices and supply constraints. Preliminary data indicates that prices have continued to rise in 2007. At present, production costs of energy for a 2 MW wind turbine range from 5.3 to 6.1 euro cents (c€) per kWh, depending on the wind resource at the chosen site. According to experience curve analyses, the cost range is expected to decline to between 4.3 and 5.5c€/kWh by 2015.

OFFSHORE DEVELOPMENT

Offshore wind (Chapter III.2) only accounts for around 1 per cent of total installed wind power capacity in the world, and development has taken place mainly around the North Sea and the Baltic Sea. At the end of 2007, there was a capacity of more than 1000 MW located offshore in five countries: Denmark, Ireland, The Netherlands, Sweden and the UK. Most of the capacity has been installed in relatively shallow water (less than 20m), and no further than 20km from the coast, so as to minimise the costs of foundations and sea cables.

The costs of offshore capacity, like those of onshore turbines, have increased in recent years. On average, investment costs for a new offshore wind farm are

Figure S.12: Calculated production cost for selected offshore wind farms, including balancing costs (2006 prices)

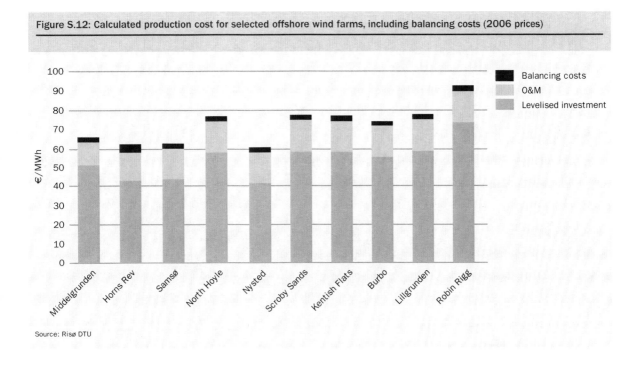

Source: Risø DTU

expected be in the range of €2.0 to €2.2 million/MW for a near-shore, shallow-depth facility. Compared to land-based turbines, the main differences in the cost structure are related to more expensive foundations, the transformer station and sea transmission cables. The cost of offshore-generated electricity ranges from approximately 6–8c€/kWh, mainly due to differences in sea depth, distance from shore and investment costs.

FINANCE

The nature of business in wind energy is changing. Although there are still many small, privately owned projects, a substantial shift towards bigger, utility-owned projects can be observed – this is discussed in Chapter III.3. This change brings new money to the industry and decreases dependence on banks for initial funding. Powerful sponsors are also arriving on the scene. Projects are increasing in size and large-scale offshore activity is taking off; since banks favour big projects, this is a change for the better. If the general

economic picture deteriorates, project finance may suffer, but the strong political and environmental support for renewable energy means that wind energy funding is still viewed as a very attractive option.

PRICES AND SUPPORT MECHANISMS

When clustering different types of support mechanism for electricity from renewables (RES-E), a distinction is made between direct and indirect policy instruments (Chapter III.4). Direct policy measures attempt to stimulate the immediate installation of RES-E technologies, whereas indirect instruments focus on improving the long-term framework conditions. As well as regulatory instruments, there are also voluntary approaches for the promotion of RES-E technologies, mainly based on the willingness of consumers to pay premium rates for green electricity. Other important classification criteria include whether policy instruments address price or quantity, and whether they support investment or generation.

When reviewing and evaluating the various RES-E support schemes, it is important to assess the success of the different policy instruments according to the following criteria:

- Effectiveness: Did the RES-E support programmes lead to a significant increase in deployment of capacities from RES-E in relation to the additional potential?
- Economic efficiency: What was the absolute support level compared to the actual generation costs of RES-E generators and what was the trend in support over time?

Regardless of whether a national or an international support system is concerned, a single instrument is usually not enough to stimulate the long-term growth of RES-E.

IMPACT OF WIND POWER ON SPOT POWER PRICES

In a number of countries, wind power is increasing its share of total power production (Chapter III.5). This is particularly noticeable in Denmark, Spain and Germany, where wind power's contribution to total power supply is 21 per cent, 12 per cent and 7 per cent respectively. In these cases, wind power is becoming an important player in the power market and can significantly influence power prices. As wind power has very low marginal cost (due to zero fuel costs), it enters near the bottom of the supply curve. This shifts the curve to the right, resulting in lower power prices, with the extent of the price reduction depending on the price elasticity of the power demand.

In general, when wind power provides a significant share of the power supply, the price of power is likely to be lower during high-wind periods and higher during low-wind periods. A study carried out in Denmark shows that the price of power to consumers (excluding transmission and distribution tariffs, and VAT and other taxes) in 2004 to 2007 would have been approximately 4–12 per cent higher if wind power had not contributed to power production. This means that in 2007, power consumers saved approximately 0.5c€/kWh due to wind power reducing electricity prices. This should be compared to consumer payments to wind power of approximately 0.7c€/kWh as feed-in tariffs. So although the cost of wind power to consumers is still greater than the benefits, a significant reduction in net expenses is certainly achieved due to lower spot prices.

The analysis involves the impacts of wind power on power spot prices being quantified using structural analyses. A reference is fixed, corresponding to a situation with zero contribution from wind power in the power system. A number of levels with increasing contributions from wind power are identified and, relating to the reference, the effect of wind power's power production is calculated. This is illustrated in the left graph of Figure S.13, where the shaded area between the two curves approximates the value of wind power in terms of lower spot power prices.

WIND POWER COMPARED TO CONVENTIONAL POWER

In general, the cost of conventional electricity production is determined by four cost components:
1. fuel;
2. CO_2 emissions (as given by the European Trading System for CO_2);
3. operation and maintenance (O&M); and
4. capital, including planning and site work.

Implementing wind power avoids the full fuel and CO_2 costs, as well as a considerable share of conventional power plants' O&M costs. The amount of capital costs avoided depends on the extent to which wind power capacity can displace investments in new conventional power plants; this is linked directly to how wind power plants are integrated into the power system.

Figure S.13: The impact of wind power on the spot power price in the West Denmark power system in December 2005

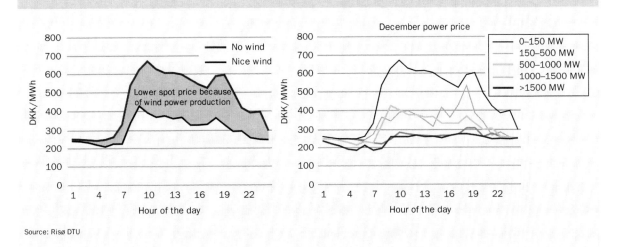

Source: Risø DTU

Studies show that the cost of integrating variable wind power is approximately 0.3 to 0.4c€/kWh of wind power generated, even at fairly high levels of wind power penetration (approximately 20 per cent, depending on the nature of the operating system). Figure S.14 shows the results of the reference case, assuming the two conventional power plants are coming on-stream in 2010.

As shown in the reference case, the cost of power generated at conventional power plants is lower than the cost of wind-generated power under the given assumptions of low fuel prices. At a European inland site, wind-generated power is approximately 33–34 per cent more expensive than natural gas- and coal-generated power (Chapter III.6).

This case is based on the *World Energy Outlook* assumptions on fuel prices, including a crude oil price of US$59/barrel in 2010. At present (mid-2008), the crude oil price has reached as high as $147/barrel. Although this oil price is combined with a lower exchange rate for the US dollar, the present price of oil is significantly higher than the forecasted IEA oil price for 2010. Therefore, a sensitivity analysis has been carried out and the results are shown in Figure S.15.

In Figure S.15, the natural gas price is assumed to double compared to the reference (equivalent to an oil price of $118/barrel in 2010), the coal price to increase by 50 per cent and the price of CO_2 to increase

Figure S.14: Costs of generated power comparing conventional plants to wind power, 2010 (constant 2006-€)

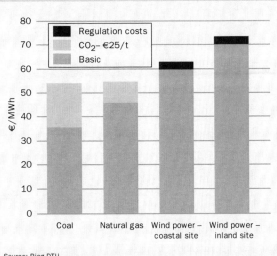

Source: Risø DTU

Figure S.15: Sensitivity analysis of costs of generated power comparing conventional plants to wind power, assuming increasing fossil fuel and CO_2 prices, 2010 (constant 2006-€)

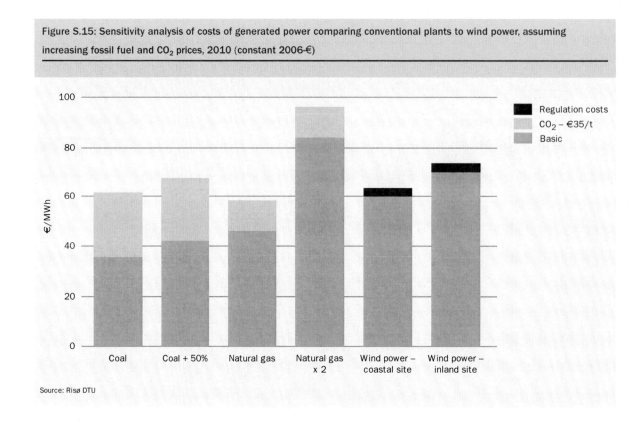

Source: Risø DTU

to €35/t from €25/t in 2008. As shown in the figure, the competitiveness of wind-generated power increases significantly, costs at the inland site becoming lower than those of the natural gas plant and only around 10 per cent more expensive than those of the coal-fired plant. On coastal sites wind power produces the cheapest electricity.

EMPLOYMENT

Wind energy companies in the EU currently employ around 108,600 people; when indirect jobs are taken into account, this figure rises to more than 150,000 (Chapter III.7). A significant share of direct wind energy employment (approximately 77 per cent) is located in three countries, Denmark, Germany and Spain, whose combined installed capacity represents 70 per cent of the EU total. However, the sector is

less concentrated now than it was in 2003, due to the opening of manufacturing and operation centres in emerging markets and to the fact that many wind-related activities, such as promotion, O&M, engineering and legal services, are now carried out at a local level. Wind turbine and component manufacturers account for most of the jobs (59 per cent).

In addition to the 108,600 direct jobs outlined in the previous section, the European wind energy sector also affects employment in sectors not directly related to wind energy. Approximately 43,000 people were indirectly employed in wind energy in 2007. In total, the EU wind energy sector therefore employed more than 150,000 in 2007. EWEA's analysis concludes that 15.1 jobs are created in the EU for each new MW installed. In addition, 0.4 jobs are created per MW of total installed capacity in operations and maintenance and other activities related to existing installations.

Table S.3: Direct employment from wind energy companies in selected European countries

Country	No of direct jobs
Austria	700
Belgium	2000
Bulgaria	100
Czech Republic	100
Denmark	23,500
Finland	800
France	7000
Germany	38,000
Greece	1800
Hungary	100
Ireland	1500
Italy	2500
The Netherlands	2000
Poland	800
Portugal	800
Spain	20,500
Sweden	2000
United Kingdom	4000
Rest of EU	400
TOTAL	**108,600**

Source: Own estimates, based on EWEA (2008a); ADEME (2008); AEE (2007); DWIA (2008); Federal Ministry of the Environment in Germany, BMU (2008)

Part IV: Industry and Markets

In 2001, the EU passed its Directive on the promotion of electricity produced from renewable energy sources in the internal electricity market. This is still the most significant piece of legislation in the world for the integration of electricity produced by renewable energies, including wind power. This directive contains an indicative target of 21 per cent of final electricity demand in the EU to be covered by renewable energy sources by 2010, and regulates the electricity markets in which they operate. It has been tremendously successful in promoting renewables, particularly wind energy, and is the key factor explaining the global success of the European renewable energy industries and the global leadership position of European wind energy companies.

The gradual implementation of the 2001 Renewable Electricity Directive in the Member States, as well as the unanimous decision made by the European Council at its Spring Summit in March 2007 for a binding 20 per cent share of renewable energy in the EU by 2020, are all steps in the right direction and indicators of increased political commitment. A new directive, based on a European Commission proposal from January 2008, was adopted by the European Parliament and Council in December 2008. It will raise the share of renewable energy in the EU from 8.5 per cent in 2005 to 20 per cent in 2020, which means that more than one-third of the EU's electricity will have to come from renewables in 2020, up from 15 per cent in 2007. It is already clear that wind energy will be the largest contributor to this increase.

THE EU ENERGY MIX

While thermal generation, totalling over 430 GW, has long served as the backbone of Europe's power production, combined with large hydro and nuclear, Europe is steadily transitioning away from conventional power sources towards renewable energy technologies (Chapter IV.1). Between 2000 and 2007, total EU power capacity increased by 200 GW, reaching 775 GW by the end of 2007. The most notable changes in the mix were the near doubling of gas capacity to 164 GW and wind energy more than quadrupling, from 13 to 57 GW.

WIND ENERGY IN THE EUROPEAN POWER MARKET

The EU is leading the way with policy measures to facilitate the move towards the deployment of renewable energy technologies. With an impressive compound annual growth rate of over 20 per cent in MW installed between 2000 and 2007 (Figure S.16), wind energy has clearly established itself as a relevant power source

Figure S.16: New power capacity, EU, 2000–2007 (in MW)

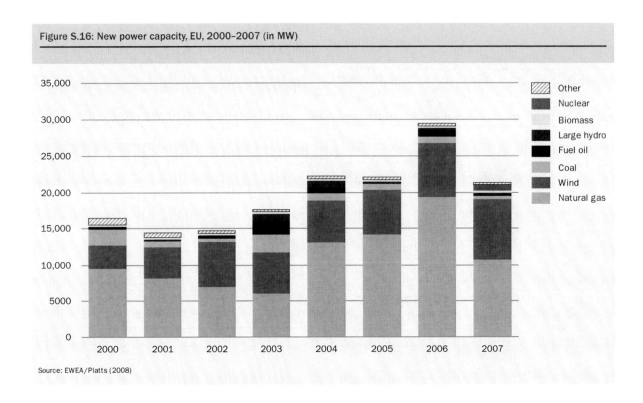

Source: EWEA/Platts (2008)

in Europe's power generation market. Thirty per cent of all power capacity installed in the EU over the five-year period has been wind power, making it the second largest contributor to installation of EU capacity over the last ten years, after natural gas (55 per cent). In 2007, 40 per cent of annual EU capacity installed was wind power, and wind power increased more than any other power-generating technology in Europe, including natural gas.

Wind power's share has jumped to over 10 per cent of total installed capacity and more than 5 per cent of national electricity demand in five European markets, Germany, Spain, Denmark, Portugal and Ireland, surpassing 10 per cent in both Spain and Denmark.

THE CURRENT STATUS OF THE EU WIND ENERGY MARKET

In the EU, installed wind power capacity has increased by an average of 25 per cent annually over the past 11 years, from 4753 MW in 1997 to 56,535 MW in 2007 (Chapter IV.2). In terms of annual installations, the EU market for wind turbines has grown by 19 per cent annually, from 1277 MW in 1997 to 8554 MW in 2007. In 2007, Spain was by far the largest market for wind turbines, followed by Germany, France and Italy. Eight countries – Germany, Spain, Denmark, Italy, France, the UK, Portugal and The Netherlands – now have more than 1000 MW installed. Germany, Spain and Denmark – the three pioneering countries of wind power – are home to 70 per cent of installed wind power capacity in the EU (see Figures S.17 and S.18).

The more than 56,000 MW of total wind power capacity installed in the EU at the end of 2007 will produce 3.7 per cent of the EU-27's electricity demand in an average wind year.

With 1080 MW by the end of 2007, offshore accounted for 1.9 per cent of installed EU capacity and 3.5 per cent of the electricity production from wind power in the EU. The market is still below its 2003

Figure S.17: 2007 Member State market shares of new capacity

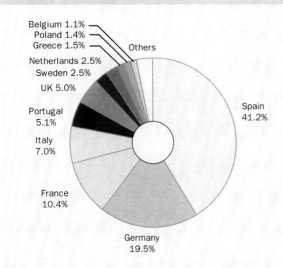

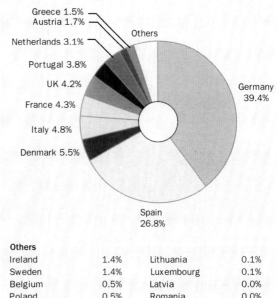

Others			
Czech Republic	0.7%	Cyprus	0.0%
Ireland	0.7%	Denmark	0.0%
Estonia	0.3%	Hungary	0.0%
Finland	0.3%	Latvia	0.0%
Bulgaria	0.4%	Luxembourg	0.0%
Austria	0.2%	Malta	0.0%
Lithuania	0.1%	Slovakia	0.0%
Romania	0.1%	Slovenia	0.0%

Source: EWEA (2008a)

Figure S.18: End 2007 Member State market shares of total capacity

Others			
Ireland	1.4%	Lithuania	0.1%
Sweden	1.4%	Luxembourg	0.1%
Belgium	0.5%	Latvia	0.0%
Poland	0.5%	Romania	0.0%
Czech Republic	0.2%	Slovakia	0.0%
Finland	0.2%	Cyprus	0.0%
Bulgaria	0.1%	Malta	0.0%
Hungary	0.1%	Slovenia	0.0%
Estonia	0.1%		

Source: EWEA (2008a)

level and development has been slower than previously anticipated.

EUROPEAN WIND INDUSTRY: PLAYERS AND INVESTMENT TRENDS

Wind's spectacular growth as a vehicle for new generation capacity investment has attracted a broad range of players across the industry value chain (Chapter IV.3). From local, site-focused engineering firms to global vertically integrated utilities, all have formed part of wind energy's European growth story.

As the region responsible for pioneering widespread, larger-scale take-up of wind power, Europe hosts the tightest competition for market share, among roughly a dozen suppliers. The European market has seen a highly stable market share distribution, with few major shifts since a round of consolidation among leading suppliers during 2003 and 2004. Between 2004 and 2007, three players averaged over a 15 per cent market share of annual MW added each, followed by four players with a 5–10 per cent share.

Supply chain management represents a key competitive driver in wind turbine supply. The relationships between turbine manufacturers and their component suppliers have become increasingly crucial, and have come under increasing stress in the past three years, as soaring demand has required faster ramp-up times,

Table S.4: Cumulative installations of wind power in the EU and projections for 2010 (in MW)

Country	Cumulative installations								
	2000	2001	2002	2003	2004	2005	2006	2007	2010
Austria	77	94	140	415	606	819	965	982	1200
Belgium	13	32	35	68	96	167	194	287	800
Bulgaria					10	10	36	70	200
Cyprus			0	0	0	0	0	0	0
Czech Republic			3	9	17	28	54	116	250
Denmark	2417	2489	2889	3116	3118	3128	3136	3125	4150
Estonia			2	2	6	32	32	58	150
Finland	39	39	43	52	82	82	86	110	220
France	66	93	148	257	390	757	1567	2454	5300
Germany	6113	8754	11,994	14,609	16,629	18,415	20,622	22,247	25,624
Greece	189	272	297	383	473	573	746	871	1500
Hungary			3	3	3	17	61	65	150
Ireland	118	124	137	190	339	496	746	805	1326
Italy	427	682	788	905	1266	1718	2123	2726	4500
Latvia			24	27	27	27	27	27	100
Lithuania			0	0	6	6	48	50	100
Luxembourg	10	15	17	22	35	35	35	35	50
Malta			0	0	0	0	0	0	0
Netherlands	446	486	693	910	1079	1219	1558	1746	3000
Poland			27	63	63	83	153	276	1000
Portugal	100	131	195	296	522	1022	1716	2150	3500
Romania			1	1	1	2	3	8	50
Slovakia			0	3	5	5	5	5	25
Slovenia			0	0	0	0	0	0	25
Spain	2235	3337	4825	6203	8264	10,028	11,623	15,145	20,000
Sweden	231	293	345	399	442	510	571	788	1665
UK	406	474	552	667	904	1332	1962	2389	5115
EU accumulated*	12,887	17,315	23,098	28,491	34,372	40,500	48,031	56,535	80,000

Note: *From 2004 EU-25; from 2007 EU-27.

Source: EWEA (2008a)

larger investments and greater agility to capture value in a rapidly growing sector.

Furthermore, Europe's wind energy value chain is seeing dynamic shifts, as asset ownership is redistributed, growth is sought in maturing markets and players seek to maximise scale on an increasingly pan-European stage. The proliferation of players looking to develop, own or operate wind plants has pushed competition to a new level, underlining the key elements of local market knowledge, technical expertise and financial capacity as crucial to positioning on the value chain.

KEY PLAYER POSITIONING

Europe's shifting distribution of wind power asset ownership clearly illustrates the industry's scaling up and geographic expansion. From an industry concentrated in Denmark and Germany with single, farmer-owned turbines at the end of the 1990s, wind power ownership now includes dozens of multinational players, owning several GWs of installed capacity. Five main blocks of ownership now characterise the structure of the European market:

1. utilities;
2. Europe's largest independent power producers (IPPs);
3. other Spanish IPPs;
4. German investors;
5. other European investors/IPPs.

Over the past five years, the most salient trend has been the increased participation of utilities in the industry. Utility share of total wind power installed increased from 17 per cent in 2002 to 25 per cent in 2007. The biggest jump took place between 2005 and 2006, as the region's top wind utilities saw annual additions of well over 500 MW.

PLANNED FUTURE INVESTMENT

For the 2007 to 2010 timeframe, Europe's top 15 utilities and IPPs in terms of MW owned declared construction pipelines totalling over 18 GW, which translates into well over €25 billion in wind plant investment, based on current cost estimates per MW installed. Overall, the European wind market is expected to grow at a rate of over 9 GW installed annually through to 2010, which translates into annual investments pushing past €10 billion to nearly €16 billion.

The overall European wind power market environment is coming of age with the technology's steady emergence into the overall power market. Although wind has become an integral part of the generation mix, alongside conventional power sources, of markets such as Germany, Spain and Denmark, it continues to face the double challenge of competing against other renewable technologies while proving to be a strong energy choice for large power producers seeking to grow and diversify their portfolios.

THE STATUS OF THE GLOBAL WIND ENERGY MARKETS

In its best year yet, the global wind industry, discussed in Chapter IV.4, installed 20,000 MW in 2007. This development was lead by the US, China and Spain, and it brought the worldwide installed capacity to 94,122 MW. This was an increase of 31 per cent compared with the 2006 market, and represented an overall increase in global installed capacity of about 27 per cent.

The top five countries in terms of installed capacity are Germany (22.3 GW), the US (16.8 GW), Spain (15.1 GW), India (7.8 GW) and China (5.9 GW). In terms of economic value, the global wind market in 2007 was worth about €25 billion (US$37 billion) in new generating equipment, and attracted €34 billion (US$50.2 billion) in total investment.

Europe remains the leading market for wind energy — its new installations represented 43 per cent of the global total and European companies supplied 66 per cent of the world's wind power capacity in 2007.

US AND CHINESE MARKETS CONTINUE TO BOOM

The US reported a record 5244 MW installed in 2007, more than double the 2006 figure, accounting for about 30 per cent of the country's new power-producing capacity in 2007. Overall US wind power generating capacity grew by 45 per cent in 2007, with total installed capacity now standing at 16.8 GW. While wind energy in the EU covered some 4 per cent of 2008 electricity demand, however, US wind farms will generate around 48 billion kWh of electricity in 2008, representing just over 1 per cent of US electricity supply.

China added 3449 MW of wind energy capacity during 2007, representing market growth of 156 per cent over 2006, and now ranks fifth in total installed wind energy capacity, with over 6000 MW at the end of 2007. However, experts estimate that this is just the beginning, and that the real growth in China is still to come. European manufacturers are well positioned to exploit this market opportunity.

ADMINISTRATIVE AND GRID ACCESS BARRIERS

Today, integration of electricity from renewable energy sources into the European electricity market faces multiple barriers. Chapter IV.5 takes a developer's point of view and observes the barriers occurring during the process of acquiring building permits, spatial planning licences and grid access, using the example of four EU Member States.

Barriers are encountered if the procedures with which a project developer has to comply are not set out in a coherent manner; these include a lack of

transparency and excessive administrative requirements. Every European Member State faces such barriers, but their impact on the deployment of renewable energy differs for each country. Barriers at the administrative, social and financial levels, as well as in relation to grid connection, are a serious obstacle to investment and the competitiveness of wind energy in the European and global markets.

Part V: Environment

Not all energy sources have the same negative environmental effects or natural resources depletion capability. Fossil energies exhaust natural resources and are mostly responsible for environmental impacts. On the other hand, renewable energies in general, and wind energy in particular, produce few environmental impacts, and these are significantly lower than those produced by conventional energies.

ENVIRONMENTAL BENEFITS

Chapter V.1 describes the life-cycle assessment (LCA) methodology for emissions and environmental impact assessments and, based on relevant European studies, shows the emissions and environmental impacts derived from the electricity production of onshore and offshore wind farms throughout their life cycles. The avoided emissions and environmental impacts from wind electricity compared to the other fossil electricity generation technologies are also examined.

ENVIRONMENTAL IMPACTS

Although the environmental impacts of wind energy are much lower in intensity than those created by conventional energies, they still have to be assessed. The possible negative influences on fauna and nearby populations have been analysed for both onshore and offshore schemes. Specific environmental impacts, such

LM Glasfiber

as those on landscape, noise, bird and marine organism populations, and electromagnetic interference, are examined in Chapter V.2.

Wind energy has a key role to play in combating climate change by reducing CO_2 emissions from power generation. The emergence of international carbon markets, which were spurred by the flexible mechanisms introduced by the Kyoto Protocol, as well as improved regional emissions trading schemes, such as the EU Emissions Trading System (ETS), could provide an additional incentive for the development and deployment of renewable energy technologies, specifically wind energy.

POLICY MEASURES TO COMBAT CLIMATE CHANGE

Wind energy has the potential to make dramatic reductions in CO_2 emissions from the power sector. Chapter V.3 gives an overview of the development of the international carbon markets, assesses the impact of the Clean Development Mechanism and Joint Implementation on wind energy, and outlines the path towards a post-2012 climate regime. It also gives an outline of the EU ETS, discussing the performance to date, the allocation method and proposals for the post-2012 period.

EXTERNALITIES OF WIND COMPARED TO OTHER TECHNOLOGIES

The electricity markets do not currently take account of external effects and the costs of pollution. Therefore it is important to identify the external effects of different electricity generation technologies and calculate the related external costs. External costs can then be compared with the internal costs of electricity, and competing energy systems, such as conventional electricity generation technologies, can be compared with wind energy (Chapter V.4).

Science began to study the external costs of electricity generation in the late 1980s, notably with the Externalities of Energy study (ExternE), which attempted to develop a consistent methodology to assess the externalities of electricity generation technologies. Work and methodologies on the ExternE project are updated on a regular basis. This project values the external costs of wind energy at less than 0.26c€/kWh, with those of conventional (fossil fuel-based) electricity generation being significantly higher.

In Chapters V.4 and V.5, *Wind Energy – The Facts* presents the results of empirical analyses of the emissions and external costs avoided by the replacement of conventional fossil fuel-based electricity generation by wind energy in each of the EU27 Member States in 2007, 2020 and 2030. Wind energy avoided external costs of more than €10 billion in 2007, and this figure is expected to increase as penetration of wind energy increases over the coming decades (Table S.5).

Table S.5: Avoided external energy costs			
	2007	**2020 (estimated)**	**2030 (estimated)**
Wind energy's contribution to avoided external costs (€ billion, 2007 prices)	10.2	32.9	69.2

Note: A precondition for full implementation of the environmental benefits estimated for 2020 and 2030 is continuous adaptation of financial support instruments and the removal of barriers to market integration of wind energy.

SOCIAL ACCEPTANCE OF WIND ENERGY AND WIND FARMS

Experience with wind energy in the EU shows that social acceptance is crucial for the successful development of wind energy (Chapter V.6). Social research on wind energy has focused on three main points:

1. assessment of the levels of public support for wind energy (public acceptance);
2. identification of and understanding the social response at the local level (community acceptance); and
3. analysis of the key issues involved in social acceptance by key stakeholders and policymakers (stakeholder acceptance).

The way in which wind farms are developed and managed, as well as the way in which the public engages with them, may be more important in shaping public reactions to new projects than the purely physical or technical characteristics of the technology. Such factors significantly affect the relationships between hosting communities, developers and the authorities. There are no fixed rules for the management of social acceptance on technological matters, but proper consideration of this wide range of issues may help promoters and authorities learn from past experiences and find mechanisms to maintain and expand public engagement in wind development.

Part VI: Scenarios and Targets

The European Commission's 1997 White Paper on renewable sources of energy set the goal of doubling the share of renewable energy in the EU's energy mix from 6 to 12 per cent by 2010. It included a target of 40,000 MW of wind power in the EU by 2010, producing 80 TWh of electricity and saving 72 million tonnes (Mt) of CO_2 emissions per year. The 40,000 MW target was reached in 2005.

The 40,000 MW goal from the European Commission's White Paper formed EWEA's target in 1997, but three years later, due to the strong developments in the German, Spanish and Danish markets for wind turbines, EWEA increased its target by 50 per cent, to 60,000 MW by 2010 and 150,000 MW by 2020 (Chapters VI.1 and VI.2). In 2003, EWEA once again increased its target, this time by 25 per cent to 75,000 MW by 2010 and 180,000 MW by 2020. Due to the expansion of the EU with 12 new Member States, EWEA increased its reference scenario for 2010 to 80,000 MW, while maintaining its 2020 target of 180,000 MW and setting a target of 300,000 MW by 2030 (Figure S.19).

If the reference scenario is reached, wind power production will increase to 177 TWh in 2010, 477 TWh in 2020 and 935 TWh in 2030 (Chapter VI.3). The European Commission's baseline scenario assumes an increase in electricity demand of 33 per cent between 2005 and 2030 (4408 TWh). Assuming that EU electricity demand develops as projected by the European Commission, wind power's share of EU electricity consumption will reach 5 per cent in 2010, 11.7 per cent in 2020 and 21.2 per cent in 2030.

If political ambitions to increase energy efficiency are fulfilled, moreover, wind power's share of future electricity demand will be greater than the baseline scenario. In 2006, the European Commission released new scenarios to 2030 on energy efficiency and renewables. If EU electricity demand develops as projected in the European Commission's 'combined high renewables and efficiency' (RE&Eff) case, wind energy's share of electricity demand will reach 5.2 per cent in 2010, 14.3 per cent in 2020 and 28.2 per cent in 2030 (see Table S.6).

Since 1996, the European Commission has changed its baseline scenario five times. Over the 12-year period, targets for wind energy in 2010 and 2020 have been increased almost tenfold, from 8000 MW to 71,000 MW (2010) and from 12,000 MW to 120,000 MW (2020) in the European Commission's latest baseline scenario from 2008.

Figure S.19: EWEA's three wind power scenarios (in GW)

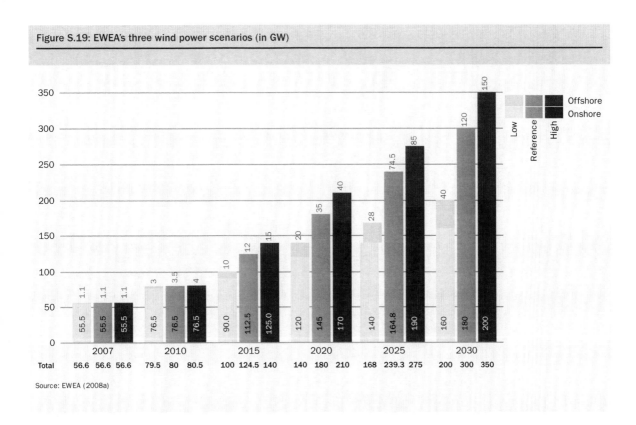

Source: EWEA (2008a)

Somewhat surprising, the baseline scenario from 2008 gives significantly lower figures for wind energy than the baseline scenario from 2006. The 71,000 MW projection for 2010 implies that the wind energy market in Europe will decrease by approximately 50 per cent over the next three years with respect to the present market. In light of the current market achievements, growth trends and independent market analyses, the European Commission's baseline scenario seems out of touch with current reality and clearly underestimates the sector's prospects in the longer term.

Both the European Commission and IEA's baseline scenarios for wind energy assume that market growth will slow significantly – the European Commission by as much as 50 per cent (compared to the EWEA scenario), to reach its 71 GW target for 2010. Their advanced scenarios, however, are in line with EWEA's target for 2010, while the European Commission's 2006 scenario even exceeds EWEA's 180 GW target for 2020.

Turbine prices have increased since 2005; however, one of the significant advantages of wind power is that

Table S.6: Wind power's share of EU electricity demand

	2000	2007	2010	2020	2030
Wind power production (TWh)	23	119	177	477	935
Reference electricity demand (TWh)*	2577	3243	3568	4078	4408
RE&Eff case electricity demand (TWh)*	2577	3243	3383	3345	3322
Wind energy share (reference)	0.9%	3.7%	5.0%	11.7%	21.2%
Wind energy share (RE&Eff case)	0.9%	3.7%	5.2%	14.3%	28.2%

*Sources: Eurelectric, EWEA and European Commission

Table S.7: Savings (in billions of €) made depending on the price of fuel and CO_2 (per tonne)

Totals (fuel prices equivalent to oil at \$90; CO_2 €25)	2008–2010	2011–2020	2021–2030	2008–2020	2008–2030
Investment	31,062	120,529	187,308	151,591	338,899
Avoided CO_2 cost	21,014	113,890	186,882	134,904	321,786
Avoided fuel cost	51,165	277,296	455,017	328,462	783,479
Totals (fuel prices equivalent to oil at \$50; CO_2 €10)	2008–2010	2011–2020	2021–2030	2008–2020	2008–2030
Investment	31,062	120,529	187,308	151,591	338,899
Avoided CO_2 cost	8,406	45,556	74,753	53,962	128,714
Avoided fuel cost	30,456	165,057	270,843	195,513	466,356
Totals (fuel prices equivalent to oil at \$120; CO_2 €40)	2008–2010	2011–2020	2021–2030	2008–2020	2008–2030
Investment	31,062	120,529	187,308	151,591	338,899
Avoided CO_2 cost	33,623	182,223	299,011	215,846	514,857
Avoided fuel cost	67,002	363,126	595,856	430,128	1,025,984

Source: EWEA (2008)

the fuel is free. Therefore the total cost of producing wind energy throughout the 20- to 25-year lifetime of a wind turbine can be predicted with great accuracy. Neither the future prices of coal, oil or gas nor the price of carbon will affect the cost of wind power production. This, as Chapter VI.4 points out, is probably wind energy's most significant competitive advantage in the global energy market.

Cumulative investments in wind energy over the three decades from 2000 to 2030 will total €390 billion. According to EWEA's reference scenario, approximately €340 billion will be invested in wind energy in the EU-27 between 2008 and 2030.

As can be seen from Table S.7, changing the CO_2 and the fuel price assumptions has a dramatic impact on the resulting fuel and CO_2 costs that are avoided by installing wind power capacity. With low CO_2 prices (€10/t) and fuel prices (equivalent of \$50/barrel of oil) throughout the period, the wind power investments over the

next 23 years avoid €466 billion instead of €783 billion in fuel and CO_2 costs. With high prices for CO_2 (€40/t) and fuel (equivalent to \$120/barrel of oil), wind power would avoid fuel and CO_2 costs equal to more than €1 trillion over the three decades from 2000 to 2030.

Table S.7 shows the different savings made depending on the price of oil (per barrel) and CO_2 (per tonne).

The Global Wind Energy Council (GWEC) predicts that the global market for wind turbines will grow by over 155 per cent from 94 GW in 2007 to reach 240.3 GW of total installed capacity by 2012 (Chapter VI.5). In particular, the US and Chinese markets are expected to expand dramatically.

Depending on the increase in demand, wind power could cover 11.5 to 12.7 per cent of global electricity consumption in 2020, according to GWEC, and as much as 20.2 to 24.9 per cent – in other words between a fifth and a quarter of the world's electricity needs – in 2030 (Chapter VI.6).

WIND ENERGY - THE FACTS

PART I

TECHNOLOGY

Acknowledgements

Part I was compiled by Paul Gardner, Andrew Garrad, Lars Falbe Hansen, Peter Jamieson, Colin Morgan, Fatma Murray and Andrew Tindal of Garrad Hassan and Partners, UK; José Ignacio Cruz and Luis Arribas of CIEMAT, Spain; Nicholas Fichaux of the European Wind Energy Association (EWEA).

We would like to thank all the peer reviewers for their valuable advice and for the tremendous effort that they put into the revision of Part I.

Part I was carefully reviewed by the following experts:

Nicolas Fichaux	European Wind Energy Association
Henning Kruse	Siemens
Takis Chaviaropoulos	CRES
Angeles Santamaria Martin	Iberdrola
Erik Lundtang Petersen	Risø DTU National Laboratory
Jos Beurskens	ECN
Josep Prats	Ecotècnia
Eize de Vries	Planet
Flemming Rasmussen	Risø DTU National Laboratory
Simon Watson	Loughborough University
Félix Avia	CENER
Murat Durak	Turkish Wind Energy Association
Jørgen Højstrup	Suzlon Energy A/S

I.1 INTRODUCTION

Electricity can be generated in many ways. In each case, a fuel is used to turn a turbine, which drives a generator, which feeds the grid. The turbines are designed to suit the particular fuel characteristics. The same applies to wind-generated electricity: the wind is the fuel, which drives the turbine, which generates electricity. But unlike fossil fuels, it is free and clean.

The politics and economics of wind energy have played an important role in the development of the industry and contributed to its present success, but the engineering is still pivotal. As the wind industry has become better established, the central place of engineering has become overshadowed by other issues, but this is a tribute to the success of engineers and their turbines. Part I of this volume addresses the key engineering issues:

- the wind – its characteristics and reliability; how it can be measured, quantified and harnessed;
- the turbines – their past achievements and future challenges, covering a range of sizes larger than most other technologies, from 50 W to 5 MW and beyond;
- the wind farms – the assembly of individual turbines into wind power stations or wind farms; their optimisation and development; and
- going offshore – the promise of a very large resource, but with major new technical challenges.

Part I provides a historical overview of turbine development, describes the present status and considers future challenges. This is a remarkable story, which started in the 19th century and accelerated over the last two decades of the 20th, on a course very similar to the early days of aeronautics. The story is far from finished, but it has certainly started with a vengeance.

Wind must be treated with great respect. The wind speed on a site has a very powerful effect on the economics of a wind farm, and wind provides both the fuel to generate electricity and, potentially, loads that can destroy the turbines. This part describes how it can be quantified, harnessed and put to work in an economic and predictable manner. The long- and short-term behaviour of the wind is described. The latter can be successfully forecasted to allow wind energy to participate in electricity markets.

The enormous offshore wind resource offers great potential, but there are major engineering challenges, especially regarding reliability, installation and access.

In short, Part I explores how this new, vibrant and rapidly expanding industry exploits one of nature's most copious sources of energy – the wind.

I.2 WIND RESOURCE ESTIMATION

Introduction

The wind is the fuel for the wind power station. Small changes in wind speed produce greater changes in the commercial value of a wind farm. For example, a 1 per cent increase in the wind speed might be expected to yield a 2 per cent increase in energy production.

This chapter explains why knowledge of the wind is important for each and every stage of the development of a wind farm, from initial site selection to operation.

Europe has an enormous wind resource, which can be considered on various levels. At the top level, the potential resource can be examined from a strategic standpoint:

- Where is it?
- How does it compare to the EU and national electricity demands?
- What regions and areas offer good potential?

At the next level, it is necessary to understand the actual wind resource on a site in great detail:

- How is it measured?
- How will it change with time?
- How does it vary over the site?
- How is it harnessed?

It is at this stage that commercial evaluation of a wind farm is required, and robust estimates must be provided to support investment and financing decisions. Once the wind speed on the site has been estimated, it is then vital to make an accurate and reliable estimate of the resulting energy production from a wind farm that might be built there. This requires wind farm modelling and detailed investigation of the environmental and ownership constraints.

As its contribution to electricity consumption increases, in the context of liberalised energy markets, new questions are beginning to emerge, which are critically linked to the nature of the wind:

- How can wind energy be consolidated, traded and generally integrated into our conventional electricity systems?
- Will an ability to forecast wind farm output help this integration?

These questions, and more, are addressed in this chapter. The first section looks at the strategic 'raw' resource issues, and the following sections provide a detailed step-by-step evaluation of the assessment process. A worked example of a real wind farm is then provided and, finally, recommendations are made about the important matters that need to be tackled in the near future to help wind energy play its full part.

Regional Wind Resources

Naturally, wind energy developers are very interested in the energy that can be extracted from the wind, and how this varies by location. Wind is ubiquitous, and in order to make the choice of potential project sites an affordable and manageable process, some indication of the relative size of the 'wind resource' across an area is very useful. The wind resource is usually expressed as a wind speed or energy density, and typically there will be a cut-off value below which the energy that can be extracted is insufficient to merit a wind farm development.

ON-SITE MEASUREMENT

The best, most accurate indication of the wind resource at a site is through on-site measurement, using an anemometer and wind vane (described in detail later in this chapter). This is, however, a fairly costly and time-consuming process.

COMPUTER MODELLING

On a broader scale, wind speeds can be modelled using computer programs which describe the effects on the wind of parameters such as elevation, topography and ground surface cover. These models must be primed with some values at a known location, and usually this role is fulfilled by local meteorological station measurements or other weather-related recorded data, or data extracted from numerical weather prediction models, such as those used by national weather services.

Typically, these wind-mapping programs will derive a graphical representation of mean wind speed (for a specified height) across an area. This may take the form of a 'wind atlas', which represents the wind speed over flat homogeneous terrain, and requires adjustments to provide a site-specific wind speed prediction to be made with due consideration of the local topography. In some areas, 'wind maps' may be available; these include the effects of the terrain and ground cover. Wind atlases and wind maps have been produced for a very wide range of scales, from the world level down to the local government region, and represent the best estimate of the wind resource across a large area. They do not substitute for anemometry measurements – rather they serve to focus investigations and indicate where on-site measurements would be merited.

As a further stage in investigations, theoretical wind turbines can be placed in a chosen spacing within a geographical model containing wind speed values as a gridded data set. This is usually computed in a geographical information system (GIS). Employing assumptions on the technology conversion efficiency to units of energy, it is possible to derive an energy estimate that corresponds to a defined area. This is typically expressed as Region X having a wind energy potential of Y units of energy per year.

CONSTRAINTS

Most wind energy resource studies start with a top-level theoretical resource, which is progressively reduced through consideration of so-called 'constraints'. These are considerations which tend to reduce the area that in reality will be available to the wind energy developer. For instance, they can be geographically delineated conservation areas, areas where the wind speed is not economically viable or areas of unsuitable terrain. Areas potentially available for development are sequentially removed from the area over which the energy resource is summed.

Different estimates of the potential energy resource can be calculated according to assumptions about the area that will be available for development. The resource without constraints is often called the 'theoretical' resource; consideration of technical constraints results in an estimation of a 'technical' resource; and consideration of planning, environmental and social issues results in the estimation of a so-called 'practical' resource. Such studies were common in the 1980s and 1990s, when wind energy penetration was relatively low, but have been overtaken somewhat by events, as penetrations of wind energy are now substantial in many European countries.

Wind Atlases

ONSHORE

Figure I.2.1 shows the onshore wind energy resource as computed on a broad scale for the European Wind Atlas. The map shows different wind speed regions. The wind speeds at a 50 m height above ground level within the regions identified may be estimated for different topographic conditions using the table below the figure.

The wind speed above which commercial exploitation can take place varies according to the specific market conditions. While countries such as Scotland clearly have exceptional potential, with rising fuel prices and consequently increasing power prices, every European country has a substantial technically and economically exploitable wind resource.

Figure I.2.1: European Wind Atlas, onshore

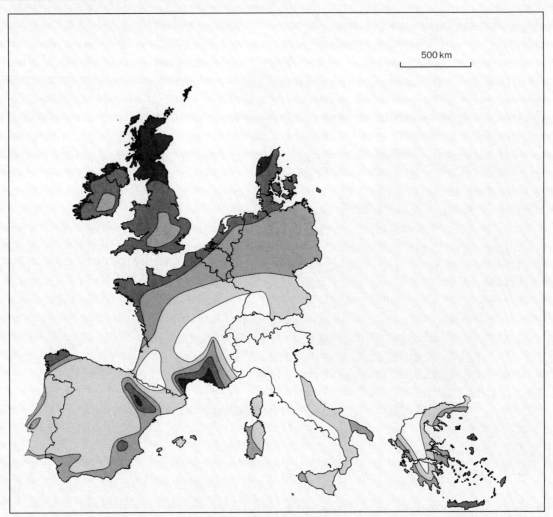

Wind resources at 50 metres above ground level for five different topographic conditions										
	Sheltered terrain		Open terrain		At a sea coast		Open sea		Hills and ridges	
	m/s	W/m²	m/s	W/m²	m/s	W/m²	m/s	W/m²	m/s	W/m²
	>6.0	>250	>7.5	>500	>8.5	>700	>9.0	>800	>11.5	>1800
	5.0–6.0	150–250	6.5–7.5	300–500	7.0–8.5	400–700	8.0–9.0	600–800	10.0–11.5	1200–1800
	4.5–5.0	100–150	5.5–6.5	200–300	6.0–7.0	250–400	7.0–8.0	400–600	8.5–10.0	700–1200
	3.5–4.5	50–100	4.5–5.5	100–200	5.0–6.0	150–250	5.5–7.0	200–400	7.0–8.5	400–700
	<3.5	<50	<4.5	<100	<5.0	<150	<5.5	<200	<7.0	<400

Source: Risø DTU (see Appendix A for colour version)

The European Wind Atlas employs meteorological data from a selection of monitoring stations, and shows the distribution of wind speeds on a broad scale. It has been used extensively by developers and governments in estimating the resource and regional variations. It is possible to map wind speeds at a higher resolution, using, for instance, more detailed topographical data and a larger sample size of meteorological data, in order to show more local variations in wind speed. This can be used by developers looking for sites in a particular country.

There are many examples of national, regional and local wind atlases, for Europe and the rest of the world, but they are far too numerous to list here. When investigating a particular region or country regarding its development potential, one of the first questions is always 'Is there a wind atlas for this area?'.

A review of national wind atlases for European countries has been undertaken for this edition of *Wind Energy – The Facts*, the results of which are shown in Table I.2.1. Where obtained and permission granted, map reproductions are contained in Appendix A. The European Wind Atlas resulted in the development of a wind-mapping tool called WAsP, and this is used widely for both broad-scale wind mapping and more site-specific applications. Table I.2.1 distinguishes between the use of WAsP and 'other' wind mapping methods.

OFFSHORE

Wind atlases for offshore are covered in Chapter I.5 (pages 107–124).

Local Wind Resource Assessment and Energy Analysis

The previous section presented wind maps for Europe and considered the wind resource at a strategic level. The purpose of this section is to consider the resource

Table I.2.1: Wind atlases, Europe

Country	Coverage in the European Wind Atlas	Other WAsP application	Other model
EU-27			
Austria		✓ [1]	
Belgium	✓		
Bulgaria			✓ [14]
Cyprus			
Czech Republic		✓ [1]	
Denmark	✓	✓ [2]	✓ [3]
Estonia		✓ [15]	✓ [16, 17]
Finland		✓ [4]	
France	✓		
Germany	✓ [5]		
Greece	✓		✓ [6]
Hungary		✓ [1]	
Ireland	✓	✓ [7]	✓ [8]
Italy	✓		✓ [9, 10]
Latvia		✓ [15]	[18]
Lithuania		✓ [15]	
Luxembourg	✓		
Malta			
The Netherlands	✓		
Poland			✓ [19]
Portugal	✓		
Romania			✓ [20]
Slovakia		✓ [1]	
Slovenia		✓ [1]	
Spain	✓		
Sweden		✓ [11]	
UK	✓		✓ [12]
Other countries			
Turkey			✓ [21]
Armenia			✓ [22]
Croatia		✓ [1]	
Georgia		✓ [23]	
Norway			✓ [13]
Russia		✓ [24]	
Switzerland			✓ [25]

Sources: Mortensen et al. (2007); for sources 1–25, see the section at the end of References

assessment and modelling at a local, wind farm, level. To the wind farm developer, the regional wind maps are valuable tools for site finding, but are not accurate enough to justify the financing of the development. Here it will be shown that the single most important characteristic of a site is its wind speed, and that the performance of a wind farm is very sensitive to uncertainties and errors in the basic wind speed estimate.

For the majority of prospective wind farms, the developer must undertake a wind resource measurement and analysis programme. This must provide a robust prediction of the expected energy production over its lifetime. This section discusses the issues that are pertinent to recording an appropriate set of site wind data, and the methodologies that can be used to predict the expected long-term energy production of a project. It is noted that a prediction of the energy production of a wind farm is possible using methods such as the wind atlas methodology within WAsP, using only off-site data from nearby meteorological stations. However, where the meteorological stations used have only data from low elevations, such as 10 m height, and/or the stations are located far from the site, such analyses are generally used only to assess the initial feasibility of wind farm sites. It is also possible to make predictions of the wind speed at a site using a numerical wind atlas methodology, based on a data source such as the 'reanalysis' numerical weather model data sets. Again, such data are usually used more for feasibility studies than final analyses. The text below describes an analysis where on-site wind speed and direction measurements from a relatively tall mast are available.

Figure I.2.2 provides an overview of the whole process. The sections below describe this process step by step. Appendix C provides a worked example of a real wind farm, for which these techniques were used to estimate long-term energy production forecast and compares this pre-construction production estimate with the actual production of the wind farm over the

first year of operation. It is noted that this example is from a wind farm constructed several years ago, so the turbines are relatively modest in size compared with typical current norms. Also, some elements of the analysis methods have altered a little – for example, a more detailed definition of the wind farm loss factor is now commonly used.

Figure I.2.2 represents a simplification of the process. In reality it will be necessary to also iterate the turbine selection and layout design process, based on environmental conditions such as turbine noise, compliance with electrical grid requirements, commercial considerations associated with contracting for the supply of the turbines and detailed turbine loading considerations.

THE IMPORTANCE OF THE WIND RESOURCE

Wind energy has the attractive attribute that the fuel is free and that this will remain the case for the project lifetime and beyond. The economics of a project are thus crucially dependent on the site wind resource. At the start of the project development process, the long-term mean wind speed at the site is unknown. To illustrate the importance of the long-term mean wind speed, Table I.2.2 shows the energy production of a 10 MW project for a range of long-term annual mean wind speeds.

It can be seen that when the long-term mean wind speed is increased from 6 to 10 m/s, about 67 per cent, the energy production increases by 134 per cent. This range of speeds would be typical of Bavaria at the low end and hilltop locations in Scotland or Ireland at the high end. As the capital cost is not strongly dependent on wind speed, the sensitivity of the project economics to wind speed is clear. Table I.2.2 illustrates the importance of having as accurate a definition of the site wind resource as possible.

The sensitivity of energy yield to wind speed variation varies with the wind speed. For a low wind speed

Figure I.2.2: Overview of the energy prediction process

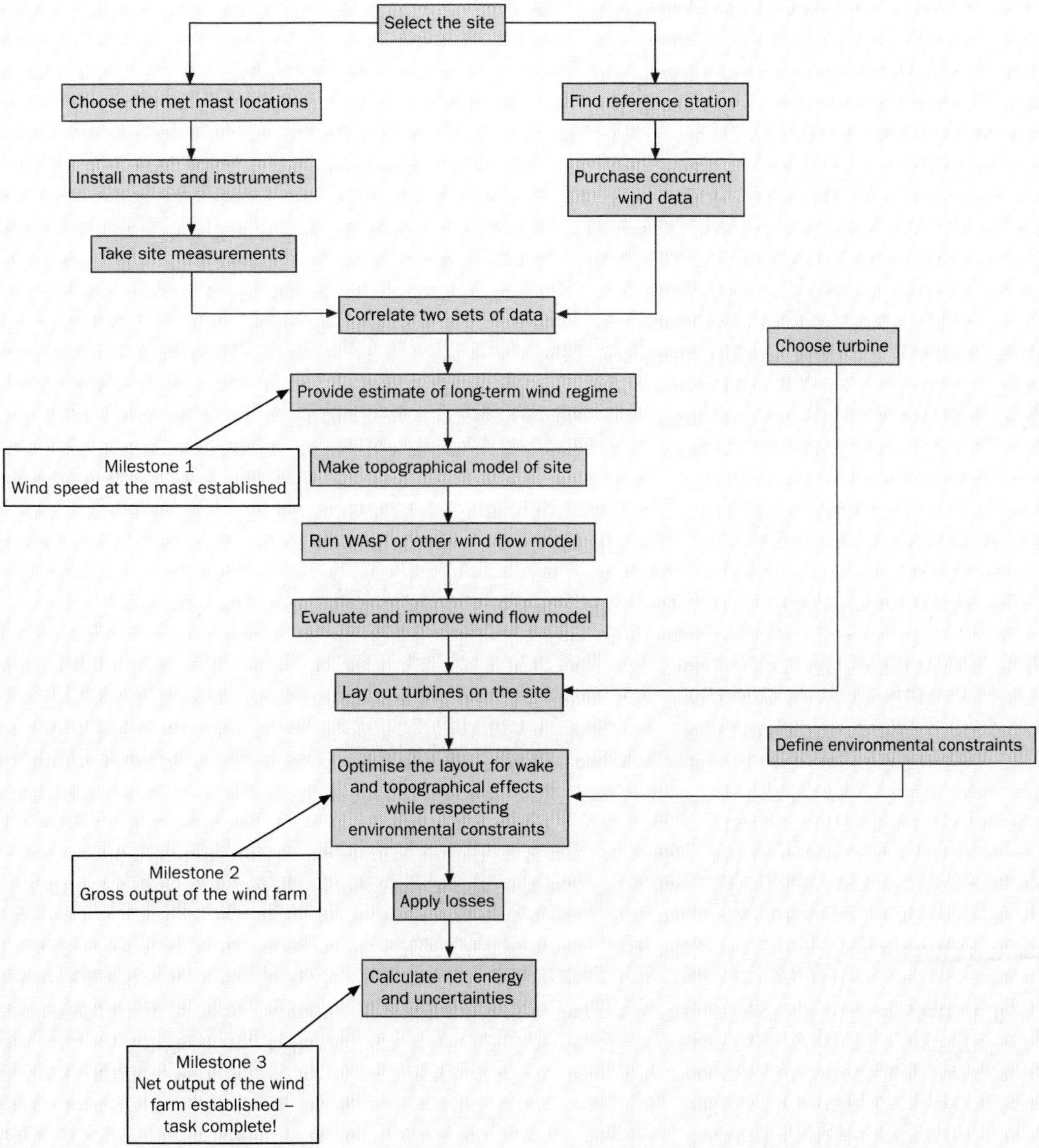

Note: For some sites, no suitable reference station is available. In such cases, only site data is used and longer on-site data sets are desirable.

Source: Garrad Hassan

Table I.2.2: Sensitivity of wind farm energy production to annual mean wind speed

Wind speed (m/s)	Wind speed normalised to 6 m/s (%)	Energy production of 10 MW wind farm (MWh/ annum)*	Energy production normalised to 6 m/s site (%)	Capital cost normalised to 6 m/s site (%)
5	83	11,150	63	100
6	100	17,714	100	100
7	117	24,534	138	102
8	133	30,972	175	105
9	150	36,656	207	110
10	167	41,386	234	120

Note: *Assumes typical turbine performance, air density of 1.225 kg/m^3, total losses of 12 per cent and Rayleigh wind speed distribution.

Source: Garrad Hassan

site, the sensitivity is greater than for a high wind speed site. For example, at a low wind speed site, a 1 per cent change in wind speed might result in a 2 per cent change in energy, whereas for a high wind speed site the difference might be only 1.5 per cent. Table I.2.2 is in fact a simplification of the reality of the situation, where different specifications of turbine model would typically be selected for low and high wind speed sites, but it serves to illustrate the importance of wind speed to energy production.

The commercial value of a wind farm development is therefore crucially dependent on the energy yield, which in turn is highly sensitive to the wind speed. A change of wind speed of a few per cent thus makes an enormous difference in financial terms for both debt and equity.

In summary, the single most important characteristic of a wind farm site is the wind speed. Thus every effort should be made to maximise the length, quality and geographical coverage across the wind farm site of the data collected. However, measurement is undertaken at the very beginning of the project, and some compromise is therefore inevitable.

BEST PRACTICE FOR ACCURATE WIND SPEED MEASUREMENTS

The results shown here illustrate the importance of having an accurate knowledge of the wind resource. A high-quality site wind speed measurement campaign is therefore of crucial importance in reducing the uncertainty in the predicted energy production of a proposed project. The goal for a wind measurement campaign is to provide information to allow the best possible estimate of the energy on the site to be provided. The question of how many masts to use and how tall they should be then arises.

Number and Height of Meteorological Masts

For a small wind farm site, it is likely that one meteorological mast is sufficient to provide an accurate assessment of the wind resource at the site. For medium wind farms, say in excess of 20 MW, located in complex terrain, it is likely that more than one mast will be required to give an adequate definition of the wind resource across the site. For large projects of around 100 MW, located in complex terrain, it is particularly important to take great care in 'designing' a monitoring campaign to record the necessary data for a robust analysis in a cost-effective way.

In simple terrain, where there is already a lot of experience and close neighbouring wind farms, the performance of these wind farms can be used instead of a measurement campaign. North Germany and Denmark are obvious examples. A great many turbines have been sited in this way. However, great caution must be exercised in extending this approach to more complex areas.

The locations and specifications of the mast or masts need to be considered on a site-specific basis, but generally, if there are significant numbers of turbines more than 1 km from a meteorological mast in terrain that is either complex or in which there is significant forestry, it is likely that additional masts will be required.

In such circumstances, discussion with the analyst responsible for assessing the wind resource at the site is recommended at an early stage.

Turning now to the height of the masts, it is known that the wind speed generally increases with height, as illustrated in Figure I.2.3.

Figure I.2.3 schematically shows the way in which the wind speed grows. This characteristic is called 'shear' and the shape of this curve is known as the 'wind shear profile'. Given the discussion above about the importance of accurate wind speed measurements, it is clear that it will be important to measure the wind speed as near to the hub height of the proposed turbine as possible. If a hub height measurement is not made, then it will be necessary to estimate the shear profile. This can be done, but it creates uncertainties. Commercial wind turbines typically have hub heights in the 60–120 m range. The costs of meteorological masts increase with height. Tilt-up guyed masts may be used up to heights of 60 or 80 m. Beyond such heights, cranes are required to install masts, which increases costs. If 'best practice' of a hub height mast is not

followed, then a reasonable compromise is to ensure that masts are no less than 75 per cent of the hub height of the turbines.

Specification of Monitoring Equipment and Required Signals

A typical anemometry mast will have a number of anemometers (devices to measure wind speed) installed at different heights on the mast, and one or two wind vanes (devices to measure wind direction). These will be connected to a data logger, at the base of a mast, via screened cables. It is unusual for there to be a power supply at a prospective wind farm site, so the whole anemometry system is usually battery operated. Some systems have battery charging via a solar panel or small wind turbine. For some systems, particularly in cold climates, the measurement of the temperature is important to assist with the detection of icing of the anemometers. In such circumstances, the use of heated or 'ice-free' anemometers is beneficial; however, their use without an external power source is usually impractical. Measurement of the pressure at the site is desirable but often not essential.

Remote sensing techniques are now being used to measure the wind speeds at wind farm sites. The technology available for this is being developed rapidly, and although the use of remote sensing devices on wind farms is not currently widespread, such devices are expected to become more widely used in the near future.

Remote sensing devices are essentially ground-based devices which can measure wind speeds at a range of heights without the need for a conventional mast. There are two main sorts of devices:

1. Sodar (SOund Detection And Ranging), which emits and receives sound and from this infers the wind speed at different heights using the doppler shift principle; and
2. Lidar (LIght Detection And Ranging), which also uses the doppler shift principle, but emits and receives light from a laser.

Figure I.2.3: The atmospheric boundary layer shear profile

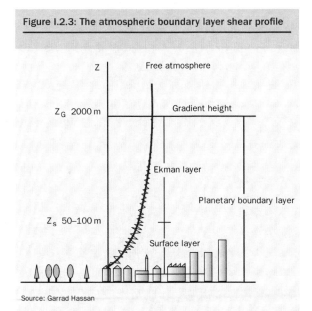

Source: Garrad Hassan

Sodar has been used for assessing wind farm sites for some years, particularly in the US and Germany. It is often used in combination with conventional anemometry and, historically, the results have been used to provide more information to better understand the patterns of the wind regime at a site, rather than necessarily using the data in a direct, quantitative way. Recently, some wind energy-specific Sodar products have come onto the market and experience is currently being gained from these new devices.

Lidar devices have made an entry into the wind market over the last one to two years, and two main commercial models are currently available. Published papers on the devices show they are capable of achieving impressive accuracy levels, and it is expected that their use in wind energy applications will increase.

The clear merit of remote sensing devices is that they do not need a mast. However, Lidar devices, in particular, are relatively expensive to purchase and both devices draw significantly more power than conventional anemometry, so for remote sites a local, off-grid power supply solution would be needed.

It is recommended that in-house or external experts are used to help make an informed decision about when and how to use remote sensing devices at potential wind farm sites.

Signals that would typically be recorded for each sensor, with a ten-minute averaging period, are as follows:
- mean wind speed;
- maximum three-second gust wind speed;
- true standard deviation of wind speed;
- mean wind direction;
- mean temperature; and
- logger battery voltage.

In recent years, it has become standard practice to download data remotely, via either modem or a satellite link. This approach has made managing large quantities of data from masts, on a range of prospective sites, significantly more efficient than manual downloading. It also has the potential to improve data coverage rates.

Recommendations provided by the International Electrotechnical Committee (IEC), the International Energy Agency (IEA) and the International Network for Harmonised and Recognised Wind Energy Measurement (MEASNET), provide substantial detail on minimum technical requirements for anemometers, wind vanes and data loggers. It is strongly recommended that anyone intending to make 'bankable' wind measurements should refer to these documents. Historically, a notable deviation from best practice, as defined in the IEC and IEA documents, is the use of anemometers that have not been individually calibrated for the assessment of the wind resource at the site. Each sensor will have a slightly different operational characteristic, as a result of variations in manufacturing tolerances. The use of individually calibrated anemometers has a direct impact on reducing the uncertainty in the predicted wind speed at a site and is therefore to be recommended.

Over the past decade, perhaps the most significant shortcoming of wind speed measurements at prospective wind farm sites has been the poor mounting arrangement of the sensors. There is an increasing body of measured data which has demonstrated that if the separation of anemometers from the meteorological mast, booms and other sensors is not sufficient, then the wind speed recorded by the sensor is not the true wind speed. Instead, it is a wind speed that is influenced by the presence of the other objects. The effect of the mast structure on the flow field around the mast top is illustrated in Figure I.2.4. The figure shows that there is a complicated flow pattern, which must be accommodated when mounting the anemometry.

These results have been predicted using a commercial computational fluid dynamics (CFD) code. It is important to be aware of the potential influence of the support structure on the measured data. Detailed guidance is provided in the International Energy Agency's Annex XI (1999), on specific separation distances which are required to reduce the influence of the support structure on the measurement to

Figure I.2.4: Predicted wind speed distribution around and above a meteorological mast using CFD

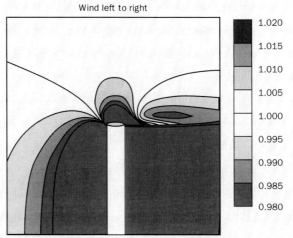

Wind into the page   Wind left to right

1.020
1.015
1.010
1.005
1.000
0.995
0.990
0.985
0.980

Source: Garrad Hassan

acceptable levels. Illustrative examples that demonstrate good and poor mounting arrangements are presented in Figure I.2.5.

If the guidance presented above is followed, a high-quality set of wind data should become available, in time, from a prospective site. The absolute minimum requirement is for data to cover one year, so that any seasonal variation can be properly captured. In addition to specifying and installing appropriate equipment, vigilance is required in the regular downloading and checking of data, to ensure that high levels of data coverage are achieved. It will be necessary to demonstrate, either internally or externally, the provenance of the data on which important financial decisions are being made. Therefore it is important to maintain accurate records regarding all aspects of the specification, calibration, installation and maintenance of the equipment used.

THE ANNUAL VARIABILITY OF WIND SPEED

A 'wind rose' is the term given to the way in which the joint wind speed and direction distribution is defined. An example is given in Figure I.2.6. The wind rose can be thought of as a wheel with spokes, spaced, in this example, at 30 degrees. For each sector, the wind is considered separately. The duration for which the wind comes from this sector is shown by the length of the

Figure I.2.5: Summary of good practice (left) and poor practice (right) mounting arrangements (arrow indicates dominant wind direction)

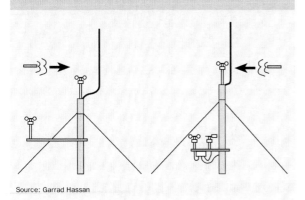

Source: Garrad Hassan

Figure I.2.6: Wind rose

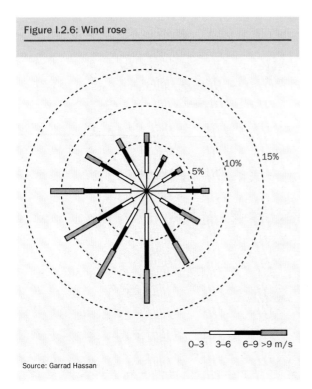

0–3   3–6   6–9 >9 m/s

Source: Garrad Hassan

spoke and the speed is shown by the thickness of the spoke. The design of a wind farm is sensitive to the shape of the wind rose for the site. In some areas, particularly in areas where the wind is driven by thermal effects, the wind can be very unidirectional. For example, in Palm Springs in the US, the wind comes from a sector 10 degrees wide for 95 per cent of the year. In this type of site, the wind farms tend to be arranged in tightly packed rows, perpendicular to the wind, with large spaces downwind. In northern Europe, the wind, although predominantly from the southwest, also comes from other directions for a significant amount of time, and hence wind turbines tend to be more uniformly spaced in all directions.

The description above focused on the wind speed and wind rose. The other important parameter determining the output of a wind farm is the wind speed distribution. This distribution describes the amount of time on a particular site that the wind speed is between different levels. This characteristic can be very important, but is

often inadequately treated. The distribution is important since it is the combination of the wind speed distribution and the power curve of the proposed turbine which together determine the energy production.

Consider, as an example, two sites, A and B. At one extreme, at Site A, the wind blows at 9 m/s permanently and the wind farm would be very energetic. At the other extreme, at Site B, let us assume that the wind blows at 4 m/s (below cut-in wind speed for a typical wind turbine) for one-third of the time, at 26 m/s (above cut-out wind speed for most turbines) for one-third of the time and at 9 m/s for one-third of the time. The mean wind speed would then be $(1/3) \times (4 + 9 + 26) = 13$ m/s, much higher than Site A, but the energy yield at Site B would be only a third of that at Site A.

These two examples are both unrealistic, but serve to illustrate a point – that wind speed alone is not enough to describe the potential energy from the site. Some more realistic site wind speed distributions are shown in Figure I.2.7. In this figure, the actual wind speed distribution is shown, as well as a 'Weibull fit' to the distribution. The Weibull distribution is a mathematical expression which provides a good approximation to many measured wind speed distributions. The Weibull distribution is therefore frequently used to characterise a site. Such a distribution is described by two parameters: the Weibull 'scale' parameter, which is closely related to the mean wind speed, and the 'shape' parameter, which is a measurement of the width of the distribution. This approach is useful since it allows both the wind speed and its distribution to be described in a concise fashion. However, as can be seen from the figure, care must be taken in using a Weibull fit. For many sites it may provide a good likeness to the actual wind speed distribution, but there are some sites where differences may be significant.

The annual variability in wind rose and wind speed frequency distribution is also important in assessing the uncertainty in the annual energy production of a wind farm; this is described in detail in the sections below.

Figure I.2.7: Examples of wind speed distributions

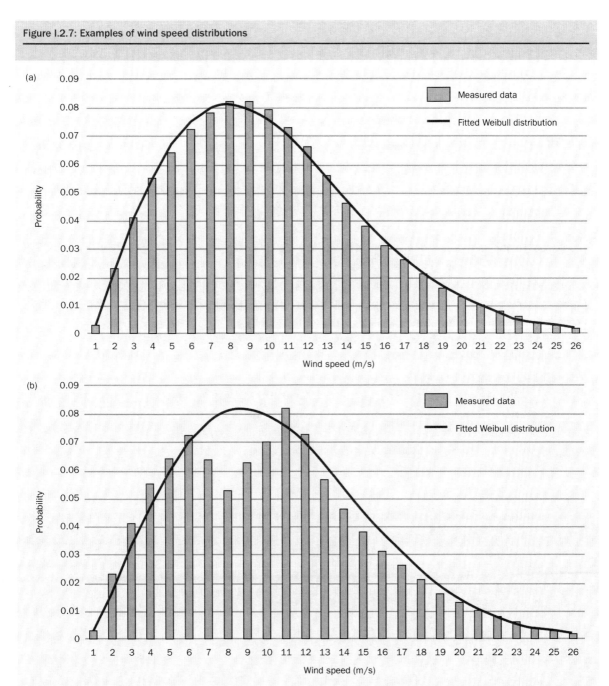

For the purpose of the example, only the variation in annual mean wind speed is considered, as the other factors usually have a secondary effect.

Variability of One-Year Periods

As discussed above, the annual variability of wind speed has a strong influence on the analysis methodologies developed for the assessment of the long-term wind resource at a site and the uncertainty in such predictions. Before describing typical methodologies used for the prediction of the long-term mean wind speed at a site, an example is given to illustrate typical levels of annual variability of wind speed. The example presented below seeks to answer the following questions:
- If there is one year of wind data available from a potential wind farm site, what error is likely to be associated with assuming that such data is representative of the long term?
- If, instead, there are three years of data available from the site, how does the picture change?

Figure I.2.8a presents the annual mean wind speed recorded at Malin Head Meteorological Station in Ireland over a 20-year period. It can be seen that there is significant variation in the annual mean wind speed, with maximum and minimum values ranging from less than 7.8 m/s to nearly 9.2 m/s. The standard deviation of annual mean wind speed over the 20-year period is approximately 5 per cent of the mean.

Table I.2.3 presents the average and annual maximum and minimum wind speeds. As an illustration, the equivalent annual energy productions for the example 10 MW wind farm case described earlier are also presented in the table.

Table I.2.3 shows that, had wind speed measurements been carried out on the site for just one year, on the assumption that this year would be representative of the long term, then the predicted long-term wind speed at the site could have had a 10 per cent margin of error. It is often the case that little on-site data is available and hence this situation can arise. In terms of energy production, it is evident that the predicted energy production could be in error by some 14 per cent if the above assumption had been made. For a lower wind speed site, a 10 per cent error in wind speed could easily have a 20 per cent effect on energy production, owing to the higher sensitivity of energy

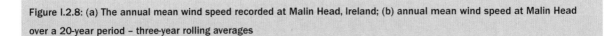

Figure I.2.8: (a) The annual mean wind speed recorded at Malin Head, Ireland; (b) annual mean wind speed at Malin Head over a 20-year period – three-year rolling averages

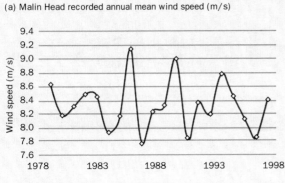

(a) Malin Head recorded annual mean wind speed (m/s)

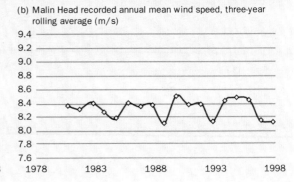

(b) Malin Head recorded annual mean wind speed, three-year rolling average (m/s)

Source: Garrad Hassan

Table I.2.3: Wind speeds and energy production for the average, lowest and highest wind speed year in the period 1979 to 1998, based on a nominal 10 MW wind farm at Malin Head

	Annual mean wind speed (m/s)	Percentage of average year (%)	Energy production (MWh/annum)	Percentage of average year (%)
Lowest wind speed year (1987)	7.77	93.3	29,491	89.8
Average year	8.33	100.0	32,847	100.0
Highest wind speed year (1986)	9.16	110.0	37,413	113.9

production to changes in wind speed at lower wind speeds.

Variability of Three-Year Periods

Figure I.2.8b illustrates the same data as presented in Figure I.2.8a, but in this instance, a three-year rolling average of the data has been taken. It is immediately apparent that the variability in the mean wind speed over three-year periods is substantially reduced compared with that of one-year periods.

The results presented in Table I.2.3 are reproduced in Table I.2.4, this time based on the highest and lowest three-year averages.

Table I.2.4 illustrates that, if three years' worth of data are available from a site, the maximum deviations in wind speed and energy production over these periods, from long-term averages, is substantially reduced. The deviations of 10 and 14 per cent in wind speed and energy for the analysis based on one-year data sets reduces to deviations of 3 and 4 per cent respectively when three-year periods are considered.

While the results presented here are site-specific, they are broadly representative of any wind farm in Europe. The reliability of long-term data and the consistency of the wind is a matter that is central to the commercial appraisal of a wind farm. Substantial work has been undertaken to try and identify some key characteristics of the long-term behaviour of wind. This effort consisted of the identification of reliable long-term data sets around the world and attempting to identify common characteristics. One of the results of this approach is illustrated in Figure I.2.9. Data sets of around 30 years in duration have been assembled, and for each site the mean of the 30 annual figures was calculated, together with their standard deviation. The ratio of the standard deviation to the mean was then calculated and it was found that the ratio varied very little from location to location. This feature was observed in many areas across the world – data was collected from Australia, Japan and the US, as well as Europe. This result is useful in determining how much variation in wind is to be expected.

In summary, this work indicated that the annual variability of long-term mean wind speeds at sites across Europe tends to be similar, and can reasonably be characterised as having a normal distribution with a standard deviation of 6 per cent. This result plays an

Table I.2.4: Wind speeds and energy production for the average, lowest and highest three-year periods within the period 1979 to 1998

	Annual mean wind speed (m/s)	Percentage of average year (%)	Energy production (MWh/annum)	Percentage of average year (%)
Lowest wind speed year (1989)	8.10	97.2	31,540	96.0
Average year	8.33	100.0	32,847	100.0
Highest wind speed year (1990)	8.51	102.2	33,871	103.1

Figure I.2.9: Wind map of Europe – Inter-annual variations shown as standard deviation as a percentage of mean

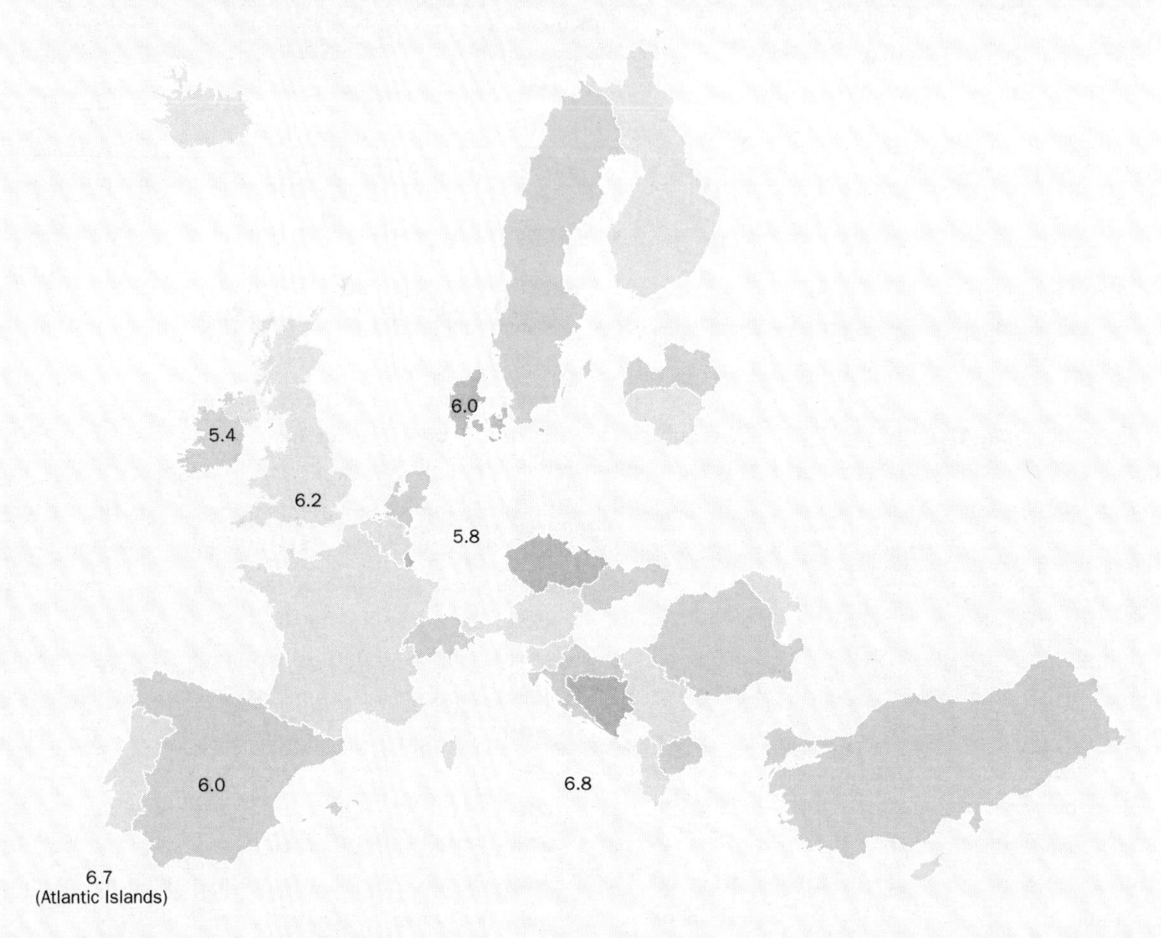

Source: Garrad Hassan

important role in the assessment of the uncertainty in the prediction of wind farm energy production.

ANALYTICAL METHODS FOR THE PREDICTION OF THE LONG-TERM WIND REGIME AT A SITE

From the above, it is clear that the key element of the assessment of the energy production of a proposed wind farm site is the prediction of the long-term wind regime at the site. The outcome of the analyses described in this section is a long-term wind speed distribution, together with the wind rose. Other meteorological inputs to the energy production analysis are the long-term site air density and site turbulence intensity, a measurement of the 'gustiness' of the wind. These, while still important, are of secondary influence to the energy production of the wind farm, and therefore their derivation is not considered in detail here. It should be noted, however, that the turbulence intensity is very important in determining the loading on a wind turbine, and hence its life expectancy.

Overview

When assessing the feasibility of a potential wind farm site or where, for strategic purposes, an indication of the variation of wind speed over an area is required, it is unlikely that any wind data from a relatively tall meteorological mast will be available. If there is no on-site data available, the 'wind atlas method' (Troen and Petersen, 1989) is commonly used. This method uses modelling techniques to translate the long-term reference station data to the site. This method can be quite accurate in many cases, but should not replace on-site measurements for more formal wind farm energy assessment. It is also possible to make predictions of the wind speed at a site using a numerical wind atlas methodology, based on a data source such as the 'reanalysis' numerical weather model data sets. Again, such analyses are generally used to assess the feasibility of a site or sites for development.

There are essentially two methods that can be used for the prediction of the long-term wind resource at a site where on-site measurements are available:

- Method 1: Correlate on-site wind data with wind data recorded at a long-term reference station; and
- Method 2: Use only on-site wind data.

Unless a long data set is already available from a site, it is desirable to use Method 1 for the prediction of the long-term wind resource at a site. Typically, a reliable result can be obtained with as little as one year of site data. As illustrated by the example presented for Malin Head Meteorological Station above, if Method 1 cannot be used and Method 2 is used with only one year of data, the uncertainty in the assumption that the year of data recorded is representative of the long term is substantial. Therefore, it is normal practice to find a suitable source of longer-term data in the vicinity of the wind farm site. This allows a correlation analysis to be undertaken and, if only relatively short data sets are available from the site, it is likely to result in an analysis with significantly less

uncertainty than that resulting from use of the site data alone. However, before a data set from a long-term reference station can be used in an analysis, it is vital that thorough checks on the validity of the data for the analysis are undertaken.

Before discussing the details of this approach, it may be helpful to consider the broader picture. It would be ideal if every site benefited from a long-term data set of, say, ten years, measured at hub height. Now and again this happens, but it is very rare. It is, therefore, necessary either to use limited on-site data or to try and use other data to gain a long-term view. The correlation approach can be thought of in the following way. Data is gathered on the site using good quality calibrated equipment. This data provides absolute measurements of the wind speed on the site during the measurement period. If it can be established that there is a close relationship (a good correlation) between the site data and a reference mast, then it will be possible by using the long-term reference data and the relationship to recreate the wind speeds on the site. Thus it is possible to 'pretend' that the long-term wind speed records exist on the site. If a good correlation exists, this is a very powerful technique, but if the correlation is weak, it can be misleading and hence must be used with caution.

Necessary conditions for an off-site wind data set to be considered as a long-term reference are set out below:

- The reference data set includes data which overlaps with the data recorded on site.
- It can be demonstrated that the data has been recorded using a consistent system over the period of both the concurrent and longer-term data. This should include consideration not just of the position and height of the mast and the consistency of equipment used, but also potential changes in the exposure of the mast. For example, the construction of a new building at an airport or the erection of a wind farm near an existing mast will corrupt the data. The absolute values recorded at the

reference station are not important, but any changes to it, in either process or surrounding environment, will render it useless as a reference site. This investigation is therefore very important and is usually done by a physical visit to the site, together with an interview with site staff.

- The exposure of the reference station should be good. It is rare that data recorded by systems in town centres, or where the mean wind speed at the reference station is less than half that of the site, prove to be reliable long-term reference data sets.
- The data is well correlated with that recorded at the site.

Where there have been changes in the consistency of data sample at a reference long-term data source, or where a reliable correlation cannot be demonstrated, it is important that the use of a prospective source of long-term data is rejected. If no suitable reference meteorological station can be found, then the long-term wind resource can only be derived from the data recorded at the site itself. It is likely that longer data sets of two or more years are required to achieve similar certainty levels to those that would have been obtained had a high-quality long-term reference data set been available.

Experience of the analysis of wind energy projects across Europe has indicated that the density of public sources of high-quality wind data is greater in northern Europe than in southern Europe. This observation, combined with the generally more complex terrain in much of southern Europe, often leads to analyses in southern Europe being based on only the data recorded at the wind farm site, or other nearby wind farm sites. In contrast, for analyses in northern Europe, correlation of site data to data recorded at national meteorological stations is more common. Clearly, this observation is a generalisation, however, and there are numerous exceptions. Thus the establishment of a good set of long-term reference masts, specifically for wind energy use in areas of Europe where wind energy projects are likely to be developed, would be an extremely valuable asset. An EU-wide network of this sort would be highly beneficial.

Correlation Methodologies

The process of comparing the wind speeds on the site with the wind speeds at the reference station, and using the comparison to estimate the long-term wind speed on the site, is called 'measure correlate predict' (MCP). This process is also described in some detail in Appendix D, along with a more detailed discussion of the merits of different methodologies. It is difficult to provide definitive guidance on how poor the quality of a correlation can be before the reference station may no longer be reliably used within an analysis. However, as a general rule, where the Pearson coefficient (R^2) of an all-directional monthly wind speed correlation is less than 0.8, there is substantial uncertainty in using long-term data from the reference station to infer long-term wind conditions at the wind farm site.

Once the MCP process has been completed, an estimate exists of the long-term wind speed at the site. This stage – Milestone 1 in Figure I.2.2 – is a very important one, since the position has now been reached at which we have knowledge of the long-term wind speed behaviour at a single point (or points if there are multiple masts) on the site. This estimate will contain both the mean long-term expected value and the uncertainty associated with that value. So far, however, we know nothing of the distribution of the wind speed across the site and neither have we considered the way in which the wind speed values can be converted into energy.

THE PREDICTION OF THE ENERGY PRODUCTION OF A WIND FARM

In order to predict the energy production of the wind farm it is necessary to undertake the following tasks:
- predict the variation in the long-term wind speed over the site at the hub height of the machines, based on the long-term wind speeds at the mast locations;

- predict the wake losses that arise as a result of one turbine operating behind another – in other words in its wake; and
- calculate or estimate the other losses.

Information Required for an Analysis

In addition to the wind data described in the earlier sections, inputs to this process are typically the following:
- wind farm layout and hub height;
- wind turbine characteristics, including power curve (the curve which plots the power output of a turbine as a function of the wind speed) and thrust curve (the equivalent curve of the force applied by the wind at the top of the tower as a function of wind speed);
- predicted long-term site air density and turbulence intensity (the turbulence intensity is the 'gustiness' of the wind);
- definition of the topography over the site and surrounding area; and
- definition of the surface ground cover over the site and surrounding area.

Energy Production Prediction Methodologies

Typically, the prediction of the variation in wind speed with height, the variation in wind speed over the site area and the wake interaction between wind turbines are calculated within a bespoke suite of computer programs, which are specifically designed to facilitate accurate predictions of wind farm energy production. The use of such tools allows the energy production of different options of layout, turbine type and hub height to be established rapidly once models are set up. Such programs are commonly described as 'wind farm design tools' (WFDTs).

Within the WFDT, site wind flow calculations are commonly undertaken, using the WAsP model, which has been widely used within the industry over the past decade. There are also other commercial models, physically similar to WAsP, that are sometimes used. This area of the wind farm energy calculation is in need of the greatest level of fundamental research and development. Flow models that can be used in commercial wind farm development have to be quick to execute and have to be reliable and consistent. At present, the industry generally opts for simple but effective tools such as WAsP. However, in the last few years, use of computational fluid dynamics (CFD) codes is increasing, although CFD tools are typically used in addition to and not instead of the more simple tools, to investigate specific flow phenomena at more complex sites. CFD tools need to be used with care, as the results are sensitive to modelling assumptions. CFD codes are also substantially more computationally onerous to run. Typical use of CFD tools is, first, to give another estimate of the local acceleration effects or variability at the site, and second, to identify hot spots, in other words areas where the wind conditions are particularly difficult for the wind turbines. In particular, such tools are starting to be used to assist the micro-siting of wind turbines on more complex terrain sites.

Thus, the challenge is to take a topographical map and the long-term wind rose at a known point on the map, and use this information to calculate the long-term wind speed at all the points on the map where the intention is to place wind turbines. A typical set of topographical input and wind contour output (normalised to the wind speed at the location of an on-site mast) is shown in Figure I.2.10 for a hilly area of approximately 6 by 4 km.

The WAsP model does have shortcomings under certain topographical and flow conditions, as the model is designed for specific wind flow conditions. For slopes steeper than approximately 20°, where the flow will separate, the model is beyond its formal bounds of validity. It therefore needs to be used with care and experience and not as a 'black box'. As indicated above, it does not include 'viscous effects', which cause the wind to 'separate' as it flows over a sharp change in topography. The WAsP model will follow the

Figure I.2.10: Input topography (left) and output-normalised wind speed (right)

 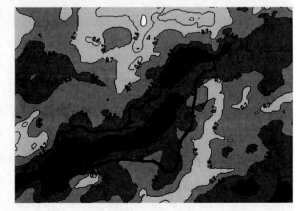

Source: Garrad Hassan

terrain, whereas the real wind will behave as shown in Figure I.2.11. Therefore in complex terrain, the manual interpretation of calculation results is required.

There is an enormous amount of work in progress in all aspects of engineering, quite distinct from the wind energy industry, in which developments are being made in the numerical prediction of complicated flows. Most notably, these efforts centre on aerospace problems – accurate flow over aircraft wings and fuselage, for example, or internal flows in turbo-machinery. Efforts are now being made to apply such models to the arbitrary terrain that defines a wind farm (complex topographies characterising wind farms). There is still a long way to go before these models can be considered sufficiently reliable, however, to replace rather than complement conventional tools.

The energy estimate is only as good as its weakest link, and hence its accuracy is largely defined by this step – the topographical wind model. Data sets now exist that can be used for the validation of new codes, and developments are expected. The task is, nevertheless, a demanding one, and accurate calculation of the flow over steep terrain is challenging. At present, it is necessary to use a mixture of computation and

human insight. For such sites, there is currently no alternative to a comprehensive monitoring campaign to provide data that may be used to initiate localised flow modelling.

Once the topographical effects on the flow have been computed, it is necessary to determine how the individual turbines affect one another – the wake effects. If a turbine is working downstream of another (in its wake), then it will see less wind than it would if it were in the free stream. For some types of wind rose and wind farm design this effect can be significant, of the order of 15 per cent in lost energy, and hence needs to be carefully calculated. The models that

Figure I.2.11: An example of flow separation over a hill

Flow separation

Source: Garrad Hassan

estimate this loss are known as 'wake models'. Different complexities of wake model are used in the various commercially available WFDTs. These tools are now well validated for a variety of different types of wind farm layout. However, it is well known that they do not work well for very tightly packed wind farms, such as those described above at Palm Springs, and further fundamental work is required to improve the modelling in this area. Also, with the advent of data from large offshore wind farms, it is apparent that some adjustments are required to conventional wake models for very large wind farms.

It is important to appreciate that as the distance of the turbine from the meteorological mast increases, the uncertainty in the prediction also increases. This increase in uncertainty is typically more rapid in complex terrain than in simple terrain. Experience of the decrease in accuracy with distance from the mast, when using models such as WAsP, is inherent in making recommendations regarding the appropriate number of meteorological masts for a wind farm site, as discussed above. The WFDTs also allow environmental constraints to be included – areas of the site which may not be used because of rare flora or fauna, noise constraints, visual intrusion, shadow flicker or land ownership, for example. These considerations are discussed in more detail below.

The ability of WFDTs to provide an integrated model of a wind farm also allows them to be used to optimise the wind farm design. This task is performed in an automatic process. It moves the turbines to provide the best possible compromise between concentration of the wind turbines at the maximum wind speed areas of the site, which maximises wake losses (loss of energy by turbulence effects of a wind turbine on its neighbours), and the spreading out of the turbines all over the site to minimise wake loss. This process can be undertaken successfully at the same time as observing the environmental and ownership constraints. The development of these tools has been a significant development in wind farm design. The successful completion of an energy calculation using a WFDT may be considered as Milestone 2 of Figure I.2.2.

Wind Farm Energy Loss Factors

When WFDTs have been used to predict the output of a wind farm, it is necessary to estimate or calculate a range of potential sources of energy loss. There may be considered to be six main sources of energy loss for wind farms, each of which may be subdivided into more detailed loss factors:

1. the wake effect;
2. availability;
3. electrical efficiency;
4. turbine performance;
5. environmental losses; and
6. curtailments.

A comprehensive list of potential losses is presented in Table I.2.5 below. Several of the loss factors will not be relevant to most projects, but they are listed here for the sake of completeness. Wind farm availability and the influence of tree growth on energy production may be time-dependent factors. As the values in the table below are site-specific, example values have not been presented, although in aggregate the total losses for a wind farm site would typically be in the 10–20 per cent range.

The loss factors used to derive the wind farm net energy output prediction are described below. For each loss factor, a general description is given.

Gross Energy Output

The gross energy production is the energy production of the wind farm obtained by matching the predicted free stream hub height wind speed distribution at each turbine location and the manufacturer's supplied turbine power curve. In defining the gross energy output, it is assumed that there are no wake interactions between the turbines and none of the loss factors

Table I.2.5: Comprehensive list of loss factors

	Wind farm rated power	##.#	MW
	Gross energy output	##.#	GWh/annum
1	**Wake effect**		
1a	Wake effect internal	##.#	%
1b	Wake effect external	##.#	%
1c	Future wake effect	##.#	%
2	**Availability**		
2a	Turbine availability	##.#	%
2b	Balance of plant availability	##.#	%
2c	Grid availability	##.#	%
3	**Electrical efficiency**		
3a	Operational electrical efficiency	##.#	%
3b	Wind farm consumption	##.#	%
4	**Turbine Performance**		
4a	Generic power curve adjustment	##.#	%
4b	High wind speed hysteresis	##.#	%
4c	Site-specific power curve adjustment	##.#	%
5	**Environmental**		
5a	Performance degradation – non-icing	##.#	%
5b	Performance degradation – icing	##.#	%
5c	Icing shutdown	##.#	%
5d	Temperature shutdown	##.#	%
5e	Site access	##.#	%
5f	Tree growth (may be time-dependant)	##.#	%
6	**Curtailments**		
6a	Wind sector management	##.#	%
6b	Grid curtailment	##.#	%
6c	Noise, visual and environmental curtailment	##.#	%
	Net energy output	##.#	GWh/annum

Note: Sample values for each variable are not included as there are wide ranges of possible values.

listed in the remainder of the energy table are applied. This result includes adjustments to the power curve to account for differences between the predicted long-term annual site air density and the air density to which the power curve is referenced. It also includes the effect of the terrain on the flow.

Wake Effect

Wind turbines extract energy from the wind and downstream there is a wake from the wind turbine, where wind speed is reduced. As the flow proceeds downstream, there is a spreading of the wake and the wake recovers towards free stream conditions. The wake effect is the aggregated influence on the energy production of the wind farm which results from the changes in wind speed caused by the impact of the turbines on each other.

Wake Effect Internal

This is the effect that the wind turbines within the wind farm being considered have on each other.

Wake Effect External

This is the effect that the wind turbines from neighbouring wind farms (if any) have on the wind farm being considered.

Future Wake Effect

Where future wind farms are to be constructed in the vicinity of the project under consideration, the wake effect of these has to be estimated and taken into account. If appropriate, this factor can be derived as a profile over the project lifetime.

Availability

Wind turbines, the balance of plant infrastructure and the electrical grid will not be available the whole time. Estimates are included for likely levels of availability for these items, averaged over the project lifetime.

Turbine Availability

This factor defines the expected average turbine availability of the wind farm over the life of the project. It represents, as a percentage, the factor which needs to be applied to the gross energy to account for the loss of energy associated with the amount of time the turbines are unavailable to produce electricity.

Balance of Plant (BOP) Availability

This factor defines the expected availability of the turbine transformers, the on-site electrical infrastructure and the substation infrastructure up to the point of connection to the grid of the wind farm. It represents, as a percentage, the factor that needs to be applied to the gross energy to account for the loss of energy associated with the downtime of the balance of plant.

Grid Availability

This factor defines the expected grid availability for the wind farm in mature operation. It is stressed that this factor relates to the grid being outside the operational parameters defined within the grid connection agreement, as well as actual grid downtime. This factor also accounts for delays in the wind farm coming back to full operation following a grid outage. It represents, as a percentage, the factor that needs to be applied to the gross energy to account for the loss of energy associated with the downtime of the grid connection.

Electrical Efficiency

There will be electrical losses experienced between the low voltage terminals of each of the wind turbines and the wind farm point of connection, which is usually located within a wind farm switching station.

Operational Electrical Efficiency

This factor defines the electrical losses encountered when the wind farm is operational and which will be manifested as a reduction in the energy measured by an export meter at the point of connection. This is presented as an overall electrical efficiency, and is

based on the long-term average expected production pattern of the wind farm.

Wind Farm Consumption

This factor defines the electrical efficiency including electrical consumption of the non-operational wind farm, such as consumption by electrical equipment within the turbines and substation. In most cases, this issue is omitted from the list of loss factors, and is instead considered as a wind farm operational cost. However, for some metering arrangements, it may be appropriate to include this as an electrical efficiency factor, rather than an operational cost, and so this factor is included within the table.

Turbine Performance

In an energy production calculation, a power curve supplied by the turbine supplier is used within the analysis.

Generic Power Curve Adjustment

It is usual for the supplied power curve to represent accurately the power curve that would be achieved by a wind turbine on a simple terrain test site, assuming the turbine is tested under an IEC power curve test. For certain turbine models, however, there may be reason to expect that the supplied power curve does not accurately represent the power curve that would be achieved by a wind turbine on a simple terrain site under an IEC power curve test. In such a situation, a power curve adjustment is applied.

High Wind Hysteresis

Most wind turbines will shut down when the wind speed exceeds a certain limit. High wind speed shutdown events can cause significant fatigue loading. Therefore, to prevent repeated start-up and shutdown of the turbine when winds are close to the shutdown threshold, hysteresis is commonly introduced into the turbine control algorithm. Where a detailed description of the wind turbine cut-in and cut-out parameters are available, this is used to estimate the loss of production due to high wind hysteresis, by repeating the analysis using a power curve with a reduced cut-out wind speed.

Site-Specific Power Curve Adjustment

Wind turbine power curves are usually based on power curve measurements, which are made on simple terrain test sites. Certain wind farm sites may experience wind flow conditions that materially differ from the wind flow conditions seen at simple terrain test sites. Where it is considered that the meteorological parameters in some areas of a site differ substantially from those at a typical wind turbine test station, then the impact on energy production of the difference in meteorological parameters at the site compared with a typical power curve test site is estimated. This may be undertaken where turbulence or up-flow angle are considered to be substantially different at the wind farm site to that which is experienced at a typical test site, and sufficient data is available to make the appropriate adjustments.

Environmental

In certain conditions, dirt can form on the blades, or over time the surface of the blade may degrade. Also, ice can build up on a wind turbine. These influences can affect the energy production of a wind farm in the ways described below. Extremes of weather can also affect the energy production of a wind farm; these are also described below.

Performance Degradation – Non-icing

The performance of wind turbines can be affected by blade degradation, which includes the accretion of dirt (which may be washed off by rain from time to time), as well as physical degradation of the blade surface over prolonged operation.

Performance Degradation – Icing

Small amounts of icing on the turbine blades can change the aerodynamic performance of the machine, resulting in loss of energy.

Icing Shutdown

As ice accretion becomes more severe, wind turbines will shut down or will not start. Icing can also affect the anemometer and wind vane on the turbine nacelle, which may also cause the turbine to shut down.

Temperature Shutdown

Turbines are designed to operate over a specific temperature range. For certain sites, this range may be exceeded, and for periods when the permissible temperature range is exceeded, the turbine will be shut down. For such sites, an assessment is made to establish the frequency of temperatures outside the operational range, and the correlation of such conditions with wind speed. From this, the impact on energy production is estimated.

Site Access

Severe environmental conditions can influence access to more remote sites, which can affect availability. An example of this might be an area prone to severe snow drifts in winter.

Tree Growth/Felling

For wind farm sites located within or close to forests or other areas of trees, the impact of how the trees may change over time and the effect that this will have on the wind flow over the site, and consequently the energy production of the wind farm, must be considered. The impact of the future felling of trees, if known, may also need to be assessed.

Curtailments

Some or all of the turbines within a wind farm may need to be shut down to mitigate issues associated with turbine loading, export to the grid or certain planning conditions.

Wind Sector Management

Turbine loading is influenced by the wake effects from nearby machines. For some wind farms with particularly close machine spacing, it may be necessary to shut down certain turbines for certain wind conditions. This is referred to as wind sector management, and will generally result in a reduction in the energy production of the wind farm.

Grid Curtailment

Within certain grid connection agreements, the output of the wind farm is curtailed at certain times. This will result in a loss of energy production. This factor also includes the time taken for the wind farm to become fully operational following grid curtailment.

Noise, Visual and Environmental Curtailment

In certain jurisdictions, there may be requirements to shut down turbines during specific meteorological conditions to meet defined noise emission or shadow flicker criteria at nearby dwellings or due to environmental requirements related to birds or bats.

DEFINITION OF UNCERTAINTY IN THE PREDICTED ENERGY PRODUCTION

Uncertainty analysis is an important part of any assessment of the long-term energy production of a wind farm. Although an uncertainty analysis needs to be considered on a site-specific basis, the process can be described as follows:

- identify the different inputs to and processes within the analysis;
- assign an uncertainty to each of these elements, both in terms of the magnitude of the uncertainty and the shape of the distribution;
- convert each of the uncertainties into common units of energy;
- combine the various uncertainties to define a total uncertainty for the entire prediction; and
- present uncertainty statistics at requested levels.

Research work reported in Raftery et al. (1999) defined a comprehensive risk register for wind power projects and included detailed Monte Carlo-based analysis techniques to assess the uncertainty in the results obtained. Based on the results of this work, use of an uncertainty analysis with a number of simplifying assumptions can be justified. The main simplifying assumptions are that it is reasonable to consider a relatively small number of key uncertainties and that these individual uncertainties can be assumed to be normally distributed. Making these assumptions, it is possible to define energy production levels with a defined probability of those levels being exceeded.

It is common to present uncertainty results for both a long future period of, say, ten years and also for a shorter future period of one year. It is now normal practice for such figures, in parallel with a central estimate for the production of a wind farm, to be used to inform investment decisions for projects.

The uncertainty analyses presented within energy assessments typically assume that the turbines will perform exactly to the defined availability and power performance levels. The power performance and availability levels are usually covered by specific warranty arrangements, and hence any consideration of the uncertainty in these parameters needs machine-specific and contract-specific review, which is generally outside the scope of a 'standard' energy analysis. However, it is increasingly the norm to assign a moderate uncertainty to the estimated availability, loss factor and power performance factors, to reflect that small deviations from expected availability and power performance levels may not be sufficient to trigger damage payments under the warranty. Uncertainty in the energy estimates is a vital part of the result.

Forecasting

This chapter has so far considered only the industry's ability to estimate long-term energy production for a wind farm. Usually, this is the most important task, since, to date, most of the power purchase agreements are 'take or pay', meaning that the utility or other customer is obliged to buy all the energy produced by the wind farm. As the penetration of wind power generation increases (in terms of the overall energy mix), fluctuations in energy output (caused by variations in wind speed) will be more visible on the electrical system. Transmission system operators (TSOs) working to balance supply and demand on regional or national grid systems will need to predict and manage this variability to avoid balancing problems. The point at which this is required changes from system to system, but it has been observed as becoming important when penetration of wind energy reaches about 5 per cent of installed capacity.

As the level of penetration of wind energy into individual grids increases, it will become necessary to make wind farms appear much more like conventional plants and hence it will be necessary to forecast, at short to medium timescales (one hour to seven days), how much energy will be produced. In some countries, forecasting is already required. New wind farms in California are required to 'use the best possible means available' to forecast the output, and send such estimates to Cal ISO (the California independent system operator). In European countries where there is already a high level of penetration, such as Spain, Germany and Denmark,

operators, managers and TSOs are routinely forecasting the output from their wind farms. These forecasts are used to schedule the operations of other plants, and are also used for trading purposes.

Forecasting the wind energy production will grow in importance as the level of installed capacity grows. The wind industry must expect to do its very best to allow the TSOs to use wind energy to its best effect, which means aggregated output forecasts from wind farms must be accurate. In the UK, where the market is already deregulated, energy traders are using sophisticated forecasts to trade wind energy on the futures market.

At the same time as improving the predictability of the output of wind energy plants through improvements in forecasting techniques, awareness of the true behaviour of conventional plants should be considered. In order to provide the best mix of plants and technologies, it will be important for all the different energy forms to be considered on an equitable basis. Therefore the proper, formal statistical analysis of both renewable and conventional plants is important. This task should be considered as an essential element of a wind energy development strategy and must be conducted on a total power system approach.

As a result of its strategic importance as described above, forecasting has been the focus of considerable technical attention in recent years. A good source of general review materials, as well as detailed papers, can be found in Landberg et al. (2003). Although there are a variety of different techniques being used, they all share similar characteristics. It is therefore possible to provide a generic description of the techniques presently being used, whereby data is provided by a weather forecast, and production data is provided by the wind farms. The two sets of data are combined to provide a forecast for future energy production. Figure I.2.12 provides a schematic picture of typical forecasting approach.

To integrate wind energy successfully into an electricity system at large penetration levels of more than

Figure I.2.12: A schematic representation of a forecasting approach

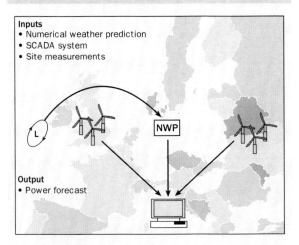

Inputs
- Numerical weather prediction
- SCADA system
- Site measurements

NWP

L

Output
- Power forecast

Source: Garrad Hassan

NWP model, and are often termed 'storm-scale' or 'convective-scale'; they aim to model local thermal and terrain effects that are not apparent at the coarse scale.

The following aspects of the physical model approach need to be considered:

- skill – the implementation of local meso-scale models requires the skill and competency of a meteorologist; there is always the possibility of a poorly formed model introducing further errors; and
- computational requirements – the formulation and execution of the models are computationally expensive.

OVERVIEW OF THE METHOD

There are several groups (see, for example, Landberg et al., 2003; Martf et al., 2000; Moehrlen et al., 2000) working in this area and they all have slightly different approaches. However, in all cases, the creation of power output forecasts is a two-stage process. First, there is the creation of site-specific meteorological forecasts (for some predefined reference point, such as a site met mast). These meteorological forecasts are then transformed, via site-specific power models, to power output forecasts. This process is shown schematically in Figure I.2.13.

To enable the meteorological model to be both autoregressive and adaptive, feedback data from the site is also required. In other words, the method needs to know what is happening at the site where predictions are being made, as well as using the forecast. If it does so, it can 'learn' and also can adjust the forecasts, depending on the degree of success.

EXAMPLES OF TIME SERIES POWER PREDICTION RESULTS

An example of time series plots for two separate forecast horizons is shown in Figure I.2.14. It shows

10 per cent, accurate wind energy predictions are needed.

The numerical weather prediction (NWP) models run by national institutes are typically of continental, if not global, scale. Consequently, their resolutions tend to be too coarse for wind energy needs. The model run by the Danish Meteorological Institute for northwest Europe, for example, has a minimum horizontal resolution of approximately 5 km.

The methods of achieving the transformation between coarse NWP forecasts and site-specific ones are varied. Despite this variation, they can largely be grouped into two main types:

1. physical models; and
2. statistical models.

Physical models primarily aim to improve the resolution of the 'original' NWP model. The models used to achieve this can include:

- simple linear-flow models such as WAsP; and
- high-resolution NWP models: these are essentially local (nested) meso-scale versions of the original

Figure I.2.13: Method overview

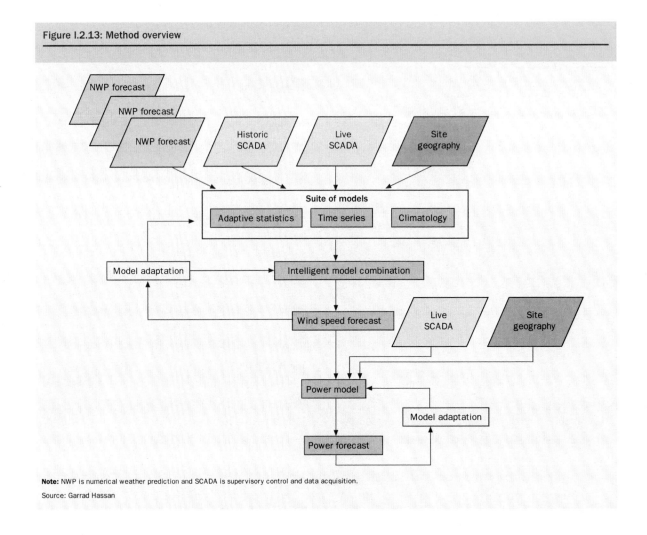

Note: NWP is numerical weather prediction and SCADA is supervisory control and data acquisition.

Source: Garrad Hassan

how well the model predicted the power output of the wind farm, both 1 hour (T + 1) and 12 hours (T + 12) in advance. Also shown within the figure are 'P10', 'P25', 'P75' and 'P90' error bars associated with the prediction: for example the P75 level is a power level with a 75 per cent probability of being exceeded.

To an 'engineering eye' this prediction looks good. It is clear that the forecast has captured the shape of the profile rather well. To the 'commercial eye', however, the situation is not so good. For example, on 1 February, there was greater volatility in the power output than predicted. For hourly trading purposes, such a prediction would then be poor. Whether or not the forecast is 'good' or 'bad' therefore depends very strongly on its precise purpose. For scheduling plant maintenance it is acceptable, whereas for hourly trading it is poor. The purpose of the forecast thus needs to be very carefully defined; this is a strategic as well as a technical question. That said, forecasts of the accuracy of that defined in Figure I.2.14 are currently substantially increasing the value of wind energy in some markets and where suitable commercial trading strategies are employed, and in

Figure I.2.14: Time series of power forecast, T + 1 (left) and T + 12 (right)

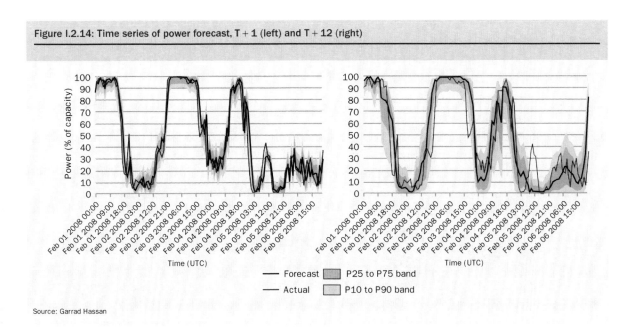

Source: Garrad Hassan

other markets such forecasts help TSOs balance the system.

EXAMPLE STATISTICAL ACCURACY OF FORECASTS

To better understand the accuracy of a forecast and set the 'snap shot' time series results presented above into a more statistical long-term context, it is necessary to look at a particular measure of accuracy over time. There are several different measures that can be used, but a commonly used measure is to present the mean absolute error of the forecast, normalised by the rated capacity of the wind farm, against forecast horizon. Forecast accuracy is discussed further in Madsen et al. (2005). The accuracy achieved is, of course, dependent on the specifics of the wind regime and the complexity of the site, as well as the country where the wind farm is located. An example of the typical range of accuracies achieved by state-of-the-art forecasting methods is presented in Figure I.2.15.

PORTFOLIO EFFECTS

As the geographical spread of an electricity system increases, the wind speeds across it become less correlated. Some areas will be windy; some will not.

Figure I.2.15: Typical range of forecast accuracy for individual wind farms

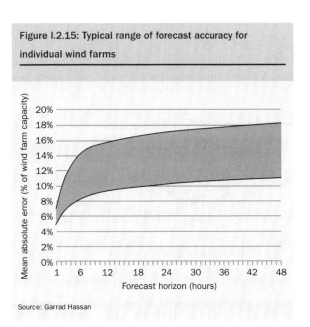

Source: Garrad Hassan

Some areas will have rising power output; some will have falling power output. The effect of aggregation on wind farm power fluctuations, and on capacity credit issues, has been looked at by several analysts, and forecasting of the output of large numbers of wind farms for the system operator is now commonplace in both Denmark and Germany.

As can be imagined, forecast accuracy for portfolios of wind farms is better than for individual wind farms. An example time series plot of the forecast accuracy of a portfolio of seven geographically dispersed wind farms is presented in Figure I.2.16. It can be seen that the average deviation of the forecast is significantly lower than that for the individual wind farm example presented in Figure I.2.14, despite the portfolio being forecast 24 hours in advance rather than 12 hours. It would be typical for the mean absolute error of the forecast 24 hours in advance to be 3 to 5 per cent lower for a portfolio of seven wind farms than that for one individual wind farm. As forecasts are extended to include all wind farms in a region, such as a part of Germany or Denmark, where densities of turbines are high, then mean absolute errors of only 5–7 per cent are seen even 24 to 48 hours in advance.

CONCLUSIONS

There is now a range of sophisticated modelling tools which can provide robust short-term forecasts of the output of wind farms and portfolios of wind farms. These models are being used in various markets for a range of different purposes. Key areas of use are by TSOs to facilitate balancing of the grid, by energy traders to trade energy from wind farms on futures markets and by wind farm owners to optimise O&M arrangements.

The use of wind energy forecasts in many countries is at an embryonic stage and there are substantial benefits to be gained from the wider adoption of current state-of-the-art forecasting in all of the areas mentioned above. In addition, changes need to be made in the way the power systems are operated and the markets are functioning. The tools are available; they just need to be more widely used.

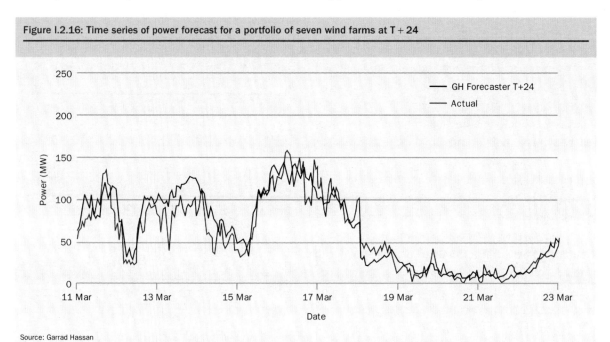

Figure I.2.16: Time series of power forecast for a portfolio of seven wind farms at T + 24

Source: Garrad Hassan

There is substantial ongoing work aimed at improving the forecast tools. There are two main aspects to this work. First, global-level weather models, which are behind all sophisticated short-term forecast models, are continuously improving, due to:

- the availability of greater computing power;
- a better understanding of the physics; and
- more numerous input data from sources, such as satellites and aeroplanes.

Second, and in parallel with such advances, the short-term forecasters specifically serving the wind industry are continually improving models. A key practical way in which models are improving is due to improvements in wind farm SCADA (supervisory control and data acquisition) systems, meaning an increase in the quality of the information flowing back from the wind farm to be used in the forecast.

Future Developments

Wind speed and energy prediction are, and will remain, a very critical part of the development of a wind farm. Enormous investments are made based on the estimates provided. Lender and investor confidence must be maintained or improved. Improvement in these techniques is, therefore, an important part of European and global wind energy development. Below is a list of important topics for future development.

- An intention of this chapter is to introduce some of the challenges of predicting the future output of wind farms to non-wind analysis specialists involved in the wind industry. Speaking in general terms, for a typical site, the energy production level with a 90 per cent chance of being exceeded (P90 level) may well be 15 per cent below the central estimate energy production (P50) level. Put another way, one in ten projects analysed may be expected to have an energy production in the long term more than 15 per cent below the central estimate level. Of course, in a similar way, statistically one in ten

projects may be expected to exceed the central estimate production level by 15 per cent. The challenge is therefore to improve the accuracy of the models and methods.

- In addition to this, the production will vary substantially from year to year, due to annual variations in the wind regime. Also, the relatively high levels of availability met by most modern wind farms only happen if the right O&M structures and budgets are in place. Neglect in these areas can cause wind farms to produce significantly less than their potential output.

- To a high degree, the accuracy of the pre-construction prediction of the energy production of a wind farm is in the hands of the owner. The better the site wind measurements, the lower the uncertainty and the lower the likelihood of surprises once the wind farm is operational. It is recommended that in-house or external experts are involved to carefully design a monitoring strategy for all wind farm developments, and that this area of a development is adequately funded.

- Assessment methods for wind farms are continuously improving and key areas of development include:
  - the use of increasingly sophisticated flow modelling techniques;
  - further validation of wake models;
  - optimisation of the use of data sources that are not on the wind farm site to adjust site data to be representative of the long term;
  - the use of remote sensing techniques to measure wind speed;
  - improved estimates of 'loss factors'; and
  - more sophisticated approaches to uncertainty analysis.

- A central principle of the development of improved scientific predictions is the refinement and validation of models against measured data. Wind farm SCADA systems are recording huge volumes of data. It is considered that the quality of SCADA data recorded at wind farm sites is improving, but needs further improvement. Also, SCADA data from wind farms contains information that will allow the further validation of models and assumptions used in energy assessment. The challenge for the industry is to focus on understanding what that data tells us and, where appropriate, amending models, techniques and assumptions.

- The wind industry needs to work more closely with climate change scientists to better understand how weather patterns may change in the future.

High-quality short-term forecasting techniques are available now. The initial challenge for the industry is to extract the maximum value from the existing state-of-the-art forecast models. At present, the use of such techniques to address some of the fundamental challenges for the integration of wind energy into the grid is patchy. In parallel with the increased use of existing tools, continued effort needs to be focused on improving forecasting technology.

EWEA/Winter

I.3 WIND TURBINE TECHNOLOGY

Evolution of Commercial Wind Turbine Technology

The engineering challenge for the wind industry is to design an efficient wind turbine to harness wind energy and turn it into electricity. In this chapter, the evolution of wind turbines is discussed, their present status described and the future challenges identified.

The evolution of modern wind turbines is a story of engineering and scientific skill, coupled with a strong entrepreneurial spirit. In the last 20 years, turbines have increased in size by a factor of 100 (from 25 kW to 2500 kW and beyond), the cost of energy has reduced by a factor of more than five and the industry has moved from an idealistic fringe activity to an acknowledged component of the power generation industry. At the same time, the engineering base and computational tools have developed to match machine size and volume. This is a remarkable story, but it is far from finished: many technical challenges remain and even more spectacular achievements will follow.

THE TECHNICAL CHALLENGE OF A UNIQUE TECHNOLOGY

The concept of a wind-driven rotor is ancient, and electric motors were widely disseminated, both domestically and commercially, in the latter half of the 20th century. Making a wind turbine may seem simple, but it is a big challenge to produce a wind turbine that:

- meets specifications (frequency, voltage, harmonic content) for standard electricity generation, with each unit operating as an unattended power station;
- copes with the variability of the wind (mean wind speeds on exploitable sites range from 5 to 11 m/s, with severe turbulence in the Earth's boundary layer and extreme gusts up to 70 m/s); and
- competes economically with other energy sources.

The traditional 'Dutch' windmill (Figure I.3.1) had proliferated to the extent of about 100,000 machines throughout Europe at their peak in the late 19th century. These machines preceded electricity supply and were indeed wind-powered mills used for grinding grain. Use of the wind for water pumping also became common. The windmills were always attended, sometimes inhabited and largely manually controlled. They were also characterised by direct use of the mechanical energy generated on the spot. They were integrated within the community, designed for frequent replacement of certain components and efficiency was of little importance.

In contrast, the function of a modern power-generating wind turbine is to generate high-quality, network frequency electricity. Each wind turbine must function as an automatically controlled independent 'mini-power station'. It is unthinkable for a modern wind turbine to be permanently attended, and uneconomic for it to be frequently maintained. The development of the microprocessor has played a crucial role in enabling cost-effective wind technology. A modern wind turbine is required to work unattended, with low maintenance, continuously for in excess of 20 years.

Figure I.3.1: Traditional 'Dutch' windmill

Garrad Hassan

Stall

Although most of the largest wind turbines now employ active pitch control, in the recent history of wind turbine technology, the use of aerodynamic stall to limit power has been a unique feature of the technology. Most aerodynamic devices (aeroplanes and gas turbines, for example) avoid stall. Stall, from a functional standpoint, is the breakdown of the normally powerful lifting force when the angle of flow over an aerofoil (such as a wing section) becomes too steep. This is a potentially fatal event for a flying machine, whereas wind turbines can make purposeful use of stall as a means of limiting power and loads in high wind speeds.

The design requirements of stall regulation led to new aerofoil developments and also the use of devices, such as stall strips, vortex generators, fences and Gurney flaps, for fine-tuning rotor blade performance. Even when not used to regulate power, stall still very much influences aerofoil selection for wind turbines. In an aircraft, a large margin in stall angle of attack, compared to the optimum cruising angle, is very desirable. On a wind turbine, this may be undesirable and lead to higher extreme loads.

Fatigue

The power train components of a wind turbine are subject to highly irregular loading input from turbulent wind conditions, and the number of fatigue cycles experienced by the major structural components can be orders of magnitude greater than for other rotating machines. Consider that a modern wind turbine operates for about 13 years in a design life of 20 years and is almost always unattended. A motor vehicle, by comparison, is manned, frequently maintained and has a typical operational life of about 160,000 km, equivalent to four months of continuous operation.

Thus in the use rather than avoidance of stall and in the severity of the fatigue environment, wind technology has a unique technical identity and unique R&D demands.

THE DEVELOPMENT OF COMMERCIAL TECHNOLOGY

An early attempt at large-scale commercial generation of power from the wind was the 53 m diameter, 1.25 MW Smith Putnam wind turbine, erected at Grandpa's Knob in Vermont, US, in 1939. This design brought together some of the finest scientists and engineers of the time (the aerodynamic design was by von Karman and the dynamic analysis by den Hartog). The wind turbine operated successfully for longer than some megawatt machines of the 1980s.

It was a landmark in technological development and provided valuable information about quality input to design, machine dynamics, fatigue, siting and sensitivity. However, preceding the oil crisis of the 1970s, there was no economic incentive to pursue the technology further in the post-war years.

The next milestone in wind turbine development was the Gedser wind turbine. With assistance from Marshall Plan post-war funding, a 200 kW, 24 m diameter wind turbine was installed during 1956 and 1957 in the town of Gedser in the southeast of Denmark. This machine operated from 1958 to 1967 with about a 20 per cent capacity factor.

In the early 1960s, Professor Ulrich Hütter developed high tip speed designs, which had a significant influence on wind turbine research in Germany and the US.

1970 to 1990

In the early 1980s, many issues of rotor blade technology were investigated. Steel rotors were tried but rejected as too heavy, aluminium as too uncertain in the context of fatigue endurance, and the wood-epoxy system developed by Gougeon Brothers in the US was employed in a number of both small and large wind turbines. The blade manufacturing industry has, however, been dominated by fibreglass polyester construction, which evolved from the work of LM Glasfiber, a boat-building company, and became thoroughly consolidated in Denmark in the 1980s.

By 1980 in the US, a combination of state and federal energy and investment tax credits had stimulated a rapidly expanding market for wind in California. Over the 1980–1995 period, about 1700 MW of wind capacity was installed, more than half after 1985 when the tax credits had reduced to about 15 per cent.

Tax credits attracted an indiscriminate overpopulation of various areas of California (San Gorgonio, Tehachapi and Altamont Pass) with wind turbines, many of which were ill-designed and functioned poorly. However, the tax credits created a major export market for European (especially Danish) wind turbine manufacturers, who had relatively cost-effective, tried and tested hardware available. The technically successful operation of the later, better-designed wind turbines in California did much to establish the foundation on which the modern wind industry has been built. The former, poor quality turbines conversely created a poor image for the industry, which has taken a long time to shake off.

1990 to Present

The growth of wind energy in California was not sustained, but there was striking development in European markets, with an installation rate in Germany of around 200 MW per annum in the early 1990s. From a technological standpoint, the significant outcome was the development of new German manufacturers and of some new concepts. The introduction of innovative direct drive generator technology by the German manufacturer Enercon is particularly noteworthy. Subsequently, a huge expansion of the Spanish market

occurred, including wind farm development, new designs and new manufacturers.

Over this period there have been gradual, yet significant, new technological developments in direct drive power trains, in variable speed electrical and control systems, in alternative blade materials, and in other areas. However, the most striking trend in recent years has been the development of ever larger wind turbines, leading to the current commercial generation of multi-megawatt onshore and offshore machines.

DESIGN STYLES

Significant consolidation of design has taken place since the 1980s, although new types of electrical generators have also introduced further diversification.

Vertical Axis

Vertical axis wind turbine (VAWT) designs were considered, with expected advantages of omnidirectionality and having gears and generating equipment at the tower base. However, they are inherently less efficient (because of the variation in aerodynamic torque with the wide range in angle of attack over a rotation of the rotor). In addition, it was not found to be feasible to have the gearbox of large vertical axis turbines at ground level, because of the weight and cost of the transmission shaft.

The vertical axis design also involves a lot of structure per unit of capacity, including cross arms in the H-type design (Figure I.3.2). The Darrieus design (Figure I.3.3) is more efficient structurally. The blade shape is a so-called 'troposkein curve' and is loaded only in tension, not in bending, by the forces caused as the rotor spins. However, it is evident that much of the blade surface is close to the axis. Blade sections close to the axis rotate more slowly and this results in reduced aerodynamic efficiency. The classic 'egg-beater' shaped Darrieus rotors also suffered

Figure I.3.2: H-type vertical axis wind turbine

Figure I.3.3: Darrieus type vertical axis wind turbine

designs have also been made for large-scale offshore applications.

from a number of serious technical problems, such as metal fatigue-related failures of the curved rotor blades. These disadvantages have caused the vertical axis design route to disappear from the mainstream commercial market. FlowWind, previously the main commercial supplier of vertical axis turbines, stopped supplying such machines over a decade ago.

Although there is not yet any substantial market penetration, there has recently been a remarkable resurgence of innovative VAWT designs in the category of small systems for diverse applications, especially on rooftops of buildings, and also some innovative

Number of Blades

Small-scale, multi-bladed turbines are still in use for water pumping. They are of relatively low aerodynamic efficiency but, with the large blade area, can provide a high starting torque (turning force). This enables the rotor to turn in very light winds and suits a water pumping duty.

Most modern wind turbines have three blades, although in the 1980s and early 1990s some attempt was made to market one- and two-bladed wind turbine designs.

The single-bladed design (Figure I.3.4) is the most structurally efficient for the rotor blade, as it has the

Figure I.3.4: Single-bladed wind turbine

Figure I.3.5: Two-bladed wind turbines of Carter Wind Turbines Ltd

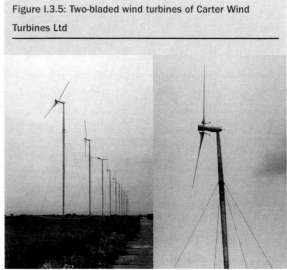

greatest blade section dimensions with all the installed blade surface area in a single beam. It is normal to shut down (park) wind turbines in very high winds, in order to protect them from damage. This is because they would generally experience much higher blade and tower loads if they continued to operate. The one-bladed design allows unique parking strategies – with the single blade acting as a wind vane upwind or downwind behind the tower – which may minimise storm loading impact. However, there are a number of disadvantages. With a counterweight to balance the rotor statically, there is reduced aerodynamic efficiency and complex dynamics requiring a blade hinge to relieve loads. The designs of Riva Calzoni, MAN, Messerschmidt and others were also of too high a tip speed to be acceptable in the modern European market from an acoustic point of view.

The two-bladed rotor design (Figure I.3.5) is technically on a par with the established three-bladed design. In order to obtain a potentially simpler and more efficient rotor structure with more options for rotor and nacelle erection, it is necessary either to accept higher cyclic loading or to introduce a teeter hinge, which is often complex. The teeter hinge allows the two blades of the rotor to move as a single beam through typically ±7° in an out-of-plane rotation. Allowing this small motion can much relieve loads in the wind turbine system, although some critical loads return when the teeter motion reaches its end limits. The two-bladed rotor is a little less efficient aerodynamically than a three-bladed rotor.

In general, there are small benefits of rotors having increasing number of blades. This relates to minimising losses that take place at the blade tips. These losses are, in aggregate, less for a large number of narrow blade tips than for a few wide ones.

In rotor design, an operating speed or operating speed range is normally selected first, taking into account issues such as acoustic noise emission. With the speed chosen, it then follows that there is an optimum total blade area for maximum rotor efficiency.

The number of blades is, in principle, open, but more blades imply more slender blades for the fixed (optimum) total blade area.

Note also that it is a complete misconception to think that doubling the number of blades would double the power of a rotor. Rather, it would reduce power if the rotor was well designed in the first instance.

It is hard to compare the two- and three-bladed designs on the basis of cost-benefit analysis. It is generally incorrect to suppose that, in two-bladed rotor design, the cost of one of three blades has been saved, as two blades of a two-bladed rotor do not equate with two blades of a three-bladed rotor. Two-bladed rotors generally run at much higher tip speed than three-bladed rotors, so most historical designs would have noise problems. There is, however, no fundamental reason for the higher tip speed, and this should be discounted in an objective technical comparison of the design merits of two versus three blades.

The one-bladed rotor is perhaps more problematic technically, whilst the two-bladed rotor is basically acceptable technically. The decisive factor in eliminating the one-blade rotor design from the commercial market, and in almost eliminating two-bladed design, has been visual impact. The apparently unsteady passage of the blade or blades through a cycle of rotation has often been found to be objectionable.

Pitch Versus Stall

This section discusses the two principal means of limiting rotor power in high operational wind speeds – stall regulation and pitch regulation. Stall-regulated machines require speed regulation and a suitable torque speed characteristic intrinsic in the aerodynamic design of the rotor. As wind speed increases and the rotor speed is held constant, flow angles over the blade sections steepen. The blades become increasingly stalled and this limits power to acceptable levels, without any additional active control. In stall control, an essentially constant speed is achieved through the connection of the electric generator to the grid. In this respect, the grid behaves like a large flywheel, holding the speed of the turbine nearly constant irrespective of changes in wind speed.

Stall control is a subtle process, both aerodynamically and electrically. In summary, a stall-regulated wind turbine will run at approximately constant speed in high wind without producing excessive power and yet achieve this without any change to the rotor geometry.

The main alternative to stall-regulated operation is pitch regulation. This involves turning the wind turbine blades about their long axis (pitching the blades) to regulate the power extracted by the rotor. In contrast to stall regulation, pitch regulation requires changes of rotor geometry by pitching the blades. This involves an active control system, which senses blade position, measures output power and instructs appropriate changes of blade pitch.

The objective of pitch regulation is similar to stall regulation, namely to regulate output power in high operational wind speeds. A further option, active stall regulation, uses full span pitching blades. However, they are pitched into stall in the opposite direction to the usual fine pitching where the aerofoil sections are rotated leading edge into wind direction. This concept, like the conventional fine pitch solution, uses the pitch system as a primary safety system, but also exploits stall regulation characteristics to have much reduced pitch activity for power limiting.

Variable Speed Versus Fixed Speed

Initially, most wind turbines operated at fixed speed when producing power. In a start-up sequence the rotor may be parked (held stopped), and on release of the brakes would be accelerated by the wind until the required fixed speed was reached. At this point, a connection to the electricity grid would be made and then the grid (through the generator) would hold the speed constant. When the wind speed increased beyond the level at which rated power was generated, power would be regulated in either of the ways described above, by stall or by pitching the blades.

Subsequently, variable speed operation was introduced. This allowed the rotor and wind speed to be matched, and the rotor could thereby maintain the best flow geometry for maximum efficiency. The rotor could be connected to the grid at low speeds in very light winds and would speed up in proportion to wind speed. As rated power was approached, and certainly after rated power was being produced, the rotor would revert to nearly constant speed operation, with the blades being pitched as necessary to regulate power. The important differences between variable speed operation as employed in modern large wind turbines and the older conventional fixed speed operation are:

• Variable speed in operation below rated power can enable increased energy capture.
• Variable speed capability above rated power (even over quite a small speed range) can substantially relieve loads, ease pitch system duty and much reduce output power variability.

The design issues of pitch versus stall and degree of rotor speed variation are evidently connected. In the 1980s, the classic Danish, three-bladed, fixed speed, stall-regulated design was predominant. Aerodynamicists outside the wind industry (such as for helicopters and gas turbine) were shocked by the idea of using stall. Yet, because of the progressive way in which stall occurs over the wind turbine rotor, it proved to be a thoroughly viable way of operating a wind turbine. It is one of the unique aspects of wind technology.

Active pitch control is the term used to describe a control system in which the blades pitch along their axis like a propeller blade. Superficially, this approach seemed to offer better control than stall regulation, but it emerged through experience that pitch control of a fixed speed wind turbine at operational wind speeds that are a lot higher than the rated wind speed (minimum steady wind speed at which the turbine can produce its rated output power) could be quite problematic. The reasons for this are complex, but in turbulent (constantly changing) wind conditions, it is demanding to keep adjusting pitch to the most appropriate angle and under high loads, and excessive power variations can result whenever the control system is 'caught out' with the blades in the wrong position.

In view of such difficulties, which were most acute in high operational wind speeds (of say 15–25 m/s), pitch control in conjunction with a rigidly fixed speed became regarded as a 'challenging' combination. Vestas initially solved this challenge by introducing OptiSlip (which allows a certain degree of variable speed using pitch control in power-limiting operations, in the range of 10 per cent speed variation using a high slip induction generator). Suzlon presently use a similar technology, Flexslip, with a maximum slip of 17 per cent. Speed variation helps to regulate power and reduces demand for rapid pitch action.

Variable speed has some attractions, but also brings cost and reliability concerns. It was seen as a way of the future, with expected cost reduction and performance improvements in variable speed drive technology. To some extent these have been realised. However, there was never a clear case for variable speed on economic grounds, with small energy gains being offset by extra costs and also additional losses in the variable speed drive. The current drive towards variable speed in new large wind turbines relates to greater operational flexibility and concerns about the power quality of traditional stall-regulated wind turbines. Two-speed systems emerged during the 1980s and 1990s as a compromise, improving the energy capture and noise emission characteristics of stall-regulated wind turbines. The stall-regulated design remains viable, but variable speed technology offers better output power quality to the grid, and this is now driving the design route of the largest machines. Some experiments are underway with the combination of variable speed and stall regulation, although variable speed combines naturally with pitch regulation. For reasons related to the methods of power control, an electrical variable speed system allows pitch control to be effective and not overactive.

Another significant impetus to the application of pitch control, and specifically pitch control with independent pitching of each blade, is the acceptance by certification authorities that this allows the rotor to be considered as having two independent braking systems acting on the low speed shaft. Hence, only a parking brake is required for the overall safety of the machine.

Pitch control entered wind turbine technology primarily as a means of power regulation which avoided stall when, from the experience of industries outside wind technology, stall was seen as problematic if not disastrous. However, in combination with variable speed and advanced control strategies, stall offers unique capabilities to limit loads and fatigue in the wind turbine system and is almost universally employed in new large wind turbine designs. The load-limiting capability of the pitch system improves the power to weight ratio of the wind turbine system and compensates effectively for the additional cost and reliability issues involved with pitch systems.

DESIGN DRIVERS FOR MODERN TECHNOLOGY

The main design drivers for current wind technology are:
• low wind and high wind sites;
• grid compatibility;
• acoustic performance;
• aerodynamic performance;
• visual impact; and
• offshore.

Although only some 1.5 per cent of the world's total installed capacity is currently offshore, the latest developments in wind technology have been much influenced by the offshore market. This means that,

in the new millennium, the technology development focus has been mainly on the most effective ways to make very large turbines. Specific considerations are:
• low mass nacelle arrangements;
• large rotor technology and advanced composite engineering; and
• design for offshore foundations, erection and maintenance.

A recent trend, however, is the return of development interest to new production lines for the size ranges most relevant to the land-based market, from 800 kW up to about 3 MW. Of the other main drivers, larger rotor diameters (in relation to rated output power) have been introduced in order to enhance exploitation of low wind speed sites. Reinforced structures, relatively short towers and smaller rotor diameters in relation to rated power are employed on extremely high wind speed sites.

Grid compatibility issues are inhibiting further development of large wind turbines employing stall regulation. Acoustic performance regulates tip speed for land-based applications and requires careful attention to mechanical and aerodynamic engineering details. Only small improvements in aerodynamic performance are now possible (relative to theoretical limits), but maximising performance without aggravating loads continues to drive aerodynamic design developments. Visual impact constrains design options that may fundamentally be technically viable, for example two-bladed rotors.

ARCHITECTURE OF A MODERN WIND TURBINE

Many developments and improvements have taken place since the commercialisation of wind technology in the early 1980s, but the basic architecture of the mainstream design has changed very little. Most wind turbines have upwind rotors and are actively yawed to preserve alignment with the wind direction.

The three-bladed rotor proliferates and typically has a separate front bearing, with low speed shaft connected to a gearbox that provides an output speed suitable for the most popular four-pole (or two-pole) generators. This general architecture is shown in Figure I.3.6. Commonly, with the largest wind turbines, the blade pitch will be varied continuously under active control to regulate power in higher operational wind speeds. For future large machines, there appears to be a consensus that pitch regulation will be adopted.

Support structures are most commonly tubular steel towers tapering in some way, both in metal wall thickness and in diameter from tower base to tower top. Concrete towers, concrete bases with steel upper sections, and lattice towers are also used but are much less prevalent. Tower height is rather site-specific and turbines are commonly available with three or more tower height options.

The drive train of Figure I.3.6 shows the rotor attached to a main shaft driving the generator through the gearbox. Within this essentially conventional architecture of multi-stage gearbox and high-speed generator, there are many significant variations in structural support, in rotor bearing systems and in general layout. For example, a distinctive layout (Figure I.3.7) has been developed by Ecotècnia

Figure I.3.6: Typical nacelle layout of a modern wind turbine

Nordex

Figure I.3.7: Ecotecnia 100 nacelle layout

Ecotècnia

(Alstom), which separates the functions of rotor support and torque transmission to the gearbox and generator. This offers a comfortable environment for the gearbox, resulting in predictable loading and damping of transients, due to its intrinsic flexibility. Among the more innovative of a large variety of bearing arrangements is the large single front bearing arrangement adopted by Vestas in the V90 3 MW design (Figure I.3.8). This contributes to a very compact and lightweight nacelle system.

Whilst rotor technology is set amongst the leading commercial designs and the upwind three-bladed rotor prevails generally, more unconventional trends in nacelle architecture are appearing. The direct drive systems of Enercon are long established, and many direct drive designs based on permanent magnet generator (PMG) technology have appeared in recent years. A number of hybrid systems, such as Multibrid, which employ one or two gearing stages, and multipole generators have also appeared. These developments are discussed in 'Technology Trends' below. It is far from clear which of the configurations is the optimum. The effort to minimise capital costs and

Figure I.3.8: Vestas V90 nacelle layout

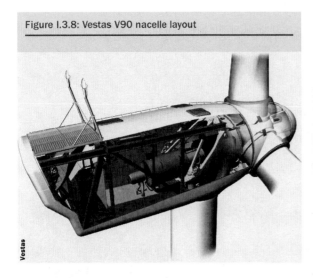

Figure I.3.9: Modern wind technology

maximise reliability continues – the ultimate goal is to minimise the cost of electricity generated from the wind.

GROWTH OF WIND TURBINE SIZE

Modern wind technology is available for a range of sites: low and high wind speeds and desert and arctic climates can be accommodated. European wind farms operate with high availability (97 per cent) and are generally well integrated with the environment and accepted by the public.

At the start of the millennium, an ever increasing (in fact mathematically exponential) growth in turbine size with time had been documented by manufacturers, such as Siemens Wind Power (earlier Bonus AS), and was a general industry trend. In the past three or four years, although interest remains in yet larger turbines for the offshore market, there has been a levelling of turbine size at the centre of the main, land-based market and a focus on increased volume supply in the 1.5 to 3 MW range.

The past exponential growth of turbine size was driven by a number of factors. The early small sizes, around 20–60 kW, were very clearly not optimum for system economics. Small wind turbines remain much more expensive per kW installed than large ones, especially if the prime function is to produce grid quality electricity. This is partly because towers need to be higher in proportion to diameter to clear obstacles to wind flow and escape the worst conditions of turbulence and wind shear near the surface. But it is primarily because controls, electrical connection to grid and maintenance are a much higher proportion of the capital value of the system.

Also, utilities have been used to power in much larger unit capacities than the small wind turbines, or even wind farm systems of the 1980s, could provide.

When wind turbines of a few hundred kW became available, these were more cost-effective than the earlier smaller units, being at a size where the worst economic problems of very small turbines were avoided. However, all the systems that larger wind turbines would require were also needed and the larger size was the most cost-effective. It also became apparent that better land utilisation could often be realised with larger wind turbine units, and larger unit sizes were also generally favourable for maintenance cost per kW installed. All these factors, the psychology of 'bigger is better' as a competitive element in manufacturers' marketing and a focus of public research funding programmes on developing larger turbines contributed to the growth of unit size through the 1990s.

Land-based supply is now dominated by turbines in the 1.5 and 2 MW range. However, a recent resurgence in the market for turbines around 800 kW size is interesting, and it remains unclear, for land-based projects, what objectively is the most cost-effective size of wind turbine.

The key factor in maintaining design development into the multi-megawatt range has been the development of an offshore market. For offshore applications, optimum overall economics, even at higher cost per kW in the units themselves, requires larger turbine units to limit the proportionally higher costs of infrastructure (foundations, electricity collection and subsea transmission) and lower the number of units to access and maintain per kW of installed capacity.

Figure I.3.10 summarises the history of sizes of leading commercial wind turbines up to the present (2008) and illustrates a few concepts for the largest turbines of the near future. The future challenges in extending the conventional three-bladed concept to size ranges above 5 MW are considerable, and are probably as much economic as engineering issues. REpower, exploiting reserve capacity in design margins, has up-rated its 5 MW wind turbine to 6 MW and

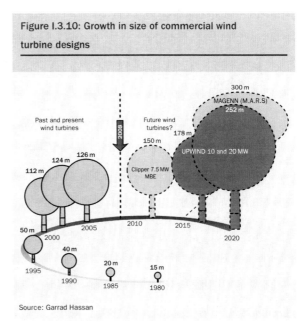

Figure I.3.10: Growth in size of commercial wind turbine designs

Source: Garrad Hassan

BARD Engineering has announced a similar up-rating of its 5 MW design to 6 MW and later 7 MW. Clipper Windpower has announced a 7.5 MW prototype to be purchased by the UK Crown Estates (an unprecedented type of investment for them), with no specific timeline for development, but suggestions of production by around 2012. The interest in yet larger wind turbines, especially for offshore markets, is reflected in the UpWind project. This major project of the EU 6 Framework Programme addresses a wide range of wind energy issues, including up-scaling, by evaluating the technical and economic issues in developing unit wind turbines of 10 and 20 MW capacity. The Magenn airborne wind energy concept is one of a number of speculative new concepts for large capacity wind energy systems that is reviewed later in the chapter ('Future Innovations').

LARGE COMMERCIAL WIND TURBINES

The 'top ten' wind turbine manufacturers, as measured by global market share in 2007, and some salient features of the technology of some of their flagship designs, are listed in Table I.3.1.

Vestas

Vestas has long been the world's leading supplier of wind turbines. Key volume products are the V80 and V90 series. Vestas technology is generally particularly lightweight. Blades made using high strength composites in the form of prepregs, and innovation in nacelle systems design, has contributed to this characteristic.

According to industry sources, Vestas is developing a new offshore wind turbine model.

GE Energy

GE is now focusing increasingly on their 2.5 MW 2.5XL series, which entered series production in the summer of 2008. This is seen as the next generation of turbines to succeed the proven 1.5 MW series and be produced at high volumes. It is interesting to note the change in design to a permanent magnet generator with full converter retains a high-speed generator with a multi-stage planetary gear system. Doubly fed induction generator (DFIG) technology has been challenged recently by more stringent network requirements, the 'fault ride-through' requirement in particular, and adaptations have been made to respond to these issues. The DFIG solution is undoubtedly cheaper in capital cost than systems with full converters. However, the implications for part-load efficiency can make it inferior in cost of energy to efficient systems with PMG and full converter, and the value of capitalised losses should not be underestimated. This may in part be the motivation for GE's change in design route and adoption of a PMG generator in their latest turbine series. Note also that a synchronous PMG can be applied without design hardware modifications in both 50 Hz and 60 Hz network regions. This greatly increases flexibility for international developers operating in multiple wind markets.

Gamesa Eolica

The latest design from Gamesa Eolica, the Gamesa G10X, a 4.5 MW, 128 m diameter prototype, is presently being developed. Key features of this design include:

- a two-bearing arrangement, integrated with a two-stage planetary type gearbox;

Table I.3.1: Design choices of leading manufacturers

		Share (%)	Model	Drive train	Power rating (kW)	Diameter (m)	Tip speed (m/s)	Power conversion
1	Vestas	22.8	V90	Geared	3000	90	87	Asynchronous
2	GE Energy	16.6	2.5XL	Geared	2500	100	86	PMG converter
3	Gamesa	15.4	G90	Geared	2000	90	90	DFIG
4	Enercon	14.0	E82	Direct	2000	82	84	Synchronous
5	Suzlon	10.5	S88	Geared	2100	88	71	Asynchronous
6	Siemens	7.1	3.6 SWT	Geared	3600	107	73	Asynchronous
7	Acciona	4.4	AW-119/3000	Geared	3000	116	74.7	DFIG
8	Goldwind	4.2	REpower750	Geared	750	48	58	Induction
9	Nordex	3.4	N100	Geared	2500	99.8	78	DFIG
10	Sinovel	3.4	1500 (Windtec)	Geared	1500	70		

Source: Garrad Hassan

Figure I.3.11: Nacelle of the Gamesa G10X 4 MW wind turbine

Figure I.3.12: Gamesa G87 (2 MW) at Loma de Almendarache

- a low mass sectional blade (inserts bonded into carbon pultruded profiles are bolted on site);
- a hybrid tower with concrete base section and tubular steel upper section; and
- an attached FlexFit™ crane system that reduces the need for large external cranes.

Gamesa is also using state-of-the art control and converter technology in this design. The main shaft is integrated into a compact gearbox, limited to two stages and providing a ratio of 37:1. It would appear that their system will have a synchronous generator with fully rated power converter. The nacelle of the G10X is shown in Figure I.3.11. Figure I.3.12 shows presently operational Gamesa G87 2 MW wind turbines in the wind farm at Loma de Almendarache.

Enercon

Enercon has dominated supply of direct drive turbines (Figure I.3.13). They have favoured wound-rotor generator technology in their designs, although permanent magnet technology is now the choice of most manufacturers developing new direct drive designs. A direct drive generator, with a wound field rotor is more complex, requiring excitation power to be passed to the rotor, but it benefits from additional controllability.

Enercon has perhaps the aerodynamic design that gives most consideration to flow around the hub area, with their blade p rofile smoothly integrated with the hub cover surface in the fine pitch position. Their latest designs achieve a very high rotor aerodynamic efficiency, which may be due both to the management of flow in the hub region and tip winglets (blade tip ends curved out of the rotor plane), which can inhibit tip loss effects.

Enercon has quite diverse renewable energy interests, which include commercially available wind desalination and wind-diesel systems. In addition, it

Figure I.3.13: Enercon E-82

has involvement in hydro energy systems and Flettner rotors for ship propulsion.

Suzlon Energy and REpower

Suzlon produces wind turbines in a range from 350 kW to 2.1 MW. It has developed its technology through acquisitions in the wind energy market and is targeting a major share in the US market.

Recent additions to the range include the S52, a 600 kW turbine for low wind speed Indian sites, and S82, a 1.50 MW turbine. The S52 employs a hydraulic torque converter that can allow up to 16 per cent slip, thus providing some of the benefits of variable speed operation.

In 2007, after a five-month takeover battle with the French state-owned nuclear company Areva, Suzlon took a controlling stake in REpower, with 87 per cent of the German wind company's voting rights.

REpower, which had previously acquired the blade supplier Abeking & Rasmussen, continues to expand its manufacturing facilities in Germany and also rotor blade production in Portugal. The company cooperates with Abeking & Rasmussen ROTEC, manufacturing rotor blades at Bremerhaven in a joint venture, PowerBlades GmbH.

Siemens Wind Power

Siemens Wind Power (formerly Bonus) is among the few companies that are becoming increasingly successful in the offshore wind energy market. Its 3.6 MW SWT turbines of 107 m diameter (Figure I.3.14) are now figuring prominently in offshore projects. Senior management in Siemens has indicated the end of a trend of exponential growth in turbine size (considering, year by year, the turbine design at the centre of their commercial supply). The stabilisation of turbine size has been a significant trend in the past three or four years and, although there is much discussion of larger machines and developments on the drawing board and a few prototypes, there is some evidence that, at least for land-based projects, turbine size is approaching a ceiling.

Figure I.3.14: Siemens Wind Power (Bonus)

Acciona Energy

In a four-year period, Acciona has become the seventh ranked world manufacturer in terms of MW supplied. The company presently has four factories, two located in Spain and one each in the US and China. In total, this amounts to a production capacity of 2625 MW a year.

Acciona's latest design is a new 3 MW wind turbine to be commercially available in 2009 and delivered to projects in 2010. The new turbine is designed for different wind classes (IEC Ia, IEC IIa and IEC IIIa). It will be supplied with a concrete tower of 100 or 120 m hub height and will have three rotor diameter options of 100, 109 and 116 m, depending on the specific site characteristics. The rotor swept area for the 116 m diameter is 10,568 m^2, the largest in the market of any 3 MW wind turbine, which will suit lower wind speed sites.

Electricity is generated at medium voltage (12 kV), aiming to reduce production losses and transformer costs. The main shaft is installed on a double frame to reduce loads on the gearbox and extend its working life. The AW-3000 operates at variable speed, with independent blade pitch systems.

In North America, the AW-3000 will be manufactured at the company's US-based plant, located in West Branch, Iowa. The AW-1500 (Figure I.3.15) and AW-3000 machines will be built concurrently. The present production capacity of 675 MW/year is planned to be increased to 850 MW/year.

Goldwind

Goldwind is a Chinese company in the wind industry providing technology manufactured under licences from European suppliers. Goldwind first licensed REpower's 48 kW to 750 kW turbine technology in 2002, and then acquired a licence in 2003 from Vensys Energiesysteme GmbH (Saarbrücken, Germany) for the Vensys 62 1.2 MW turbine. When Vensys developed a low wind speed version, with a larger 64m diameter rotor that increased output to 1.5 MW,

Figure I.3.15: Acciona new AW1500

Goldwind also acquired the licence for this turbine and is currently working with Vensys to produce 2.0 MW and 2.5 W turbines. Goldwind won a contract to supply 33 wind turbines (1.5 MW Vensys 77 systems) for the 2008 Olympic Games in Beijing. The company operates plants in Xinjiang, Guangdong, Zhejiang and Hebei Provinces and is building plants in Beijing (Figure I.3.16) and Inner Mongolia. In 2008, Goldwind signed a six-year contract with LM Glasfiber (Lunderskov, Denmark) to supply blades for Vensys 70 and 77 turbines (Figure I.3.17) and develop blades for Goldwind's

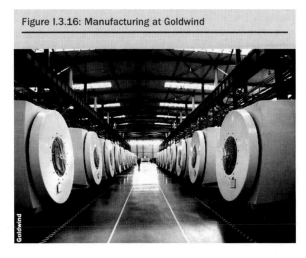

Figure I.3.16: Manufacturing at Goldwind

Figure I.3.17: Vensys

Figure I.3.18: Sinovel 1.5 MW wind turbines

next generation of 2 MW and larger turbines at LM's factory in Tianjin.

Nordex

Nordex is developing new control techniques and has a condition monitoring system, which monitors component wear, also incorporating ice sensors and an automatic fire extinguishing system.

Sinovel

In March 2007, the AMSC (American Superconductor Corporation) signed a multi-million dollar contract with Sinovel Wind, under which 3 and 5 MW wind turbines would be developed. Sinovel is continuing to manufacture and deploy the 1.5 MW wind turbines (Figure I.3.18) it began producing in 2005. The 1.5 MW wind turbines also utilise core electrical components produced by AMSC.

Earlier in 2007, AMSC had acquired the Austrian company Windtec to open opportunities for them in the wind business. Windtec has an interesting history, originating as Floda, a company that developed groundbreaking variable speed wind turbines in the latter part of the 1980s. Based in Klagenfurt, Austria,

Windtec now designs a variety of wind turbine systems and licenses the designs to third parties.

In June 2008, AMSC received a further $450 million order from Sinovel for core electrical components for 1.5 MW wind turbines. The contract calls for shipments to begin in January 2009 and increase in amount year by year until the contract's completion in December 2011. According to AMSC, the core electrical components covered under this contract will be used to support more than 10 GW of wind power capacity, nearly double China's total wind power installed base at the end of 2007.

Technology Trends

LARGER DIAMETERS

Figure I.3.19 shows trends by year of the typical largest turbine sizes targeted for mainstream commercial production. There were megawatt turbines in the 1980s, but almost all were research prototypes. An exception was the Howden 55 m 1 MW design (erected at Richborough in the UK), a production prototype, which was not replicated due to Howden withdrawing from the wind business in 1988. Although there is much more active consideration of larger designs than indicated

Figure I.3.19: Turbine diameter growth with time

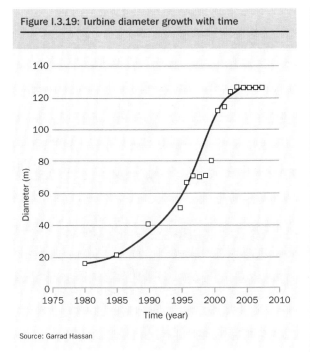

Source: Garrad Hassan

Figure I.3.20 Enercon E-126

in Figure I.3.19, there has been a definite pause in the appearance of any larger turbines since 2004.

The world's largest wind turbine is currently the Enercon E-126 (Figure I.3.20) installed in Emden, Germany, in February 2008. The E-126 is a development from the E-112, which had been up-rated to 6 MW. The new E-126 has a rating of 6 MW and may be up-rated to 7 MW. The Enercon E-126 extends the design of the E-112 and, although this design represents an increase in rated power of the world's largest wind turbines, the physical size of the rotor is similar to the REpower 5 MW design. Thus there has been no significant increase in rotor size since 2004.

TIP SPEED TRENDS

There is no fundamental reason for tip speed to change with scale. However, for turbines on land, restrictions on acoustic noise emission increase as a power function to the tip speed, and so often limit how fast the tip can go. This was especially the case when turbines

predominantly operated in fixed speed mode. Variable speed wind turbines have greater operational flexibility and can benefit from a high rated speed, but still operate at reduced speed (at night, for example) in noise-sensitive areas. Higher tip speed has the advantage that, for a given output power, the torque on the drive train is reduced and therefore the drive train mass and cost also decrease.

Offshore, there is a clear potential benefit in higher tip speeds, and less constraint on acoustic emission levels. However, with increasing tip speed, blade solidity decreases (in an optimised rotor design), and blades will tend to become more flexible. This can be beneficial for system loads but problematic for maintaining the preferred upwind attitude, with adequate tower clearance of the blade tips in extreme loading conditions. Listed data on design tip speed (Figure I.3.21) shows an increase with scale, albeit with great scatter in the data. It is evident that the present ceiling is around 90 m/s, but also that, to restrict tower top mass, very large offshore turbines will not adopt design tip speeds much below 80 m/s.

PITCH VERSUS STALL

There has been an enduring debate in the wind industry about the merits of pitch versus stall regulation. These alternatives were discussed in the earlier section on 'The technical challenge of a unique technology'.

Until the advent of MW-scale wind turbines in the mid-1990s, stall regulation predominated, but pitch regulation is now the favoured option for the largest machines (Figure I.3.22). There are now about four times as many pitch-regulated turbine designs on the market than stall-regulated versions. The dominance of pitch-regulated turbines will be still greater in terms of numbers of machines installed. The prevalence of pitch regulation is due to a combination of factors.

Overall costs are quite similar for each design type, but pitch regulation offers potentially better output power quality (this has been perhaps the most significant factor in the German market), and pitch regulation with independent operation of each pitch actuator allows the rotor to be regarded as having two independent braking systems for certification purposes.

There has been some concern about stall-induced vibrations (vibrations which occur as the blade enters stall), being particularly problematic for the largest machines. However, there has in fact been little evidence of these vibrations occurring on a large scale. There were specific problems of edgewise vibrations of stall-regulated rotor blades (or in the nacelle, tuned to the edgewise rotor whirling modes experienced by the nacelle), associated with loss of aerodynamic damping in deep stall, but this was addressed by introducing dampers in the rotor blades.

SPEED VARIATION

Operation at variable speed offers the possibility of increased 'grid friendliness', load reduction and

Figure I.3.21: Tip speed trends

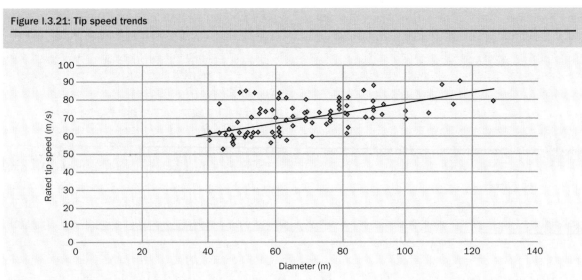

Source: Garrad Hassan

Figure I.3.22: Ratio of pitch-regulated to stall-regulated designs of ≥1 MW rating

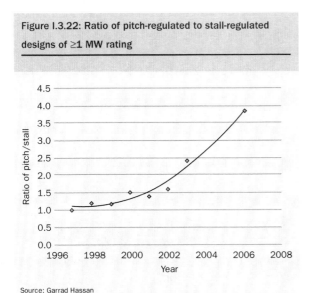

Source: Garrad Hassan

The most popular system is currently the doubly fed induction generator (DFIG), also called the wound rotor induction generator (WRIG). This provides almost all the benefits of full-range variable speed drives, but only a proportion, perhaps one-third, of the power passes through the converter. The power converter thus has approximately a third of the size, cost and losses of a conventional variable speed drive. In this concept, the stator of the electrical machine is connected directly to the network, and the rotor circuit is connected via the power converter. This is a modern version of the classical Kramer or Scherbius system. The DFIG has a more limited speed range than the conventional variable speed drive (approximately 1.5–2:1, compared to 2.5:1 or more). This speed range, however, is sufficient to provide the benefits listed above. The conventional option of a power converter, with the same rating as the generator, is unlikely to compete with the DFIG until the cost of power electronic converters falls substantially and the efficiency improves. There is evidence that this point may have been reached, with some manufacturers moving over to fully rated converters. In this respect, the potential for improved efficiency in avoiding the DFIG route may come to outweigh cost differentials. Also, some of the benefit of a DFIG system has been eroded by more stringent network requirements impacting on DFIG system cost.

The so-called 'squirrel-cage induction generator' may be used with a fully rated converted, as in some Siemens designs. Other novel generator configurations have been proposed for wind turbine applications, including the switched reluctance machine (SR, also known as variable reluctance). All rely on full-size power converters, and are therefore also at a disadvantage relative to the DFIG. The DFIG configuration used at present requires slip-rings to transfer power to and from the rotor circuit. There is an alternative method, which in effect transfers the rotor power magnetically, called the brushless doubly fed induction generator (BDIG), which avoids the use of slip-rings.

some minor energy benefits. Among wind turbines over 1 MW, out of 95 distinct models from 29 different manufacturers, 9 were fixed speed, 14 had two-speed systems and 72 employed variable speed. This shows that it is almost mandatory for MW-scale turbines to have some degree of speed variation and that continuously variable speed is the predominant choice.

Variable speed operation is realised in many ways, each differing in significant details. Direct drive systems have a natural capability for a very wide speed range, although even there some restriction on minimum speed may reduce the cost of power electronics. The 'conventional' variable speed concept, using a geared drive train, connects the generator to the network through a power electronic converter and enables systems that may have wide or narrow speed ranges. The electrical energy is generated at variable frequency – a frequency related to the rotational speed of the rotor – and then converted, by the converter or inverter (both power electronic devices), to the frequency of the grid. There are several possible configurations, based on both synchronous and induction generators.

However, at least one generator manufacturer has concluded that such machines are inherently larger and more expensive than the slip-ring option. There is no commercial turbine using the BDIG. As the experience of WRIG with slip-rings is good in wind turbines, this remains the preferred option. Slip-ring maintenance intervals of six months are achieved, and may be stretched to yearly.

DRIVE TRAIN TRENDS

Bearing Arrangements

A wide variety of rotor bearing arrangements has been tried within the context of the established conventional drive train, with multiple stage gearbox and high-speed generator. The basic requirement is to support rotor thrust and weight, whilst communicating only torque to the gearbox. Early highly modular drive trains, with well-separated twin rotor bearings and a flexible connection to the gearbox, could achieve this. However, especially with the increase in turbine size, there has been increasing interest in systems that reduce the weight and cost of the overall system. A large single slew-ring type bearing has been employed in the V90 as the main rotor bearing. Two- and three-point bearing systems are adopted in many designs, with no clear sign of consolidation of design choices. The general trend towards more integrated nacelle systems means that bearing designs are highly interactive with the complete systems concept. This is substantially true for the conventional drive train and even more so for direct drive and related concepts.

Direct Drive and Permanent Magnet Generators (PMGs)

There has been a significant trend towards innovative drive train systems, and direct drive or hybrid systems with reduced gearing figure in many new designs.

Table I.3.2 shows direct drive turbines commercially available in the wind energy market. Enercon has long pioneered direct drive and is the only company with a large market share delivering this technology.

Design trends are now predominantly towards PMGs and the design struggle is to realise the potentially better efficiency and reliability of a direct drive system without cost or weight penalties. It may be noted that the trend towards PMG technology is much wider than in the context of direct drive alone. Clipper Windpower (after research into systems with multiple induction generators), Northern Power Systems (after initially adopting a wound rotor direct drive design) and GE Energy, in their recent 2.5XL series, have all adopted PMG systems.

Table I.3.2: Direct drive wind turbines		
Turbine	**Diameter (m)**	**Power (kW)**
ENERCON E-112	114.0	4500
ENERCON E-70 E4	71.0	2300
Harakosan (Zephyros) Z72	70.0	2000
ENERCON E-66	70.0	1800
MTorres TWT 1650/78	78.0	1650
MTorres TWT 1650/70	70.0	1650
VENSYS 70	70.0	1500
VENSYS 77	77.0	1500
Leitwind LTW 77	77.0	1350
VENSYS 64	64.0	1200
VENSYS 62	62.0	1200
Leitwind LTW 61	61.0	1200
ENERCON E-58	58.0	1000
ENERCON E-48	48.0	800
Jeumont J53	53.0	750
Jeumont J48	48.0	750
ENERCON E-33	33.4	330
Subaru 22/100 (FUJI)	22.0	100
Northern Power NW 100/19	19.1	100
Unison U50 750 kW	50.0	750

The Unison U50 750 kW wind turbine (Figure I.3.23) is among recent designs emerging in East Asia that endorse the direct drive PMG concept. The PMG is overhung downwind of the tower, balancing the rotor weight and giving the nacelle its characteristic shape.

Hybrid

Designs with one or two stages of gearing, but not involving high-speed two- or four-pole generators, are classified here as 'hybrid'. Such designs from Clipper Windpower, WinWinD and Multibrid are reviewed in 'Alternative drive train configurations'. These designs also maintain the trend towards PMG technology.

Figure I.3.23: Unison U50

Fully Rated Converters

The DFIG route was discussed in 'Speed variation'. It has been predominant and will probably endure for some time in the high volume markets of Asia, for example, but is being challenged elsewhere by more stringent network requirements, the efficiency benefits of PMG technology and advances in power converter technology.

HUB HEIGHT

When wind turbines were designed exclusively for use on land, a clear average trend of hub height, increasing linearly in proportion to diameter, had been evident, although there is always very large scatter in such data, since most manufacturers will offer a range of tower heights with any given turbine model, to suit varying site conditions.

Figure I.3.24 shows that the trend in hub height with scale is now less than in proportion to diameter. This trend has resulted, quite naturally, from the largest machines being for offshore, where there is reduced wind shear. Offshore, the economic penalties of increased foundation loads and tower cost will typically outweigh any small energy gains from a much increased hub height.

ROTOR AND NACELLE MASS

Rotor mass trends are always complicated by quite different material solutions, choice of aerofoils and design tip speed, all of which can impact very directly on the solidity (effectively surface area) and mass of a blade. Table I.3.3 shows the blade mass of very large wind turbines.

The introduction into Enercon's E-126 design of a jointed blade with a steel spar on the inner blade is a clear example of where blade technology is radically different from most other large blades.

Figure I.3.24: Hub height trends

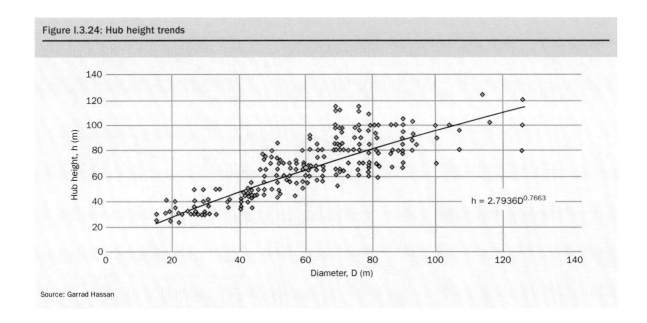

$$h = 2.7936D^{0.7663}$$

Source: Garrad Hassan

BARD has also made an interesting decision in blade design, eliminating carbon-fibre reinforcement from their blades. Their blade design preserves a very large chord on the inboard section. On most blade designs, maximum chord is usually limited to less than the aerodynamic optimum, in order to facilitate manufacturing, handling and transport. BARD will directly ship their blades from a dockside manufacturing site, avoiding the land transport issues with very large blades. LM also avoids carbon reinforcement in their latest LM 61.5 blade.

For any given design style, nacelle mass is very much determined by turbine torque rating, which scales as the cube of diameter. This implies that, with consistent design at the same level of technology development, the scaling exponent of nacelle mass will be cubic. Considering data from various public sources, it appears (Figure I.3.25) that nacelle mass scales approximately as the square of diameter. It is clear, however, that the largest turbines deviate substantially from the trend line and, considering only modern large turbines above 80m, the exponent is seen to be approximately cubic (Figure I.3.26).

Manufacturers are continually introducing new concepts in drive train layout, structure and components to reduce mass and cost, but avoiding the cubic scaling (or worse when gravity loads begin to dominate), increases in system mass and cost linked to up-scaling, a concept which is currently being studied in depth in the UpWind project. This project includes an exploration of the technical and economic feasibility of 10 and 20 MW wind turbines.

Table I.3.3: Blade mass of very large wind turbines

	Diameter (m)	Blade mass (tonnes)	Normalised mass* (tonnes)
BARD VM	122	26	28
E-112	114	20	24
M5000	116	17	20
REpower 5M	126	18	18

Note: *The 'normalised mass' is an approximate adjustment to bring all the designs to a common point of reference relative to any one design (in this case REpower 5M), taking account of different diameters and different relations between rated power and diameter.

TRANSPORT AND INSTALLATION

Erection of wind farms and systems for handling ever larger components has progressed since the

Figure I.3.25: Scaling of wind turbine nacelle system mass

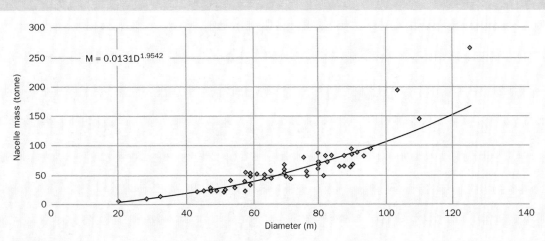

Source: Garrad Hassan

early commercial projects of the 1980s. For a period up to the mid-1990s, the allowable mass of components to be lifted to hub height was determined by available cranes. Subsequently, there has been a shift, indicative of the maturity and growth of the wind industry, where crane manufacturers are producing designs specially suited to wind farm installation.

Often complete rotors are lifted onto nacelles. Sometimes hub and blades are lifted individually.

The Enercon E-126 exploits a jointed blade design to facilitate transport, handling and erection of the rotor components. The blade joints are made up in the air. Thus rotor erection practice moved through size ranges where complete rotors were crane lifted, to size ranges where hub and blades were lifted

Figure I.3.26: Scaling of nacelle system mass (turbines >80 m diameter)

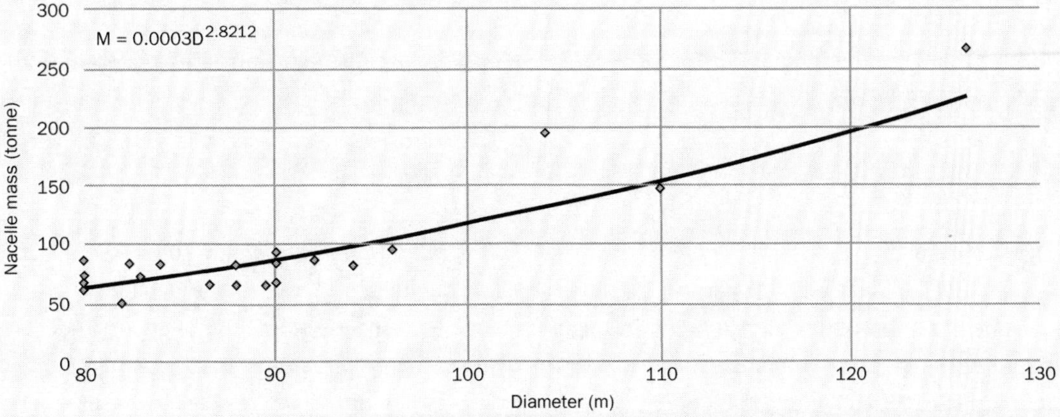

$$M = 0.0003D^{2.8212}$$

Source: Garrad Hassan

Figure I.3.27: Erection of Nordex wind turbines

Figure I.3.28: LM Glasfiber blade test (LM 61.5 m blade for 126 m rotor diameter)

individually, and now to the E-126, where blade parts are lifted individually.

A wind farm of Nordex N100 wind turbines (the largest Nordex wind generators in the US) was erected (Figure I.3.27) over a five mile long ridge south of Wadena, Minnesota.

This project made first use of the new 2007 DEMAG CC2500-1, a 550 ton crawler, a crane with 126 m of main boom with jib combination to 168 m. Transport of the CC2250-1 with maximum boom and counterweights requires 36 truckloads. This assembly approach for a wind farm involving whole rotor lifts of 100m diameter rotors contrasts strongly with the erection strategy of the E-126.

Current Developments

ROTOR BLADE DEVELOPMENT

Large Blade Design

Development in materials for rotor blades is ongoing. Testing ranging from characterisation of constituent materials, through blade sub-components to whole blades is vital for the integrity of new designs.

As the wind industry has matured, proof testing, ultimate load testing and fatigue testing of new rotor blade designs has become the norm. Figure I.3.28 shows blade testing at LM Glasfiber, which has long been the world's largest independent blade supplier. The commitment of LM Glasfiber to blade development for wind turbines is shown by a recent investment in their own specially designed wind tunnel, which can provide tunnel speeds up to 105 m/s and test an airfoil with 0.90 m chord at Reynolds number up to 6 million.

The handling and transport issues with very large blades have caused some manufacturers to revisit ideas for jointed blades.

The world's largest wind turbine, the Enercon E-126, adopts a jointed blade design. In the E-126 blade, an essentially standard outer blade section with a conventional blade root attachment is bolted to a steel inner blade spar. The trailing edge of the inner blade is a separate composite structure.

Gamesa has also developed a jointed blade design for the G10X.

Enercon has been pushing the limits in maximum attainable aerodynamic performance of a horizontal axis wind turbine rotor, achieving a measured rotor power performance coefficient in excess of 0.5. It may be noted that its recent designs take great care to manage the flow regime in the region of the hub and also employ winglets, which are suggested by research at Risø DTU to reduce power loss associated with the blade tip effect.

Pitch Systems

The Genesys 600 kW direct drive wind turbine, on which the Vensys designs are based, employed almost wear-free toothed belts in the pitch drives, instead of the more common hydraulic cylinders or geared electric pitch motors. One key advantage claimed for this innovation is insensitivity to shock loads, since such impact forces are distributed over multiple meshing teeth pairs. A second advantage is that the drive system does not require grease lubrication and is almost maintenance free. Toothed belts are widely used in industrial and automotive applications. Vensys have maintained this feature in their present designs.

ALTERNATIVE DRIVE TRAIN CONFIGURATIONS

Whilst the rotor configuration of large wind turbines as three composite blades in upwind attitude with full span pitch control has consolidated in the past ten years, there is increasing variety in current drive train arrangements. A few of the considerable variety of layout and bearing options of the drive train, with gearbox and conventional high-speed generator, were discussed in 'Architecture of a modern wind turbine'. Other drive train options can be classified as direct drive (gearless) or as hybrid, in other words with some level of

gearing (usually one or two stages) and one or more multi-pole generators.

Direct Drive

The motivation for direct drive is to simplify the nacelle systems, increase reliability, increase efficiency and avoid gearbox issues. A general trend towards direct drive systems has been evident for some years, although there are considerable challenges in producing technology that is lighter or more cost-effective than the conventional geared drive trains. Although these developments continue, direct drive turbines have not, as yet, had a sizeable market share. The exception is Enercon, which has long supplied direct drive generators employing a synchronous generator and having an electrical rotor with windings rather than permanent magnets. Most other direct drive designs are based on PMG technology, using high-strength Neodymium magnets. In July 2008, Siemens installed the first of two new 3.6 MW direct drive turbines to assess whether direct drive technology is competitive with geared machines for large turbines. The two turbines, which have a rotor diameter of 107 m, use a synchronous generator and permanent magnets.

MTorres started wind industry activities in 1999, leading to development of the TWT-1500, a 1500 kW wind turbine with a multi-pole synchronous generator. The nacelle layout of the MTorres wind turbine is indicated in the schematic diagram above (Figure I.3.29).

The Netherlands manufacturer Lagerwey supplied small wind turbines for a number of years and, at a later stage, developed wind turbines of 52, 58 and 72 m diameter with direct drive generators. The LW 52 and LW 58 were wound rotor synchronous machines like Enercon's. Lagerwey then sought to develop a larger 1.5 MW direct drive turbine with Zephyros, the Zephyros LW 72. The first installation, at a site in The Netherlands, used a permanent magnet generator

Figure I.3.29: MTorres wind turbine design

Figure I.3.30: Northern Power Systems 100 kW drive train concept

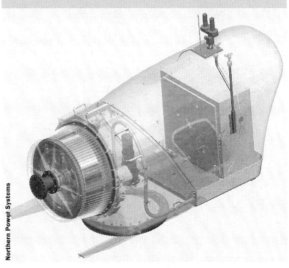

design and generation at medium voltage (3 to 4 kV). Subsequently, Zephyros separated from Lagerwey and was acquired by Harakosan.

Another notable development in direct drive has come from the Vensys designs, which derive from the Genesys 600 kW prototype of 1997, developed at Saarbrucken University. Vensys turbines may see increasing market presence through the interests of the Chinese developer Goldwind.

Northern Power Systems (NPS) developed the Northwind 100 wind turbine. Several hundred of their 100 kW turbines have been installed, often in remote locations. Their direct drive generator originally employed a salient pole wound-rotor technology, but, in line with most new direct drive designs, they have since developed a permanent magnet generator design and an innovative power converter design (Figure I.3.30).

Hybrid Systems

Hybrid systems are a middle route between the conventional solution with three stages of gearing at megawatt scale and direct drive solutions, which generally demand rather a large diameter generator. The intention is to have a simpler and more reliable gearbox, with a generator of comparable size, leading to a dimensionally balanced and compact drive train.

This design route was launched in the Multibrid concept licensed by Aerodyn. The inventor, George Bohmeke, has pursued that technology with the Finnish company WinWind.

A characteristic of the system is the more balanced geometry of gearbox and generator, leading to a compact arrangement. The nacelle need not extend much aft of tower centre-line (Figure I.3.31), as is generally appropriate for offshore machines, unless it will be accommodating electrical power equipment, such as the converter and transformer.

The structural economy achieved with such an integrated design is well illustrated in Figure I.3.31, with the main nacelle structure tending towards an open shell structure, a broadly logical result since, rather like the hub, it also connects circular interfaces (yaw bearing and main rotor bearing) that have substantial angular spacing.

Figure I.3.31: WinWind 3 MW

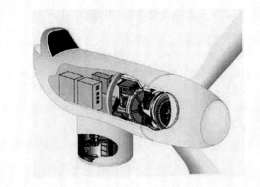

Source: WinWind

Clipper Windpower (Figure I.3.32) manufactures 2.5 MW wind turbines, with a hybrid drive train of very distinctive design. After initial research into systems with multiple induction generators, Clipper developed a system with an innovative gearbox with outputs to four PMGs. As with other hybrids, this again leads to a very compact drive train.

Prokon Nord Energiesysteme GmbH, based in Leer, acquired the previous Multibrid company in 2003. The prototype M5000 (Figure I.3.33) was installed in Bremerhaven and commissioned in 2005. The Multibrid technology was subsequently acquired by Areva in June 2008.

Distinctive features of the M5000 include a highly compact integrated slow rotating drive system,

Figure I.3.32: Clipper Windpower Liberty wind turbine with Multi-PMG system

Figure I.3.33: Multibrid 5 MW wind turbine

comprising a single main bearing (no main shaft), a single-stage gearbox and a medium speed PMG (58–147 rpm). With a tower-head mass of 310 tonnes, the M5000 is apparently the lightest wind turbine rated around 5 MW.

Other Drive Train Developments

Hydraulic components have figured in drive train design for some time in motors, brakes, fluid couplings or torque limiting systems. Hydraulic drives comprising pump(s) and motor(s) for main power transmission were employed in the unsuccessful Bendix 3 MW prototype of the early 1980s, but this design route was not pursued. Key problems were inadequate capacity, efficiency, reliability and life of existing commercial hydraulic components – the lack of components specifically designed for the needs of efficient wind power generation.

The Scottish company Artemis has addressed this and has developed a high-efficiency, long-life, ring cam pump, with electronically controlled poppet valves to suit wind turbine applications. Ring cam pumps are very rugged and reliable. Those, for example, made by the Scottish supplier MacTaggart Scott are welded into the hulls of submarines for life. Development work which will subsequently consider wind power transmission systems in the 5 to 10 MW range is progressing with funding assistance from the UK Carbon Trust. Artemis claims a 20 per cent mass reduction in nacelle systems can result, commenting that the power density of hydraulic machines is at least three times higher than the most advanced electric motor.

Another recent use of fluidic systems is in the Voith transmission system, adopted by De Wind (now owned by Composites Technology Inc.). This is essentially a way of realising a variable speed in the gearbox, thereby allowing direct connection of a synchronous generator to the output and hence avoiding the need for an electrical power converter.

The Voith WinDrive system uses a hydrodynamic torque converter to provide the variable speed relationship between the output shafts. WinDrive is essentially a mechanical solution to variable speed operation, based on a torque converter in combination with a planetary gear system. As a fluid machine, the torque converter is well matched to the wind turbine rotor and, via the fluid in the converter, the system decouples the input and output shafts absorbing input torque spikes and providing damping of vibrations.

With WinDrive, added mechanical complexity and cost in the gear system is compensated by elimination of the cost, mass and losses of an electrical power converter. The damping and compliance, intrinsic in the hydrodynamic coupling, ensures that a synchronous generator can be used. The Voith technology is long established in industrial drives, but the wind power application presents new challenges, especially in fatigue life and efficiency, which Voith have been addressing.

CONTROLLER CAPABILITIES

In the early days, wind turbine controllers had simple sequential control tasks to perform: start-up, controlled shutdown, and the monitoring of temperatures and other status indications from important components. At the academic level, it was realised early on that more advanced control could reduce the mechanical loads on the turbine, and thereby allow mass to be reduced. This has now been implemented to some extent in some turbine designs, principally by controlling individual blade pitch and generator torque (via the variable-speed electronic power converter). Additional inputs, such as nacelle acceleration, are required. The control algorithm then has a complex optimisation task to perform, and the controller principles and algorithms are considered highly confidential by turbine manufacturers.

This trend is likely to continue as experience is gained. The computation required is not great compared to control tasks in other industries. Therefore, the rate of development is likely to be set by the rate at which experience is gained with prototypes and the rate at which models of the turbines and their environments can be developed and validated.

Further progress is expected, as means are developed to provide further reliable measurements of mechanical loads: measurement of blade loads through optical fibre strain gauges appears to be a hopeful development.

NETWORK OPERATOR REQUIREMENTS

As noted in previous sections, the technical requirements of network operators are becoming more onerous. The principal areas are:

- the ability to stay connected and perhaps also contribute to system stability during disturbances on the electricity system (including 'fault ride-through');
- the ability to control reactive power generation/ consumption in order to contribute towards control of voltage; and
- a general aim for the wind farm (not necessarily each wind turbine) to respond similarly to conventional thermal generation, where possible.

These issues are considered in more detail in Part II. However, the net effect is to increase the arguments for variable speed operation, and in particular for concepts using a fully rated electronic converter between the generator and the network. Concepts with mechanical or hydraulic variable speed operation and a synchronous generator, connected directly to the network, are also suitable.

It should be noted that some of the required functions require fast response and communications from the wind farm SCADA (supervisory control and data acquisition) system.

It is also feasible to meet the requirements for reactive power control and for fault ride-through by using fixed-speed or DFIG wind turbines, and additional power electronic converters (commonly called statcoms or static VAR compensators). This can be done by a single converter at the wind farm point of connection to the grid, or by smaller units added to each turbine.

TESTING, STANDARDISATION AND CERTIFICATION

Standardisation is largely achieved through the IEC 61400 series of standards specifically for the wind industry, and their national or European equivalents. Standardisation work is driven by industry requirements, but requires consensus, and therefore takes considerable time. A process of gradual change and development is expected, rather than any radical changes.

Testing of major components, such as blades, is expected to develop, driven both by standards and by manufacturers. Certification of turbine designs by independent bodies is now well established.

Future Innovations

NEW SYSTEM CONCEPTS

Technology development in the largest unit sizes of conventional wind turbines has been particularly stimulated by the emerging offshore market, and many of the most innovative wind energy systems proposed in recent years target that market. System concepts, such as floating turbines that are intrinsically for offshore, are reviewed in Chapter I.5. Systems involving innovations that may operate from land or offshore are presented below. Some of these systems may be the way of the future; some will undoubtedly disappear from the scene. At the very least, they illustrate the huge stimulus for creative engineering which has arisen from the challenges in harnessing renewable energy sources and from the establishment of wind energy technology in the power industry.

Electrostatic Generator

An electrostatic generator is being investigated in the High Voltage Laboratory at TU Delft. It is presently at the laboratory feasibility stage, being tested at milliwatt scale. The considerable attraction is to have a system with few mechanical parts. It is seen as being potentially suitable for buildings, due to minimal noise and vibration, or for offshore, on account of simplicity and the potential to be highly reliable.

In the EWICON design, charged droplets are released, and transported and dispersed by the wind to create DC current in collector wires. One key issue with this system is the level of power consumption used in charging the droplets. Up-scaling to useful power capacities from micro laboratory size is thought to benefit performance but has yet to be demonstrated.

AIRBORNE TURBINES

Airborne turbine concepts have appeared, at least in patent documents, for many years, and such concepts are presently generating increased interest. The designs are broadly classified as:

- systems supported by balloon buoyancy;
- kites (lifting aerofoils); and
- tethered auto gyros.

Some systems combine these features. Key issues are:

- controllability, in the case of kites;
- general problems associated with maintaining the systems in flight for long periods; and
- reliable power transmission to land/sea level without heavy drag-down from the power cables or tethers.

Just as limits on land-based sites and better offshore resources have justified the extension of wind technology offshore, much is made of the increased wind energy density at higher altitudes and reduced energy collection area per MW of capacity, as compensation for the obvious challenges of airborne technology.

The Kite Gen concept (Figure I.3.34) is to access wind altitudes of 800 to 1000 m using power kites in the form of semi-rigid, automatically piloted, efficient aerofoils. All generating equipment is ground-based, and high-strength lines transmit the traction of the kites and control their position. Compared to a rotor, the figure-of-eight path swept by the kite can be beneficial for energy capture, as the inner parts of a conventional wind turbine rotor blade travel at comparatively low speeds through some of the swept area.

In the early 1990s, the Oxford-based inventor Colin Jack patented airborne wind energy ideas, based on the autogyro principle, where some of the lift of the rotor is used to support its weight. This was prompted by the realisation that a balloon capable of supporting the weight of a wind turbine rotor in power producing mode requires much more buoyancy than one able to support the dead weight in calm conditions. This arises

Figure I.3.34: Kite Gen

Figure I.3.35: Sky Windpower flying electric generator

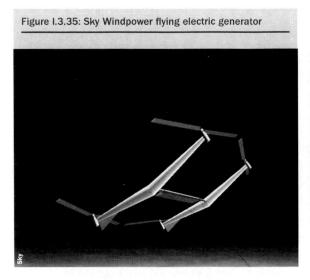

Figure I.3.36: Magenn

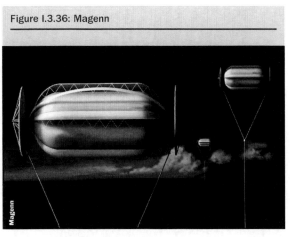

from the 'drag-down' effect of rotor thrust when the turbine is operating. A recent concept, using auto gyro rotors to be self-supporting as well as generating, is illustrated in Figure I.3.35.

In November 2007, Google announced a strategic initiative to develop electricity from renewable energy sources and, in particular, are known to be providing R&D funding for Makani Power to develop innovative airborne technology. As yet, no details of the concept are available.

The Magenn Air Rotor System (MARS) is an airborne tethered horizontal axis wind turbine system, with its rotor in the form of a helium balloon. Generation is at an altitude of around 300m, with power transmission down the tether cable. The generators are at each end of the rotor (Figure I.3.36), with a direct output power connection to the twin cables. Outboard of the generators at each end of the rotor are wind vane stabilisers in the form of conical wheels. The Magnus effect associated with the rotor rotation also provides additional lift, which stabilises the rotor position, causing it to pull up overhead rather than drift downwind on its tether.

MAGLEV

Maglev is an innovative vertical axis turbine concept. Construction began on a large production site for Maglev wind turbines in central China in November 2007. According to a press release, Zhongke Hengyuan Energy Technology has invested 400 million yuan in building this facility, which will produce Maglev wind turbines with capacities ranging from 400 to 5000 kW.

Using magnetic levitation, the blades of the turbine are suspended on an air cushion, and the energy extracted by linear generators with minimal friction losses. The major advantage of Maglev is claimed to be that it reduces maintenance costs and increases the lifespan of the generator.

I.4 WIND FARM DESIGN

Introduction

Previous chapters have discussed wind turbine technology and the wind resource. This chapter presents a brief summary of the design of a wind farm as a whole. The chapter is based on onshore wind farms, though many of the topics are also relevant for offshore. Offshore wind farms are covered in Chapter I.5. Environmental factors that affect wind farm design are discussed in this chapter. Further environmental issues are covered in detail in Part V.

Factors Affecting Turbine Location

Once a site has been identified and the decision has been taken to invest in its development, the wind farm design process begins. The fundamental aim is to maximise energy production, minimise capital cost and operating costs, and stay within the constraints imposed by the site. As the constraints and costs are all subject to some level of uncertainty, the optimisation process also seeks to minimise risk. The first task is to define the constraints on the development:

- maximum installed capacity (due to grid connection or power purchase agreement terms);
- site boundary;
- 'set back' – distances from roads, dwellings, overhead lines, ownership boundaries and so on;
- environmental constraints;
- location of noise-sensitive dwellings, if any, and assessment criteria;
- location of visually-sensitive viewpoints, if any, and assessment criteria;
- location of dwellings that may be affected by 'shadow flicker' (flickering shadows cast by rotating blades) when the sun is in particular directions, and assessment criteria;
- turbine minimum spacings, as defined by the turbine supplier (these are affected by turbulence, in particular); and

- constraints associated with communications signals, for example microwave link corridors or radar.

These constraints may change as discussions and negotiations with various parties progress, so this is inevitably an iterative process.

When the likely constraints are known, a preliminary design of the wind farm can be produced. This will allow the size of the development to be established. As a rough guide, the installed capacity of the wind farm is likely to be of the order of 12 MW per km^2, unless there are major restrictions that affect the efficient use of the available land.

For the purpose of defining the preliminary layout, it is necessary to define approximately what sizes of turbine are under consideration for the development, as the installed capacity that can be achieved with different sizes of turbine may vary significantly. The selection of a specific turbine model is often best left to the more detailed design phase, when the commercial terms of potential turbine suppliers are known. Therefore at this stage it is either necessary to use a 'generic' turbine design, defined in terms of a range of rotor diameters and a range of hub heights, or alternatively to proceed on the basis of two or three layouts, each based on specific wind turbines.

The preliminary layout may show that the available wind speed measurements on the site do not adequately cover all the intended turbine locations. In this case it will be necessary to consider installing additional anemometry equipment. The preliminary layout can then be used for discussions with the relevant authorities and affected parties. This is an iterative process, and it is common for the layout to be altered at this stage.

The factors most likely to affect turbine location are:

- optimisation of energy production;
- visual influence;
- noise; and
- turbine loads.

OPTIMISATION OF ENERGY PRODUCTION

Once the wind farm constraints are defined, the layout of the wind farm can be optimised. This process is also called wind farm 'micro-siting'. As noted above, the aim of such a process is to maximise the energy production of the wind farm whilst minimising the infrastructure and operating costs. For most projects, the economics are substantially more sensitive to changes in energy production than infrastructure costs. It is therefore appropriate to use energy production as the dominant layout design parameter.

The detailed design of the wind farm is facilitated by the use of wind farm design tools (WFDTs). There are several that are commercially available, and others that are research tools. Once an appropriate analysis of the wind regime at the site has been undertaken, a model is set up that can be used to design the layout, predict the energy production of the wind farm, and address issues such as visual influence and noise.

For large wind farms it is often difficult to manually derive the most productive layout. For such sites a computational optimisation using a WFDT may result in substantial gains in predicted energy production. Even a 1 per cent gain in energy production from improved micro-siting is worthwhile, as it may be achieved at no increase in capital cost. The computational optimisation process will usually involve many thousands of iterations and can include noise and visual constraints. WFDTs conveniently allow many permutations of wind farm size, turbine type, hub height and layout to be considered quickly and efficiently, increasing the likelihood that an optimal project will result. Financial models may be linked to the tool so that returns from different options can be directly calculated, further streamlining the development decision-making process.

An example screen dump from a typical WFDT is presented in Figure I.4.1. The darker shaded areas represent the areas of highest wind speed. The turbines are represented by the small markers with a number underneath. The spacing constraint is illustrated by the ellipses (this constraint is discussed in the turbine loads section).

VISUAL INFLUENCE

'Visual influence' is the term used for the visibility of the wind turbines from the surrounding area.

In many countries the visual influence of a wind farm on the landscape is an important issue, especially in regions with high population density. The use of computational design tools allows the zone of visual influence (ZVI), or visibility footprint, to be calculated to identify from where the wind farm will be visible. It is usually necessary to agree a number of cases with the permitting authorities or other interested parties, such as:

- locations from which 50 per cent of turbine hubs can be seen;
- locations from which at least one hub can be seen; and
- locations from which at least one blade tip can be seen.

Figure I.4.2 shows an example ZVI for a wind farm, generated using a WFDT. The variation in colour represents the proportion of the wind farm that is visible from viewpoints anywhere in the wind farm vicinity.

Such maps tend to exaggerate the actual visual effect of the wind farm, as they do not clearly indicate the effect of distance on visual appearance. They are also difficult for non-specialists to interpret. Therefore it is also common to generate 'visualisations' of the appearance of the wind farm from defined viewpoints. These can take the form of 'wireframe' representations of the topography. With more work, photomontages can be produced in which the wind turbines are superimposed upon photographs taken from the defined viewpoints.

Figure I.4.1: Example wind speed map from a WFDT

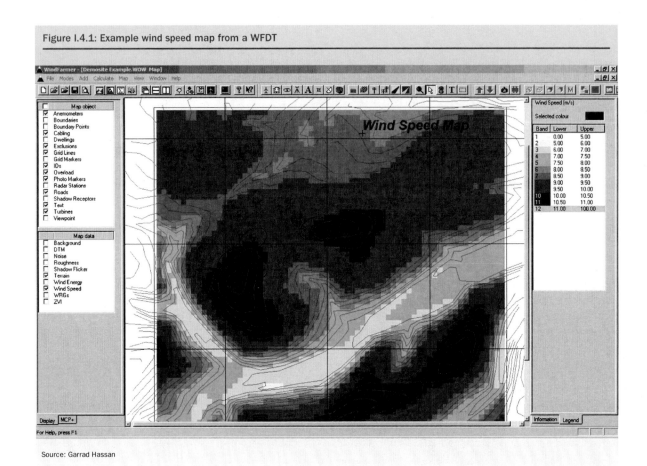

Source: Garrad Hassan

Figure I.4.3 shows an example wireframe representation of a wind farm generated using a WFDT. Figure I.4.4 shows the same image after rendering software has been applied and Figure I.4.5 shows the same image as a photomontage.

Other factors also affect the visual appearance of a wind farm. Larger turbines rotate more slowly than smaller ones, and a wind farm of fewer larger turbines is usually preferable to a wind farm of many smaller ones. In some surroundings, a regular area or straight line may be preferable compared to an irregular layout.

NOISE

In densely populated countries, noise can sometimes be a limiting factor for the generating capacity that can be installed on any particular site. The noise produced by operating turbines has been significantly reduced in recent years by turbine manufacturers, but is still a constraint. This is for two main reasons:

1. Unlike most other generating technologies, wind turbines are often located in rural areas, where background noise levels can be very low, especially overnight. In fact the critical times are when wind speed is at the lower end of the turbine operating range, because then the wind-induced background noise is lowest.

2. The main noise sources (blade tips, the trailing edge of the outer part of the blade, the gearbox and generator) are elevated, and so are not screened by topography or obstacles.

Figure I.4.2: A wind farm ZVI generated using a WFDT

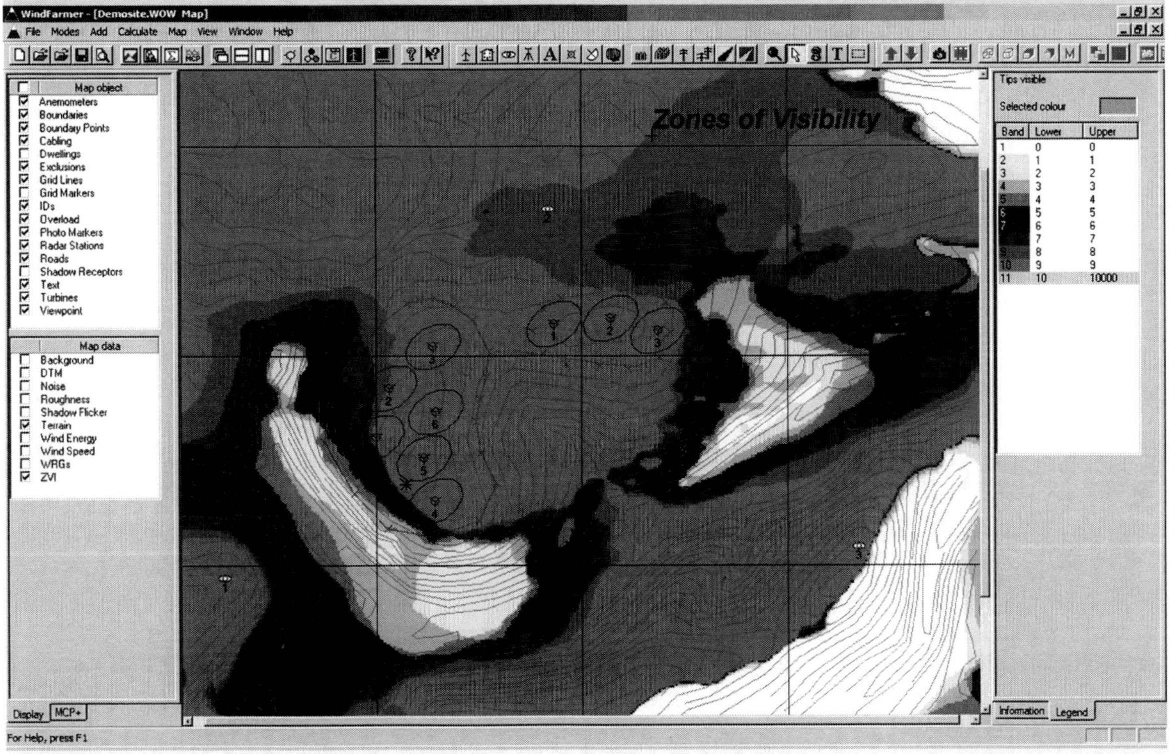

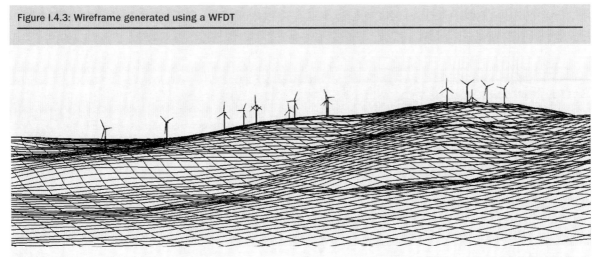

Source: Garrad Hassan

Figure I.4.3: Wireframe generated using a WFDT

Source: Garrad Hassan

Figure I.4.4: Rendered wireframe generated using a WFDT

Turbine manufacturers may provide noise characteristic certificates, based on measurements by independent test organisations to agreed standards. The internationally recognised standard that is typically referred to is 'Wind turbine generator systems: Acoustic noise measurement techniques' (IEC 61400 Part 11 of 2003).

Standard techniques, taking into account standard noise propagation models, are used to calculate the expected noise levels at critical locations, which are usually the nearest dwellings. The results are then compared with the acceptable levels, which are often defined in national legislation. An example of a 'noise map' that can be generated by a WFDT is shown in Figure I.4.6. The noise contours shown represent the modelled noise at any point in the vicinity of the wind farm.

The internationally recognised standard for such calculations is 'Acoustics – Attenuation of sound during propagation outdoors; Part 2: General method of calculation' (ISO 9613-2).

Sometimes the permitting authorities will require the project to conform to noise limits, with penalties if it can be shown that the project does not comply. In turn the turbine manufacturer could provide a warranty for the noise produced by the turbines. The warranty may be backed up by agreed measurement techniques in case it is necessary to undertake noise tests on one or more turbines.

TURBINE LOADS

A key element of the layout design is the minimum turbine spacing used. In order to ensure that the turbines

Figure I.4.5: Photomontage generated using a WFDT

are not being used outside their design conditions, the minimum acceptable turbine spacing should be obtained from the turbine supplier and adhered to.

The appropriate spacing for turbines is strongly dependent on the nature of the terrain and the wind rose for a site. If turbines are spaced closer than five rotor diameters (5D) in a frequent wind direction, it is likely that unacceptably high wake losses will result. For areas with predominantly unidirectional wind roses, such as the San Gorgonio Pass in California, or bidirectional wind roses, such as Galicia in Spain, greater distances between turbines in the prevailing wind direction and tighter spacing perpendicular to the prevailing wind direction will prove to be more productive.

Tight spacing means that turbines are more affected by turbulence from the wakes of upstream turbines. This will create high mechanical loads and requires approval by the turbine supplier if warranty arrangements are not to be affected.

Separately from the issue of turbine spacing, turbine loads are also affected by:
• 'Natural' turbulence caused by obstructions, topography, surface roughness and thermal effects; and
• Extreme winds.

Defining reliable values for these parameters, for all turbine locations on the site, may be difficult. Lack of knowledge is likely to lead to conservative assumptions and conservative design.

Within the wind industry there is an expectation that all commercial wind turbines will be subject to independent certification in accordance with established standards or rules. A project-specific verification of the suitability of the certification for the proposed site should be carried out, taking into

Figure I.4.6: A wind farm noise map generated using a WFDT

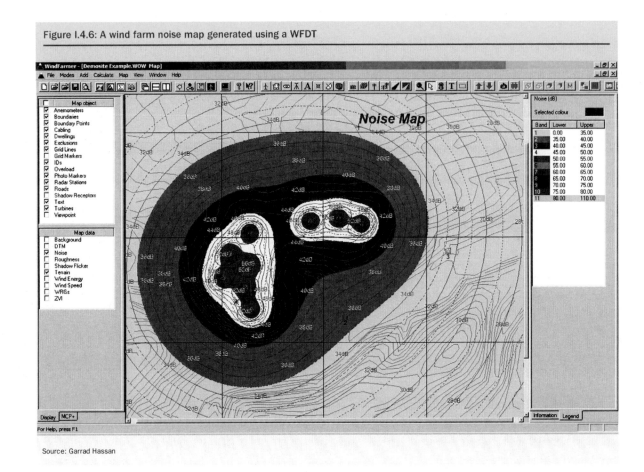

Source: Garrad Hassan

account the turbine design specifications and the expected climatic conditions of the site.

Infrastructure

The wind farm infrastructure consists of:

- Civil works:
  - roads and drainage;
  - wind turbine foundations;
  - met mast foundations (and occasionally also the met masts); and
  - buildings housing electrical switchgear, SCADA central equipment, and possibly spares and maintenance facilities.

- Electrical works:
  - equipment at the point of connection (POC), whether owned by the wind farm or by the electricity network operator;
  - underground cable networks and/or overhead lines, forming radial 'feeder' circuits to strings of wind turbines;
  - electrical switchgear for protection and disconnection of the feeder circuits;
  - transformers and switchgear associated with individual turbines (although this is now commonly located within the turbine and is supplied by the turbine supplier);
  - reactive compensation equipment, if necessary; and
  - earth (grounding) electrodes and systems.

- Supervisory control and data acquisition (SCADA) system:
  - central computer;
  - signal cables to each turbine and met mast;
  - wind speed and other meteorological transducers on met masts; and
  - electrical transducers at or close to the POC.

The civil and electrical works, often referred to as the 'balance of plant' (BOP), are often designed and installed by a contractor or contractors separate from the turbine supplier. The turbine supplier usually provides the SCADA system.

As discussed above, the major influence on the economic success of a wind farm is the energy production, which is principally determined by the wind regime at the chosen site, the wind farm layout and the choice of wind turbine. However, the wind farm infrastructure is also significant, for the following reasons:

- The infrastructure constitutes a significant part of the overall project cost. A typical cost breakdown is given in Figure I.4.7.

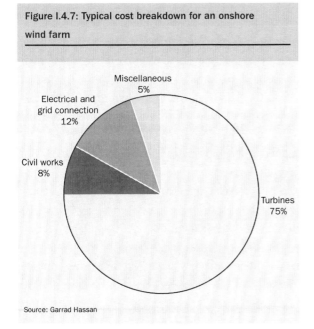

Figure I.4.7: Typical cost breakdown for an onshore wind farm

Miscellaneous 5%

Electrical and grid connection 12%

Civil works 8%

Turbines 75%

Source: Garrad Hassan

- The civil works present significant risks to the project costs and programme. It is not unknown for major delays and cost overruns to be caused by poor understanding of ground conditions, or the difficulties of working on sites that, by definition, are exposed to the weather and may have difficult access.
- The major electrical items (transformers, switchgear) have long lead times. At the time of writing, a large HV/MV power transformer may have a lead time of several years.
- The grid connection works may present a significant risk to the programme. It is likely that works will need to be undertaken by the electricity network operator, and the programme for these works is effectively out of the control of the wind farm developer. It is very unusual for electricity network operators to accept liability for any delay.

CIVIL WORKS

The foundations must be adequate to support the turbine under extreme loads. Normally the design load condition for the foundations is the extreme, 'once in 50 years' wind speed. In Europe this wind speed is characterised by the 'three-second gust'. This is a site-specific parameter, which will normally be determined as part of the wind speed measurements and energy production assessment for the site. For most sites this will lie between 45 and 70 m/s. At the lower end of this range it is likely that the maximum operational loads will be higher than the loads generated by the extreme gust and will therefore govern the foundation design.

The first step towards the proper design of the foundations is therefore the specification of a load. The turbine supplier will normally provide a complete specification of the foundation loads as part of a tender package. As the turbine will typically be provided with reference to a generic certification class (see page 94), these loads may also be defined with reference to the generic classes, rather than site-specific load cases.

Although extremely important, the foundation design process is a relatively simple civil engineering task. A typical foundation will be perhaps 13 m across a hexagonal form and might be 1–2 m deep. It will be made from reinforced concrete cast into an excavated hole. The construction time for such a foundation, from beginning to end, can easily be less than a week.

The site roads fall well within normal civil engineering practice, provided the nature of the terrain and the weather are adequately dealt with.

For wind farms sited on peat or bogs, it is necessary to ensure that the roads, foundations and drainage do not adversely affect the hydrology of the peat.

The wind farm may also need civil works for a control building to house electrical switchgear, the SCADA central computer, welfare facilities for maintenance staff and spare parts. There may also be an outdoor electricity substation, which requires foundations for transformers, switchgear and other equipment. None of this should present unusual difficulties.

For upland sites, it is often beneficial to locate the control building and substation in a sheltered location. This also reduces visual impact.

ELECTRICAL WORKS

The turbines are interconnected by a medium voltage (MV) electrical network, in the range 10 to 35 kV. In most cases this network consists of underground cables, but in some locations and some countries overhead lines on wooden poles are adopted. This is cheaper but creates greater visual influence. Overhead wooden pole lines can also restrict the movement and use of cranes.

The turbine generator voltage is normally classed as 'low', in other words below 1000 V, and is often 690 V. Some larger turbines use a higher generator voltage, around 3 kV, but this is not high enough for economical direct interconnection to other turbines. Therefore, it is necessary for each turbine to have a transformer to step up to MV, with associated MV switchgear. This

equipment can be located outside the base of each turbine. In some countries these are termed 'padmount transformers'. Depending on the permitting authorities and local electricity legislation, it may be necessary to enclose the equipment within GRP or concrete enclosures. These can be installed over the transformers, or supplied as prefabricated units complete with transformers and switchgear.

However, many turbines now include a transformer as part of the turbine supply. In these cases the terminal voltage of the turbine will be at MV, in the range 10 to 35 kV, and can connect directly to the MV wind farm network without the need for any external equipment.

The MV electrical network takes the power to a central point (or several points, for a large wind farm). A typical layout is shown in Figure I.4.8. In this case the central point is also a transformer substation, where the voltage is stepped up again to high voltage (HV, typically 100 to 150 kV) for connection to the existing electricity network. For small wind farms (up to approximately 30 MW), connection to the local MV network may be possible, in which case no substation transformers are necessary.

The MV electrical network consists of radial 'feeders'. Unlike industrial power networks, there is no economic justification for providing ring arrangements. Therefore a fault in a cable or at a turbine transformer will result in all turbines on that feeder being disconnected by switchgear at the substation. If the fault takes considerable time to repair, it may be possible to reconfigure the feeder to allow all turbines between the substation and the fault to be reconnected.

Figure I.4.8 shows two possible locations for the POC. Definitions of the POC vary from country to country (and are variously called delivery point, point of interconnection or similar), but the definitions are similar: it is the point at which responsibility for ownership and operation of the electrical system passes from the wind farm to the electricity network operator. More complex division of responsibilities is possible (for example, the wind farm developer may build and install

Figure I.4.8: A typical electrical layout

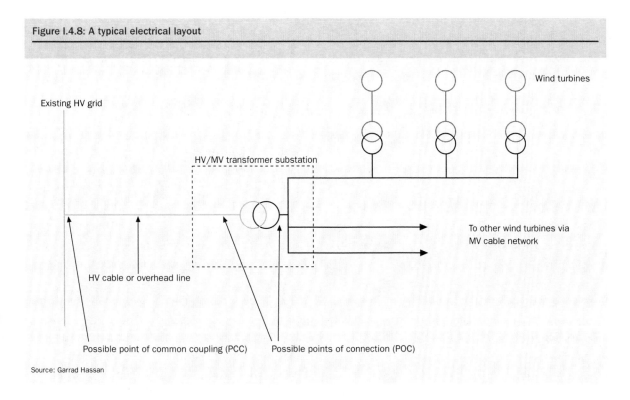

Wind turbines

Existing HV grid

HV/MV transformer substation

To other wind turbines via
MV cable network

HV cable or overhead line

Possible point of common coupling (PCC)   Possible points of connection (POC)

Source: Garrad Hassan

equipment, which then is taken over by the network operator), but this is unusual.

The revenue meters for the wind farm will usually be located at or close to the POC. In some cases, where the POC is at HV, the meters may be located on the MV system to save costs. In this case it is usual to agree correction factors to account for electrical losses in the HV/MV transformer.

Figure I.4.8 also shows a possible location of the 'point of common coupling' (PCC). This is the point at which other customers are (or could be) connected. It is therefore the point at which the effect of the wind farm on the electricity network should be determined. These effects include voltage step changes, voltage flicker and harmonic currents. Grid issues are discussed in detail in Part II. Often the PCC coincides with the POC.

The design requirements for the wind farm electrical system can be categorised as follows:

- It must meet local electrical safety requirements and be capable of being operated safely.

- It should achieve an optimum balance between capital cost, operating costs (principally the electrical losses) and reliability.
- It must ensure that the wind farm satisfies the technical requirements of the electricity network operator.
- It must ensure that the electrical requirements of the turbines are met.

The technical requirements of the electricity network operator are set out in the connection agreement, or a 'grid code' or similar document. This is discussed further in Part II.

SCADA AND INSTRUMENTS

A vital element of the wind farm is the SCADA system. This system acts as a 'nerve centre' for the project. It connects the individual turbines, the substation and meteorological stations to a central computer. This computer and the associated communication system allow the operator to supervise the behaviour of all the wind turbines and also the wind farm as a whole. It keeps a record on a ten-minute basis of all the activity, and allows the operator to determine what corrective action, if any, needs to be taken. It also records energy output, availability and error signals, which acts as a basis for any warranty calculations and claims. The SCADA system also has to implement any requirements in the connection agreement to control reactive power production, to contribute to network voltage or frequency control, or to limit power output in response to instructions from the network operator.

The SCADA computer communicates with the turbines via a communications network, which almost always uses optical fibres. Often the fibre-optic cables are installed by the electrical contractor, then tested and terminated by the SCADA supplier.

The SCADA system is usually provided by the turbine supplier, for contractual simplicity. There is also a market for SCADA systems from independent suppliers. The major advantages of this route are claimed to be:

- identical data reporting and analysis formats, irrespective of turbine type; this is important for wind farm owners or operators who have projects using different wind turbines; and
- transparency of calculation of availability and other possible warranty issues.

In addition to the essential equipment needed for a functioning wind farm, it is also advisable, if the project size can warrant the investment, to erect some permanent meteorological instrumentation on met masts. This equipment allows the performance of the wind farm to be carefully monitored and understood. If the wind farm is not performing according to its budget, it will be important to determine whether this is due to poor mechanical performance or less-than-expected wind resource. In the absence of good quality wind data on the site, it will not be possible to make this determination. Large wind farms therefore usually contain one or more permanent meteorological masts, which are installed at the same time as the wind farm.

Construction Issues

A wind farm may be a single machine or it may be a large number of machines, possibly many hundreds. The design approach and the construction method will, however, be almost identical whatever the size of project envisaged. The record of the wind industry in the construction of wind farms is generally good. Few wind farms are delivered either late or over budget.

Newcomers to the wind industry tend to think of a wind farm as a power station. There are, however, some important differences between these two types of power generation. A conventional power station is one large machine, which will not generate power until it is complete. It will often need a substantial and complicated civil structure, and construction risk will be an important part of the project assessment. However, the construction of a wind farm is more akin to the purchase of a fleet of trucks than it is to the construction of a power station. The turbines will be purchased at a fixed cost agreed in advance and a delivery schedule will be established exactly as it would be for a fleet of trucks. In a similar way the electrical infrastructure can be specified well in advance, again probably at a fixed price. There may be some variable costs associated with the civil works, but this cost variation will be very small compared to the cost of the project as a whole. The construction time is also very short compared to a conventional power plant. A 10 MW wind farm can easily be built within a couple of months.

To minimise cost and environmental effects, it is common to source material for roads from on-site quarries or 'borrow pits', where suitable. It may be necessary to seek permission for this from the permitting authorities.

Costs

Wind farm costs are largely determined by two factors: the complexity of the site and the likely extreme loads. The site may be considered complex if the ground conditions are difficult – hard rock or very wet or boggy ground, for example – or if access is a problem. A very windy site with high extreme loads will result in a more expensive civil infrastructure as well as a higher specification for the turbines.

The cost of the grid connection may also be important. Grid connection costs are affected by:

- distance to a suitable network connection point;
- the voltage level of the existing network; and
- the network operator's principles for charging for connections and for the use of the electricity system.

Costs are covered in Part III.

Commissioning, Operation and Maintenance

Once construction is completed, commissioning will begin. The definition of 'commissioning' is not standardised, but generally covers all activities after all components of the wind turbine are installed. Commissioning of an individual turbine can take little more than two days with experienced staff.

Commissioning tests will usually involve standard electrical tests for the electrical infrastructure as well as the turbine, and inspection of routine civil engineering quality records. Careful testing at this stage is vital if a good quality wind farm is to be delivered and maintained.

The long-term availability of a commercial wind turbine is usually in excess of 97 per cent. This value means that for 97 per cent of the time, the turbine will be available to work if there is adequate wind. This value is superior to values quoted for conventional power stations. It will usually take a period of some six months for the wind farm to reach full, mature, commercial operation and hence, during that period, the availability will increase from a level of about 80–90 per cent after commissioning to the long-term level of 97 per cent or more.

It is normal practice for the supplier of the wind farm to provide a warranty for between two and five years. This warranty will often cover lost revenue, including downtime to correct faults, and a test of the power curve of the turbine. If the power curve is found to be defective, then reimbursement will be made through the payment of liquidated damages. For modern wind farms, there is rarely any problem in meeting the warranted power curves, but availability, particularly for new models, can be lower than expected in the early years of operation. During the first year of operation of a turbine some 'teething' problems are usually experienced. For a new model this effect is more marked. As model use increases, these problems are resolved and availability rises.

After commissioning, the wind farm will be handed over to the operations and maintenance crew. A typical crew will consist of two people for every 20 to 30 wind turbines in a wind farm. For smaller wind farms there may not be a dedicated O&M crew but arrangements will be made for regular visits from a regional team. Typical routine maintenance time for a modern wind turbine is 40 hours per year. Non-routine maintenance may be of a similar order.

There is now much commercial experience with modern wind turbines and high levels of availability are regularly achieved. Third party operations companies are well established in all of the major markets, and it is likely that this element of the industry will develop very much along the lines associated with other rotating plant and mechanical/electrical equipment.

The building permits obtained in order to allow the construction of the wind farm may have some ongoing environmental reporting requirements, for example the monitoring of noise, avian activity, or other flora or fauna issues. Similarly there may, depending on the local regulations, be regulatory duties to perform in connection with the local electricity network operator.

Therefore, in addition to the obvious operations and maintenance activity, there is often a management role to perform in parallel. Many wind farms are funded through project finance and hence regular reporting activities to the lenders will also be required.

EWEA/Winter

I.5 OFFSHORE

Introduction

Previous chapters have covered the fundamental technical aspects of wind energy, largely from the point of view of onshore installations. This chapter covers those technical issues that are different for offshore wind. The potential for offshore wind is enormous in Europe and elsewhere, but the technical challenges are also great. The capital costs are higher than onshore, the risks are greater, the project sizes are greater and the costs of mistakes are greater.

Offshore wind technology and practice has come a long way in a short time, but there is clearly much development still to be done. Although the fundamentals of the technology are the same onshore and offshore, it is clear that offshore wind technology is likely to diverge further from onshore technology. Methods of installation and operation are already very different from onshore wind generation, with great attention being given to reliability and access.

Wind Resource Assessment Offshore

This section describes the differences in wind flow, monitoring and data analysis offshore in comparison to onshore. It also highlights the key differences associated with the assessment of the offshore wind resource and the energy production of offshore wind farms when compared with onshore wind farms. Many of the elements of the analyses are common to onshore and offshore projects and it is therefore recommended that this section is read in parallel with the chapters on onshore wind.

FUNDAMENTALS

Onshore, topographic effects are one of the main driving forces of the wind regime. With no topographic effects offshore, other factors dominate the variation in wind speed with height.

The surface roughness (a parameter used to describe the roughness of the surface of the ground, referred to as Z_0) is low, which results in a steeper boundary layer profile (also referred to as the wind shear profile), characterised by the symbol α. A range of typical values for Z_0 and α are illustrated in Table I.5.1. Offshore, the surface roughness length is dependent on sea state, increasing with the local wave conditions, which are in turn influenced by wind conditions. However, this relationship is complex, as the sea surface, even when rough, does not present fixed roughness elements such as trees, hills and buildings, as tends to be the case onshore.

The low surface roughness also results in low turbulence intensity. This serves to reduce mechanical loads. It also may increase the energy capture compared to an identical wind turbine at an onshore location with identical mean wind speed.

The coastal zone, where the properties of the boundary layer are changeable, extends away from the shore for varying distances, and this can result in variations in wind speed, boundary layer profiles and turbulence across the wind farm.

Stable flow conditions are also evident offshore. In these situations, air flows with different origins and air

Table I.5.1: Typical values for Z_0 and α		
Type of terrain	**Z_0 (m)**	**α**
Mud flats, ice	0.00001	
Smooth sea	0.0001	
Sand	0.0003	0.10
Snow surface	0.001	
Bare soil	0.005	0.13
Low grass, steppe	0.01	
Fallow field	0.03	
Open farmland	0.05	0.19
Shelter belts	0.3	
Forest and woodland	0.5	
Suburb	0.8	
City	1	0.32

temperatures can be slow to mix. This can manifest itself as unusual boundary layer profiles, and in some rare situations wind speed may even reduce with height.

A further factor influencing offshore winds can be the tide level in areas with a high tidal range. The rise and fall of the sea level effectively shifts the location of the turbine in the boundary layer. This can have impacts in variation of mean wind speed within a period of approximately 12 hours, and also on the variation in mean winds across the turbine rotor itself.

Temperature-driven flows due to the thermal inertia of the sea initiate localised winds around the coastal area. Compared to the land, the sea temperature is more constant over the day. During the day, as the land heats up, the warmer air rises and is replaced by cooler air from over the sea. This creates an onshore wind. The reverse effect can happen during the night, resulting in an offshore wind. The strength and direction of the resulting wind is influenced by the existing high-level gradient wind, and in some situations the gradient wind can be cancelled out by the sea breeze, leaving an area with no wind. Finally, as all sailors are aware, close to the coast there are 'backing' and 'veering' effects.

MEASUREMENT OFFSHORE

Offshore wind farms typically use the largest available wind turbines on the market. Their size presents several issues, including understanding the characteristics of the boundary layer up to and above heights of 100 m. Measurements offshore are costly, with costs driven to a large extent by the cost of constructing the support structure for the meteorological mast. These masts cost some €1–5 million, depending on site location and specification, which is perhaps 100 times that required for equivalent onshore work. Offshore monitoring towers are un-guyed, and therefore need to be wider, which can mean measurements are more susceptible to wind flow effects from the tower. Anemometry equipment is otherwise standard.

If high-quality wind measurements are not available from the site or nearby, there are other sources of information that can be utilised to determine the approximate long-term wind regime at the wind farm location. There are offshore databases for wind data, including meteorological buoys, light vessels and observation platforms. Additionally, meso-scale modelling (based on global reanalysis data sets) and Earth-observation data play a role in preliminary analysis and analysis of spatial variability. None of these are suitable for a robust financing report, however.

WIND ANALYSIS OFFSHORE

Depending on the amount of data available, different analysis methods can be employed. A feasibility study can be carried out based on available wind data in the area. WAsP can be used from coastal meteorological stations to give a prediction offshore, which is aided by its latest tool, the coastal discontinuity model (CDM) (see 'Offshore wind resource assessment in European Seas, state-of-the-art' in Sempreviva et al., 2003).

Existing offshore measurements can also be used. There are problems associated with using long-distance modelling, however, especially around the coast, due to the differences in predominant driving forces between onshore and offshore breezes and the variation in the coastal zone in between.

For a more detailed analysis, measurements offshore at the site are necessary. 'Measure correlate predict' (MCP) methods from a mast offshore to an onshore reference station can be used. With several measurement heights and attention to measurement, more accurate modelling of the boundary layer will help extrapolate to heights above the monitoring mast. A photograph of an offshore mast is presented in Figure I.5.1.

Figure I.5.1: An offshore meteorological mast

is therefore less mixing of the air behind the turbine, which results in a slower re-energising of the slow-moving air, and the wake lasts longer. Observations from the largest current offshore wind farms have identified shortcomings in the classic wind farm wake modelling techniques, due to the large size of the projects and perhaps due to specific aspects of the wind regime offshore. Relatively simple amendments to standard wake models are currently being used to model offshore wake effects for large projects, but further research work is ongoing to better understand the mechanisms involved and to develop second generation offshore wake models.

There is likely to be more downtime of machines offshore, primarily due to difficult access to the turbines. If a turbine has shut down and needs maintenance work, access to it may be delayed until there is a suitable window in the weather conditions. This aspect of offshore wind energy is a critical factor in the economic appraisal of a project. Increasingly sophisticated Monte Carlo-based simulation models are being used to assess the availability of offshore wind farms, which include as variables the resourcing of servicing crews, travel time from shore, the turbine technology itself and sea state.

ENERGY PREDICTION

The energy prediction step is essentially the same as for onshore predictions. There is generally only minor predicted variation in wind speed over a site. Given the absence of topography offshore, measurements from a mast can be considered representative of a much larger area than would be possible onshore.

For large offshore sites, wake losses are likely to be higher than for many onshore wind farms. The wake losses are increased due to the size of the project and also due to lower ambient turbulence levels – the wind offshore is much smoother. There

Wind Turbine Technology for Offshore Locations

AVAILABILITY, RELIABILITY AND ACCESS

High availability is crucial for the economics of any wind farm. This depends primarily on high system reliability and adequate maintenance capability, with both being achieved within economic constraints on capital and operational costs. Key issues to be addressed for good economics of an offshore wind farm are:

• minimisation of maintenance requirements; and
• maximisation of access feasibility.

The dilemma for the designer is how best to trade the cost of minimising maintenance by increasing reliability – often at added cost in redundant systems or greater design margins – against the cost of systems for facilitating and increasing maintenance capability. Previous studies within the EU research programmes, such as OptiOWECS, have considered a range of strategies from zero maintenance (abandonment of faulty offshore turbines) to highly facilitated maintenance.

Access is critical as, in spite of the direct cost of component or system replacement in the difficult offshore conditions, lost production is often the greatest cost penalty of a wind turbine fault. For that reason much attention is given to access. Related to the means of access is the feasibility of various types of maintenance activities and whether or not support systems (cranes and so on) and other provisions are needed in the wind turbine nacelle systems.

Impact on Nacelle Design

The impacts of maintenance strategy on nacelle design relate to:

- Provision for access to the nacelle;
- Systems in the nacelle for handling components; and
- The strategic choice between whether the nacelle systems should be (a) designed for long life and reliability in an integrated design that is not particularly sympathetic to local maintenance and partial removal of subsystems or (b) designed in a less cost-effective modular way for easy access to components.

Location of Equipment

Transformers may be located in the nacelle or inside the tower base. Transformer failures have occurred in offshore turbines, but it is not clear that there is any fundamental problem with location in either the nacelle or the tower base.

Importance of Tower Top Mass

The tower top mass is an important influence on foundation design. In order to achieve an acceptable natural frequency, greater tower top mass may require higher foundation stiffness, which could significantly affect the foundation cost for larger machines.

Internal Cranes

One option is to have a heavy duty internal crane. Siemens and Vestas have adopted an alternative concept, which in general consists of a lighter internal winch that can raise a heavy duty crane brought in by a maintenance vessel. The heavy duty crane may then be hoisted by the winch and set on crane rails provided in the nacelle. Thus it may be used to lower major components to a low-level platform for removal by the maintenance vessel.

Critical and difficult decisions remain about which components should be maintained offshore in the nacelle, which can be accessed, handled and removed to shore for refurbishment or replacement, and when to draw a line on component maintenance capability and accept that certain levels of fault will require replacement of a whole nacelle.

Means of Access

The costs of turbine downtime are such that an effective access system offshore can be relatively expensive and still be justified. Helicopter access to the nacelle top has been provided in some cases. The helicopter cannot land, but can lower personnel. Although having a helipad that would allow a helicopter to land is a significantly different issue, the ability to land personnel only on the nacelle top of a wind turbine has very little impact on nacelle design. Although adopted for the Horns Rev offshore wind farm, helicopter access is probably too expensive as a routine method of transporting personnel to and from offshore wind turbines, assuming current project sizes and distance from shore.

Access Frequency

At Horns Rev, which is the first major offshore wind farm in the North Sea, a vast number of worker transfers have taken place since construction, and this is a concern for the health and safety of personnel. It is expected (and essential) that the required number of transfers for the establishment and commissioning of offshore wind plant will reduce as experience is gained.

Access Impediments

In the Baltic Sea especially, extensive icing occasionally takes place in some winters. This changes the issues regarding access, which may be over the ice if it is frozen solid or may use ice-breaking ships. Also, the ice in general is in motion and may be quite unstable. Lighthouses have been uprooted from their foundations and moved by pack ice. The wind turbine foundation design used by Bonus in the Middelgrunden offshore wind farm, situated in shallow water between Denmark and Sweden, provides for a section at water level with a bulbous shape. This assists in ice breaking and easing the flow of ice around the wind turbine, thereby reducing loads that would tend to move the whole foundation.

In the European sites of the North Sea, the support structure design conditions are more likely to relate to waves than ice, and early experience of offshore wind has shown clearly that access to a wind turbine base by boat is challenging in waves of around 1 m height or more.

Currently most standard boat transfers cannot – and should not – be performed in sea states where the significant wave height is greater than 1.5 m and wind conditions are in excess of 12 m/s. This sea state constraint is generally not an onerous parameter for wind farms located in the Baltic region. However, in more exposed locations, such as in UK and Irish waters, the average number of days where the wave height is greater than 1.5 m is considerably greater.

Feasibility of Access

Operating in concert with the wave height restrictions are restrictions from the water depth, swell and underwater currents. As an example, UK wind farms are generally sited on shallow sandbanks, which offer advantages in easier installation methods, scaled reductions in foundation mass requirements and the tendency of shallow sandbanks to be located in areas away from shipping channels. However, shallow waters, particularly at sandbank locations where the seabed topography can be severe, amplify the local wave height and can significantly change the wave form characteristics. Generally speaking, the turbine in a wind farm that is located in the shallowest water will present the most access problems.

Wave data that is representative of UK offshore wind farm sites shows that access using a standard boat and ladder principle (significant wave heights up to 1.5 m) is generally possible for approximately 80 per cent of the available time. However, this accessibility rate is too low for good overall wind farm availability. In winter, accessibility is typically worst when there is the greatest likelihood of turbine failures; yet at these times there are higher winds and hence potentially higher levels of production loss. Accessibility can be improved to above 90 per cent if access is made possible in significant wave heights between 2.0 and 2.5 m. Providing access in yet more extreme conditions is probably too challenging, considering cost, technical difficulty and safety. A safety limit on sea conditions has to be set and rigidly adhered to by the wind farm operator. This implies that 100 per cent accessibility to offshore wind plant will not be achievable and 90 per cent accessibility seems a reasonable target. Improvements in availability thereafter must be achieved through improved system reliability.

Safe personnel access is currently one of the most important topics under discussion in offshore wind energy. For example, the British Wind Energy Association, in consultation with the UK Health and Safety Executive, has produced guidelines for the wind energy industry (BWEA, 2005). These guidelines were issued as general directions for organisations operating or considering operating wind farms.

Access Technology Development

There may be much benefit to be gained from the general knowledge of offshore industries that are already developed, especially the oil and gas industry. However, there are major differences between an offshore wind farm and, for example, a large oil rig. The principal issues are:

- There are multiple smaller installations in a wind farm and no permanent (shift-based) manning, nor the infrastructure that would necessarily justify helicopter use.
- Cost of energy rules wind technology, whereas maintenance of production is much more important than access costs for oil and gas.

Thus, although the basis of solutions exists in established technology, it is not the case that the existing offshore industry already possesses off-the-shelf solutions for wind farm construction and maintenance. This is evident in the attention being given to improved systems for access, including the development of special craft.

Conclusions Regarding Access Issues

There appears to be a clear consensus on offshore wind turbine access emerging for UK sites. Purpose-built aluminium catamaran workboats are currently in use for the several wind farms. Catamarans generally provide safe access in sea conditions with a maximum significant wave height up to 1.5 m. On occasion this figure has been exceeded by skippers experienced in offshore wind transfers on a particular site.

In UK waters it is generally accepted that the standard boat and ladder access principle is practicable for approximately 50–80 per cent of the available service time, depending on the site. However, when this accessibility figure is considered in concert with the overall wind farm availability equation, there is scope for improvement. The main reason for improvement is that winter accessibility rates are typically much worse than for the summer period. This is compounded with a higher likelihood of turbine failures in winter and also higher winds, hence higher levels of production loss.

With some effort this accessibility figure can be improved markedly – that is to say, where access can be made possible in significant wave heights of between 1.5 and 2.0 m. Providing access above 2.0 m becomes an economically and technically challenging decision. It is likely that significant expenditure and technical resources would be necessary to gain modest incremental improvements in access rates above the 2.5 m significant wave threshold. Furthermore, it may be proved, using analytical techniques, that around 80 per cent of the development and capital costs would only net, at most, gains of 20 per cent in accessibility.

In short, it is doubtful from a pragmatic and safety perspective whether personnel should be expected to attempt any transport and transfer in wave conditions in excess of 2–2.5 m. A safety limit on sea conditions has to be set and rigidly adhered to by a wind farm operator, with acceptance that a 100 per cent accessibility rate will never be achievable in an offshore wind context. Project economics based on the premise of accessibility rates around 90 per cent of the available time is a reasonable target. Improvements in availability thereafter must be achieved through improved reliability of machinery.

In summary, there is an appreciable benefit in increasing accessibility rates upwards from the current 1.2 to 1.5 m threshold to 2.0 m. This has proven to be achievable using catamarans. There are a number of alternative vessels at the concept stage that are being designed to allow safe transfer in 2 m significant waves. It may be possible to achieve further improvements in accessibility using specialised workboats fitted with flexible gangways that can absorb the wave energy that cannot be handled by the vessel. Alternatively, larger vessels with a greater draught, which are more inherently stable in rougher sea conditions, must be employed. However, the drawback for offshore wind energy is that foundation technology and the economics for many projects dictate that turbines should be installed in shallow banks, hence shallow draught service vessels are obligatory.

LIGHTNING RISK OFFSHORE

Lightning has been more problematic offshore than expected. The answer, however, lies with providing wind turbine blades with better methods of lightning protection, as used on the more problematic land-based sites, rather than looking for systems to ease handling and replacement of damaged blades. The consequences of lightning strikes can be severe if systems are not adequately protected.

MAINTENANCE STRATEGY – RELIABILITY VERSUS MAINTENANCE PROVISION

Regarding cost of energy from offshore wind, a general view is emerging that it is a better to invest in reliability to avoid maintenance than in equipment to facilitate it. Also, expenditure on maintenance ships and mobile gear is generally more effective than expenditure per machine on added local capability such as nacelle cranes. In selected cases, a strategy to facilitate in-situ replacement of life-limited components (such as seals) may be advised.

Wind Farm Design Offshore

Designing an offshore wind farm is a staged process involving:
- data-gathering;
- preliminary design and feasibility study;
- site investigation;
- concept development and selection;
- value engineering;
- specification; and
- detailed design.

Close interaction throughout the design process is required, including interaction about the grid connection arrangements and constraints introduced through the consenting process. Key aspects of the design, and the process of arriving at that design, are described in the following sections.

The capital cost of offshore projects differs markedly from those onshore, with perhaps 50 per cent of the capital cost being due to non-turbine elements, compared to less than 25 per cent in onshore projects. Figure I.5.2 shows a typical breakdown of cost for an offshore wind farm.

SITE SELECTION

The selection of the site is the most important decision in the development of an offshore wind farm. It is best accomplished through a short-listing process that draws together all known information on the site options, with selection decisions driven by feasibility, economics and programme, taking account of information on consenting issues, grid connection and other technical issues discussed below.

WIND TURBINE SELECTION

Early selection of the wind turbine model for the project is typically necessary so that the design process

Figure I.5.2: Capital cost breakdown for a typical offshore wind farm

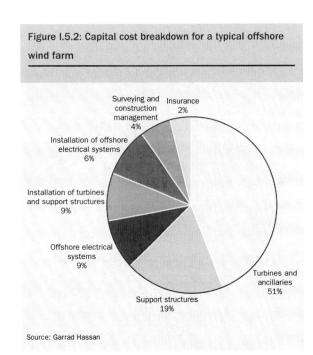

Source: Garrad Hassan

for support structures (including site investigations), electrical system and grid connection can progress. Offshore projects require use of the larger wind turbines on the market, meaning that there is often limited

choice; hence, securing the wind turbine model may be necessary to start up the project programme.

LAYOUT

The process for designing the layout of an offshore wind farm is similar to the process for an onshore wind farm, albeit with different drivers. Once the site is secured by a developer, the constraints and known data on the site are evaluated and input in the layout design, as shown in Figure I.5.3.

One driver that often dominates onshore wind farm design is the noise footprint of the project, which is not usually an issue for offshore projects.

The layout design process evaluates and compares layout options in relation to technical feasibility, overall capital cost and the predicted energy production.

Determining the optimum layout for an offshore wind farm involves many trade-offs. One example of such a trade-off is related to the array spacing, where a balance must be struck between array losses, that is to say, energy production, and electrical system costs and efficiency. Several other such trade-offs exist.

Figure I.5.3: Layout design process

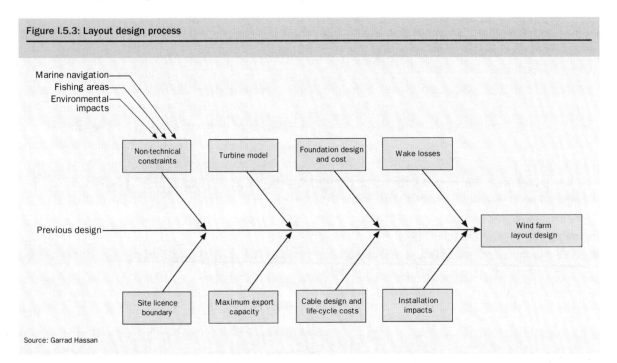

Source: Garrad Hassan

The experience to date is that for sites of homogeneous depth and soil properties, revenue, and hence production, has a dominant impact on the cost of energy, and as a result the design of the layout is dictated by energy production. However, where the water depth and soil properties vary widely across a site, more complex trade-offs must be made between production, electrical system costs and support structure costs, including installation costs.

OFFSHORE SUPPORT STRUCTURES

Support structures for offshore wind turbines are highly dynamic, having to cope with combined wind and hydrodynamic loading and complex dynamic behaviour from the wind turbine. It is vital to capture the integrated effect of the wind and wave loads and the wind turbine control system, as this is a situation where the total loading is likely to be significantly less than the sum of the constituent loads. This is because the loads are not coincident and because the aerodynamic damping provided by the rotor significantly damps the motions due to wave loading.

Structures to support wind turbines come in various shapes and sizes; the most common are illustrated in Table I.5.2 and Figures I.5.4 to I.5.7. Monopiles have been chosen for most of the installed offshore wind farms to date. Concrete gravity base structures have also been used on several projects. As wind turbines get larger, and are located in deeper water, tripod or jacket structures may become more attractive.

The design process for offshore wind turbine support structures is illustrated in Figure I.5.8.

Table I.5.2: Support structure options

Structure	Examples	Use	Notes
Monopile	Utgrunden, SE; Blyth, UK; Horns Rev, DK; North Hoyle, UK; Scroby Sands, UK; Arklow, Ireland; Barrow, UK; Kentish Flats, UK	Shallow to medium water depths	• Made from steel tube, typically 4–6 m in diameter • Installed using driving and/or drilling method • Transition piece grouted onto top of pile
Jacket	Beatrice, UK	Medium to deep water depths	• Made from steel tubes welded together, typically 0.5–1.5 m in diameter • Anchored by driven or drilled piles, typically 0.8–2.5 m in diameter
Tripod	Alpha Ventus, DE	Medium to deep water depths	• Made from steel tubes welded together, typically 1.0–5.0 m in diameter • Transition piece incorporated onto centre column • Anchored by driven or drilled piles, typically 0.8–2.5 m in diameter
Gravity base	Vindeby, DK; Tuno Knob, DK; Middlegrunden, DK; Nysted, DK; Lilgrund, SE; Thornton Bank, BE	Shallow to medium water depths	• Made from steel or concrete • Relies on weight of structure to resist overturning; extra weight can be added in the form of ballast in the base • Seabed may need some careful preparation • Susceptible to scour and undermining due to size
Floating structures	None	Very deep water depths	• Still under development • Relies on buoyancy of structure to resist overturning • Motion of floating structure could add further dynamic loads to structure • Not affected by seabed conditions

Figure I.5.4: Monopile

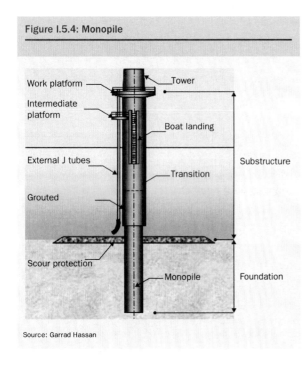

Source: Garrad Hassan

Figure I.5.6: Jacket structure

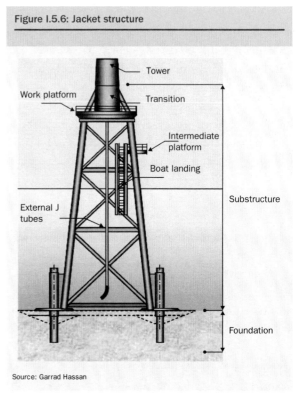

Source: Garrad Hassan

Figure I.5.5: Gravity base structure (indicative only: actual implementations can vary significantly)

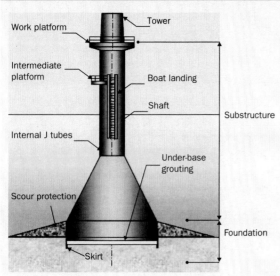

Source: Garrad Hassan

The structure designs are strongly influenced by met-ocean site conditions and site investigations. Metocean conditions are determined by detailed hydrodynamic analysis based on long-term hindcast model data and calibrated against short-term site wave measurements.

Site investigations are major tasks in their own right, requiring careful planning to achieve optimum results within programme and financial constraints. These involve a combination of geophysical and geotechnical measurements. The geotechnical investigations identify the physical properties of the soils into which foundations are to be placed and are achieved using cone penetrometer or borehole testing. Geophysical tests involve measurement of water depth and of the seismic properties of the underlying soil layers and can be used to interpolate the physical findings of geotechnical tests. The type and extent of geotechnical tests is dependent on the soil type occurring at the site and the homogeneity of those site conditions.

Figure I.5.7: Tripod structure (third leg hidden)

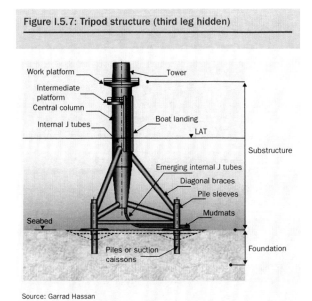

Source: Garrad Hassan

Design of the secondary structures, such as decks, boat landings and cable J-tubes, is typically developed in the detailed design phase. These details have a major impact on ease of construction, support structure maintenance requirements, accessibility of the wind turbines and safety of personnel during the operations phase. Hence a significant design period is recommended.

Where an offshore substation is required, this is likely to require a substantial support structure, although as it does not have the complexities of wind turbine loading, it is a more conventional offshore structure to design. As discussed in the following section, an offshore substation may range from a unit of below 100 MW with a small single deck structure to a large, multi-tier high voltage DC (HVDC) platform.

ELECTRICAL SYSTEM

An offshore wind farm electrical system consists of six key elements:

1. wind turbine generators;
2. offshore inter-turbine cables (electrical collection system);
3. offshore substation (if present);
4. transmission cables to shore;
5. onshore substation (and onshore cables); and
6. connection to the grid.

Figure I.5.9 illustrates these schematically and the following subsections describe them in more detail.

The design of the electrical system is determined by the characteristics of the wind turbine generators and

Figure I.5.8: Typical primary support structure design inputs

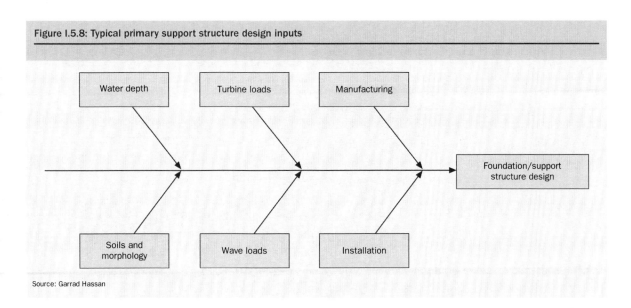

Source: Garrad Hassan

Figure I.5.9: Typical single line diagram

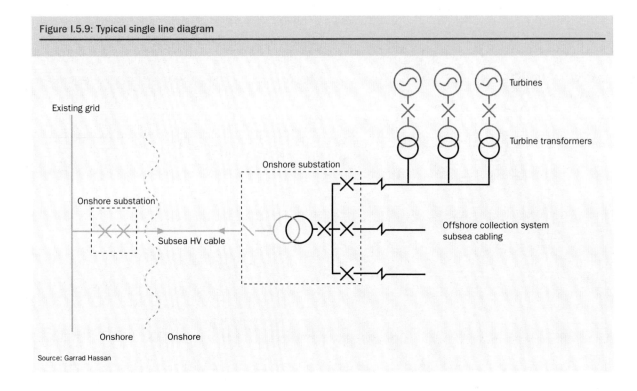

Source: Garrad Hassan

of the network to which the project is to be connected, as well as regulations imposed upon it, notably through grid codes. The network operator controls the grid to meet its operational objectives and also requires a degree of control over large generators (which may include offshore wind farms). Additionally, the wind farm must be designed to respond appropriately to grid faults. These demands can be expected for any large wind farm located offshore (see Part II).

Wind turbine control and electrical systems are constantly evolving to provide improved characteristics and fault response for the purpose of grid integration. Nevertheless, the wind farm electrical system can be expected to have additional functional requirements in addition to the basic transmission from turbines to the grid connection point.

Offshore Substations

Offshore substations are used to reduce electrical losses by increasing the voltage and then exporting the power to shore. Generally a substation does not need to be installed if:

- the project is small (~100 MW or less);
- it is close to shore (~15 km or less); or
- the connection to the grid is at collection voltage (for example 33 kV).

Most early offshore wind projects met some or all of these criteria, so were built without an offshore substation. However, most future offshore wind farms will be large and/or located far from shore, and so will require one or more offshore substations.

Offshore substations typically serve to step up the voltage from the site distribution voltage (30–36 kV) to a higher voltage (say 100–220 kV), which will usually be the connection voltage. This step-up dramatically reduces the number of export circuits (subsea cables) between the offshore substation and the shore. Typically, each export circuit may be rated in the range 150–200 MW.

Such substations may be configured with one or more export circuits. Future units will be larger and more complex. To date, no standard substation layout has yet evolved.

For projects located far from the grid connection point, or of several hundred megawatts in capacity, AC transmission becomes costly or impossible, due to cable-generated reactive power using up much of the transmission capacity. In such cases, HVDC transmission is becoming an option. Such a system requires an AC/DC converter station both offshore and onshore; both stations are large installations.

Onshore Substations

Design of the onshore substation may be driven by the network operator, but there will be some choices to be made by the project developer. Generally, the onshore substation will consist of switchgear, metering, transformers and associated plant. The onshore substation may also have reactive compensation equipment, depending on the network operator requirements and the design of the offshore network.

Subsea Cables

Subsea cables are of well-established design. Each circuit runs in a single cable containing all three phases and optical fibre for communications, with a series of fillers and protective layers and longitudinal water blocking to prevent extensive flooding in the event of the external layers failing.

Inter-turbine (array) cables are typically rated at 30 to 36 kV and installed in single lengths from one turbine to its neighbour, forming a string (collection circuit) feeding the substation. Each collection circuit is usually rated up to 30 MW. Export cables are of similar design but for higher voltage, typically 100 to 220 kV. Cables are terminated at each structure through a vertical tube from seabed to above water level (J-tube or I-tube) and into conventional switchgear.

Long-term reliability of the subsea cables is a major concern, addressed mainly by ensuring the safe burial of the cables at a depth that avoids damage from trawlers and anchors and the exposure of cables to hydrodynamic loading.

INSTALLATION

The installation of the wind turbines and their support structures is a major factor in the design of offshore wind farms, with the specific challenge of having to perform multiple repeated operations in difficult offshore locations. As well as being a significant contributor to the capital cost, the installation process may drive the selection of support structure technology.

The substation installation is also a major operation, albeit a single one, unlike the wind turbine and wind turbine foundation installation, and therefore much more conventional in the offshore industry. A large crane vessel is likely to be required – either a shear-leg crane or other heavy lift unit. Alternatively, self-installing substation designs may find an increasing role in the future.

Cable installation is a significant industry sector, with specialist design and planning, and installation vessels and equipment. The design and planning of the cable installation is an early activity, covering:

- identification of hazards to cables;
- site investigation to identify seabed properties (geophysical survey, vibrocore sampling, cone penetrometer tests, boreholes);
- development of burial protection indices;
- scour protection;
- cable route selection;
- cable transport; and
- vessel and equipment selection.

Future Trends for Offshore Wind

GENERAL

The offshore wind sector remains relatively immature, and despite the first demonstration project
being built in 1991, the total installed capacity
only breached the 1000 MW barrier in 2007/2008.
Added to this, experience has shown that the sector presents unique technical challenges that must
be addressed through research and development
efforts:

- any project involves multiple distributed installations, spread over much larger areas and in much
  larger numbers than other offshore industries;
- nearshore shallow water (for most projects) siting –
  unlike oil and gas rigs, sea defence works, and
  ports and harbours; and
- more stringent economics than oil and gas.

These factors combine so that there is limited bor-
rowing available from other sectors, and technology
has had to evolve within a short timescale on a small
number of projects, leaving significant scope for further
maturing. This issue covers all parts of the industry,
including:

- wind turbines;
- wind turbine support structures;
- modelling tools;
- electrical infrastructure;
- assembly and installation; and
- operations and maintenance.

Two drivers cut across all these areas: safety of personnel and the public, and environmental protection.

EWEA has led the EU Wind Energy Technology
Platform (see Chapter I.7) and has convened a
working group to identify necessary future technical
initiatives for offshore wind. The outcomes are
discussed in Chapter I.7.

WIND TURBINES

Wind turbine technology in general is discussed in
Chapter I.3, with some future innovative wind energy
conversion systems that may be exploited on land or
offshore reviewed under 'Future innovations'.

It has long been acknowledged that some of the
design drivers for a wind turbine installed offshore are
fundamentally different from those installed onshore,
specifically:

- the non-turbine elements of an offshore project represent a much higher proportion of the capital cost,
  with that cost element only partially scaling with
  turbine size;
- acceptable noise levels are much higher offshore;
  and
- better reliability is required offshore.

These drivers have already influenced the design of
wind turbines used offshore, and this is leading to the
development of wind turbines specifically designed for
offshore use with features such as:

- larger rotors and rated power;
- higher rotor tip speeds;
- sophisticated control strategies; and
- electrical equipment designed to improve grid connection capability.

Wind resource assessment offshore provides more
background, but other technological innovations that
may be deployed in future offshore turbines include:

- two-bladed rotors;
- downwind rotors;
- more closely integrated drive trains;
- multi-pole permanent magnet generators;
- high temperature superconductors (in generators);
  and
- high voltage output converters (eliminating the
  need for turbine transformers).

WIND TURBINE SUPPORT STRUCTURES

As shown in Figure I.5.2, support structures form a significant proportion of offshore wind development costs. It is expected that there will be both innovation and value engineering of structure designs and improved manufacturing processes to improve the economics and meet the demands for more challenging future sites and wind turbines.

Such developments are likely to include modifications to conventional designs, scale-up of manufacturing capacity and processes, and more novel design concepts. Such innovative designs may include:

- suction caisson monotowers;
- use of suction caissons as the foundation of jacket or tripod structures;
- application of screw piles;
- floating structures; and
- braced supports to monopiles.

There are two key aspects to the maturing of offshore support structures:

- acquisition of data on the behaviour of the existing structures in order to support research into the development of improved design tools and techniques and better design standards; this will be used to extend the life of structures, to reduce costs and to develop risk-based life-cycle approaches for future designs; and
- the build-up of scale and speed in production in order to achieve cost reduction and the capacity necessary to supply a growing market

The development of floating structures, while long-term, will be a major advance if successful. This is discussed in 'Offshore support structures' above.

MODELLING TOOLS

The offshore wind sector will deploy turbines of greater size and in greater numbers than has been done previously. Understanding of the engineering impacts of this is achieved through modelling, and this increase in scale requires development and validation of the industry's modelling tools. Associated with this is the refinement of design standards. The priority areas that must be addressed for large offshore wind farms are:

- development of wind turbine wakes within the wind farm;
- meso-scale modifications to the ambient flow in the immediate environs of the wind farm;
- downstream persistence of the modification to ambient flow, and therefore the impact of neighbouring wind farms upon each other; and
- dynamic loads on wind turbines deep within wind farms.

ELECTRICAL INFRASTRUCTURE

Incremental development in electrical equipment (switchgear, transformers and reactive power compensation equipment) is to be expected, driven by the wider electricity supply industry. The offshore wind business will soon be the largest market for subsea cables and so some innovations there may be driven by the specific requirements of the sector, although cables are a relatively mature technology. Voltage-source high voltage DC transmission is a relatively new commercial technology, and one that will find extensive application in offshore wind.

The major electrical impact of the offshore sector will be the reshaping of the transmission network of the countries involved in order to serve these major new generating plants. Also to be expected is an increase in the interconnection of countries to improve the firmness of national power systems, which may also involve providing an international offshore transmission network dedicated to serving offshore wind projects.

ASSEMBLY AND INSTALLATION

Future technical developments in the construction process are likely to be:

- improvements in harbour facilities that are strategically located for the main development regions;
- construction of further purpose-built installation equipment for the installation of wind turbines, support structures and subsea cables: vessels and also piling hammers, drilling spreads and cable ploughs; and
- development of safe, efficient, reliable and repeatable processes to reduce costs, minimise risks, guarantee standards and deliver investor confidence.

OPERATIONS AND MAINTENANCE

Successful performance of O&M is most critically dependent on service teams being able to access the wind farm as and when needed. Good progress has been made on this in recent years and accessibility has improved significantly. This has been achieved by incremental improvements in:

- vessels used;
- landing stages on the wind turbine structures; and
- procedures.

Future offshore wind farms offer new access challenges, being larger and much further offshore. This will result in increased use of helicopters for transferring service crews, larger vessels to give fast comfortable transit from port to site and the use of offshore accommodation platforms, combined with evolution of strategies to perform O&M.

FLOATING SYSTEMS

The US Department of Energy (DoE) has hosted conferences on 'deepwater' solutions in recent years. In both the EU and the US for over ten years, there has been exploratory research of floating offshore systems and preliminary development of design tools for modelling a wind turbine system on a dynamically active support that is affected by wave climate. Until recently, such technology, even at the level of a first demonstration, was considered rather far in the future. However, interest has accelerated and demonstration projects have been announced.

The main drivers for floating technology are:

- access to useful resource areas that are in deep water yet often near the shore;
- potential for standard equipment that is relatively independent of water depth and seabed conditions;
- easier installation and decommissioning; and
- the possibility of system retrieval as a maintenance option.

The main obstacle to the realisation of such technology is:

- development of effective design concepts and demonstration of cost-effective technology, especially in respect of the floater and its mooring system.

StatoilHydro-Siemens

StatoilHydro and Siemens Power Generation entered into an agreement to cooperate on technology to develop floating wind turbines, based on StatoilHydro's Hywind concept. StatoilHydro is aiming to build the world's first full-scale floating wind turbine and test it over a two-year period offshore near Karmøy, an island to the southwest of Norway. The company has announced an investment of approximately NOK400 million for a planned start-up in autumn 2009.

A Siemens 2.3 MW wind turbine (80 m rotor diameter) is set on a floating column of the spar-buoy type, a solution long established in oil and gas production platforms and other offshore floating systems.

The flotation element is proposed to have a draught of 100 m below the sea surface, and to be moored to the seabed using three anchor points. The system can be employed in waters depths ranging from 120 to 700 m.

The combination of two established technology solutions in the wind turbine and spar buoy may be considered a prudent approach to the development of offshore floating systems.

StatoilHydro has also acquired a substantial share in the technology company Sway, which is developing a highly innovative solution for system support. The SWAY® system (Figure I.5.10) is a floating foundation capable of supporting a 5 MW wind turbine in water depths ranging from 80 m to more than 300 m in challenging offshore locations.

In the SWAY® system, the tower is stabilised by elongation of the floating tower to approximately 100 m under the water surface and by around 2000 tonnes of ballast in the bottom. A wire bar gives sufficient strength to avoid tower fatigue. Anchoring is secured with a single tension leg between the tower and the anchor.

The tower takes up an equilibrium tilt angle (typically around 5 to 10°) due to the wind thrust on the

Figure I.5.10: The SWAY® concept

rotor. During power production in storm conditions, there is expected to be a further variation of only ±0.5 to 1.0° from the equilibrium tilt angle due to wave action.

The concept exploits active control of rotor thrust, and claims to achieve substantial cost savings over competing technology for deep water applications. The first full-scale wind turbine is expected to be built and installed in 2010–2012.

Blue H

A prototype installation using a concept similar to the tension-leg platform developed in the oil and gas industry was launched in late 2007 by Blue H Technologies of The Netherlands. The installation carries a two-bladed wind turbine, and is due for full installation and testing in 2008 in a water depth of over 100 m, approximately 17 km offshore from Puglia, Italy.

I.6 SMALL WIND TURBINES

Introduction

Small wind turbines (SWTs) are used in two main areas:

1. 'autonomous' electrical systems (also called 'stand-alone', 'grid-isolated' or 'off-grid'), in other words those that are not connected to any larger electrical system and are therefore solely responsible for the control of voltage and frequency; and
2. 'distributed generation', in other words systems with small generators connected to a larger public distribution network, where there is a network operator responsible for overall control (this is also often called 'grid-connected' or 'on-grid' generation).

Despite the attention given to multi-megawatt wind farms, the markets for autonomous electrical systems and distributed generation using small wind turbines can be attractive if prices of conventional electricity and fossil fuels are sufficiently high, or in many developing countries, where hundreds of millions of people live without access to electricity.

However, despite the maturity reached in the development of the large and medium-sized wind technology for wind farms, the state-of-the-art for small wind turbines is far from technological maturity and economic competitiveness. Average costs for current stand-alone wind turbines vary from €2500 to €6000 per installed kW, while in distributed generation, a small wind turbine can vary from €2700 to €8000 per installed kW, the additional cost mainly due to the power converter required for grid connection. Both these figures contrast with the costs of large wind turbines, which are in the region of €1500/kW.

Concerning the performance analysis for small wind turbines, the average power density is around 0.15 to 0.25 kW/m^2, because of the limited wind potential in sites where the energy is required, compared to typical sites for large wind turbines in wind farms.

The technology of SWTs is clearly different from that used in large wind turbines. These differences affect all of the subsystems: mainly the control and electrical systems, but also the design of the rotor. Most of the SWTs existing on the market are machines that have developed in an almost 'hand-crafted' way, with less mature technology compared to that achieved by large wind turbines.

SWTs have great potential, but some challenges have to be addressed to produce reliable machines. IEC standards do exist for SWTs (IEC 61400-2 for design requirements for SWT) and there are applicable standards from large wind, such as power performance or noise emissions measurements; however, something more has to be done in order to develop more appropriate standards and simpler ways to display the results obtained to end users.

In spite of these barriers, the market in developed countries is promising for grid-connected and off-grid applications, due to promotion policies (such as capital cost buy-down, feed-in tariffs and net metering), and even more so for developing countries, because of the continuing decrease in specific costs and the increasing need for energy.

Table I.6.1 gives a useful categorisation of commercial SWT ranges by rated power, from a few watts to 100 kW.

The values that define the ranges for this classification have been chosen from the norms and legislation affecting SWTs. The value of 40 m^2 was the limit established in the first edition of the IEC 61400-2 Standard and is the range intended at the present time for integration of SWTs into the built environment; the 200 m^2 limit was established in the second edition of the above-mentioned IEC 61400-2 Standard in 2006, and includes most SWT applications. Finally, the limit of 100 kW is defined in many countries as the maximum power that can be connected directly to the low voltage grid. The pico-wind range is commonly accepted as those SWTs smaller than 1 kW.

Table I.6.1: Classification of SWTs

Rated power (kW)	Rotor swept area (m²)	Sub-category
$P_{rated} < 1$ kW	$A < 4.9\,m^2$	Pico-wind
$1\,kW < P_{rated} < 7$ kW	$A < 40\,m^2$	Micro-wind
$7\,kW < P_{rated} < 50$ kW	$A < 200\,m^2$	Mini-wind
$50\,kW < P_{rated} < 100$ kW	$A < 300\,m^2$	(No clear definition adopted yet)

Source: CIEMAT

Markets and Applications for SWTs

The different applications for which SWTs are especially suitable have been summarised in Table I.6.2 for the main two markets identified: off-grid applications and grid-connected applications.

Table I.6.2 tries to reflect the possible combinations, according to market, application and size of SWT, in the ranges defined in the introduction to this chapter (Table I.6.1). The table shows how isolated systems with SWTs offer solutions for almost any application whenever there is enough wind resource at the site. Depending on the size of the system, the three technological solutions for isolated systems are:

1. wind home systems;
2. hybrid systems; and
3. wind-diesel systems.

The status of these options will be commented on under 'Isolated applications'.

In the case of on-grid applications, the possibilities are also numerous, depending on the available space for the installation and on legal and economic constraints. Some options have also been identified, including integration into the built environment, and single and multiple wind turbine installations. The status of these other options is discussed under 'Grid-connected applications'.

Table I.6.2: Applications of small wind turbines

Rated power/system	Sailboats	Signalling	Street lamp	Remote houses/ dwellings	Farms	Water pumping	Seawater desalination	Village power	Mini-grid	Street lamp	Buildings rooftop	Dwellings	Public centres	Car parking	Industrial	Industrial	Farms
	Wind home system →									Build integrated →							
				Wind hybrid →						Single wind turbine →							Wind mini-farm
					Wind-diesel →												
$P < 1$ kW	X	X	X	X	X	X	X			X	X	X	X				
$1\,kW < P < 7\,kW$	X	X	X	X	X	X	X	X		X	X	X	X	X	X	X	
$7\,kW < P < 50\,kW$				X	X	X	X	X					X	X	X	X	X
$50\,kW < P < 100\,kW$								X	X						X	X	X
Small wind systems applications	Sailboats	Signalling	Street lamp	Remote houses/ dwellings	Farms	Water pumping	Seawater desalination	Village power	Mini-grid	Street lamp	Buildings rooftop	Dwellings	Public centres	Car parking	Industrial	Industrial	Farms
	Off-grid applications									On-grid applications							

Source: CIEMAT

Evolution of Commercial SWT Technology

SWTs have traditionally been used for remote small off-grid applications, this being the bulk of the market both in the developed and in the developing world. Only in the last few years has this trend changed, due to the growth of grid connections from SWTs. The potential market for grid-connected SWTs is accelerating the development of SWT technology as the anticipated large-scale production justifies the higher financial investments required for development of the technology.

ISOLATED APPLICATIONS

As mentioned above, most of the existing systems that include SWTs are isolated applications, as this has been the most traditional use for them. Among the possible isolated applications, the most common are rural electrification, professional applications (telecommunications and so forth) and pumping.

From the technological point of view, three groups of isolated applications using SWT can be distinguished; these are described in the following sections.

Very Small Systems

Very small systems usually have a generating capacity smaller than 1 kW. The best-known applications for these configurations are mobile applications, such as boats and caravans, and wind home systems (WHSs), the wind version of solar home systems (SHSs) used for rural electrification. This configuration is based on DC connection, where the battery (usually a lead-acid battery) is the main storage and control component. Usually the system supplies DC loads, as the consumed energy is very low. In terms of number of systems, this is the most frequently used configuration. Manufacturers include Marlec, Ampair and Southwest.

Hybrid Systems

The term 'hybrid' has different meanings in the context of off-grid systems with renewable energy. In this case, 'hybrid systems' refers to systems including wind generation and other generation sources (usually photovoltaic, PV). The power generation capacity for this configuration is in most cases less than 50 kW. A diesel generator (gen-set) is used in many systems in this configuration to supply backup power. Traditionally, these systems have also been based on DC connection (Proven, Bornay, Windeco), with a battery (lead-acid in most cases) also playing the role of storage and control, and an inverter to generate AC power for the loads (common practice is to use only AC loads in this configuration). However, in recent years some solutions have been proposed using AC connection. This solution has been possible through the development of bidirectional converters (SMA, Conergy, Xantrex) that allow the flow from the DC bus to the AC bus and vice versa using only one stage of power electronics. Developers using this second solution have proposed the use of SWTs with asynchronous generators directly connected to the AC bus (SMA, Conergy), which was a concept rarely used for these systems (except for Vergnet). The trend of technology for these systems is mainly in the development of modular and flexible power electronics able to provide both the power quality and the supervisory control of the system.

Wind-Diesel Systems

Even though some hybrid systems include only wind and diesel generation, the configuration described as 'wind-diesel' (W/D) refers to those systems where the gen-set plays a key role, not only as a backup source but also as an essential component for the correct control and functioning of the system. This configuration is typical for larger isolated applications (>50 kW), and some systems in the MW range have been reported. The storage system this configuration uses (if any) is a short-term storage one, commonly batteries or flywheels, which is used for power quality and control purposes only, not for long-term energy balance.

Three different types of wind-diesel systems can be distinguished, according to the proportion of wind use in the system:

1. low penetration W/D systems, which do not require additional modifications to the diesel-only grid (usually an existing system) as the diesel engine runs continuously and its controls can cope with the control of the system in the W/D mode of operation without significant modification;
2. medium penetration W/D systems, which require the inclusion of some control capabilities (usually the regulation of deferrable loads or the regulation of the wind generation) for the moments when wind generation is higher than load consumption; and
3. high penetration W/D systems, which require the addition of complex control strategies and devices in order to guarantee the stability of the system in the wind-only mode (in other words when the diesel gen-set has been shut off).

Low penetration systems can be found at the commercial level, whereas solutions for high penetration W/D systems are at the demonstration level. Technology trends for this configuration include the development of robust and proven control strategies. Prospects for this configuration (mainly for high penetration W/D systems) are very promising, as the cost of fuel has recently increased dramatically.

GRID-CONNECTED APPLICATIONS

Another market with great potential for SWTs is in grid-connected applications for residential, industrial or even, lately, urban environments. The so-called distributed wind applications are poised for rapid market growth in response to continuing energy price increases and increased demand for on-site power generation. However, in order for distributed wind to reach its mainstream market potential, the industry must overcome several hurdles, primarily in system costs, quality of design, grid interconnection and installation restrictions.

Presently, the major share of development of this market is in the US, Canada and Australia, in parallel with new trends in the development of distributed generation systems. This emerging market provides a new impulse to the development of SWT technology.

Wind power can also be used to generate electricity in an urban environment. This trend has mainly been seen in Europe, where the integration of SWTs in the built environment is being actively discussed. New wind turbines are under development for this application, which is looking mainly for quiet and efficient devices under turbulent and skewed wind flow. As well as the installation of wind turbines around and on buildings, there is also interest in 'building-integrated' wind turbines, where the turbine is part of the building structure or façade.

Market Development

ISOLATED SYSTEMS

The following description of the development of the market for isolated systems follows the above divisions under 'Isolated applications'. The market for very small systems represents the most active sector for wind off-grid systems, especially due to the boat and caravan market (thousands of units per year). The use of WHSs for rural electrification is far from the generalised use of SHSs, but some current developments can be noted. For example, the use of WHSs is a traditional practice in Inner Mongolia, where around 250,000 SWTs (adding up to 64 MW) have already been installed, with a manufacturing capability of 40,000 units per year. Apart from this huge local market, another interesting project is PERMER (Renewable Energies for Rural Electrification) in Argentina, where, after a pilot phase of 115 WHSs, an implementing phase of 1500 WHSs (in two configurations, 300 or 600 W) has been approved.

The market for hybrid systems is widely spread in single system configurations throughout the world. Experiences of planned global rural electrification programmes that include hybrid systems are:

- China, where wind/PV hybrid stations were included as an option, with the participation of some of the existing hybrid systems' developers and manufacturers (such as SMA and Bergey); and
- the Rural Electrification Program in Chile, which included hybrid systems mainly for small islands; this programme is still running, and some systems are still being developed.

Finally, the market for W/D systems is closely related to the cost of producing power with diesel engines. As mentioned above, the recent fuel price increases open a new era for this solution. There is experience with W/D systems all over the world, with Alaska, Canada and Australia as the main near-term markets, ranging from low penetration to high penetration systems. Until now, high penetration W/D systems have been installed mainly in cold climates, where the surplus energy can be used for heating. Chile has also included the W/D systems solution in its electrification programme.

GRID-CONNECTED SYSTEMS

Different markets exist in the world for small wind turbines, depending mainly on the current state of development of the country and household characteristics, as affected by the residential or built environment.

Europe has an extensive electric grid, so there is little need for off-grid wind energy systems. However, there is some potential for small grid-connected systems, which many Europeans would find attractive. The high concentration of population in urban areas provides a great opportunity for on-site distributed generation from wind power by installing small wind turbines on rooftops, even though the roughness of the urban environment can mean a reduced and more turbulent wind flow. Because of this, distributed generation based on small wind energy in residential and industrial areas is under development, and urban wind integration is an emerging application that seems to help reduce electricity power demand. Some countries have policies for the promotion of these applications.

The UK is the European leader on micro-generation, which includes small wind energy as one of the important contributors to national targets for renewable energy. The British Wind Energy Association (BWEA) claims that it would be possible to install enough micro and small wind turbines by 2020 to generate up to 1200 MW.

Currently in the UK there are nine companies manufacturing 17 commercial small wind turbine models and more than 11 prototypes. Most of these are horizontal axis wind turbines (HAWT), though some are vertical axis wind turbines (VAWT), for installation on or around buildings. The market is very dynamic, with over 3500 micro and small wind turbines installed in 2007 alone, and high expected domestic and export market growth (more than 120 per cent forecast for 2007/2008). The on-grid/off-grid market share ratio was 50:50 in 2007; however, the on-grid market is expected to increase strongly over the next two years. The building-mounted/free-standing market share ratio was 25:75 in 2007.

The BWEA has adopted a standard for performance, safety, reliability and sound emissions by which small wind turbines will be tested in order to be eligible for incentive programmes. Little information on actual as-installed performance is currently available.

Also in the UK, Proven Energy has launched a project called WINDCROFTING™, available to any landowner with a grid connection. In return for a 25-year lease, the landowner will receive rent on turbines installed on their land, and may also buy a subsidised turbine at the same time to provide electricity. A subsidiary company installs, operates and maintains the SWT.

The Netherlands is conducting studies on SWTs for grid connection:

- in the field of local attitudes, testing turbines for one year, at the end of which the owner will be able to buy the turbine for 50 per cent of the retail price; and
- in the field of actual performance, measuring the performance of 12 SWTs with a maximum power of 5 kW.

Portugal passed a new law in 2007, with feed-in tariffs for micro-generation systems below 3.68 kW, which includes small wind. This new law encourages renewable energy self-consumption, limiting the maximum amount of energy for which a premium price is paid to 4 MWh/year. The first target of the plan is 10 MW, and the duration of the support is 15 years.

Spain is promoting the SWT sector through different initiatives. Currently, the SWT market is covered by three domestic manufacturers, but new manufacturers are working on larger SWT developments and wind turbines for urban environment integration. In addition, a working group has been created inside the Association of Renewable Energy Producers (APPA) devoted to SWT generation issues. The objectives of this working group are twofold:

1. to inform the public about this technology; and
2. to act as the voice of the SWT industry to public and private entities, trying to achieve the necessary favourable conditions for the development of the technology, both from the financial and the legal points of view.

The US is the main market for SWTs, with more than 100,000 in operation in the 90 W to 25 kW size range, totalling more than 35 MW. During 2007, more than 900 SWTs were sold, 98 per cent by US manufacturers. The US SWT market is growing at about 15 to 20 per cent per year. Rural households are the most frequent applications for small wind systems, with net metering as the principal fiscal incentive as Production Tax Credits (PTCs, the main US support mechanism for grid-connected wind farms) are not available for this technology. By 2009, most US states will require turbines to be Small Wind Certification Council certified in order to be eligible for their incentive programmes.

The Canadian SWT market is also significant, with between 1.8 and 4.5 MW installed, and approximately CA$4.2 million in annual sales (2005). There are net metering policies in several provinces, and a feed-in tariff of CA$0.11/kWh available in Ontario. At least 17 manufacturers are based in Canada; these are working on a new product certification programme adapted to Canadian interests from the existing US and UK certification programmes.

Technology Trends and Recent Developments

Most of the SWTs that are currently deployed around the world have three blades, but there are also models with two, four or more at the micro-scale. Rotor diameter is below 20 m and most of the commercial small wind turbines have a rotor diameter below 10 m. These turbines are mounted typically on 12 to 24 m towers.

For the rotor, technology trends are towards advanced blade manufacturing methods based mainly on alternative manufacturing techniques such as injection moulding, compression moulding and reaction injection moulding. The advantages are shorter fabrication time, lower parts cost, and increased repeatability and uniformity, but tooling costs are higher.

Most of the current SWTs use a synchronous permanent magnet generator based on rare earth permanent magnets as the electromechanical converter, for the following reasons:

- Rare earth permanent magnets are now taking over from ferrite magnets: they have superior magnetic properties, and there has been a steady decline in prices.

- They result in more compact and lighter-weight generators.

An important characteristic to achieve in permanent magnet generators is a reduced generator 'cogging' torque, which enhances low wind speed start-up.

Some manufacturers still continue to use induction generators. However, in the recent past, no turbines of less than 50 kW rated power have used induction generators directly connected to the grid. Currently, though, designs utilising induction generators are re-emerging to avoid power electronics in order to achieve reduced cost and improved reliability.

A costly component for grid-connected SWTs is the inverter, or DC/AC converter. Most of the inverters used come from the PV market and are being adapted for use with wind turbines, installed downstream of voltage control devices.

Lately, wind turbine-specific inverters have also started to appear in both single- and three-phase configurations. These can be certified against International Power Quality and EMC (electromagnetic compatibility) standards.

As a general tendency, SWTs are currently being designed for low wind speeds, which means larger rotors, taller towers and precise regulation devices for gust events. Usually the turbine is protected against high winds by yawing or 'furling', in other words the rotor is turned out of the wind passively, by aerodynamic forces. Some alternatives to furling, such as stall control, dynamic brakes, mechanical brakes and pitch control (both centrifugal and active) have also been developed.

In order to reduce noise emissions, reduced operating and peak rotor speeds are being pursued. Because of this, the typical design tip speed ratio is 5:1.

New standards for SWT design (IEC 61400-02, second edition) were published in 2006 for turbines with a rotor area <200 m^2 (~16 m rotor diameter). There is slowly increasing use of these standards by the industry.

The industry is diverse and manufacturers vary widely in degree of maturity. Over 300 different models (in various stages of development) exist worldwide, of which 100 are engineered by US manufacturers.

The most recent developments in the field of small wind turbines can be summarised as follows:

- active pitch controls to maintain energy capture at very high wind speeds;
- vibration isolators to dampen sound;
- advanced blade design and manufacturing methods;
- alternative means of self-protection in extreme winds;
- adapting a single model to either on-grid or off-grid use;
- software and wireless display units;
- inverters integrated into the nacelle (rotor hub);
- electronics designed to meet stronger safety and durability standards;
- systems wired for turnkey interconnection;
- attempts to make SWTs more visually attractive; and
- integrating turbines into existing tower structures, such as utility or lighting poles.

Technology Status

A review of the technology status is given here, arranged by components. A colour code has been used to show the popularity of various technical options for all cases that have been analysed and for which there was information available. The colour code shows the estimated frequency of occurrence of each option, as a percentage (see Table I.6.3).

Table I.6.3: Colour code for estimated frequency of occurrence of technical options				
0%	0–25%	25–50%	50–75%	75–100%

This analysis has been made for each of the SWT power ranges defined in Table I.6.1.

ROTOR

The rotor technology status is summarised in Table I.6.4.

The results for the analyses of the generator and the power electronics used in SWTs are summarised in Table I.6.5. Brief comments have been included for the different sizes of SWT.

CLAIMED EFFICIENCY

Another aspect in which SWTs are different to large grid-connected wind turbines is the generation efficiency. First, the efficiency for SWTs is not well

Table I.6.4: Rotor and related issues

	P < 1 kW	1 kW < P < 7 kW	7 kW < P < 50 kW	50 kW < P < 100 kW
Number of blades				
2				
3				
>3				
Type of rotor				
Horizontal (HAWT)				
Vertical (VAWT)				
For HAWT				
Upwind				
Downwind				
Blade material				
Composite				
Wood + epoxy				
Composite + epoxy				
Mechanical overspeed protection				
No control	x	x	x	x
Passive pitch		x	x	-
Active pitch	-	-	-	x
Centrifugal stall		x	x	-
Mechanical maximum power regulation				
No control	x	x	x	x
Autofurl	x	x	x	-
Tilt	x	x	x	-
Stall	-	x	x	x
Rotational speed				
Very high				
High				
Medium				
Low				

Table I.6.5: Generator, power electronics and other comments

	P < 1 kW	1 kW < P < 7 kW	7 kW < P < 50 kW	50 kW < P < 100 kW
Generator				
PMG				
Axial flux				
Radial flux				
Synchronous generator				
Asynchronous generator				
Power electronics Charge controller for SWT with PM or synchronous generator				
Parallel regulator (+ dump load)				
Series regulator				
PMG short circuit available				
Maximum power point tracking				
Other comments	Robust designs; Most common range in terms of sales	Lack of inverters for grid connection	Lack of inverters for grid connection	Lack of manufacturers

known, as there is not much information available; second, the values are usually significantly lower than those for large wind turbines. Aerodynamics has something to do with this, but principally, designs for SWTs are not optimised. Figure I.6.1 shows a graph of manufacturers' claimed efficiencies for SWTs as a function of rated power at rated wind speed.

It should be noted that IEC standards for power curve measurement are not as rigorously applied by SWT manufacturers as they are in the large wind

Figure I.6.1: Claimed efficiency as a function of rated power

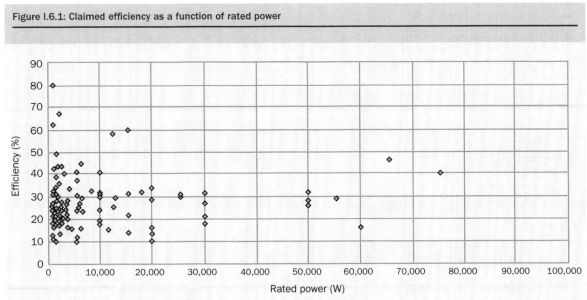

Source: CIEMAT

turbine field, and so comparison is difficult (see Appendix E for further details).

Omitting the models that appear to break the fundamental Betz limit, some comments can be made:

- In general, the efficiency for SWT generation is lower than for large wind turbines (as previously noted, this is in part due to aerodynamics but also due to the lack of optimised designs).
- In reality, the actual efficiencies tend to be even lower (values between 10 and 25 per cent are common).

COST ANALYSIS

It is common to use the cost per kilowatt for the analysis of the cost of different generating technologies. For the case of SWTs, the picture obtained is given in Figure I.6.2.

Values are greatly scattered for the low power range, but the trend is that cost per kW diminishes as the rated power increases.

However, an overall cost analysis for SWTs must be preceded by a comment on the definition of the rated power of an SWT. In wind energy generation, there are no universally accepted standard test conditions (STC is the term used for photovoltaic generation) to which all of the characterisations of the devices are referred. So it is the manufacturer who chooses the conditions (rated wind speed) for which to define the rated power of the SWT. The situation is clarified to some extent in Figure I.6.3, where the chosen rated wind speeds are shown for the different rated power values of the SWT under study.

The large variation in the rated wind speed value means that the specific parameters related to rated power cannot be compared directly, as they do not refer to the same conditions. For large wind turbines there are no defined standard conditions either, but the higher maturity of the market has led to a much lower dispersion.

Another parameter, which is objective, not subjective like rated power, is the swept rotor area. Figure I.6.4 shows the variation of power rating as a function of the diameter.

Even though for higher diameters some dispersion appears, the swept rotor area seems to be a better representation of the power of the SWT than the rated

Figure I.6.2: Cost per kW as a function of rated power

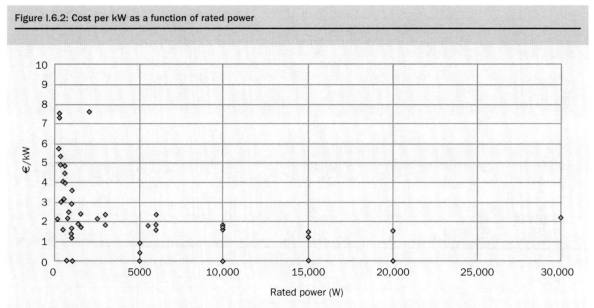

Source: CIEMAT

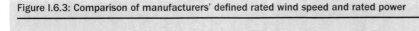

Figure I.6.3: Comparison of manufacturers' defined rated wind speed and rated power

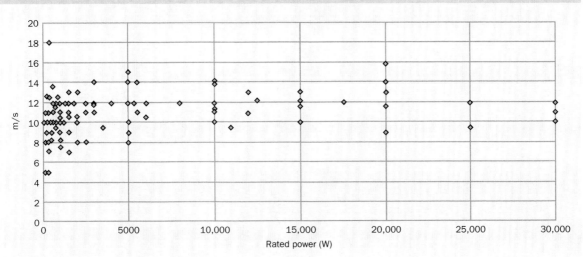

Source: CIEMAT

Figure I.6.4: Manufacturers' defined rated power as a function of rotor diameter

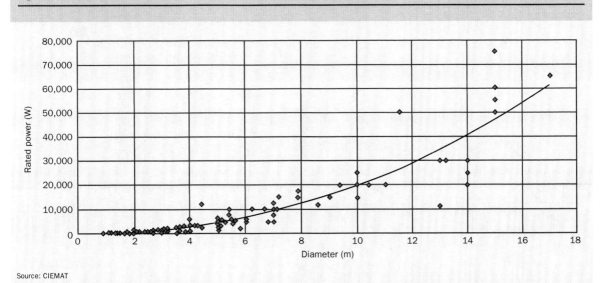

Source: CIEMAT

power itself, and it is definitively more representative of the total energy generated by the SWT. This is also the reason for using cost per square metre, as shown in Figure I.6.5, for the cost analysis.

Even though there is still a significant scattering in the low power range, which is a sign of lack of maturity of the market, the trend of lower costs per square metre is maintained as the size of the SWT increases,

Figure I.6.5: Cost per m² as a function of rated power

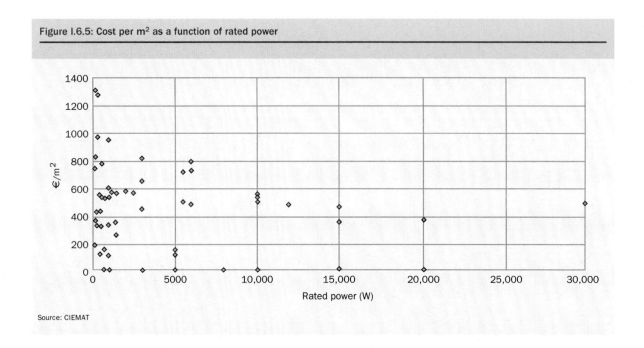

Source: CIEMAT

which is an advantage when compared with PV generation, a direct competitor for SWTs.

Future Trends

Future trends are outlined in the following sections.

BLADE AND ROTOR DESIGN

- new aerofoil design:
  - Improved aerofoil shapes, variable chord distributions and variable twist distributions, such that the overall performance of the SWT can be improved significantly.
  - Noise emission reduction. Noise is still a constraint on the further expansion of wind energy in certain countries. The current trend towards integrating SWTs closer to populated areas and on buildings enhances this problem. Several design parameters can change, such as aerofoil shape, tip speed and angle of attack, but modify-

ing design parameters always means compromising between noise reduction and optimal performance.
  - Low Reynolds number aerodynamics. The Reynolds number is an important parameter in fluid mechanics, and tends to be lower for SWTs. Low Reynolds numbers make the problem of aerofoil design difficult because the boundary layer is much less capable of handling an adverse pressure gradient without separation. Thus, very low Reynolds number designs do not have severe pressure gradients, and therefore the maximum lift capability is restricted.
  - Deforming blades.
- new materials such as thermoplastics (nylon);
- new SWT regulation methods in order to avoid the use of furling systems. Furling systems are unattractive because of the high acoustic noise emission and vibrations when the SWT furls. New cheap, reliable pitch systems may be developed;
- magnetic bearings to reduce losses in SWTs.

GENERATOR

- solutions for cogging torque reduction, such as asymmetric poles; and
- low rotational speed SWT, based on hybrid planetary gear and PMG.

MANUFACTURING SPECIALISATIONS

- blade manufacturers: new methods based on, for example, pultrusion process and filament winding;
- open-use power electronics: power converters capable of working in both stand-alone and grid-connection mode; and
- light tower manufacturing.

CONCENTRATION OF MANUFACTURERS

Standardisation, Legislation and Characterisation

- wider use of existing standards;
- further development of standards; and
- commonly accepted means to characterise performance.

OFF-GRID

- storage technologies;
- improved sizing tools;
- improved understanding of hybrid systems and W/D systems;
- wider offer of components (sizes, manufacturers);
- communication protocols; and
- new applications.

GRID-CONNECTED

- interconnecting power electronics;
- communication protocols;

- standardised interfaces ('plug and play' devices); and
- simpler, more uniform and better-understood technical requirements of network operators.

UNDERSTANDING OF WIND RESOURCE IN AREAS WHERE SWT ARE INSTALLED

- near buildings;
- on buildings; and
- within buildings (building-integrated).

Concluding Remarks and Future R&D Needs

SWTs play an important role in off-grid projects, where in windy locations they can provide a relatively economical power supply, since alternatives such as diesel generators have high fuel costs when used for continuous power supply. This can also be true for grid-connected installations, despite the fact that their production cost per kWh is often higher than that of large wind turbines.

Internationally accepted IEC standards (IEC 61400) relevant to the SWT industry already exist, but are not much used in practice. Some effort is required to develop the existing standards for SWTs, in order to increase their use. For instance, the IEC 61400-2 Standard 'Design requirements for small wind turbines', which applies to wind turbines with a rotor swept area smaller than 200 m^2 and generating at a voltage below 1000 VAC (volts alternating current) is difficult and costly to apply. The power performance standard IEC 61400-12 includes directions for small turbine power performance testing for battery charging, but does not include power performance characterisation for grid-connected SWTs. Finally, while the intent of including noise measurements in the standard rating system is laudable, the test procedure outlined is imperfect.

All components of SWTs – blades, generators, regulation systems, power converters and so on – could be improved.

New designs for integration in the urban environment should be efficient and aesthetic. It goes without saying that they must be extremely quiet and robust.

The market for SWTs is promising. There are an increasing number of SWT manufacturers all over the world, and even manufacturers of large wind turbines are beginning to assess this sector, attracted by the emerging possibilities of the new market.

I.7 RESEARCH AND DEVELOPMENT

Introduction

This chapter reviews research and development (R&D) for wind energy, concentrating on programmes and priorities within Europe.

Added Value of R&D

The wind energy market outperforms its own record every year. Market growth rates are in the same range as those of high-tech technologies (internet, mobile phones and so on). Europe is the world leader in terms of installations and manufacturing, with most of the top ten manufacturers European.

A popular misconception is to consider wind energy as a mature technology where R&D efforts are not necessarily needed. As a result, there is a risk of progressive loss of European leadership, as demonstrated by recent trends in the wind industry:

- High demand has increased wind turbine delivery time and the prices of raw materials such as steel and copper have increased in recent years, meaning that the cost of wind turbines has increased.
- Although most wind turbine manufacturers are still European, two Chinese companies (GoldWind and Sinovel) and one Indian company (Suzlon) have entered the market.

In parallel, the European target of 20 per cent of energy production from renewable sources raises new challenges. In its recently published Strategic Research Agenda[1], the European Wind Energy Technology Platform, TPWind, proposed an ambitious and feasible vision for Europe. In this vision, 300 GW of wind energy capacity would be delivered by 2030, representing up to 28 per cent of EU electricity consumption. To implement this vision, an average 10 to 15 GW of additional capacity must be manufactured, delivered and implemented every year in Europe. This is equivalent to more than 20 turbines of 3 MW being installed each working day.

Moreover, TPWind's vision includes a sub-objective of offshore wind energy representing some 10 per cent of EU electricity consumption by 2030. It proposes an intermediate step of the implementation of 40 GW by 2020, compared to 1 GW installed today.

In this context, R&D is needed on two fronts:

1. an efficient implementation of the TPWind vision for wind energy, supporting the implementation of European targets; and
2. ensuring European leadership for the long term, through technological leadership.

Priority R&D Areas in Wind Energy

TPWind has established R&D priorities in order to implement its 2030 vision for the wind energy sector. Four thematic areas have been identified:

1. wind conditions;
2. wind turbine technology;
3. wind energy integration; and
4. offshore deployment and operation.

In order to implement the 2030 vision and enable the large-scale deployment of wind energy, the support of stable and well-defined market, policy and regulatory environments is essential. The following areas are considered:

- enabling market deployment;
- cost reduction;
- adapting policies;
- optimising administrative procedures;
- integrating wind into the natural environment; and
- ensuring public support.

WIND CONDITIONS

Current techniques must be improved so that, given the geographic coordinates of any wind farm (flat terrain, complex terrain or offshore; in a region covered by extensive data sets or largely unknown), predictions with an uncertainty of less than 3 per cent can be made.

Three main research objectives – resource, design conditions and short-term forecasting – are being supported by six research topics, identified by TPWind:

1. **Siting of wind turbines in complex terrain and forested areas**, in order to accurately calculate the external wind load acting on a wind turbine, and its lifetime energy production.

2. Improving the understanding of **wakes inside and between wind farms**, and using this knowledge in the design and financial analysis of offshore wind projects. The specific objectives are to increase the availability of data sets from large wind farms, improve models to predict the observed power losses from wakes and evaluate the downwind impacts of large wind farms, especially offshore.

3. **Offshore meteorology:** improving the knowledge and understanding of processes in offshore conditions. This will be used to develop new models and to extend existing ones. This is necessary in order to develop methods for determining the external design conditions, resource assessment and short-term forecasting.

4. **Extreme wind speeds:** producing a worldwide extreme wind atlas, including guidelines for the determination of the 50-year extreme wind speed and extreme statistics.

5. Investigating and modelling the behaviour of the **wind profile above 100 m**, through models, measurements and theoretical tools describing the wind profile in the entire boundary layer.

6. **Short-term forecasting** over a timeframe of one or two weeks for wind power prediction and electricity grid management.

WIND TURBINE TECHNOLOGY

The aim is to ensure that by 2030, wind energy will be the most cost-efficient energy source on the market. This can only be achieved by developing technology that enables the European industry to deliver highly cost-efficient wind turbines, and adequate grid infrastructure and changed grid operation procedures.

Research topics are categorised according to the technical disciplines and cross-sector criteria on which the integral design and operation of wind power systems are based. The seven research areas are:

1. **The wind turbine as a flow device**. With the increasing size and complexity of wind turbines, a full understanding of aerodynamic phenomena is required, including external conditions, such as the wind speed distribution on the rotor plane, for different wind turbine configurations and sites.

2. **The wind turbine as a mechanical structure/materials used to make the turbine**. The goal is to improve the structural integrity of the wind turbine through an improved estimation of design loads, new materials, optimised designs, verification of structural strength, and reliability of components such as drive trains, blades and the tower.

3. **The wind turbine as an electricity plant**. This should develop better electrical components, improve the effect of the wind turbine on grid stability and power quality, and minimise the effect of the grid on wind turbine design.

4. **The wind turbine as a control system**. This will aim to optimise the balance between performance, loading and lifetime. This will be achieved through advanced control strategies, new control devices, sensors and condition monitoring systems.

5. **Innovative concepts and integration**. This should achieve a significant reduction in the lifetime cost of energy by researching highly innovative wind turbine concepts. With the support of an integrated design approach, this will be made possible through incremental improvements in technology, together with higher risk strategies involving fundamental conceptual changes in wind turbine design.

6. **Operation and maintenance strategies**. These become more critical with up-scaling and offshore deployment of wind power systems. The objective

is to optimise O&M strategies in order to increase availability and system reliability.

7. **Developing standards for wind turbine design**. This is to allow technological development, whilst retaining confidence in the safety and performance of the technology.

WIND ENERGY INTEGRATION

Large-scale integration of wind power at low integration costs requires research in three main areas:

1. **Wind power plant capabilities**. The view is to operate wind power plants like conventional power plants as far as possible. This approach implies fulfilling grid code requirements and providing ancillary services. It requires investigating the wind power plant capabilities, grid code requirements and possible grid code harmonisation at the EU level.

2. **Grid planning and operation.** One of the main barriers to the large-scale deployment of wind technology is limited transmission capacity and inefficient grid operation procedures. The grid infrastructure and interconnections should be extended and reinforced through planning and the early identification of bottlenecks at the European level. A more efficient and reliable utilisation of existing infrastructures is also required. Review of the existing rules and methodologies for determining transmission capacity is needed. Further investigation is required for offshore to assess the necessity of offshore grids. Dynamic models are needed to assess the influence of wind generation on power system operation, such as a more coordinated supervision scheme and a better understanding and improved predictability of the state of the power system.

3. **Energy and power management**. Wind power variability and forecast errors will impact the power system's short-term reserves. At higher wind power penetration levels, all sources of power system flexibility should be used and new flexibility and reserves sought.

Additional possibilities for flexibility must be explored, by both generation and demand-side management, together with the development of storage. In the context of variable production, variable demand and variable storage capacity, probabilistic decision methods should be promoted.

Also, a more centrally planned management strategy would mean that available grid capacities could be used more effectively and reinforcements could be planned more efficiently. The emphasis should be put on developing good market solutions for the efficient operation of a power system with large amounts of renewable generation.

OFFSHORE DEPLOYMENT AND OPERATIONS

The objectives are to achieve the following:

- coverage of more than 10 per cent of Europe's electricity demand by offshore wind;
- offshore generating costs that are competitive with other sources of electricity generation;
- commercially mature technology for sites with a water depth of up to 50 m; and
- technology for sites in deeper water, proven through full-scale demonstration.

Five research topics have been prioritised by the European wind energy sector:

1. **Substructures**. These represent a significant proportion of offshore development costs. It is necessary to extend the lifetime of structures, reduce costs and develop risk-based life-cycle approaches for future designs. Novel substructure designs, improved manufacturing processes and materials are critical. In the near term, the major deployment issue is the development of the production facilities and equipment for manufacturing the substructures.

This will require significant investment in new manufacturing yards and in the associated supply chain. Further data are needed on the behaviour of existing structures, supporting research into improved design tools and techniques, and better design standards.

2. **Assembly, installation and decommissioning**. This must solve the challenge of transferring equipment to wind farm sites. Such transfer requires efficient transport links, large drop-off areas and good harbours. The second challenge of wind turbine installation will require specially designed vessels and equipment. Safe, efficient and reliable processes must be developed that are easy to replicate. Finally, techniques should be developed for the dismantling of offshore wind farms and for quantifying the cost of doing so.

3. **Electrical infrastructure**. The manufacturing and installation of electricity infrastructure represents a significant cost in offshore developments. The full potential of offshore wind can only be realised through the construction of interconnected offshore grid systems and regulatory regimes that are better able to manage the variability of wind power generation. This will require significant investment in cable equipment and in vessels.

4. **Larger turbines**. The economics of offshore wind favour large machines. The key factors affecting the deployment of offshore wind are the current shortage of turbines and their reliability. New designs might be developed that address the challenge of marine conditions, corrosion and reliability issues. The development of testing facilities is a crucial issue.

5. **Operations and maintenance** (O&M). Strategies that maximise energy production while minimising O&M costs are essential. Better management systems and condition monitoring systems will be required. Effective access systems will be essential for the operation of the offshore facilities and the safety of personnel involved.

In addition to the five research topics, there are three common themes that underpin each of these topics and that are critical to delivering an offshore wind industry in Europe that is a world leader. These are:

1. **Safety**. Safe operation of offshore facilities and the safety of the staff involved are vital. It requires the examination and review of turbine access systems, and escape and casualty rescue.

2. **The environment**. This covers two main areas:
   i. The construction of substantial infrastructure in the seas around Europe must be done responsibly with minimal adverse ecological impacts; and
   ii. More knowledge about the offshore environment is needed, including collecting and understanding climatic, meteorological, oceanic and geotechnical data.

3. **Education** is critical for delivering safety. Moreover, more trained people with the necessary skills to develop the industry are needed. These will range from skilled workers needed to manufacture, build and operate the facilities to graduates that understand the technical, commercial and social context of the industry.

Market Deployment Strategy

The market deployment strategy developed by the European wind energy sector under the framework of TPWind is outlined in 'Priority R&D areas in wind energy'.

ENABLING MARKET DEPLOYMENT

Thematic priorities on these aspects are:

- **Removing electricity market barriers** by implementing market rules that promote demand-side management and flexibility and that provide much shorter gate closure times to reduce balancing costs. Improved forecasting tools, such as tools for developing assumptions on future market fluctuations in order to secure investments. In these integrated markets – at both local and international levels – wind power should be considered as an adapted market commodity that is tradable, exchangeable and transparent, like other forms of energy. The market impacts of a large penetration of wind energy on the current electricity markets should be evaluated.

- **Securing revenues** by developing a stable market with clear wind power objectives and stable incentive schemes. Markets should make use of all possible power system flexibility to keep imbalance costs low. This includes markets for ancillary services, such as bringing virtual power plants to the market, and effective balancing markets.
- **Creating a level playing field** by integrating wind power's positive externalities, such as contribution to energy independence and climate change mitigation. This would lead to recognition that wind energy deserves to be categorised as a public interest investment that is largely independent of fuel prices and has very low external costs.
- **Adapting the grid infrastructure** to make wind as manageable and as cost-efficient as possible for network operators. One approach is for wind power plants to be operated, as far as possible, like conventional power plants, to develop the electricity grid infrastructure needed, to implement common market policies and align existing markets, to develop large-scale energy storage solutions, demand-side management, and to adapt grid codes.
- **Removing policy and administrative barriers to grid development** through strategic planning, strong political leadership and adapted consenting processes.

COST REDUCTION

In the past few years, energy demand has gone up significantly and the price of fossil fuel has increased. In the case of wind energy, after a period in which costs decreased in line with experience, there has recently been an increase in wind energy costs due to very high global demand and rising commodity prices.

Other reasons for the rising costs include the increase in the overall price of materials and supply chain bottlenecks. The wind energy sector and policy-makers should focus on reducing the cost of investment, which would lead to reductions in the lifetime cost of energy, making wind more competitive. These priorities are:

- **Investment costs**. These are influenced by bottlenecks in the supply chain, which limit the effects of economies of scale on costs. Moreover, due to high demand, logistics and service suppliers will also suffer from supply bottlenecks, leading to higher investment costs. Uncertainties remain regarding the future cost of raw materials and the subsequent impact on the cost of wind energy. Finally, as installed capacity increases, wind power will move to more challenging environments, needing technological improvements to reduce costs.
- **Operating costs**. These account for a significant proportion of the overall lifetime costs. They can be reduced substantially by improving reliability, optimising operational services and component supply, and developing specific offshore systems.
- **The cost of capital**. This is closely linked to the financial sector's confidence in the technology, future revenue and market sustainability. Reducing exposure to risk in different categories will in turn reduce the cost of capital.

ADAPTING POLICIES

In order for wind energy markets to develop further, ambitious wind energy targets need to be set and appropriate measures taken in the EU Member States. Policy has to be consistent, stable and long-term, to allow for the most efficient investment and cheapest electricity for the consumer.

Long-term, legally binding wind and renewables targets should be implemented at national level, and a breakdown for the power, heat and transport sector is recommended. Penalties should be imposed on Member States that do not comply with the national targets or action plans. Stable and long-term support schemes are essential.

OPTIMISING ADMINISTRATIVE PROCEDURES

Key issues exist with the current administration of applications for wind farms and auxiliary infrastructures in many parts of Europe. These include inconsistencies and uncertainties in the requirements and judgements of administrative authorities and delays in the consenting process.

Despite policies supporting wind energy at European and national levels, it is difficult to obtain planning permits in many Member States. In some Member States, there is a lack of clarity on the administration requirements and processes for wind farm applications, particularly in relation to applications for repowering existing wind farms. Repowering projects create new challenges over and above those of developing new wind farms on a site.

INTEGRATING WIND POWER INTO THE NATURAL ENVIRONMENT

In extreme cases, wind farm projects are rejected by consenting authorities because of minor adverse local effects, as their positive impact on the global environment is not considered.

At present, environmental impact assessments (EIAs) are inefficient and the results of EIAs are not widely disseminated. A consensus has yet to be reached on whether EIAs could use studies that have already been carried out, rather than doing full and separate studies each time an EIA is performed.

The results of existing environmental monitoring across Europe need to be centrally gathered and reviewed in order to identify existing gaps in knowledge for future research.

Post-operational monitoring is currently not comprehensively evaluated against the areas of the EIA. The results are rarely shared widely or referenced in future assessments. As a consequence, much of the value of post-operational monitoring is not realised.

Priority development zones should be identified for the strategic planning of wind farms, as a policy priority.

ENSURING PUBLIC SUPPORT

Compared to conventional sources of energy generation, wind energy is popular with the general public. The industry can help to sustain this by further implementing best practices based on public consultation; by remaining willing to address public acceptance issues; and by demonstrating improvements that reduce or mitigate impacts of public concern.

Whilst there is large-scale support for wind energy, wind farm applications can be delayed or blocked by real or perceived resistance from communities at the local level.

It is essential to involve local communities in the process of wind farm developments in their area, and to ensure that they also reap some benefits. Decision-makers need to be kept well informed of the real level and nature of public support, not just the perceived level.

R&D Funding for Wind Energy

Wind energy contributes to the priorities set out by the Lisbon Strategy (2000). This strategy sets European Union goal of becoming 'the most competitive and dynamic knowledge-based economy in the world, capable of sustainable economic growth with more and better jobs and greater social cohesion' by 2010.

These objectives were complemented in 2002 at the Barcelona European Council, where Heads of State agreed that research and technological development (RTD) investment in the EU must be increased with the aim of reaching 3 per cent of GDP by 2010, up from 1.9 per cent in 2000 (1.84 per cent in 2006 for EU-27).

If the Barcelona 3 per cent objective is to be fulfilled, wind energy R&D investment would have to represent an average of €430 million per year. Two-thirds of this budget should be invested by the private sector and one-third by the public sector.

Average public annual support would then be €143 million per year. If 50 per cent of this support is provided by national (Member State) programmes and 50 per cent from EC programmes, the average contribution from the EC and from the national programmes should each be €72 million per year and increase with market turnover. The following section investigates past and current funding levels for wind energy R&D.

SUPPORT AT EC LEVEL

Historically, research and development funding for wind energy and other renewable energy technologies has been a fraction of the funding for conventional energies. According to the IEA, over the period from 1974 to 2002, nuclear energy research financing was approximately three and a half times greater than that dedicated to renewable energy. The technology achievements in wind energy are even more impressive given that the sector has received a mere 1 per cent of energy research funding in the IEA countries in 1974–2002. In the same period, nuclear energy received 58 per cent or US$175 billion, and fossil fuels 13 per cent (see Figure I.7.1).

Energy research funding in the EU has decreased dramatically and is currently at one-quarter of the level in 1980, according to the European Commission. Furthermore, the dominant part of EU research funding continues to be allocated to nuclear energy. In its first

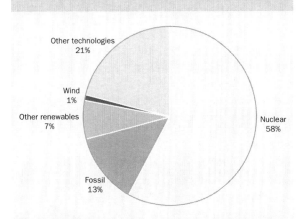

Figure I.7.1: Total energy R&D shares in IEA countries, 1974–2002 (US$)

Other technologies 21%
Wind 1%
Other renewables 7%
Nuclear 58%
Fossil 13%

Source: Renewable Energy, Market and Policy Trends in IEA Countries, OECD/IEA (2004) in *Prioritising Wind Energy Research: Strategic Research Agenda of the Wind Energy Sector*, EWEA, July 2004

review of EU energy policies, in September 2008, the IEA called for the EU to change its priorities:

The current Framework Programme allocates €1.95 billion, or almost 40 per cent of the energy funding, to nuclear fusion, a technology that is only expected to contribute past 2050. It will be important for the achievement of the EU climate change targets that this funding allocation is revised at the earliest possible opportunity, and that funding for non-nuclear energy research and development is increased significantly. (IEA, September 2008)

The EU FP7 Euratom programme allocates €1947 million to nuclear fusion, €287 million to nuclear fission and €517 million to nuclear research activities of the Joint Research Centre (JRC). In total, EU nuclear energy research funding totals €2.75 billion over the five-year period 2007–2011 or €550 million per year.

Non-nuclear energy research under the EUs FP7 receives €2300 million over the seven-year period 2007–2013, or €460 million per year. Over the next five years, the average annual EU research budget for energy will be €1010 million, allocated as follows:

- nuclear energy research: €550 million (54 per cent); and
- non-nuclear energy research: €460 million (46 per cent), of which approximately half to renewables and energy efficiency (€230 million – 23 per cent).

How much of the EU non-nuclear energy research budget will go to wind energy is not earmarked, but as shown in Table I.7.1, wind energy received €25 million under FP5 and €32 million under FP6, or approximately 3 per cent of the total FP7 energy research budget.

Funding for Overall Non-nuclear Energy (NNE) Research

In 2005, the European Commission's Advisory Group on Energy released a report[2] that demonstrated the full extent of the reduction in European Union funding for energy R&D through its Framework Programmes.

Regarding FP6, the Strategic Working Group of the Advisory Group on Energy pointed out that:

in face value terms, expenditure is now less than it was 25 years ago, in real-value terms it is very much less and as a percentage of the total Community R&D it is roughly six times smaller.

Two years later, the Strategic Energy Technology Plan (SET-Plan) was adopted. Again, it was pointed out that:

Public and private energy research budgets in the EU have declined substantially since peaking in the 1980s in response to the energy price shocks. This has led to an accumulated under-investment in energy research capacities and infrastructures. If EU governments were investing today at the same rate as in 1980, the total EU public expenditure for the development of energy technologies would be four times the current level of investment of around EUR 2.5 billion per year.

This state of affairs is illustrated in Figure I.7.2, showing that energy research funding as a percentage of all EU R&D funding has reduced from 66 per cent in FP1 to around 12 per cent in FP6 and 7 per cent in FP7.

Funding for Renewable Energy Research

Under FP6, €810 million was dedicated to R&D under the 'Sustainable energy systems' chapter (€405 million

Table I.7.1: EC funding levels in FP5 and FP6

Wind Technology paths (strategically important areas and topics)	EC funding in					
	FP5			FP6		
	Number of projects	Eligible costs in €m	Total EC contribution in €m	Number of projects	Eligible costs in €m	Total EC contribution in €m
Large size wind turbines	10	27.68	14.98	4	37.95	19.46
Integration and managing of wind power	7	12.99	7.15	4	7.75	4.43
Wind farm development management	3	4.02	2.23	2	34.03	7.70
Total wind	20	44.69	24.36	10	79.74	31.59

Source: European Commission, DG Research, *The State and Prospects of European Energy Research Comparison of Commission, Member and Non-Member States – R&D Portfolios*

Figure I.7.2: Energy spending in the seven Framework Programmes

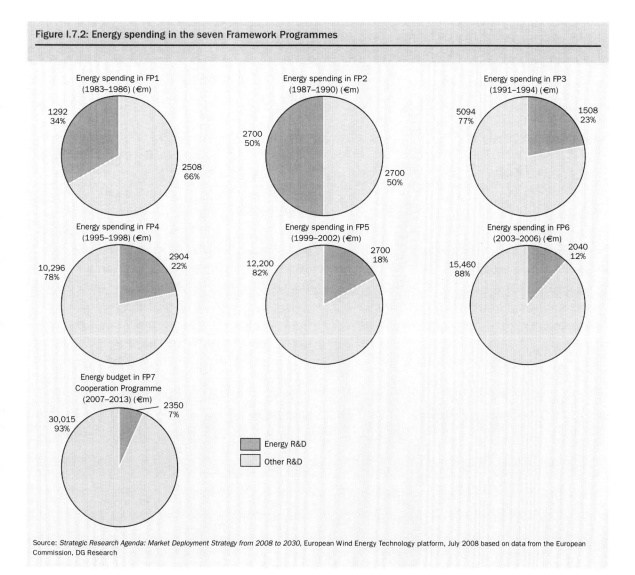

Source: *Strategic Research Agenda: Market Deployment Strategy from 2008 to 2030*, European Wind Energy Technology platform, July 2008 based on data from the European Commission, DG Research

to long-term R&D, administered by DG Research, and €405 million to short- to medium-term research, administered by DG TREN). This represented a reduction of some 20 per cent from FP5.

The name of the chapter or budget-line, 'Sustainable energy systems', engendered a lack of transparency in the funding process. The chapter included, for example, 'clean coal' technologies, focusing mainly on the sequestration of CO_2. It also included hydrogen and fuel cells, which are not energy sources.

For FP7, the lack of transparency still remains. The €2.35 billion budget available for non-nuclear energy under the Cooperation Programme includes the following chapters:

• hydrogen and fuel cells;
• renewable electricity generation;
• renewable fuel production;
• renewables for heating and cooling;
• CO_2 capture and storage technologies for zero emission power generation;

- clean coal technologies;
- smart energy networks;
- energy efficiency and savings; and
- knowledge for energy policymaking.

In 2007 the budget committed to projects was approximately €0.32 billion.

Funding for Wind Energy Research

The European Commission has provided an analysis of the evolution of the R&D budget over FP5 and FP6. Comparison between FP5 and FP6 is provided in Table I.7.1 on three main aspects:

1. large size wind turbines;
2. integration and management of wind power; and
3. wind farm development and management.

Due to two integrated projects (DOWNWIND and UpWind), the average project size increased significantly between FP5 and FP6. The EC contribution also increased by 27 per cent, reaching €31.59 million, an

average of €7 million per year – one tenth of TPWind requirements.

SUPPORT FOR WIND R&D AT MEMBER STATE LEVEL

The data discussed below have been sourced from the IEA's Energy R&D Statistics Database. R&D budgets for the period 1974–2005 are available for 19 EU countries: Austria, Belgium, Czech Republic, Denmark, Finland, France, Germany, Greece, Hungary, Ireland, Italy, Luxembourg, The Netherlands, Norway, Portugal, Spain, Sweden, Switzerland and the UK.

Figure I.7.3 illustrates the evolution of total funding for wind R&D during the period 1974–2006 (excluding EC funding). The total available budget peaked in 1985, with a significant budget available in The Netherlands, accounting for 46 per cent of the total.

The total available budget in 2003 was 48 per cent of this peak, decreasing to 37 per cent in 2004 (though for 2004, data from The Netherlands are not available). In 2005, a significant budget increase can be

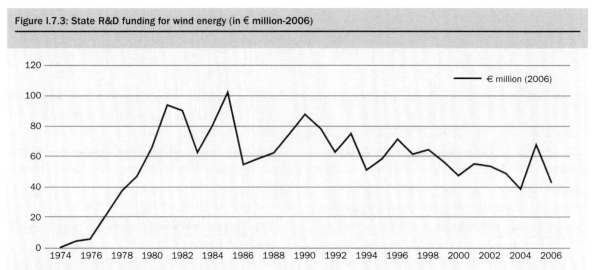

Figure I.7.3: State R&D funding for wind energy (in € million-2006)

Note: For some dates, data are not available for certain countries.

Source: EWEA, based on IEA data

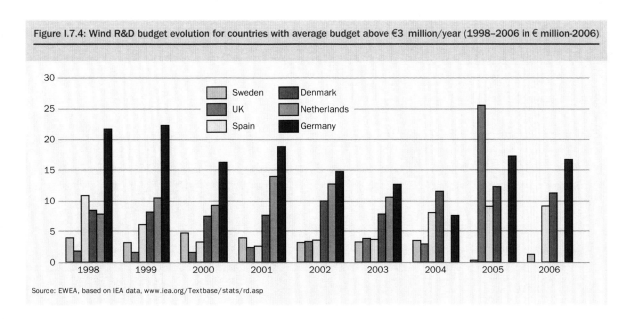

Figure I.7.4: Wind R&D budget evolution for countries with average budget above €3 million/year (1998–2006 in € million-2006)

Source: EWEA, based on IEA data, www.iea.org/Textbase/stats/rd.asp

noticed, with the R&D budget in the UK increasing by a factor of 10. In 2006, the total available budget for wind R&D was 41 per cent of the 1985 maximum and 60 per cent of TPWind's requirements.

For the period 1998–2006, these budget variations are highlighted in Figure I.7.4 for the six main contributing countries. Germany was the main contributor to R&D funding for wind energy, which is consistent with its world-leading position in installed wind power capacity and world manufacturing capacity. After a decrease in funding by a third in 2004, Germany made a significant effort in 2005 and 2006. For 2005, the budget contribution of the UK, however, far exceeded any other.

These strong budget variations prevent the sector from relying on the material research support scheme. Ambitious long-term research programmes, involving heavy research facilities, are therefore risky.

Figure I.7.5 shows the average R&D budget for the period 1998–2006 for countries with a significant budget (the seven countries above €1 million on average). Only six of these countries had an average budget that exceeded €3 million a year and only these six main contributors were able to set up research laboratories

and test facilities that were recognised worldwide and/or had world-leading turbine or component manufacturers. These figures clearly demonstrate that a high-quality research structure is built on long-term, high-level R&D budgets.

CURRENT EFFORT FROM THE PRIVATE SECTOR

Collecting and analysing the available information on R&D investment from wind turbine and component manufacturers is far from straightforward. Some manufacturers merge figures for wind into their overall R&D investment data, while data from other manufacturers is not available for public consultation.

Figures for 2006 are available for some wind turbine manufacturers,[3] making a total of €186 million – 65 per cent of TPWind requirements. This figure does not, however, include all manufacturers or sub-suppliers. The ratio of R&D expenditure to net sales varies significantly, from 0.4 to 3.1 per cent. Clipper Windpower (89 per cent) is an exception, as the manufacturing capacity is located in the US.

Figure I.7.5: Average funding, 1998–2006, for countries with an average budget above €1 million/year (in € million-2006)

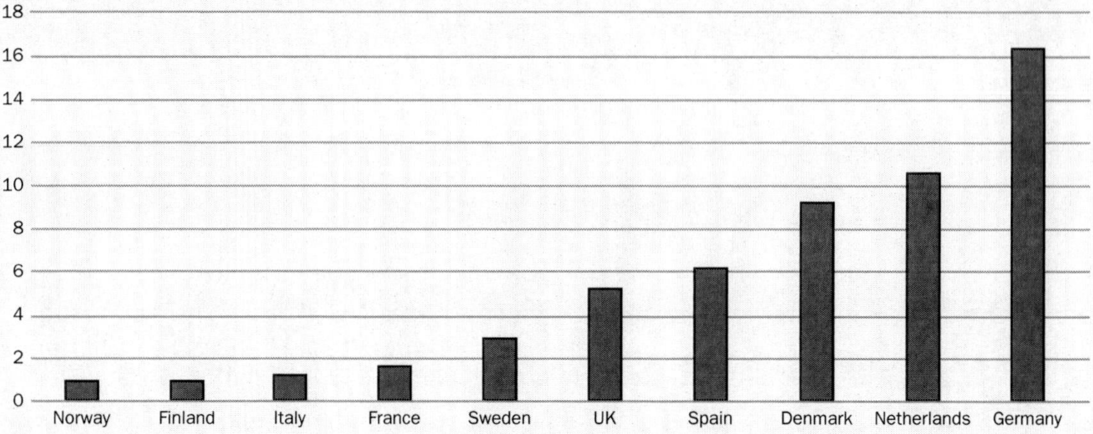

Source: EWEA, based on IEA data, www.iea.org/Textbase/stats/rd.asp in *Strategic Research Agenda Market Deployment Strategy from 2008 to 2030*, European Wind Energy Technology Programme, July 2008

Table I.7.2: R&D investment by the private sector

Company	Country	R&D Investment 2006 €m	Net Sales 2006 €m	R&D/net Ratio 2006 %	Sales 2005 %
Acciona	Spain	22.60	6272	0.4	0.1
Gamesa	Spain	33.12	2391	1.4	1.6
Nordex	Germany	11.25	514	2.2	2.9
REpower Systems	Germany	14.02	459	3.1	3.1
Vestas Wind Systems	Denmark	88.60	3854	2.3	2.4
Nordex	Germany	11.25	514	2.2	2.9
Clipper Windpower	UK	5.33	6	88.8	193.0

Source: *Strategic Research Agenda Market Deployment Strategy from 2008 to 2030*, European Wind Energy Technology Programme, July 2008

CONCLUSION

Wind energy will be a main contributor to the implementation of the EU objectives on renewable energy production. However, the current R&D efforts for wind energy are insufficient – at all levels – to respond to the energy challenges faced by the EU. The risk is therefore of failure in reaching the EU objectives for energy production from renewable sources (and therefore on reduction of CO_2 emissions) and in implementing the European strategy for growth and jobs.

A critical component is the contribution of the EU, which, in order to achieve the objectives of the Lisbon Strategy, should lead by example. A strong and clear signal from the EU would act as catalyst at Member State level in strongly supporting renewables and wind in particular.

The problem Europe faces is not a lack of technical solutions but a lack of time. 2020 is tomorrow. The longer it takes to adapt the EU energy system, the more difficult and costly it will be, with an unknown impact on the environment.

In 2007, the Strategic Energy Technology Plan set a new agenda for energy research and innovation in Europe, with the core aim of speeding up the deployment progress of energy technologies. One of the proposed key areas of action is that of industrial initiatives, among which is the European Wind Initiative. This initiative should be a major part of European research and innovation in wind energy technology. It should lead to an adapted European energy mix that is less reliant on imports, is based on zero CO_2 emissions and creates employment opportunities.

Part I Notes

[1] For more information on this, please visit the TPWind website: www.windplatform.eu

[2] European Commission (2005) *Towards the European Energy Research Area: Recommendations by the ERA Working Group of the Advisory Group on Energy*, 2005, Directorate General for Research, Luxembourg

[3] European Commission (2007b). *Monitoring Industrial Research: The 2007 EU Industrial R&D Investment Scoreboard.*

EWEA/Winter

GRID INTEGRATION

Acknowledgements

Part II was compiled by Frans Van Hulle of EWEA and Paul Gardner of Garrad Hassan and Partners.

We would like to thank all the peer reviewers for their valuable advice and for the tremendous effort that they put into the revision of Part II.

Part II was carefully reviewed by the following experts:

J. Charles Smith	UWIG, USA
Alex De Broe	3E
Ana Estanquiero	INETI
Guillaume DUCLOS	ENR

II.1 SETTING THE SCENE

Wind Energy Penetration and Integration

In Part II, we consider the large-scale integration of wind energy in the context that wind will meet a substantial share of the European electricity demand in the future. While wind energy will cover around 4 per cent of electricity demand in 2008, EWEA targets for 2020 and 2030 estimate penetration levels of up to 14 per cent and up to 28 per cent respectively (EWEA, 2008a).

Europe's wind power resources are enormous and could easily cover a larger share of the electricity demand. This is already the case, notably in a few regions in Germany, Denmark and Spain. The key issue is how to develop the future power system so that wind power can be integrated efficiently and economically. Since integration efforts, such as costs and decision-making, are related directly to the penetration level of wind power, it is essential to have a commonly defined term. Wind energy penetration can be defined in a number of ways.

WIND ENERGY PENETRATION

This looks at the percentage of demand covered by wind energy in a certain region, normally on an annual basis (see Figure II.1.1).

Wind energy penetration (%)
= Total amount of wind energy produced (TWh) /
Gross annual electricity demand (TWh)

WIND POWER CAPACITY PENETRATION

This looks at how the total installed wind power capacity in a certain region is related to the peak load in this region over a certain time period.

Wind capacity penetration (%)
= Installed wind power capacity (MW) /
Peak load (MW)

MAXIMUM SHARE OF WIND POWER

This looks at the power balance in a certain region, taking into account the minimum demand, the maximum wind power generated and the exchange with neighbouring regions or countries. This figure must remain below 100 per cent to ensure the correct power balance in the region; the nearer to 100 per cent, the closer the system is to its limits (when wind power would need to be curtailed).

Maximum share of wind power
= Maximum wind power generated (MW) /
Minimum load (MW) + power exchange
capacity (MW)

Throughout Part II, when reference is made to wind power penetration, the first definition will be used unless specified otherwise.

As shown in Figure II.1.1, the wind energy penetration levels vary throughout Europe. For the EU-27, the overall penetration in 2020 will be around 12–14 per cent according to present EWEA and European Commission (EC) targets.

Table II.1.1 shows the wind power capacity penetration values (second definition), at the beginning of 2007 for a number of countries in the Union for the Co-ordination of Transmission of Electricity (UCTE) area. These values are related to the reference load, as set out in the UCTE System Adequacy Forecast (January 2007). The installed wind power capacity refers to the situation at the end of 2006.

The share of wind power (third definition) is already high in certain areas of Europe, for example West Denmark (57 per cent) and the German state of

Figure II.1.1: Overview of wind energy penetration levels in Europe at the end of 2007

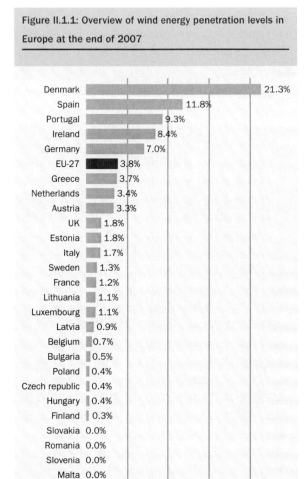

Denmark	21.3%
Spain	11.8%
Portugal	9.3%
Ireland	8.4%
Germany	7.0%
EU-27	3.8%
Greece	3.7%
Netherlands	3.4%
Austria	3.3%
UK	1.8%
Estonia	1.8%
Italy	1.7%
Sweden	1.3%
France	1.2%
Lithuania	1.1%
Luxembourg	1.1%
Latvia	0.9%
Belgium	0.7%
Bulgaria	0.5%
Poland	0.4%
Czech republic	0.4%
Hungary	0.4%
Finland	0.3%
Slovakia	0.0%
Romania	0.0%
Slovenia	0.0%
Malta	0.0%
Cyprus	0.0%

Source: EWEA (2008a)

Table II.1.1: Wind power capacity penetrations in various European countries

	Reference load (GW)	Wind power capacity (GW)	Capacity penetration
West Denmark	3.8	2.5	66%
Germany	74.0	20.6	28%
Spain	43.0	11.6	27%
Portugal	8.5	1.7	20%
The Netherlands	16.1	1.6	10%
France	80.0	1.6	2%

Source: EWEA

Schleswig-Holstein (44 per cent), but the system can absorb additional wind power before it reaches full capacity. However, with increasing amounts of wind power installed, improvements are required in the power exchange capacities between various countries. This will be discussed in more detail later in Part II.

European Policy Framework Relevant for Wind Power Integration

The electricity system in Europe needs to be modified in order to ensure security of supply, a fair and low electricity price for the consumer, and sustainable and climate-friendly electricity generation. These objectives form the basis of European energy policy. As wind power will play an ever more important role in the electricity supply, this section looks at some of the important policy developments at the EU level which are vital for the process of grid integration of wind power in Europe.

RENEWABLE ENERGY DIRECTIVE

A new Renewables Directive, agreed by the European Union in December 2008, sets a 20 per cent target for the EU as a whole for the share of energy demand to be covered by renewables by 2020. In order to achieve this target, the European Commission estimates that wind power will have to cover 12 per cent of total European electricity demand by 2020, although the individual national and sectoral targets have yet to be established. The Renewables Directive:

- stipulates that Member States take 'the appropriate steps to develop transmission and distribution grid infrastructure ... to accommodate the further development' of renewable electricity; and
- includes a clause relating to priority or guaranteed access and priority dispatch for wind power and other renewables, on condition that the reliability and safety of the grid is maintained.

INTERNAL ELECTRICITY MARKET LEGISLATION

A series of legal measures (the so-called 'Third Liberalisation Package') were proposed in 2008. The intention is to create a single electricity market in Europe, with more coordinated regulation, improved system operation at international level and fair access for renewable energy sources (RES) generators. The measures include stronger international cooperation of transmission system operators (TSOs) under the European Network of Transmission System Operators for Electricity (ENTSO).

In principle, this could provide a changed framework for the future development of harmonised grid codes in the coming years. The implementation of the proposed Liberalisation Package could also improve the interconnection between Member States. The future Agency for the Coordination of Energy Regulators in Europe (ACER), as proposed by the Liberalisation Package, needs to ensure that TSOs submit appropriate transmission development plans and that the regulation in the market is improved, strengthened and harmonised.

THE TEN-E PROGRAMME

The Trans-European Energy Networks (TEN–E) programme addresses transmission development issues at the European level, in order to support the further development of the internal electricity market. A Green Paper on transmission issues was published in the third quarter of 2008, which will form the basis for European policies for transmission development and should give guidance on the national policy frameworks. The TEN-E programme is currently being complemented by a European Commission initiative to explore and possibly implement grid reinforcements, including offshore grid transmission lines to enable the connection of the predicted offshore wind power capacity. A European coordinator has been appointed for this purpose.

Brief Outline of the Main Integration Issues

Given the current levels of wind power connected to electricity systems, it is clearly feasible to integrate wind power to a significant extent without major system changes. The 60 GW of wind power already installed in Europe shows:

- areas of high, medium and low penetration levels;
- where different conditions exist; and
- where bottlenecks and challenges occur.

Wind power as a generation source has specific characteristics, including variability, geographical distribution, favourable economics, and, above all, abundance and environmental benefits. Large-scale integration of both onshore and offshore wind raises challenges for the various stakeholders involved, ranging from generation, transmission and distribution to power trading and consumers. In order to integrate wind power successfully, a number of issues need to be addressed in the following areas:

- **design and operation of the power system:** reserve capacities and balance management, short-term forecasting of wind power, demand-side management and storage, and optimisation of system flexibility;
- **grid infrastructure issues:** optimisation of present infrastructure, extensions and reinforcements, offshore grids, and improved interconnection;
- **grid connection of wind power:** grid codes, power quality and wind power plant capabilities;
- **market redesign issues:** market aggregation and adapted market rules increasing the market flexibility, particularly for cross-border exchange and operating the system closer to the delivery hour; and
- **institutional issues:** stakeholder incentives, non-discriminatory third party grid access and socialisation of costs.

II.2 WIND POWER VARIABILITY AND IMPACTS ON POWER SYSTEMS

A wind farm does not operate all the time, so backup capacity is needed for when it does not and differences between forecast and actual production have to be balanced. Balancing and backup come at a cost, as does building new infrastructure. These facts apply to wind energy just as they apply to other power producing technologies that we integrate into the electricity grids. But for reasons that are difficult to grasp, balancing and backup of wind energy is generally perceived to be problematic whereas balancing and backup for other technologies seems as easy as breathing. Certainly, most of the mainstream media does not find it interesting to report the complexities of balancing a constant supply of nuclear power or inflexible coal-fired power against the demand from millions of consumers, with their constantly changing and unpredictable demands for power.

There is nothing simple about operating a power grid. Delivering electricity to consumers is a logistical challenge larger than for any other product market. Transmission system operators (TSOs) are tasked with delivering an invisible product, which cannot be stored, to customers who expect to receive it at the exact same second they need it. Grid operation is just-in-time management in its most extreme form; when you think about it, it seems an unrealistic task for anybody to undertake. Nevertheless, European grid operators are simultaneously servicing 500 million fickle consumers with unpredictable behaviour every second of every hour of every day. They have done so for a hundred years with minimal supply disruption. If it was not for the fact that we experience it every day, we would say that it was impossible.

Just like an individual consumer, a wind turbine is variable in output and less predictable than most other technologies. However, from a system operations perspective, the supply behaviour of a single wind farm is just as irrelevant as the demand behaviour of a person. The collective behaviour of consumers and the collective behaviour of all generating plants are what matters. That has been the guiding principle of grid operation since its inception and is likely to remain so regardless of which technologies we use. If operating a grid is inherently difficult, we are fortunate in having system operators in Europe who understand what the rest of us find difficult to comprehend. Wind power is, admittedly, different from other power technologies, and integrating large amounts of it in the existing power system is a challenge. But whatever the generating technology, the basic principles of balancing, backing up, aggregation and forecasting still apply.

Changes to the way we construct and operate the future European electricity grids are still needed if we are to meet one-third of Europe's power demand with renewables within 12 years, as projected by the European Commission. But the challenge is by no means any greater or more costly than the one system operators faced when politicians thought that nuclear was the answer and expanded its share to 30 per cent of European demand in two decades. Today the answer happens to be wind, and of course the grid needs to be adapted to that new reality.

TSOs employ some of the most skilled people in the power sector. Nevertheless, they too need practical experience to acquire knowledge when new technologies are introduced in large amounts. European TSOs are gaining vast experience and knowledge about managing over 30 per cent wind power shares for long periods of time. In Denmark, a mind-blowing 140 per cent is sometimes managed.

Understanding Variable Output Characteristics of Wind Power: Variability and Predictability

WIND POWER: A VARIABLE OUTPUT SOURCE EMBEDDED IN A VARIABLE ELECTRICITY SYSTEM

Since wind energy is a technology of variable output, it needs to be considered as just one aspect of a variable, dynamic electricity system. At modest penetration levels, the variability of wind is dwarfed by the normal variations of the load. It is impossible to analyse wind power in isolation from other parts of the electricity system, and all systems differ. The size and the inherent flexibility of the power system are crucial aspects in determining the system's capacity to accommodate a large amount of wind power.

The variability of wind energy needs to be examined in the wider context of the power system, rather than at the individual wind farm or wind turbine level. The wind does not blow continuously at any particular site, but there is little overall impact if the wind stops blowing in a certain area, as it is always blowing elsewhere. This lack of correlation means that at the system level, wind can be harnessed to provide stable output regardless of the fact that wind is not available all the time at any particular site. So in terms of overall power supply, it is largely irrelevant to consider the curve when a wind power plant produces zero power for a time, due to local wind conditions. Moreover, until wind becomes a significant producer (supplying around 10 per cent of electricity demand), there is a negligible impact on net load variability.

Box II.2.1: Variable output versus intermittency

Wind power is sometimes incorrectly considered as an intermittent energy source; however, this is misleading. At power system level, wind power does not start and stop at irregular intervals (a characteristic of conventional generation), as is suggested by the term intermittent. Even in extreme conditions, such as storms, it takes hours for wind turbines in a system area to shut down, Moreover, periods with zero wind power production are predictable and the transition to zero power is gradual.

Also worthwhile considering is the technical availability of wind turbines, which is at a very high level (98 per cent) compared to other technologies. Another advantage of wind power in this respect is its modular and distributed installation in the power system. Breakdown of a single unit has a negligible effect on overall availability.

So, the term 'intermittent' is inappropriate for system-wide wind power and the term 'variable output' should be used instead.

Wind power varies over time, mainly under the influence of meteorological fluctuations. The variations occur on all timescales: seconds, minutes, hours, days, months, seasons and years. Understanding these variations and their predictability is of key importance for the integration and optimal utilisation of wind in the power system. Electric power systems are inherently variable in terms of both demand and supply, but they are designed to cope effectively with these variations through their configuration, control systems and interconnection.

SHORT-TERM VARIABILITY

The analysis of data available from operating wind farms and meteorological measurements at typical wind farm locations allows us to quantify the variations in net wind power output that can be expected for a given time period (within the minute or hour, or during the course of several hours). The distinction between these specific timescales is made since this type of information corresponds to the various types of power plants for balancing. The results from analyses show that the power system can handle this short-term variability well. System operators only need to deal with the net output of large groups of wind farms, and the wind power variability is viewed in relation to the level and variation in power demand.

Variations within the Minute

The fast variations (seconds to minute) of aggregated wind power output (as a consequence of turbulence or transient events) are quite small, due to the aggregation of wind turbines and wind farms, and hardly impact the system.

Variations within the Hour

The variations within an hour are much more significant for the system. However, they should always be considered in relation to demand fluctuations. Local variations are largely equal to geographical diversity, and will generally remain inside ±5 per cent of installed wind power capacity at the regional level.

The most significant variations arise from the passage of storm fronts, when wind turbines reach their storm limit (cut-out wind speed) and shut down rapidly from full to zero power. However, due to the averaging effect across a wind farm, the overall power output takes several minutes to reduce to zero. And in general, this is only significant in relatively small geographical areas, since in larger areas it takes hours for the wind power capacity to cease during a storm. For example, in Denmark – a small geographical area – on 8 January 2005, during one of the biggest storms for decades, it took six hours for the installed wind power in the West Denmark area to drop from 2000 to 200 MW (5 MW/minute) (see Figure II.2.1). The passage of a storm front can be predicted and technical solutions are available to reduce the steep gradient, such as the provision of wind turbines with storm control.[1]

These intra-hour variations will be an issue for power system reserves used for balancing when wind power penetration reaches the point at which variations in supply are equal to variations in demand (when 5–10 per cent of annual electricity demand is produced by wind power).

Variations from Hour to Hour

The variations between forecast and actual wind energy production several hours ahead affect the scheduling of the power system. For system operation, the variation in itself is not a problem; it is the uncertainty of how accurately the variation can be predicted that is significant. The uncertainty of wind power predictions should always be considered in relation to the errors in demand forecasts. There is much work being conducted in this area and it is clear that solutions are available.

LONG-TERM VARIABILITY

The slower or long-term variations of wind power relevant for integration in the power system include the seasonal and inter-annual variations, caused by climatic effects. These are not particularly important for

---

Figure II.2.1: Denmark: The storm on 8 January is recorded between hours 128 and 139

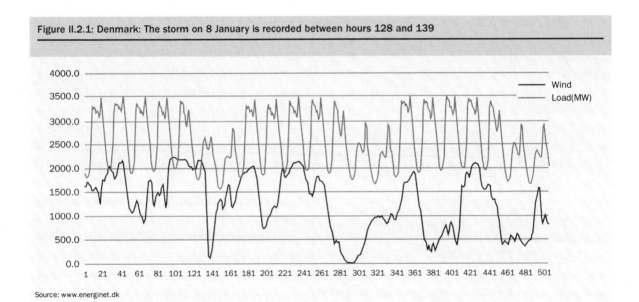

Source: www.energinet.dk

the daily operation and management of the grid, but play a role in strategic power system planning.

Monthly and Seasonal Variations

These variations are important for electricity traders that have to deal with electricity forward contracts, where wind power volume has an influence on price. They are also important for power system planning. However, it appears that for both electricity trading and system planning purposes, these deviations, resulting from annual statistics of wind power produced, can be sufficiently hedged.

Inter-annual Variations

These variations are relevant for long-term system planning, rather than daily power system operation. The annual variability of long-term mean wind speeds at sites across Europe tends to be similar, and can be characterised by a normal distribution with a standard deviation of 6 per cent. The inter-annual variability of

the wind resource is less than the variability of hydro inflow. In addition, at the power system level, the annual variations are influenced by the market growth of wind power and the projected onshore/offshore ratio.

EFFECTS OF AGGREGATION AND GEOGRAPHICAL DISPERSION

Due to the wide regional distribution of wind plants, short-term and local wind fluctuations are not correlated and therefore largely balance each other out. As a result, the maximum amplitudes of wind power fluctuations experienced in the power system are reduced. This phenomenon has been extensively studied throughout Europe.

Whereas a single wind farm can exhibit hour to hour power swings of up to 60 per cent of capacity, the maximum hourly variation of 350 MW of aggregated wind farms in Germany does not exceed 20 per cent (ISET, 2004). For larger areas, such as the Nordel system, which covers four countries, the largest hourly variations would be less than 10 per cent of installed

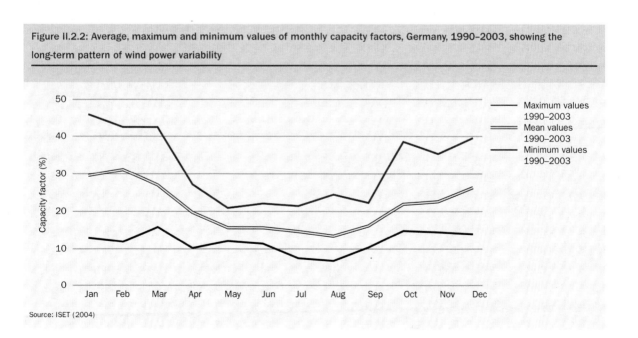

Figure II.2.2: Average, maximum and minimum values of monthly capacity factors, Germany, 1990–2003, showing the long-term pattern of wind power variability

Source: ISET (2004)

wind power capacity if the capacity was distributed throughout all the countries. The geographical spread of wind farms across a power system is a highly effective way to deal with the issue of short-term variability: the more widespread the wind farms, the lower the impact from variability on system operation.

The effect of reduced wind power variability increases with the size of the area considered. Ideally, to maximise the smoothening effect, the wind speeds occurring in different parts of the system should be as uncorrelated as possible. Due to the typical sizes of weather patterns, the scale of aggregation needed to absorb a storm front is in the order of 1500 km. By aggregating wind power over large regions of Europe, the system can benefit from the complementarities of cyclones and anticyclones over Europe (Figure II.2.3). The economic case for smoothing wind power fluctuations by utilising transmission capacity (rather than by other means) is an important area of investigation, for example in the TradeWind project.[2]

In addition to the advantage of reducing the fluctuations, the effect of geographically aggregating wind farm output is an increased amount of firm wind power capacity in the system. This will be explained further in Chapter II.6.

LOAD DURATION CURVE

One method of representing the smoothing effect of aggregation on system scale is the load duration curve of wind farms, which gives the frequency distribution of the partial load states of generated wind power (see Figure II.2.5). The effect of aggregating wind power is a flattening of the duration curve. This means that when wind power is aggregated over a large area:

- the effective number of hours when wind power is available increases; and
- the number of hours with zero or low power diminishes, while the maximum value of instantaneous aggregated power produced is decreasing.

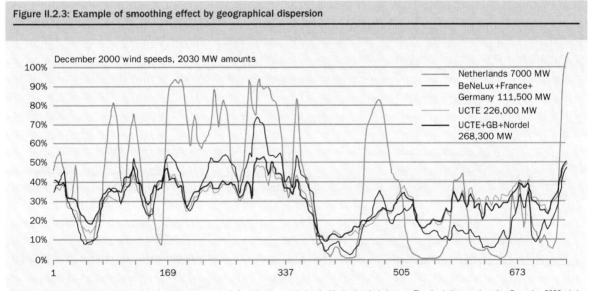

Figure II.2.3: Example of smoothing effect by geographical dispersion

Note: The figure compares the hourly output of wind power capacity in four situations, calculated with simulated wind power. The simulations are based on December 2000 wind speeds and wind power capacity estimated for the year 2030.

Source: www.trade-wind.eu

Figure II.2.4: Combined wind energy production from Europe and Northern Africa (Morocco) produces a monthly pattern that matches demand in Europe and Norway

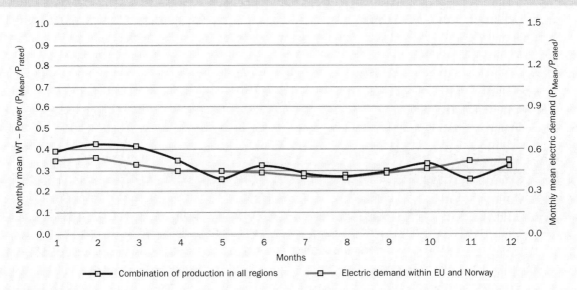

Source: Czisch (2001)

Figure II.2.5: Duration curves for the 'wind year 2000', Denmark and Nordic countries, assuming equal wind capacity in each of the four countries

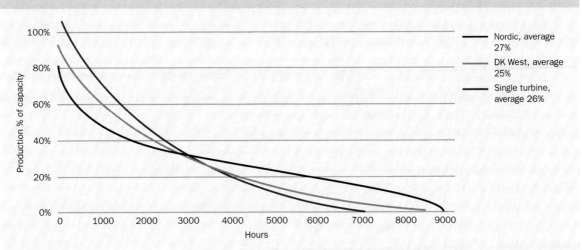

Source: Holttinen et al. (2007)

As part of the TradeWind project, a simulation was made for the EU-27, with an assumed wind capacity distribution for 2020 and 2030. The effect of geographical aggregation means that the maximum aggregated instantaneous wind power is only 60 per cent of the total capacity of 200 GW (Tande et al., 2008).

THE NEED FOR INTERCONNECTION

It is impossible to optimally aggregate large-scale wind power without a suitably interconnected grid. In this context, the grid plays a crucial role in aggregating the various wind farm outputs installed at a variety of geographical locations, with different weather patterns. The larger the integrated grid – especially beyond national borders – the more pronounced this effect becomes. This effect is equivalent to using the grid to aggregate demand over interconnected areas. In order to make best use of this effect, the present transmission system in Europe needs to be upgraded. Ideally, the interconnection capacity should be increased, and the rules governing the power exchange between countries should be adapted to ensure that interconnectors are always available for physical flow.

Variability Versus Predictability of Wind Power Production

Accurate forecasts of the likely wind power output, in the time intervals relevant for generation and transmission capacity scheduling, allow system operators and dispatch personnel to manage the variability of wind power in the system. Predictability is key to managing wind power's variability and improved accuracy of wind power prediction has a beneficial effect on the amount of balancing reserves needed, so the accurate forecasting of wind power is important for its economic integration into the power system.

Today, wind energy forecasting uses sophisticated numerical weather forecast models, wind power plant generation models and statistical methods to predict generation at 5-minute to 1-hour intervals, over periods of up to 48 to 72 hours in advance and for seasonal and annual periods.

Forecasting wind power production differs from forecasting other generation forms or forecasting the load.[3] Wind, being a natural phenomenon, is better suited to reliable statistical treatment and physical forecasting than conventional plants which are subject to physical faults.

Wind power prediction can be quite accurate for aggregated wind power, as the variations are levelled out; and the larger the area, the better the overall prediction. The extent to which prediction error decreases with the size of the region[4] considered is shown in Figure II.2.6. It should be noted that the forecast accuracy is reduced for longer prediction periods.

The quality of the short-term forecast should be considered in relation to the gate closure times in the power market. Reducing the time needed between scheduling supply to the market and actual delivery

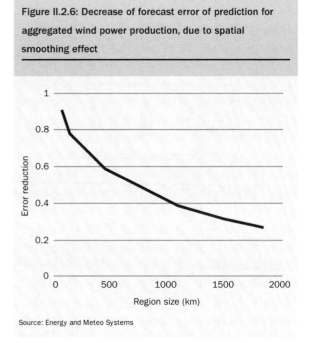

Figure II.2.6: Decrease of forecast error of prediction for aggregated wind power production, due to spatial smoothing effect

Source: Energy and Meteo Systems

(gate closure time) would allow shorter-term forecasts to be used, which could dramatically reduce unpredicted variability and lead to more efficient system operation without compromising system security. Changing from day-ahead to intraday commitments has a dramatic impact on accuracy and the cost of balancing the system. It is important to understand that for system operation, it is not just wind forecasting accuracy that is relevant for balancing the system, but also the sum of all demand and supply forecast errors relevant for system operation.

Impacts of Wind Power on Power Systems

The impacts of wind power in the electricity system depend to a large extent on the:

- level of wind power penetration;
- grid size; and
- generation mix of electricity in the system.

Wind energy penetration at low to moderate levels is a matter of cost, as demonstrated by various national and regional integration studies. And the integration costs related to the impacts listed above are fairly modest.

For low penetration levels of wind power in a system, system operation will hardly be affected. Currently (in 2008) wind power supplies less than 4 per cent of the overall EU electricity demand, but there are large regional and national differences in penetration, as shown in Figure II.1.1.

The established control methods and system reserves available for dealing with variable demand and supply are more than adequate for dealing with the additional variability at wind energy penetration levels of up to around 20 per cent, depending on the nature of a specific system. For higher penetration levels, some changes to systems and their method of operation may be required to accommodate the further integration of wind energy.

SHORT- AND LONG-TERM IMPACTS

The impacts of wind power on the power system can be categorised into short- and long-term effects. The short-term effects are caused by balancing the system at the operational timescale (minutes to hours), whereas the long-term effects are related to the contribution wind power can provide to the system adequacy (its capability to reliably meet peak load situations).

IMPACTS IN THE SYSTEM: LOCAL AND SYSTEM-WIDE

Locally, wind power plants interact with the grid voltage, just like any other power station. In this context, steady-state voltage deviations, power quality and voltage control at or near wind farm sites must all be taken into consideration. Wind power can provide voltage control and active power (frequency) control. Wind power plants can also reduce transmission and distribution losses when applied as embedded generation.

On the system-wide scale, there are other aspects to consider.

- Wind power plants affect voltage levels and power flows in the networks. These effects can be beneficial to the system, especially when wind power plants are located near load centres, and certainly at low penetration levels. For example, wind power plants can support the voltage in the system during fault (low voltage) situations. Also, wind plants that have a reactive power control system installed at the end of long radial lines benefit the system, since they support the voltage in (normally) low voltage quality parts of the grid.
- Wind power may need additional upgrades in transmission and distribution grid infrastructure, as is the case when any power plant is connected to a grid. In order to connect remote high-resource sites, such as offshore wind farms or very large wind plants in remote areas, to the load centres, new

lines need to be constructed (just as new pipelines had to be built for oil and gas). In order to maximise the smoothing effects of geographically distributed wind, and to increase the level of firm power, additional cross-border transmission is necessary to reduce the challenges of managing a system with high levels of wind power.

- Wind power requires measures for regulating control, just like any other generation technology, and, depending on the penetration level and local network characteristics, it affects the efficiency of other generators in the system (and vice versa).

- In the absence of sufficient intelligent and well-managed power exchange between regions or countries, a combination of (non-manageable) system demands and production may result in situations where wind generation has to be constrained.

- Finally, wind power plays a role in maintaining system stability and contributes to the system adequacy and security of supply.

For an overview and categorisation of the power system effects of wind power, see Table II.2.1 below.

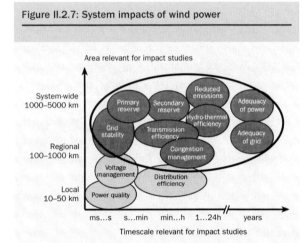

Figure II.2.7: System impacts of wind power

Note: Issues which are within the scope of Task 25 are circled in black.

Source: Holttinen et al. (2007)

A graphical overview of the various impacts of wind power in the power system is given in Figure II.2.7. It shows the local and system-wide impacts, as well as the short- and long-term impacts, for the various affected aspects of the power system, which include grid infrastructure, system reserves and system adequacy.

Table II.2.1: Power system impacts of wind power causing integration costs

	Effect or impacted element	Area	Timescale	Wind power contribution
Short-term effects	Voltage management	Local/regional	Seconds/minutes	Wind farms can provide (dynamic) voltage support (design dependent).
	Production efficiency of thermal and hydro	System	1–24 hours	Impact depends on how the system is operated and on the use of short-term forecasting.
	Transmission and distribution efficiency	System or local	1–24 hours	Depending on penetration level, wind farms may create additional investment costs or benefits. Spatially distributed wind energy can reduce network losses.
	Regulating reserves	System	Several minutes to hours	Wind power can partially contribute to primary and secondary control.
	Discarded (wind) energy	System	Hours	Wind power may exceed the amount the system can absorb at very high penetrations.
Long-term effects	System reliability (generation and transmission adequacy)	System	Years	Wind power can contribute (capacity credit) to power system adequacy.

Source: EWEA

II.3 DESIGN AND OPERATION OF EUROPEAN POWER SYSTEMS WITH LARGE AMOUNTS OF WIND POWER

In order to integrate wind power efficiently at higher levels of penetration, changes to the operating methods of various parts of the power system, such as generators and transmission systems, are required. Moreover, active management at the demand side of the power system can be used to facilitate wind power integration. Wind power, with its variable output characteristics, affects other generators in the system. As well as reducing their required output, wind power also requires other plants in the system to be scheduled differently.

In order to efficiently integrate large amounts of wind power, it is essential for the system design to be more flexible, which can be achieved by a combination of:

- flexible generating units;
- flexibility on the demand side;
- availability of interconnection capacity; and
- more flexible rules in the power market.

Balancing Demand, Conventional Generation and Wind Power

EFFECT OF WIND POWER ON SCHEDULING OF RESERVES

In this section, we outline the way in which wind affects the operation of the other generators in the system. Further information on power system operating principles is provided in Appendix H.

Primary Reserves

Wind power development will have little or no influence on the amount of primary reserves required. On second/minute timescales, rapid variations in the total wind power capacity output occur randomly, such as existing load variations. When aggregated with load and generation variations, the increase in variability due to wind is very small. Furthermore, the amount of primary reserve allocated in the power systems is dominated by potential outages of large thermal generation plants, meaning it can easily cope with these rapid variations.

Secondary and Tertiary Reserves

The impact of wind power on the need for secondary reserves will only be increasingly significant from wind energy penetrations levels of 10 per cent upwards. The main impact of wind power will be on how conventional units are scheduled to follow load (hour to day timescales). If the output from a wind plant could be accurately predicted one to two days in advance, schedulers could more easily determine units that would need to be committed. The lack of an accurate forecast adds further uncertainty to the commitment decision, on top of the uncertainty associated with load forecasting. The result is that a unit might be unnecessarily committed, or that it may not be committed when this is required. In this case, the generation mix of the power system determines scheduling in view of expected wind power production – the greater the flexibility of power units, the later unit commitment decisions can be made.

The estimate for extra reserve requirements due to wind power (EWEA, 2005a; Holttinen et al., 2007) is in the order of 2–4 per cent of the installed wind power capacity at 10 per cent penetration of gross consumption, depending on how far ahead wind power forecast errors are corrected by reserves (this is dependent on the gate closure times).

Short-term Forecasting of Wind in System Operation

Clearly, short-term forecasting becomes increasingly important for system operation as wind power penetration increases. In regions with high penetration levels, such as certain areas of Spain, Germany, Denmark and

Ireland, wind farm operators routinely forecast output from their wind farms. These forecasts are used by system operators to schedule the operation of other plant and for trading purposes. The benefits of the application of short-term forecasting depend, to a large extent, on national regulatory, technological and site-specific issues. The main advantages are cost reductions and improved system security.

ADDITIONAL BALANCING CAPACITIES AND BALANCING COSTS: OVERALL RESULTS FROM SYSTEM STUDIES

The amount of additional reserve capacity and the corresponding costs associated with increasing wind power penetration are being explored in many countries by means of system studies carried out by power engineers. This involves the simulation of system operation, whereby the effect of increasing amounts of wind power is analysed for different scenarios of generation mix. In 2006, international cooperation was established under the IEA's Task 25 to compare and analyse the outcome of different national system studies. Task 25's first report provides general conclusions, based on studies from Denmark, Finland, Norway, Sweden, Germany, Ireland, Spain, The Netherlands, Portugal, the UK and the US.

Both the allocation and the use of reserves imply additional costs. The consensus from most studies carried out so far is that the extra reserve requirements needed for larger wind power penetrations are already available from conventional power plants in the system, so in fact no new reserves are required. This means that only the increased use of dedicated reserves, or increased part-load plant requirement, will create extra costs for the energy part.

The studies calculate additional costs, compared to a situation without wind power. Most results are based on comparing the costs of system operation without wind power and then adding varying amounts of wind power into the equation (see Figure II.3.1). The costs of variability are also addressed, by

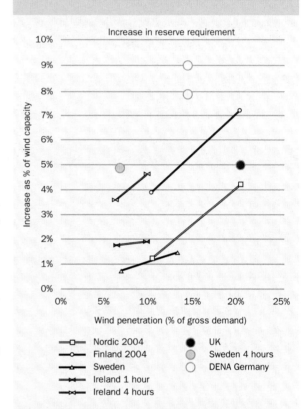

Figure II.3.1: Results for the increase in reserve requirement due to wind power, as summarised by IEA Task 25

Increase in reserve requirement

Legend:
- Nordic 2004
- Finland 2004
- Sweden
- Ireland 1 hour
- Ireland 4 hours
- UK (●)
- Sweden 4 hours (◉)
- DENA Germany (○)

Note: Major factors explaining the difference in results between various studies are assumptions with respect to forecast uncertainties (resulting from length of forecast horizon/gate closure time) and the geographical size of the area considered.

Source: Holttinen et al. (2007)

comparing simulations assuming constant (flat) wind energy to those with varying wind energy.

Estimates of the extra cost of reserves (mainly secondary load-following reserves) suggest €1–4/MWh for a wind power penetration of up to 10 per cent of gross consumption. This cost is normalised per MWh of wind energy produced. The cost per MWh at the consumption level is around €0.1–0.4/MWh at 10 per cent wind energy penetration, which is typically around 0.1 per cent of the electricity consumption price. These findings indicate that the additional system operation costs, in terms of balancing additional variability due to large-scale integration of wind power, are only a small fraction (typically less than 10 per cent) of the generation costs of wind power. The effect on the consumer price is close to zero.[5]

System Operation Aspects

TRANSMISSION LEVEL

Balancing and securing system operation by the transmission system operator (TSO) involves the use of transmission lines in the system area and interconnections to neighbouring systems. The issues include congestion management, priority access and priorities in curtailment in critical situations, such as low demand or high winds.

High penetration levels of wind power production affect the operation of the transmission system. Voltage control in the system may be required (for example near large wind farms) in order to cope with unwanted voltage changes, which might be enhanced by variable output wind power. This voltage support could be supplied by the wind farm if adequate wind energy technology were to be used; otherwise dedicated equipment would need to be installed, such as FACTS devices.[6]

Another issue is the management of power flows and possible congestion in the grid. Specific combinations of both the level and geographical location of wind power production and demand can cause changes in the size and direction of power flows in the transmission grid. This results in changing cross-border flows. In order to manage these cross-border power flows, TSOs also need high-quality wind forecasting tools. FACTS devices and phase-shifting transformers may be used for the management of power flows.

DISTRIBUTION LEVEL

Until now, the connection of wind power to the grid has usually been at distribution level. A particular feature of distribution grids is that there is no active management, for example, at transmission level. The distribution grids have to cope with greater distributed generation levels, without reducing the quality of supply to other customers.

The 'embedded generation' of wind power benefits the grid: weak grids may be supported by wind power, and users on the line may be better served, as wind power can help to control grid voltage. Power electronics of wind farms can also improve power quality characteristics in the grid. The power, if consumed within the distribution network, goes directly to the user and transmission costs can be reduced. Finally, depending on the grid code requirements of the relevant control area, wind power may maintain operations in parts of the system in the event of transmission failures, which would otherwise cause blackouts.

Adding wind power to distribution grids results in similar effects as in the transmission grid: a change in the direction and quantity of real (active) and reactive power flows, which may interact with the operation of grid control and protection equipment. The design and operation practices at the distribution level may need to be modified as additional distributed generation, such as wind power, is added. Distribution grids may have to become more 'actively managed', which would require the development of suitable equipment and design principles. However, the improved grid brings collateral benefits to the distribution grid operator.

WIND POWER CLUSTER MANAGEMENT

The pooling of several large wind farms into clusters in the GW range provides new options for optimising the integration of variable output generation into electricity supply systems. Concepts for cluster management (Rohrig et al., 2004; Estanqueiro et al., 2008) include the aggregation of geographically dispersed wind farms, according to various criteria, for optimised network management and (conventional) generation scheduling. The clusters will be operated and controlled in the same way as large conventional power plants.

The implementation of these operating methods will significantly increase wind energy's economic value to the system, by keeping additional costs for balancing to a minimum. Based on innovative wind farm operational control, a control unit between system operators and wind farm clusters (so-called 'wind farm cluster management', WFCM) will enable a profile-based generation. WFCM combines and adjusts wind plant control systems, based on forecasts, operating data, online-acquired power output and defaults from system operators.

Options for Increasing Power System Flexibility

The availability of flexible balancing solutions (generation capabilities, load management and energy storage) in power systems is an important facilitating factor for the integration of wind power. Even though power system balancing is not new, wind power provides new challenges at high penetration levels, since its variable nature requires additional flexibility in the power system – the capability to adequately respond to fast and significant net system load variations.

By increasing the flexibility of the power system, its ability to integrate variable output generation can be enhanced. In a more flexible system (for example systems with large amounts of hydro- or gas-powered electricity), the effort required to reach a certain wind energy penetration level can be lower than in a less flexible system (for example systems with a high share of nuclear power). In a system that covers a larger geographical area, a larger amount of flexibility sources are usually available. The differences in the size of power systems, dispatching principles and system flexibility explain the differences in integration costs in different countries. For example, Denmark has a high level of flexibility as it is well interconnected, thus enabling a high penetration level without significant additional costs. Portugal is another example of a flexible power system enabling easy and low-cost wind power integration, due to the large amount of fast responding, reversible hydropower plants in the system.

A serious consideration in the planning to integrate substantial amounts of wind power is the provision (flexibility sources) for additional flexibility needs in the system, compared to a situation without wind power. In the assessment of the required additional flexibility, a distinction has to be made in the different market timescales (hour/day ahead).

The main sources for additional flexibility are:

- fast markets (markets with short gate closure times);
- flexible generation (for example gas and hydro);
- demand-side management (DSM);
- energy storage; and
- interconnection.

Fast Markets

There is considerable diversity in European power market rules, but day-ahead markets exist in most countries. The day-ahead forecast error for wind has been significantly reduced in recent years, due to improved weather forecast models, but the error is still higher than for intraday forecasts. In the interest of wind power integration, gate closure times should be reduced, in order to minimise forecasting uncertainty, and therefore reducing last-minute balancing adjustments. Organising markets throughout Europe to operate faster and on shorter gate closure times (typically three hours ahead) would favour the economic integration of wind power.

A recent study (Milligan and Kirby, 2008), based on the situation in the state of Minnesota in the US, calculates the savings in balance power that could be achieved by balancing areas and assuming the presence of an energy market with a five-minute re-dispatch. In the hourly timescale, balance area consolidation reduces ramp requirements of balancing plants by 10 per cent, while in the five-minute times-cale this reduction is double – more than 20 per cent. This has considerable effects on the balancing costs, and thus on the integration of wind power.

Flexible Generation

Existing balancing solutions mostly involve conventional generation units: hydropower, pumped hydro and thermal units. Hydropower is commonly regarded as a very fast way of reducing power imbalance, due to its fast ramp-up and ramp-down rates. It also has a marginal cost, close to zero, making it a very competitive solution. Pumped hydro accumulation storage (PAC, see below) also allows energy storage, making it possible to buy cheap electricity during low-load hours and to sell it when demand and prices are higher.

Of course, thermal units are also commonly used for power system balancing (primary control and secondary control). In the category of thermal generation, gas-fired units are often considered to be most flexible, allowing rapid production adjustments. There is also potential in making existing power plants more flexible.

Storage Options

There is increasing interest in both large-scale storage implemented at transmission level and smaller-scale dedicated storage embedded in distribution networks. The range of storage technologies is potentially wide. For large-scale storage, PAC is the most common and best-known technology. PAC can also be set up underground.

Another large-scale technology option is compressed-air energy storage (CAES). On a decentralised scale, storage options include:
- flywheels;
- batteries (lead-acid and advanced), possibly in combination with electric vehicles;
- fuel cells (including regenerative fuel cells, 'redox systems');

- electrolysis (for example hydrogen for powering engine-generators or fuel cells); and
- super-capacitors.

An attractive solution would be the installation of heat boilers at selected combined heat and power (CHP) locations, in order to increase the operational flexibility of these units.

Storage involves a loss of energy. If a country does not have favourable geographical conditions for hydro reservoirs, storage is not the first solution to look at due to the poor economics at moderate wind power penetration levels (up to 20 per cent). In certain cases, it can even have an adverse effect on system operation with respect to CO_2 emissions (Ummels et al., 2008). In fact, the use of storage to balance variations at wind plant level is neither necessary nor economically viable.

Demand-side Management

With demand-side management (DSM), loads are controlled to respond to power imbalances by reducing or increasing power demand. Part of the demand can be time-shifted (for example heating or cooling) or simply switched off or on according to price signals. This enables a new balance between generation and consumption, without the need to adjust generation levels.

Today, the adjustment of generation levels is more common than DSM. The availability of this solution depends on load management possibilities (for example in industrial processes such as steel treatment) and the financial benefits offered by flexible load contracts (cost of power cuts and power increases versus lower bills). Attractive demand-side solutions in combination with decentralised storage are:
- heat pumps combined with heat boilers (at domestic or district level);
- cooling machines combined with cold storage; and
- plug-in electric vehicles.

Each of these solutions permits the separation of the time of consumption of electricity from the use of the appliance, by means of storage.

Interconnection

Available interconnection capacity for exchange of power between countries is a significant source of flexibility in a power system. However, the capacity should be technically as well as commercially available. Data on available interconnection capacities are published at www.etso-net.org. See also the section on interconnection on page 164.

II.4 GRID INFRASTRUCTURE UPGRADE FOR LARGE-SCALE INTEGRATION

European Transmission and Distribution Networks

Electricity networks can be split into two major subsections: transmission networks and distribution networks.

TRANSMISSION NETWORKS

The transmission network usually consists of high to very high voltage power lines designed to transfer bulk power from major generators to areas of demand; in general, the higher the voltage, the larger the transfer capacity. Only the largest customers are connected to the transmission network.

Transmission network voltages are typically above 100 kV. The networks are designed to be extremely robust, so they can continue to fulfil their function even in the event of several simultaneous network failures. Failure of a single element, such as a transformer or transmission line, is referred to as an 'N-1' event, and transmission systems should be capable of withstanding any such events. More complex cases of simultaneous failures of multiple elements (for example the failure of a transmission line when a parallel line has been disconnected for maintenance), are termed 'N-2' or similar. Transmission systems should also be capable of withstanding any such credible combinations.

Transmission consists mainly of overhead lines. Although underground lines offer the advantages of being less visually intrusive and raising less environmental objections, they incur higher initial investment costs and have a lower transmission capacity.

Transmission systems are operated by transmission system operators (TSOs) or independent system operators (ISOs). Responsibility for constructing or owning the physical network may belong to other organisations.

Transmission systems are actively managed through power system control centres, also known as dispatch centres. Balancing power entering and leaving the high voltage network, and reconfiguring the network to cope with planned and forced outages, is a 24-hour activity.

The European grid (Figure II.4.1) is divided into five synchronous regions and five relevant organisations: NORDEL (Organisation for the Nordic Transmission System Operators), BALTSO (Cooperation Organization of Estonian, Latvian and Lithuanian Transmission System Operators), UKTSOA (United Kingdom Transmission System Operators), ATSOI (Association of Transmission System Operators in Ireland) and UCTE (Union for the Co-ordination of Transmission of Electricity). Each of these organisations coordinates the TSOs involved at both operational and planning stages. The creation of the future European Network for Transmission System Operators for Electricity (ENTSO-E) will provide a new framework aimed at facilitating coordination between the different areas.

DISTRIBUTION NETWORKS

Distribution networks are usually below 100 kV and their purpose is to distribute power from the transmission network to customers. At present, with the exception of wind and other renewable power plants, little generation is connected to distribution networks, but this is changing rapidly, for example in Germany and Denmark.

Generation connected to distribution networks is often termed 'embedded generation' or 'distributed generation'. Distribution networks are less robust than transmission networks and their reliability decreases as voltage levels decrease. For example, a connection at 33 kV could expect to lose only a few minutes of connection per year on average, whereas a low voltage connection at 230 V for an individual domestic consumer

Figure II.4.1: Different synchronous regions in Europe

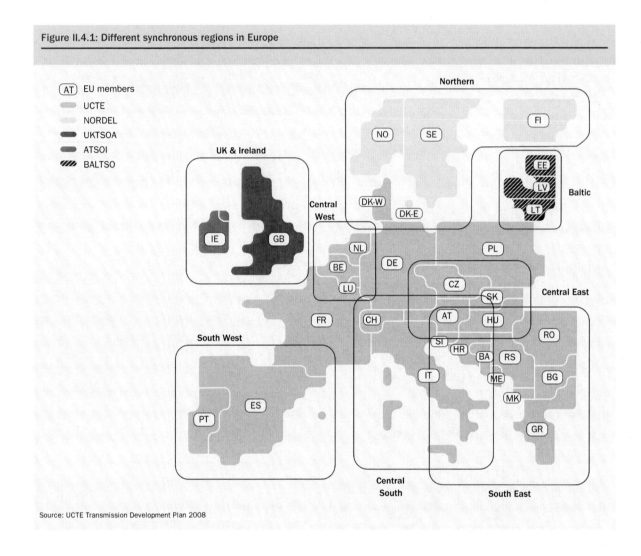

Source: UCTE Transmission Development Plan 2008

in a rural area would, on average, expect to lose at least an hour. As with transmission networks, distribution networks are operated (in some cases also owned) by distribution system operators (DSOs).

There is very little 'active' management of distribution networks. Rather, they assume a 'fit and forget' philosophy, in other words they are designed and configured on the basis of extreme combinations of circumstances (for example maximum demand in conjunction with high ambient temperatures, which reduce the capacity of overhead lines), to ensure that even in these extreme circumstances the network conditions experienced by customers are still within agreed limits.

Network Planning for Wind Power: Benefits of and Options for Increasing Transmission Capacity

THE NEED FOR IMPROVED NETWORKS

Liberalisation, market conditions, technology and the environment create fundamental changes and challenges for the European transmission and distribution networks. One of the major drivers is the emerging internal electricity market in Europe, which requires an adequate transport capacity between control regions and Member States to enable effective competition and trade of physical electricity. Therefore, enhancing the suitability of the grid for increased transnational and regional electricity transport is both in the interest of the wind industry and crucial for the development of the internal electricity market.

In addition, the specific nature of wind energy as a distributed and variable-output generation source requires specific infrastructure investments and the implementation of new technology and grid management concepts. The impacts of wind power on transmission, as described in Chapter II.2 (and see Figure II.2.1), are related to grid stability, congestion management, and transmission efficiency and adequacy. The large-scale integration of wind power requires a substantial increase in transmission capacity and other upgrade measures within and between the European Member States.

IMPROVING NETWORKS FOR INTEGRATING WIND POWER

The typical additional grid improvement measures required at increasing levels of wind power penetration can be classified into the following categories, in order of increasing effort and cost.

Soft Measures

In the short term, and at relatively low levels of wind power penetration, transmission upgrades coincide to a large extent with methods for congestion management and optimisation in the transmission system. Soft measures do not involve extensive expenditure, but rather avoid or postpone network investments.

The utilisation of existing power lines can often be increased by operating them at a higher capacity,

assisted by temperature monitoring. Improving the cross-border electricity exchange procedures, and thus the manner in which power is flowing between different countries, is also a method for alleviating congestion. If controllable power plants are available within the congested area, coordinated automatic generation control (AGC) may be applied. DSM, controlled according to the wind energy and transmission situation, is another option. Applying control systems that limit the wind power generation during critical hours should be considered as a last resort, because it is both environmentally and economically inefficient.

Investments Other Than the Construction of New Lines

At significant penetration levels, there is a need for additional voltage management in the grids, which can be achieved by devices such as FACTS[6] and also by the technical capabilities of the wind farms themselves, in particular with technologies that enable expanded MVAR capabilities.[7]

Studies in the UK (Strbac et al., 2007) have concluded that it may be preferable to insist on sufficient fault ride-through (FRT)[8] capability from large wind power plants. In certain cases, and in order to ensure power system security at higher penetration levels, this would be more economical than modifying the power system operation and not insisting on FRT capability from wind turbines.

It can be argued that the additional costs associated with the improved wind power plant capabilities at the wind farm level should also be materialised, such as in the 2008 amendment of the German Renewable Energy Law.

There are several ways in which the transmission capacity of the network can be increased (UCTE, 2008). These include:

- adding transformers to existing substations, thus enabling a higher load feed and in some cases evacuating higher generated power;

- upgrading assets: for example operating a line at higher voltage (within its design limits) or increasing the transmission capacity of a power line by tightening the conductors and reinforcing the towers;
- installing new facilities in grid substations to improve the distribution of power flows among different parallel paths and to fit better with the line capacities: for example series reactors, phase-shifting transformers or devices to increase voltage support (shunt reactive devices and static VAR compensators);
- improving the utilisation of existing assets when possible: for example replacing line conductors with high temperature conductors or adding a second circuit on an existing line (within the design limit of the towers); and
- replacing existing assets with those of a higher transmission capacity: for example replacing an existing 225 kV line with a 400 kV double-circuit line.

Construction of New Lines

Grid reinforcement is necessary to maintain adequate transmission as wind power penetration increases. This reinforcement is preceded by extensive power system analysis, including both steady-state load flow and dynamic system stability analysis. The construction of new lines is also a prerequisite for reaching regions with a high wind resource, for example offshore locations. Nowadays, in many areas of Europe, the construction of new overhead lines may take as long as 10 to 12 years from the initial concept to implementation, mainly because of lengthy planning and permission procedures.

Several studies at national and European level are now underway to back up the plans for upgrading the European transmission system in order to facilitate large-scale integration. The most important international studies are the European Wind Integration Study (EWIS) and TradeWind, which will provide

recommendations in 2009. Initiated in 2007, EWIS investigates the grid measures necessary to enable the wind power capacity foreseen for 2015 in a cooperative effort between European TSOs. The EWEA-coordinated project, TradeWind, started in 2006 and investigates the grid upgrade scenarios at the European level that would be necessary to enable wind energy penetration of up to 25 per cent, using wind power capacity scenarios up to 2030.

ENSURING ADEQUATE TRANSMISSION CAPACITY AND ACCESS FOR WIND POWER

From the above, it can be seen that in order to integrate wind power, sufficient transmission capacity needs to be available to carry the power to the demand centres. This capacity must be provided by transmission lines and a proper legal framework for accessing this capacity is required.

At the European level, two major initiatives contain basic elements of such a framework (see also 'European policy framework relevant for wind power integration' in Chapter II.1):

- The newly agreed European Renewable Energy Directive (2008) stipulates that national governments and TSOs should guarantee sufficient transmission capacity and fair access for renewables to the transmission network.
- The mandatory ownership unbundling of generation and transmission, as required by the proposed Third Energy Package (2008), should provide the legal basis to guarantee a level playing field with other generators.

In practice, the construction of the required network upgrades, especially new lines, is a very lengthy process. Also, because of the difference in speed between wind power development and transmission development, there is a need to implement fair access rules for cases where lines have to be shared between wind and other generators. As yet, there are no established rules at the European level and grid access for wind energy is presently solved in a rather pragmatic way. Some countries, such as Germany and Spain, take into account the recommendations from the 2001 RES Directive, and grant priority access to wind power to a certain extent. In practice, in case available grid capacity is limited, the principle of 'connect and manage' is often used.

INTEGRATING WIND POWER IN DISTRIBUTION NETWORKS

The addition of embedded generation, such as wind power, to distribution networks is quite common and was at the origin of wind power development in most countries. However, when wind generation reaches very high levels, it brings new challenges, for the following reasons:

- The distributed generation adds a further set of circumstances (full generation/no generation) with which the network must cope without negatively affecting the quality of supply seen by customers.
- The direction and quantity of real and reactive power flows change, which may affect the operation of network control and protection equipment at local level.
- Design and operational practices are no longer suitable and may need to be modified.

In contrast to these challenges, distributed generation systems (DGSs) also bring benefits to distribution networks, including:

- a reduction in network losses in many situations; and
- the avoidance of network reinforcement which would otherwise be required to achieve standards for quality of supply.

To address these issues, distribution networks may become more 'actively managed'. This implies cost

and requires the development of suitable equipment and design principles. Active management of the networks by DSOs may be assisted by introducing new concepts, such as 'clusters of wind farms' that aggregate and enable the monitoring of generation (see also page 161). In the future, it is expected that distributed wind generation will be fully controlled and operated as a virtual power station (VPS).

Transnational Offshore Grids

THE CASE FOR AN OFFSHORE GRID

The exploitation of the offshore wind potential in Europe brings new challenges and opportunities for European power transmission. The long-term European offshore potential amounts to up to 150 GW in 2030, according to EWEA estimates in 2008. The majority of the currently projected offshore wind plants will be situated close to the European coastlines, not further than 100 km from the shore, due to the high costs of grid connection, limited grid availability and the lack of regulatory frameworks. Looking at the North Sea alone, with a potential of hundreds of GW of wind power, an offshore grid connecting different Member States would enable the transfer of this wind power to load centres and, at the same time, facilitate the competition and the trade of electricity between countries. A multi-terminal offshore grid would reach offshore wind plants far from shore, as foreseen for German and UK waters.

The project developer Airtricity introduced the offshore Supergrid® concept in 2005. Supergrid® combines the following:

- the connection of offshore wind plants;
- the balancing of wind power variations; and
- the provision of transmission of electricity between different markets for cross-border trade.

A commercial proposal has been worked out for a first phase of 10 GW of wind power – the construction

Figure II.4.2: Vision of high voltage 'super grid' to transmit wind power through Europe

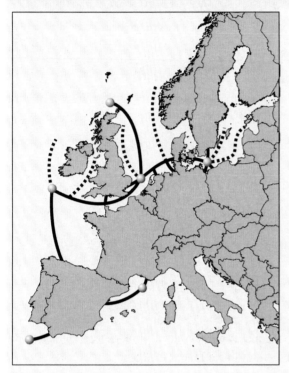

Source: Dowling and Hurley (2004)

Figure II.4.3: Offshore grid proposal by Statnett

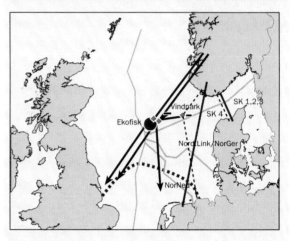

Source: Statnett (2008)

of an offshore 'super grid' which would be on a modular basis. The fact that wind farms will be able to operate collectively at variable speed and frequency, independent of the land-based grid, is expected to optimise turbine generating efficiency and offset losses incurred as a result of the increased transmission distances.

At first, large, multi-GW offshore arrays would connect to nearby networks, before being modularly extended and ultimately interconnected. A further advantage of this system will be the full controllability of power flows, eventually allowing an 'all-European' market for electricity, including 'firm' wind power.

Presently, the idea of a transnational offshore grid is being addressed by several other parties. The

Norwegian TSO Statnett proposes the progressive development of a grid linking Scandinavia with UK, Germany and The Netherlands (Figure II.4.3). On its way, it would connect offshore wind farms, as well as existing offshore oil installations that need to reduce their CO_2 emissions. The technology to be used is a HVDC VSC (see 'Ensuring adequate transmission capacity and access for wind power' above).

Greenpeace (Woyte et al., 2008) has studied the concept of an offshore grid serving electricity trade between European countries around the North Sea and at the same time providing transmission of up to 70 GW of offshore wind power capacity – a target that could be achieved between 2020 and 2030 (Figure II.4.4). The study has also evaluated the smoothing effect of aggregating the offshore wind power using such a grid. The offshore grid topology proposed seeks the maximum synergy between existing plans and reinforcements, aiming to improve the cross-border exchange between countries. For example, it includes the East Connector in the UK, which alleviates the heavy north–south congestions, as well as an offshore connection along the French, Belgian and Dutch coasts.

Figure II.4.4: Offshore grid examined in the Greenpeace study

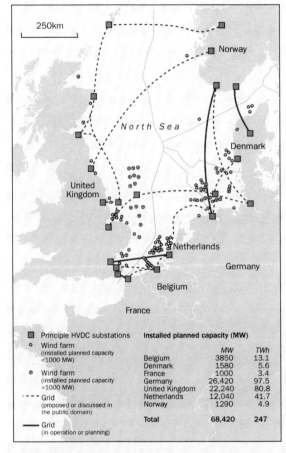

250km

Norway

North Sea

Denmark

United
Kingdom

Netherlands

Germany

Belgium

France

	Installed planned capacity (MW)		
■ Principle HVDC substations			
○ Wind farm (installed planned capacity <1000 MW)		MW	TWh
	Belgium	3850	13.1
	Denmark	1580	5.6
	France	1000	3.4
◉ Wind farm (installed planned capacity >1000 MW)	Germany	26,420	97.5
	United Kingdom	22,240	80.8
- - - Grid (proposed or discussed in the public domain)	Netherlands	12,040	41.7
	Norway	1290	4.9
—— Grid (in operation or planning)	Total	68,420	247

Source: Woyte et al. (2008)

The effects of grid configurations described above on the power flows in the European transmission system are being analysed in the TradeWind project. A common element to all these proposals is the fact that the offshore grid would provide multiple functions and serve the functioning of the European electricity market. Therefore, it should be considered as an extension of the existing onshore grid, falling under the responsibility of the various governments, TSOs and regulators involved.

TECHNICAL SOLUTIONS FOR THE OFFSHORE GRID

Compared to onshore sites, offshore wind farms will have large power capacities and be comparable in size to conventional power plants, typically in excess of 400 MW. Modern transmission technologies operating at high and extra high voltage levels will be required to transmit high levels of power over longer distances. Two main types of offshore transmission systems exist, based on either alternating or direct currents (HVAC or HVDC).

For wind farms close to shore, the HVAC system offers the best solution, as it provides the simplest, least expensive and proven technology for grid connection and is similar to the transmission network used on land. However, as transmission distances increase, the losses from the HVAC system increase significantly. To avoid ineffective operation, AC cable length should be limited to a length of approximately 120 km. HVDC technology offers a number of advantages, but has a distinct investment cost structure, as it involves the installation of expensive converter stations. The break-even distance (now around 90 km) depends on the cost developments, and will move closer to the shore as HVDC system costs decrease.

Conventional thyristor-based HVDC technology has generally been used for point-to-point power transmission. Offshore wind farm arrays would benefit from a multi-terminal transnational offshore grid system. Recent advances in HVDC technology, using insulated gate bipolar transistor (IGBT)-based converters, seem to offer a solution and facilitate the cost-effective construction of multi-terminal HVDC networks. These modern HVDC-IGBT systems offer clear technological advantages, especially in the area of controllability and efficiency. A specific advantage of HVDC systems is reactive power control capability, favouring grid integration and system stability. The technical and economic aspects of offshore transmission systems are being actively investigated by the supply industry and by electric power companies in order to be ready with the most cost-effective solutions when large-scale offshore wind power takes off.

Coordinated Network Planning at the European Level

Besides the grid upgrades required in various countries in view of the integration of wind power, there is also the need for a coordinated effort in network planning at the European level. At higher wind energy penetration levels, cross-border power flow will increase, which:

- reduces wind power variability;
- improves the predictability of wind power;
- reduces balancing costs; and
- increases wind energy's contribution to the generation adequacy.

However, this increased cross-border flow increases the need for coordinated network planning and common technical regulations. In this respect, there are several relevant ongoing initiatives and developments:

- The Trans-European Energy Networks (TEN-E) programme was set up to promote the improvement of interconnection in Europe. Coordinated by the European Commission (DG TREN), TEN-E focuses on those network aspects that improve electricity trade within Europe. Several bottlenecks in the transmission system have been identified by TEN-E as projects of European interest. Despite the launch of this programme, however, though with a few exceptions, there has been little progress in increasing interconnection capacity. In 2007, as part of the measures proposed in the Priority Interconnection Plan of the European Energy Policy, a specific European Coordinator was appointed by the European Commission with the mandate to mediate in the interconnection projects required to enable the integration of wind power in northern Europe.

- As recommended in the Third Energy Package in 2008, the European Transmission Operators founded a new organisation, the European Network of Transmission System Operators for Energy (ENTSO-E), with the aim of improving the coordination of their operations. Another Third Package requirement is the obligation for joint TSOs to publish a transmission development plan (UCTE, 2008) on a regular basis.

- Within the EWIS study (FP6), The European TSOs are currently examining the technical and market aspects that will arise from wind power integration (time horizon 2015) in the European transmission system. This should result in recommendations for transmission planning at the European level. The TradeWind project coordinated by EWEA, which runs in parallel with EWIS, will provide recommendations for interconnection and power market improvements with a time horizon of up to 2030, at which point 300 GW of wind power is expected to be integrated in Europe. Having established a common platform for the exchange of information and findings, both EWIS and TradeWind will provide quantitative input for coordinated transmission planning for the future of wind power.

II.5 GRID CONNECTION REQUIREMENTS

Regulatory and Legal Background

All customers connected to a public electricity network, whether generators or consumers, must comply with agreed technical requirements, in order for the network to operate safely and efficiently. Electricity networks rely on generators to provide many of the control functions, and so the technical requirements for generators are unavoidably more complex than for demand customers.

These technical requirements are often termed 'grid codes', though the term should be used with care, as there are often different codes, depending on the voltage level of connection, or the size of the project. Also, there may be technical requirements that are not referred to in the grid code, but which apply to the project through the connection agreement or the power purchase agreement or in some other way.

The purpose of these technical requirements is to define the technical characteristics and obligations of generators and the system operator, meaning that:

- electricity system operators can be confident that their system will be secure regardless of the generation projects and technologies applied;
- the amount of project-specific technical negotiation and design is minimised;
- equipment manufacturers can design their equipment in the knowledge that the requirements are clearly defined and will not change without warning or consultation;
- project developers have a wider range of equipment suppliers to choose from;
- equivalent projects are treated fairly; and
- different generator technologies are treated equally.

In the past, with vertically integrated utilities, the same organisation was responsible for the planning and operation of networks and generators, so the technical requirements did not need to be particularly clearly defined or fair. Nowadays, in order to avoid distortions of competition and to comply with a liberalised energy market in Europe, there is a trend towards the legal separation of generators and system owners/operators. As a result, the technical requirements governing the relationship between generators and system operators need to be more clearly defined. The introduction of renewable generation has often complicated this process significantly, as these generators have physical characteristics that are different from the directly connected synchronous generators used in large conventional power plants. In some countries, this problem has caused significant delays in the development of fair grid code requirements for wind generation.

In some countries, a specific grid code has been produced for wind farms, and in others, the aim has been to define the requirements as far as possible in a way which is independent of the power plant technology.

There are benefits to requirements that are as general as possible, such as treating all projects equally. However, this can result in small projects facing the same requirements as the largest projects, which may not be technically justifiable.

Requirements are usually written by the system operator, often overseen by the energy regulator body or government. The requirement modification process should be transparent and include consultation with generators, system users, equipment suppliers and other affected parties.

Wind Power Plant Capabilities

Wind turbine technology is discussed in detail in Part I. Turbine technology is also covered briefly in this section, to provide some background to the information that follows.

The traditional Danish stall-regulated wind turbine concept uses an induction generator. Its rotational speed is fixed by the frequency of the electricity network to which it is connected. The blades are fixed – in other words do not pitch – so the output power and structural loads in high winds are limited by passive stall regulation. Unfortunately, this concept, though cheap, simple and reliable, has several negative effects on the electricity network:

- lack of power control, meaning that system frequency cannot be controlled; this is achieved relatively simply by conventional power plants;
- limited control of reactive power, making it more difficult to control network voltages; and
- during network disturbances (such as a sudden fault on the network), a wind turbine is likely to aggravate the situation.

The fixed-speed, pitch-regulated concept (in other words the possibility to control active power output by pitching the blades) resolves the first of these issues, and the limitations of the stall-regulated wind turbine concepts can be mitigated with the addition of terminal equipment in the substation.

The development of variable-speed wind turbines, using power electronic converters, was undertaken largely to reduce mechanical loads. This introduces additional control of reactive power as a by-product, and in the majority of cases also reduces the wind turbine's effect on the network during a sudden fault.

The larger the power electronic converter (relative to the size of the wind turbine), the greater the control over reactive power. So variable-speed, pitch-regulated wind turbines, based on the full-converter principle, now allow the desired control of wind turbines within the required limits.

The currently available wind turbines do not make full use of this capability, however, and grid codes do not yet take advantage of the full capabilities. As wind penetration increases, and network operators gain experience with the behaviour of their systems, grid codes will possibly become more demanding. New technical requirements should be based on:

- a detailed assessment of requirements;
- the technical potential of all the power plant's technology; and

- the optimal way in which to meet these demands, both technically and economically.

Grid Codes and Essential Requirements for Wind Power Plants

The arrangement of the technical requirements within grid codes and related documents varies between electricity systems. However, for simplicity the typical requirements for generators can be grouped as follows:

- tolerance – the range of conditions on the electricity system for which wind farms must continue to operate;
- control of reactive power – often this includes requirements to contribute to voltage control on the network;
- control of active power;
- protective devices; and
- power quality.

It is important to note that these requirements are often specified at the point of common coupling (PCC) between the wind farm and the electricity network. In this case, the requirements are placed at the wind farm level, and wind turbines may be adapted to meet them. Often, wind turbine manufacturers specify the performance of their wind turbines, rather than that of the entire wind farm.

It is also possible for some requirements to be met by providing additional equipment, separate from the turbines; this is indicated, where relevant, in the following discussion.

TOLERANCE

The wind farm must continue to operate between minimum and maximum voltage limits. Usually, this is stated as steady-state quantities, though short-term limits are not unknown (in other words a wider range may apply for a limited duration).

The wind farm must also continue to operate between minimum and maximum frequency limits. Often there is a range which applies continuously, and several further more extreme short-term ranges.

In systems with relatively high wind penetration levels, a common requirement is for wind farms to continue to operate during severe system disturbances, during which the voltage can drop to very low levels for very short periods of time. This is termed 'fault ride-through' (FRT) or 'low voltage ride-through'. The requirements can be complex, and depend on the technical characteristics of the electricity system. Proving compliance with the requirements may be difficult and costly.

It is feasible to use wind turbines that do not comply with FRT requirements, and to meet these requirements by installing additional equipment which can produce or consume reactive power at turbine level, or centrally within the wind farm.

REACTIVE POWER CONTROL

Reactive power production and consumption by generators allows the network operator to control voltages throughout their system. The requirements can be stated in a number of ways.

The simplest is the fixed power factor. The wind farm is required to operate at a fixed power factor when generating, often equal to 1. Often, the required accuracy and integration intervals for the verification of the power factor are not stated. And the fixed value may be changed occasionally, for example for winter and summer or peak and no-load periods.

Alternatively, the wind farm may have to adjust its reactive power consumption or production in order to control the voltage to a set point. This is usually the voltage at the PCC, but other locations may be specified. There may be requirements on the accuracy of control and the speed of response. Fast control may be difficult to achieve, depending on the capabilities of the wind farm's SCADA communications system.

Some wind turbine designs can fulfil these functions, even when the wind turbine is not generating. This is potentially a very useful function for network operators, but is not yet a common requirement.

FRT requirements can be met with central reactive power compensation equipment.

ACTIVE POWER CONTROL

With pitch-regulated turbines, it is possible to reduce the output at any moment by pitching the blades. This could also be done with stall-regulated turbines, by shutting down individual turbines within a wind farm. Although this only provides relatively crude control, the output from the power system operator's point of view is effective and valuable.

All forms of active power control in a wind turbine require a reduction in output power, which means a reduction in revenue. This is less of an issue for conventional power stations, where the lost revenue will be compensated, to some extent, by a reduction in fuel cost. Therefore, system operators and energy regulators recognise that a reduction in wind farm output should be used as a last resort.

The simplest method is a cap, which means that the wind farm (or a group of wind farms) is instructed to keep its output below a certain level. A more complex version of the cap is to insist that output is kept at a fixed level (delta), below the unconstrained output available from wind.

In parallel with a cap, the wind farm may also be instructed to control ramp rate, in other words to limit the rate at which the output power can increase (due to increasing wind speed or turbines returning to service after some outage). The ramp rate is defined over periods of, for example, one minute or ten minutes. This limits the network operator's demands on other forms of generation to adjust output rapidly.

Clearly, it is not possible for wind generation to control automatically the 'negative ramp rate' if the wind drops suddenly. However, with good wind forecasting tools, it is possible to predict a reduction in wind speed in advance; the output of the wind generation can then be gradually reduced in advance of the wind speed reduction, thereby keeping the negative ramp rate at an acceptable level.

In systems with relatively high wind penetration, there is often a requirement for frequency response or frequency control. This can take many forms, but the basic principle is that, when instructed, the wind farm reduces its output power by a few per cent, and then adjusts it in response to the system frequency. By increasing power when frequency is low, or decreasing power when frequency is high, the wind farm can contribute to controlling the system frequency.

PROTECTIVE DEVICES

Protective devices, such as relays, fuses and circuit breakers are required to protect the wind farm and the network from electrical faults. Careful coordination may be needed to ensure that all conceivable faults are dealt with safely, and that correctly functioning equipment is not disconnected unnecessarily.

Short-circuit (or fault) current is a related issue. In the event of an electrical fault on the network close to the wind farm, some short-circuit current will flow from the wind turbines into the fault. Requirements may exist on the maximum or minimum permitted levels.

POWER QUALITY

This term covers several separate issues (IEC, 2008). There are usually limits on the harmonic currents that the wind farm can introduce into the network, and in this area, detailed analysis can be difficult. Ideally the existing background harmonics on the network should be established, but these are often unknown.

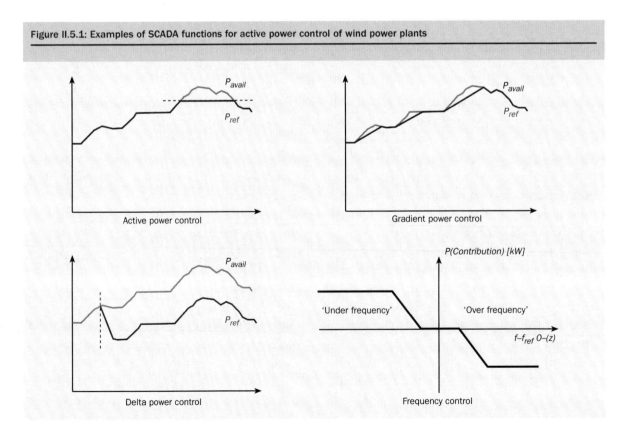

Figure II.5.1: Examples of SCADA functions for active power control of wind power plants

On weak networks, voltage steps, caused by wind turbines starting or stopping, or the energisation of transformers, can be a problem. A related problem is voltage flicker, which can be caused by wind turbines starting or stopping, or even when they are in continuous operation.

FUTURE DEVELOPMENTS

As noted above, as wind penetration increases, future technical requirements may well become more onerous.

One possible requirement is for an 'inertia function'. The spinning inertias of a conventional power plant provide considerable benefit to the power system by acting as a flywheel, and thereby reducing the short-term effects of differences in supply and demand. Variable speed wind turbines have no such equivalent effect, but theoretically their control systems could provide a function that mimics the inertia effect.

There may also be a move towards markets for control services, rather than mandatory requirements. This would make sense economically, as the generator best able to provide the service would be contracted. Also, due to the very low marginal cost of renewable energy sources, such as wind energy, it would be more environmentally and economically effective. For example, if a wind farm provided a useful service to the network operator in terms of voltage control (in other words it did more than just make up for its negative effects), then the wind farm could be paid for this service. This may be cheaper than other options available to the network operator.

HARMONISATION OF GRID CODES

The way in which grid code requirements in Europe have developed has resulted in gross inefficiencies and additional costs for consumers, manufacturers and wind farm developers. With the increasing penetration of wind energy, there is an increasing need to develop a harmonised set of grid code requirements. Harmonised technical requirements will maximise efficiency for all parties, and should be employed wherever possible and appropriate. However, it is not practical to completely harmonise technical requirements immediately, since this could lead to the unnecessary implementation of the most stringent requirements from each Member State, which would not be efficient or economically sound.

EWEA has established a Grid Code Working Group among its members. The group consists of wind turbine manufacturers, wind farm operators, service providers, certification bodies and engineering companies. There is a consensus in the industry that there is an urgent need to carry out a harmonisation exercise, as wind penetration is forecast to increase significantly in the short to medium term. The working group is working on a two-step approach:

1. a structural harmonisation exercise, with the aim of establishing a grid code template with common definitions, parameters, units and figures, as well as a common structure; and
2. a technical harmonisation exercise, with the aim of adapting existing grid code parameters to the new grid code template.

This harmonisation strategy will be of particular benefit to:
- manufacturers, who will be required to develop only common hardware and software platforms;
- developers, who will benefit from reduced costs; and
- system operators, especially those who have yet to develop their own grid code requirements for wind power plants.

The technical basis for the requirements will be further developed by TSOs and the wind power industry.

II.6   WIND POWER'S CONTRIBUTION TO SYSTEM ADEQUACY

This chapter discusses the extent to which installed wind power capacity statistically contributes to the guaranteed generation capacity at peak load. This firm capacity part of the installed wind capacity is called 'capacity credit', and is relevant since total wind power capacity will be a substantial fraction of the total generation capacity. At the European level, this will represent 30–40 per cent of total generating capacity, corresponding to the wind power targets for 2020 and 2030. However, in 2008, wind power capacity only represents around 10 per cent of European generation capacity.

Substantial amounts of new capacity need to be built in the coming decades to meet increasing demand in Europe and to replace old plants. UCTE estimates that, by 2015, generating capacity in its area will increase by 90 GW, with 60 GW coming from renewables, the majority of which will be wind power.

As for all renewable sources that cannot be stored, wind has a capacity credit that is lower than that of conventional generation technologies. However, there is a certain amount of firm wind capacity that contributes to the adequacy of the power system. Before expanding on the capacity credit of wind power, a brief explanation will be given of the current methods of estimating power system adequacy.

Security of Supply and System Adequacy

The peak demand (or peak load) of electricity in Europe is constantly increasing. Over the coming years, UCTE expects a rise in peak demand of around 1.6–1.7 per cent per year (compared with 2 per cent up to 2007) (Figure II.6.1). The peak demand is a strategic parameter, since it determines the required generating and transmission capacities. As a matter of convention, for system design purposes, peak load values at specific points during the year – in January and July – are considered.

The way in which the power system can match the evolution in electricity demand is expressed as 'system adequacy'. System adequacy measures the ability of a power system to cope with its load in all the steady states it may operate in under standard conditions. This adequacy has different components:

- the ability of the generation assets to cover the peak load, taking into account uncertainties in the generation availability and load level; and
- the ability of the transmission system to perform, considering the flexibility provided by interconnection and import and export flows.

System operators are responsible for maintaining system adequacy at a defined high level. In other words, they should ensure that the generation system is able to cover the peak demand, avoiding loss-of-load events, for a given security of supply. The various national regulations regarding this 'security of supply' range from a 99 per cent security level (in 1 out of 100 years the peak load cannot be covered, such as in Germany) to 91 per cent (1 event in 10 years, such as in the UK).

As the whole European system is interconnected, it is logical for national TSOs to harmonise their approaches towards system adequacy. This is addressed mainly by the larger systems, such as UCTE, the Nordic system, and the British and Irish systems. The assessment methods of generation adequacy can be deterministic or probabilistic, or a combination of both. Even from a national point of view, the system adequacy assessment involves transnational issues. This is because, at the moment of peak load, it may be necessary to have access to power produced by a neighbouring country, so the transmission system should be able to carry and direct these transnational power flows.

The UCTE system's adequacy is being annually reviewed over a period of ten years. Generation adequacy assessment is based on the estimation of 'remaining capacity', which can be interpreted as:

- the capacity needed by the system to cover the difference between the peak load of each country and

Figure II.6.1: Overview of growth in electricity demand in the UCTE area

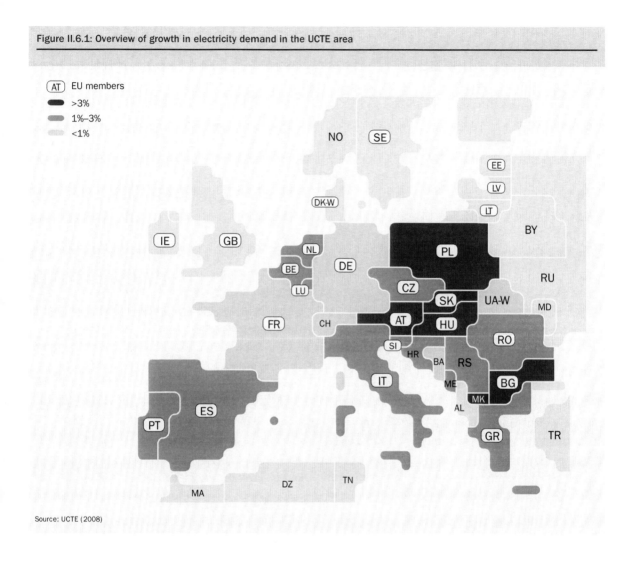

Source: UCTE (2008)

the load at the UCTE synchronous reference time ('margin against peak load'); or

- exceptional demand variation and unplanned outages that the system operators have to cover with additional reserves.

Generation adequacy assessment underscores how each country could satisfy its interior load with the available national capacity. Transmission adequacy assessment then investigates whether the transmission system is large enough to enable the potential imports and exports resulting from various national power balances, thus improving the reliability of the European power system.

In the Nordel zone, TSOs still conduct these reviews, but theoretically the electricity market price signals are considered sufficient to trigger the building of new capacity to fulfil adequacy needs. As long as the results of the reviews are positive, there is no need to keep reserves in the power system. However, in many countries, a number of contracts are drawn up to ensure that there is spare capacity available in extreme loading situations, often with older plants or loads that can be switched on or off in critical situations.

In the adequacy estimation, each power plant is assigned a typical capacity value. This takes into account scheduled and unscheduled outages. There are no plants with a capacity value of 100 per cent, since there is always the possibility that capacity will not be available when required. In its forecast, UCTE is looking at increasing shares of wind power in the coming years. It is clear from the UCTE system adequacy forecasts that there is not yet a national TSO standard for the determination of wind power's capacity credit.

Capacity Credit of Wind Power

CAPACITY CREDIT IS THE MEASURE FOR FIRM WIND POWER

The contribution of variable output wind power to system security, that is the capacity credit of wind, is estimated by determining the capacity of conventional plants displaced by wind power, whilst maintaining the same degree of system security, in other words an unchanged probability of failure to meet the reliability criteria for the system. Alternatively, it is estimated by determining the additional load that the system can carry when wind power is added, maintaining the same reliability level.

Many national wind integration studies have been giving special attention to the capacity credit of wind, as in some ways it is a 'synthetic' indicator of the potential benefit of wind as a generator in the system. Sometimes the capacity credit of wind power is measured against the outage probabilities of conventional plants.

CAPACITY CREDIT VALUES OF WIND POWER

Despite the variations in wind conditions and system characteristics among the European countries and regions, capacity credit calculations are fairly similar (Giebel, 2005).

For low wind energy penetrations levels, the relative capacity credit of wind power (that is 'firm' capacity as a fraction of total installed wind power capacity) will be equal or close to the average production (load factor) during the period under consideration, which is usually the time of highest demand. For north European countries, this is at wintertime and the load factor is typically at 25–30 per cent onshore and up to 50 per cent offshore. The load factor determining the capacity credit in general is higher than the average yearly load factor.

With increasing penetration levels of wind energy in the system, its relative capacity credit reduces. However, this does not mean that less conventional capacity can be replaced, but rather that a new wind plant added to a system with high wind power penetration levels will substitute less than the first wind plants in the system. This is illustrated in Figure II.6.2, where the relative capacity credit tails off towards a value depending mainly on the minimum load factor.

Table II.6.1 summarises the factors leading to higher or lower levels of capacity credit. Figure II.6.2, which is based on calculations in the DENA 1 study (DENA, 2005), shows the effect on capacity credit of expected improved load factors in Germany resulting from improved wind power technology (more efficient rotors) and the use of sites with higher wind speeds (offshore).

Wind power thus displaces conventional capacity in the system. The fraction of wind power that displaces conventional capacity may be limited, but the corresponding absolute capacities are significant. The aggregated capacity credit of the wind power plants in a system depends on many factors. Major decisive factors depend on the power system being considered (reliability level and flexibility of the generation mix) and the penetration level of wind power in the system. Other factors are related to wind and wind technology, such as the average capacity factor[9] and the geographical dispersion of wind plants in the system. The relative capacity credit decreases from a value approximately equal to the load factor at high load (25–35 per cent) for

Figure II.6.2: Relationship of installed wind power and capacity credit in Germany

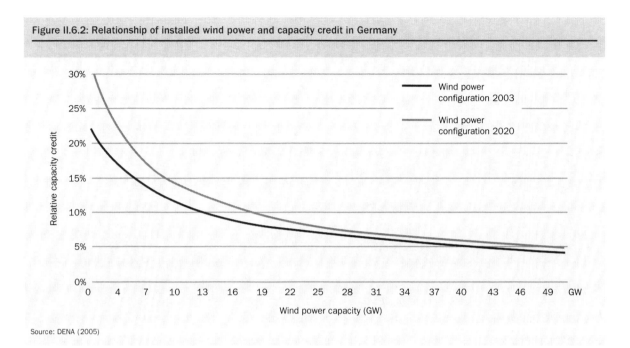

Source: DENA (2005)

Table II.6.1: Factors affecting positively and negatively the value of the capacity credit of a certain amount of wind power in the system

Higher capacity credit (%)	Lower capacity credit (%)
Low penetration of wind power	High penetration of wind
Higher average wind speed; high wind season when demand peaks	Lower average wind speeds
Lower degree of system security	High degree of system security
Higher wind power plant (aggregated) capacity factor or load factor (determined by wind climate, plant efficiency and specific rated power per m²)	Lower aggregated capacity factor (or load factor) of wind power
Demand and wind are correlated	Demand and wind uncorrelated
Low correlation of wind speeds at the wind farm sites (often related to large size of area considered)	Higher correlation of wind speeds at wind farm sites; smaller areas considered
Good wind power exchange through interconnection	Poor wind power exchange between systems

Source: EWEA

low penetrations to approximately 10–15 per cent at high penetrations.

Although wind power has a capacity credit both physically and technically, this characteristic currently has no value in the power market as wind power producers are not generally rewarded for providing firm capacity in the system.

II.7   ECONOMIC ASPECTS: INTEGRATION COSTS AND BENEFITS

The introduction of significant amounts of wind energy into the power system brings a series of economic impacts, both positive and negative. At power system level, two main factors determine wind energy integration costs:

1. balancing needs; and
2. grid infrastructure.

The additional balancing cost in a power system arises from the inherent variable nature of wind power, requiring changes in the configuration, scheduling and operation of other generators to deal with unpredicted deviations between supply and demand. Here, we demonstrate that there is sufficient evidence available from national studies to make a good estimate of such costs (see page 168). Furthermore, they are fairly low in comparison with the generation costs of wind energy and the overall balancing costs of the power system.

Network upgrade costs are necessary for a number of reasons. First, additional transmission lines and capacity need to be provided to reach and connect existing and future wind farm sites and to transport power flows in the transmission and distribution networks. These flows result both from an increasing demand and trade of electricity and from the rise of wind power. At significant wind energy penetrations, depending on the technical characteristics of the wind projects and trade flows, the networks must also be adapted to improve voltage management. Furthermore, limited interconnection capacity is often a barrier for optimally capturing the benefits of the continental nature of the wind resource, other renewable energy sources and electricity trade in general. In this respect, any infrastructure improvement will provide multiple benefits to the system, and therefore its cost should not be allocated only to wind power generation.

The cost of modifying the power system with significant amounts of wind energy increases in a linear fashion, and identifying its 'economic optimum' is not evident, as costs are accompanied by benefits. From the studies carried out so far, and the extrapolation of their results to high penetration levels, it is clear that the integration of more than 20 per cent of wind power into the EU power system would be economically as well as environmentally beneficial.

Additional Balancing and Network Costs

ADDITIONAL BALANCING COSTS

Additional balancing requirements in a system depend on a whole range of factors, including:

- the level of wind power penetration in the system, as well as the characteristic load variations and the pattern of demand compared with wind power variations;
- geographical aspects, such as the size of the balancing area, geographical spread of wind power sites and aggregation;
- the type and marginal costs of reserve plants (such as fossil and hydro);
- costs and characteristics of other mitigating options present in the system, such as storage;
- the possibility of exchanging power with neighbouring countries via interconnectors; and
- the operational routines of the power system, for example how often the forecasts of load and wind energy are updated (gate closure times) and the accuracy, performance and quality of the wind power forecast system used.

At wind energy penetrations of up to 20 per cent of gross demand, system operating cost increases by about €1–4/MWh of wind generation. This is typically 10 per cent or less of the wholesale value of wind energy. Note that these figures refer to balancing costs; costs given earlier (page 168) were for the reserves requirement.

Figure II.7.1 illustrates the costs from several studies as a function of wind power penetration. Balancing

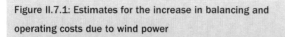

Figure II.7.1: Estimates for the increase in balancing and operating costs due to wind power

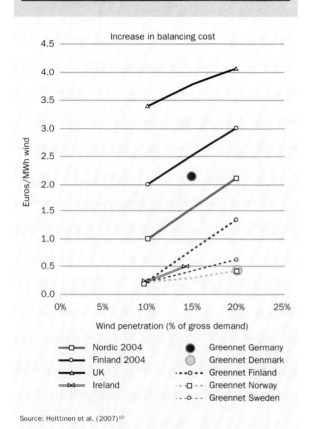

Source: Holttinen et al. (2007)[10]

costs increase on a linear basis with wind power penetration; the absolute values are moderate and always less than €4/MWh at 20 per cent level (and more often in the range below €2/MWh).

There are several major contributing factors to lower balancing costs:

- **Larger areas**: Large balancing areas offer the benefits of lower variability. They also help decrease the forecast errors of wind power, and thus reduce the amount of unforeseen imbalance. Large areas favour the pooling of more cost-effective balancing resources. In this respect, the regional aggregation of power markets in Europe is expected to improve the economics of wind energy integration. Additional

and better interconnection is the key to enlarging balancing areas. Certainly, improved interconnection will bring benefits for wind power integration, and these are presently quantified by studies such as TradeWind.

- **Reducing gate closure times**: This means operating the power system close to the delivery hour. For example, a re-dispatch, based on a four- to six-hour forecast update, would lower the costs of integrating wind power compared to scheduling based on only day-ahead forecasts. In this respect, the emergence of intraday markets is good news for the wind energy sector.

- **Improving the efficiency of the forecast systems**: Balancing costs could be decreased if the wind forecasts could be improved, leaving only small deviations to the rest of the power system. Experience in different countries (Germany, Spain and Ireland) has shown that the accuracy of the forecast can be improved in several ways, ranging from improvements in meteorological data supply to the use of ensemble predictions and combined forecasting. In this context, the forecast quality is being improved by making a balanced combination of different data sources and methods in the prediction process.

ADDITIONAL NETWORK COSTS

The consequences of adding more wind power into the grid have been analysed in several European countries (see, for example, Holttinen et al., 2007) (Table II.7.1). The national studies quantify grid extension measures and the associated costs caused by additional generation and demand in general, and by wind power production in particular. The analyses are based on load flow simulations of the corresponding national transmission and distribution grids and take into account different scenarios for wind energy integration using existing, planned and future sites.

It appears that additional grid extension/reinforcement costs are in the range of €0.1–5/MWh wind,

Table II.7.1: Grid upgrade costs from selected national system studies

Country	Grid upgrade costs, €/kW	Installed wind power capacity, GW	Remarks
Portugal	53–100	5.1	Only additional costs for wind power
The Netherlands	60–110	6.0	Specifically offshore wind
UK	45–100	8.0	
UK	62–85	26.0	20% wind power penetration
Germany	100	36.0	DENA 1 study

Source: Holttinen et al. (2007)

typically around 10 per cent of wind energy generation costs for a 30 per cent wind energy share. As for the additional balancing costs, the network cost increases with the wind penetration level. Grid infrastructure costs (per MWh of wind energy) appear to be around the same level as additional balancing costs for reserves in the system to accommodate wind power.

The costs of grid reinforcement due to wind energy cannot be directly compared, as circumstances vary significantly from country to country. These figures also tend to exclude the costs for improving interconnection between Member States. This subject is now being investigated by the TradeWind project (www.trade-wind.eu/), which investigates scenarios up to 2030.

Allocating Grid Infrastructure Costs

There is no doubt that the transmission and distribution infrastructure will have to be extended and reinforced in most EU countries when large amounts of wind power are connected. However, these adaptations are necessary not only to accommodate wind power, but also to connect other electricity sources to meet the rapidly growing European electricity demand and trade flows.

However, the present grid system is not used to its full capacity and present standards and practices of transmission lines by TSOs are still largely based on the situation before wind energy came into the picture. As wind power is producing in a whole range of partial load states, wind farms will only utilise the full rated power transmission capacity for a fraction of the time. In some cases, where there is adjustable power production (such as hydropower with reservoir), the combination of wind and hydro can use the same transmission line.

The need to extend and reinforce the existing grid infrastructure is also critical. Changes in generation and load at one point in the grid can cause changes throughout the system, which may lead to power congestion. It is not possible to identify one (new) point of generation as the single cause of such difficulties, other than it being 'the straw that broke the camel's back'. Therefore, the allocation of costs required to accommodate a single new generation plant to one plant only (for example a new wind farm) should be avoided.

In the context of a strategic EU-wide policy for long-term, large-scale grid integration, the fundamental ownership unbundling between generation and transmission is indispensable. A proper definition of the interfaces between the wind power plant itself (including the 'internal grid' and the corresponding electrical equipment) and the 'external' grid infrastructure (new grid connection and extension/reinforcement of the existing grid) needs to be discussed, especially for remote wind farms and offshore wind energy. This does not necessarily mean that the additional grid tariff components, due to wind power connection and grid extension/reinforcement, must be paid by the local/regional customers only. These costs could be socialised within a 'grid infrastructure' component at national or even EU level. Of course, appropriate accounting rules would need to be established for grid operators.

Future System Cost Developments

Assessment of the way in which integration costs beyond the present 'low to moderate' penetration level will increase depends on how the future evolution of the power system is viewed. A 'static power system' assumption becomes less plausible with increasing wind penetration, as wind serving a substantial (higher than 25 per cent) fraction of the demand will cause the system to evolve over time. Furthermore, the generation mix is likely to change significantly during this long period of wind development. For example, it is predicted that gas power generation will increase dramatically (depending on fuel costs), which will make the power system more flexible. Hence the integration costs of wind energy increase smoothly and proportionally as penetration levels increase.

Costs beyond penetration levels of about 25 per cent will depend on how the underlying system architecture changes over time, as the amount of installed wind gradually increases, together with other generating technologies. For example, in order to accommodate high amounts of wind power, a system with a generation mix dominated by fast-ramping gas turbines or hydro is much more flexible than a system dominated by nuclear or coal, as it can respond quickly to changes in supply and demand.

Up to a penetration level of 25 per cent, the integration costs have been analysed in detail and are consistently low. The economic impacts and integration issues are very much dependent on the power system in question. Important factors include:

- the structure of the generation mix and its flexibility;
- the strength of the grid;
- the demand pattern;
- power market mechanisms; and
- structural and organisational aspects.

Technically, methods that have been used by power engineers for decades can be applied for integrating wind power. But for large-scale integration (penetration levels typically higher than 25 per cent), new power system concepts may be necessary, and it would be sensible to start considering such concepts immediately. Practical experience with large-scale integration in a few regions demonstrates that this is not merely a theoretical discussion. The feasibility of large-scale penetration has already been proved in areas where wind power currently meets 20, 30 and even 40 per cent of consumption (Denmark and regions of Germany and Spain).

Wind Power will Reduce Future European Power Prices

In a 2008 study (Skytte, 2008), Econ-Pöyry used its elaborate power model to investigate the electricity price effects of increasing wind power in Europe to 13 per cent in 2020.

In a business-as-usual scenario, it is assumed that the internal power market and additional investments in conventional power will more or less levelise the power prices across Europe up until 2020 (reference scenario). However, in a large-scale wind scenario (wind covering 13 per cent of EU electricity consumption) this might not be the case.

In areas where power demand is not expected to increase very much and in areas where the amount of new deployment of wind energy is larger than the increase in power demand, wind energy will substitute the most expensive power plants. This will lower the price levels in these areas, the study shows.

In the EU, the expected price level is around 5.4c€/kWh on average in 2020 for the reference case (Figure II.7.2) with a slightly higher price on the continent than in the Nordic countries, but with smaller price differences than today.

Figure II.7.2: Price levels in 2005 and in the 2020 reference and wind scenarios

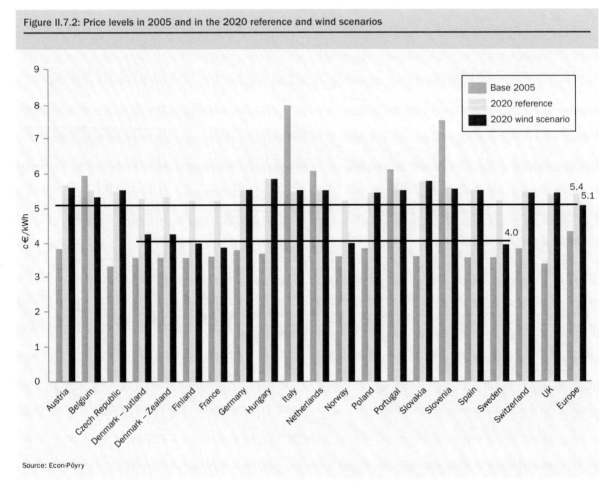

Source: Econ-Pöyry

In the wind scenario, the average price level in the EU decreases from 5.4 to 5.1c€/kWh compared to the reference scenario. However, the effects on power prices are different in the hydropower-dominated Nordic countries than in the thermal-based countries on the European continent.

In the wind scenario, wind energy reduces power prices to around 4c€/kWh in the Nordic countries. Prices in Germany and the UK remain at the higher level. In other words, a larger amount of wind power would create larger price differences between the (hydro-dominated) Nordic countries and the European continent.

One implication of price decreases in the Nordic countries is that conventional power production becomes less profitable. For large-scale hydropower the general water value decreases. In Norway, hydropower counts for the major part of the power production. However, large-scale implementation of wind creates a demand for flexible production that can deliver balancing services – opening up a window of opportunities for flexible production such as hydropower.

With large amounts of wind in the system, there will be an increased need for interconnection. This is also confirmed by the fact that, in the Econ-Pöyry model runs, with 13 per cent wind in the system compared to the reference scenario, the congestion rent (i.e. the cable income) increases on most transmission lines. This is also something one would expect: with more

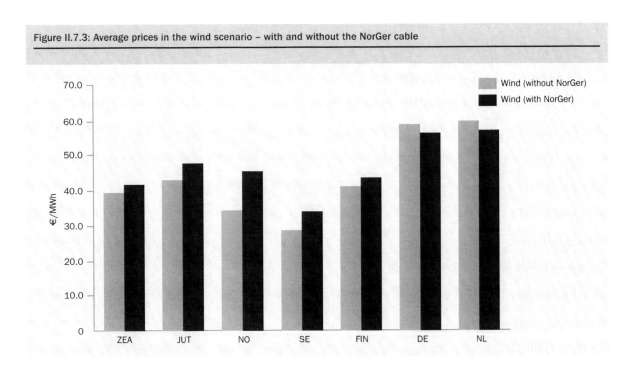

Figure II.7.3: Average prices in the wind scenario – with and without the NorGer cable

volatility in the system, there is a need for further interconnection in order to be better able to balance the system.

In order to simulate the effect of further interconnection, the same model runs were repeated, i.e. the wind and the reference scenario, but this time with a 1000 MW interconnector between Norway and Germany in place, the so-called 'NorGer cable'.[11] When running the wind scenario, Econ-Pöyry found that the congestion rent on such a cable would be around €160 million in the year 2020 in the reference scenario, while it would be around €200 million in the wind scenario.

With the cable in place it should first be observed that such a cable would have a significant effect on the average prices in the system, not only in Norway and Germany, but also in the other countries in the model. This is illustrated by Figure II.7.3. In the Nordic area the average prices are increased – the Nordic countries would import the higher prices in northern continental Europe – while in Germany (and The Netherlands) they are decreased. This is because, in the

high-price peak hours, power is flowing from Norway to Germany. This reduces the peak prices in Germany, while it increases the water values in Norway. In the off-peak low-price hours, the flow reverses, with Germany exporting to Norway in those hours where prices in Germany are very low. This increases off-peak prices in Germany and decreases water values. However, the overall effect, compared to the situation without a cable, is higher prices in Norway and lower prices in Germany. Although such effects are to be expected, this does not always have to be the case. In other cable analysis projects, Econ-Pöyry found that an interconnector between a thermal high-price area and a hydro low-price area may well reduce prices in both areas.

Concluding Remarks

Experience and studies provide positive evidence on the feasibility and solutions for integrating the expected wind power capacity in Europe for 2020, 2030 and beyond. Today, the immediate questions concern how to address a number of issues in the most cost-effective manner.

There is now a clearer understanding regarding the behaviour of aggregated wind plants at the power system level, as well as how system costs evolve with increasing wind energy penetration. In the range of expected wind energy penetration levels for the coming decades, the cost increments for additional reserves in the system, to deal with increased variability, remain moderate. Additional impacts that wind power may impose on the system, due to its variability and limited predictability, should be reduced by:

- making use of geographical aggregation;
- improved control of wind plants at both local and system levels;
- using state-of-the-art forecasting, monitoring and communication techniques; and
- reinforcing network.

Specific grid code requirements are now being imposed for wind farms, taking into account the continuous development of wind power plant capabilities. In this respect, a more proactive involvement of the wind power industry in the development of harmonised grid code requirements in Europe is essential.

A substantial upgrade of transmission networks is required, and the creation of a transnational offshore grid would bring additional benefits. This network development goes hand-in-hand with the necessary upgrades to achieve a more efficient European Internal Electricity Market and improved competition in European power markets. In this respect, a definite challenge is the creation of appropriate market rules, including incentives to make power generation and transmission evolve in a direction that facilitates variable output and decentralised generation, by increasing flexibility and providing additional network capacity, not only within the national grids, but also at the transnational level.

There is also a need for studies at the European level to provide a technical and scientific basis for grid upgrades and market organisation.

Part II Notes

[1] Instead of shutting down the wind turbine above cut-out, the power curve above cut-out wind speed (25 m/s) decreases gradually to reach zero power at 35 m/s.

[2] See www.trade-wind.eu.

[3] Except for unplanned outages of conventional plants, which by nature are not predictable. In this respect, wind power has an advantage, due to its modular nature and the low levels of capacity that are lost at one time during outages.

[4] The error reduction in the graph is defined as the ratio between the RMSE (root-mean-square-error) of regional prediction and the RMSE of a single site, based on the results of measured power production at 40 wind farms in Germany.

[5] The effect of the additional balancing costs at consumer level should take into account the benefits of wind power. Several studies show that the overall effect (resulting from costs and benefits) of integrating wind power is to reduce the price of electricity to the consumer.

[6] FACTS (Flexible AC Transmission Systems): power electronic devices locally implemented in the network, such as STATCOMs and static VAR compensators (SVCs).

[7] The MVAR capability is the capability to control reactive power by generation or controlled consumption.

[8] FRT is the ability for generators to remain stable and connected to the network when faults occur in the transmission network.

[9] Capacity factor depends on the relationship between rotor size and generator rating.

[10] The currency conversion used in this figure is EUR1 = GBP0.7 = US\$1.3. For the UK 2007 study, the average cost is presented; the range for 20 per cent penetration level is from €2.6–4.7/MWh.

[11] Classic model runs were repeated with a NorGer cable in place, in order to obtain investment figures for 2020, and to be consistent in methodology and approach. It should be noted that the NorGer cable does not have too pronounced an effect on investment levels. The size of the cable has not yet been decided, but a 1000 MW cable is probably a fair estimate and sufficient in order to simulate the effects of further interconnections.

THE ECONOMICS
OF WIND POWER

Acknowledgements

Part III was compiled by Poul Erik Morthorst of Risø DTU National Laboratory, Technical University of Denmark; Hans Auer of the Energy Economics Group, University of Vienna; Andrew Garrad of Garrad Hassan and Partners; Isabel Blanco of UAH, Spain.

We would like to thank all the peer reviewers for their valuable advice and for the tremendous effort that they put into the revision of Part III.

Part III was carefully reviewed by the following experts:

Rune Moesgaard	Danish Wind Industry Association
Hugo Chandler	IEA
Paulis Barons	Latvian Wind Energy Association
Mr. G. Bakema	Essent Netherlands

PART III   INTRODUCTION

Wind power is developing rapidly at both European and global levels. Over the past 15 years, the global installed capacity of wind power increased from around 2.5 GW in 1992 to just over 94 GW at the end of 2007, an average annual growth of more than 25 per cent. Owing to ongoing improvements in turbine efficiency and higher fuel prices, wind power is becoming economically competitive with conventional power production, and at sites with high wind speeds on land, wind power is considered to be fully commercial.

Part III of this volume focuses on the economics of wind power. The investment and cost structures of land-based and offshore turbines are discussed. The cost of electricity produced is also addressed, which takes into account the lifetime of turbines and O&M costs, and the past and future development of the costs of wind-generated power is analysed. In subsequent chapters, the importance of finance, support schemes and employment issues are discussed. Finally, the cost of wind-generated electricity is compared to the cost of conventional fossil fuel-fired power plants.

Wind power is used in a number of different applications, including grid-connected and stand-alone electricity production and water pumping. Part III analyses the economics of wind energy, primarily in relation to grid-connected turbines, which account for the vast bulk of the market value of installed turbines.

III.1   COST OF ON-LAND WIND POWER

Cost and Investment Structures

The main parameters governing wind power economics include:

- investment costs, such as auxiliary costs for foundation and grid connection;
- operation and maintenance costs;
- electricity production/average wind speed;
- turbine lifetime; and
- discount rate.

The most important parameters are turbine electricity production and investment costs. As electricity production depends to a large extent on wind conditions, choosing the right turbine site is critical to achieving economic viability.

INVESTMENT COSTS

The capital costs of wind energy projects are dominated by the cost of the wind turbine itself (ex-works).[1] Table III.1.1 shows the typical cost structure for a 2 MW turbine erected in Europe. An average turbine installed

in Europe has a total investment cost of around €1.23 million/MW. The turbine's share of the total cost is, on average, around 76 per cent, while grid connection accounts for around 9 per cent and foundations for around 7 per cent. The cost of acquiring a turbine site (on land) varies significantly between projects, so the figures in Table III.1.1 are only to be taken as examples. Other cost components, such as control systems and land, account for only a minor share of total costs.

The total cost per kW of installed wind power capacity differs significantly between countries, as shown in Figure III.1.1. The cost per kW typically varies from around €1000/kW to €1350/kW. As shown in Figure III.1.1, the investment costs per kW were found to be lowest in Denmark, and slightly higher in Greece and The Netherlands. For the UK, Spain and Germany, the costs in the data selection were found to be around 20–30 per cent higher than in Denmark. However, it should be observed that Figure III.1.1 is based on limited data, so the results might not be entirely representative for the countries involved.

Also, for 'other costs', such as foundations and grid connection, there is considerable variation between countries, ranging from around 32 per cent of total turbine costs in Portugal to 24 per cent in Germany, 21 per cent in Italy and only 16 per cent in Denmark. However, costs vary depending on turbine size as well as the country of installation.

The typical ranges of these other cost components as a share of total additional costs are shown in Table III.1.2. In terms of variation, the single most important additional component is the cost of grid connection, which, in some cases, can account for almost half of the auxiliary costs, followed by typically lower shares for foundation cost and the cost of the electrical installation. Thus these auxiliary costs may add significant amounts to the total cost of the turbine. Cost components such as consultancy and land usually only account for a minor share of the additional costs.

Table III.1.1: Cost structure of a typical 2 MW wind turbine installed in Europe (2006-€)

	Investment (€1000/MW)	Share (%)
Turbine (ex-works)	928	75.6
Foundations	80	6.5
Electric installation	18	1.5
Grid connection	109	8.9
Control systems	4	0.3
Consultancy	15	1.2
Land	48	3.9
Financial costs	15	1.2
Road	11	0.9
Total	1227	100

Note: Calculations by the author based on selected data for European wind turbine installations.

Source: Risø DTU

Figure III.1.1: Total investment cost, including turbine, foundations and grid connection, shown for different turbine sizes and countries of installation

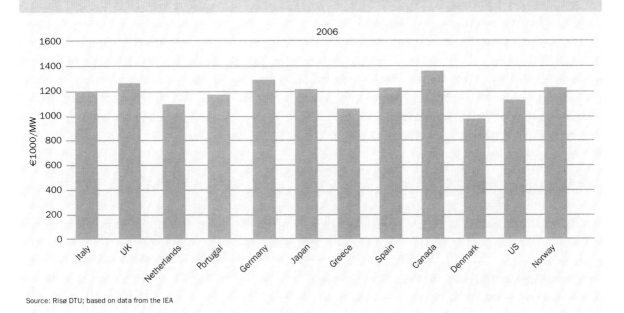

Source: Risø DTU; based on data from the IEA

For a number of selected countries, the turbine and auxiliary costs (foundations and grid connection) are shown in Figure III.1.2.

Table III.1.2: Cost structure for a medium-sized wind turbine

	Share of total cost (%)	Typical share of other costs (%)
Turbine (ex-works)	68–84	n/a
Foundation	1–9	20–25
Electric installation	1–9	10–15
Grid connection	2–10	35–45
Consultancy	1–3	5–10
Land	1–5	5–10
Financial costs	1–5	5–10
Road construction	1–5	5–10

Note: Based on a limited data selection from Germany, Denmark, Spain and the UK, adjusted and updated by the author.

Source: Risø DTU

TRENDS INFLUENCING THE COSTS OF WIND POWER

In recent years, three major trends have dominated the development of grid-connected wind turbines:

1. Turbines have become larger and taller – the average size of turbines sold on the market has increased substantially.
2. The efficiency of turbine production has increased steadily.
3. In general, the investment costs per kW have decreased, although there has been a deviation from this trend in recent years.

Figure III.1.3 shows the development of the average-sized wind turbine for a number of the most important wind power countries. It can be observed that the annual average size has increased significantly over the last 10–15 years, from approximately 200 kW in 1990 to

Figure III.1.2: Price of turbine and additional costs for foundation and grid connection, calculated per kW for selected countries (left axis), including turbine share of total costs (right axis)

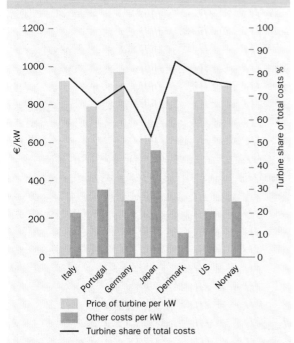

Note: The result for Japan may be caused by a different split of turbine investment costs and other costs, as the total adds up to almost the same level as seen for the other countries.

Source: Risø DTU

2 MW in 2007 in the UK, with Germany, Spain and the US not far behind.

As shown, there is a significant difference between some countries: in India, the average installed size in 2007 was around 1 MW, considerably lower than levels in the UK and Germany (2049 kW and 1879 kW respectively). The unstable picture for Denmark in recent years is due to the low level of turbine installations.

In 2007, turbines of the MW class (with a capacity of over 1 MW) had a market share of more than 95 per cent, leaving less than 5 per cent for the smaller machines. Within the MW segment, turbines with capacities of 2.5 MW and upwards are becoming increasingly important, even for on-land sites. In 2007, the

market share of these large turbines was 6 per cent, compared to only 0.3 per cent at the end of 2003.

The wind regime at the chosen site, the turbine hub height and the efficiency of production determine power production from the turbines. So just increasing the height of turbines has resulted in higher power production. Similarly, the methods for measuring and evaluating the wind speed at a given site have improved substantially in recent years and thus improved the site selection for new turbines. However, the fast development of wind power capacity in countries such as Germany and Denmark implies that the best wind sites in these countries have already been taken and that new on-land turbine capacity will have to be erected at sites with a marginally lower average wind speed. The replacement of older and smaller turbines with modern versions is also becoming increasingly important, especially in countries which have been involved in wind power development for a long time, as is the case for Germany and Denmark.

The development of electricity production efficiency, owing to better equipment design, measured as annual energy production per square metre of swept rotor area (kWh/m^2) at a specific reference site, has correspondingly improved significantly in recent years. With improved equipment efficiency, improved turbine siting and higher hub height, the overall production efficiency has increased by 2–3 per cent annually over the last 15 years.

Figure III.1.4 shows how these trends have affected investment costs, exemplified by the case of Denmark, from 1989 to 2006. The data reflects turbines installed in the particular year shown (all costs are converted to 2006 prices); all costs on the right axis are calculated per square metre of swept rotor area, while those on the left axis are calculated per kW of rated capacity.

The number of square metres covered by the turbine's rotor – the swept rotor area – is a good indicator of the turbine's power production, so this measure is a relevant index for the development in costs per kWh. As shown in Figure III.1.4, there was a substantial

Figure III.1.3: Development of the average wind turbine size sold in different countries

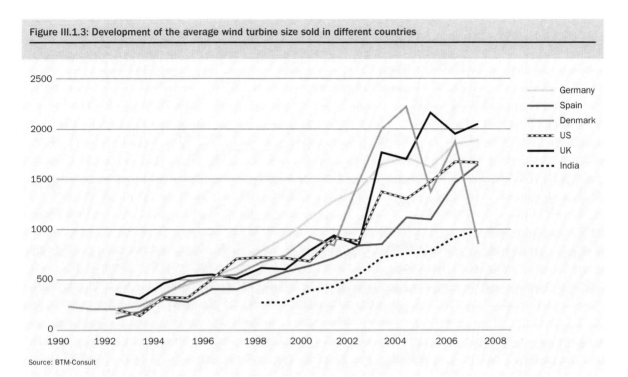

Source: BTM-Consult

Figure III.1.4: The development of investment costs from 1989 to 2006, illustrated by the case of Denmark

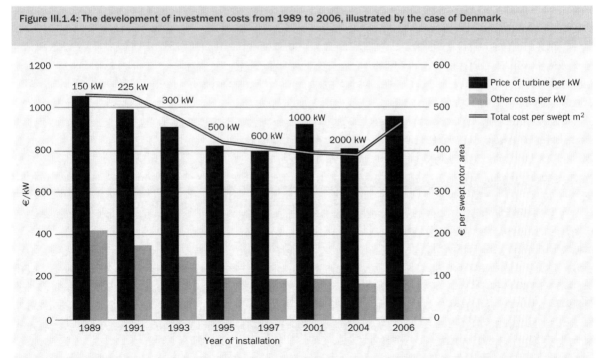

Note: Right axis – investment costs divided by swept rotor area (€/m² in constant 2006-€); Left axis – wind turbine capital costs (ex-works) and other costs per kW rated power (€/kW in constant 2006-€).

decline in costs per unit of swept rotor area in the period under consideration, except during 2006. From the late 1980s until 2004, overall investments per unit of swept rotor area declined by more than 2 per cent per annum, corresponding to a total reduction in cost of almost 30 per cent over these 15 years. But this trend was broken in 2006, when total investment costs rose by approximately 20 per cent compared to 2004, mainly due to a significant increase in demand for wind turbines, combined with rising commodity prices and supply constraints.

Looking at the cost per rated capacity (per kW), the same decline is found in the period 1989 to 2004, with the exception of the 1000 kW machine in 2001. The reason for this exception is related to the size of this particular turbine: with a higher hub height and larger rotor diameter, the turbine is equipped with a slightly smaller generator, although it produces more electricity. This fact is particularly important when analysing turbines built specifically for low and medium wind areas, where the rotor diameter is considerably larger in comparison to the rated capacity. As shown in Figure III.1.4, the cost per kW installed also rose by 20 per cent in 2006 compared to 2004.

In addition, the share of other costs as a percentage of total costs has generally decreased. In 1989, almost 29 per cent of total investment costs were related to costs other than the turbine itself. By 1997, this share had declined to approximately 20 per cent. This trend towards lower auxiliary costs continues for the last turbine model shown (2000 kW), where other costs amount to approximately 18 per cent of total costs. But from 2004 to 2006 other costs rose almost in parallel with the cost of the turbine itself.

The recent increase in turbine prices is a global phenomenon which stems mainly from a strong and increasing demand for wind power in many countries, along with constraints on the supply side (not only related to turbine manufacturers, but also resulting from a deficit in sub-supplier production capacity of wind turbine components). The general price increases for newly installed wind turbines in a number of selected countries are shown in Figure III.1.5. There are significant differences between individual countries, with price increases ranging from almost none to a rise of more than 40 per cent in the US and Canada.

Operation and Maintenance Costs of Wind-Generated Power

Operation and maintenance (O&M) costs constitute a sizeable share of the total annual costs of a wind turbine. For a new turbine, O&M costs may easily make up 20–25 per cent of the total levelised cost per kWh produced over the lifetime of the turbine. If the turbine is fairly new, the share may only be 10–15 per cent, but this may increase to at least 20–35 per cent by the end of the turbine's lifetime. As a result, O&M costs are attracting greater attention, as manufacturers attempt to lower these costs significantly by developing new turbine designs that require fewer regular service visits and less turbine downtime.

O&M costs are related to a limited number of cost components, including:

- insurance;
- regular maintenance;
- repair;
- spare parts; and
- administration.

Some of these cost components can be estimated relatively easily. For insurance and regular maintenance, it is possible to obtain standard contracts covering a considerable share of the wind turbine's total lifetime. Conversely, costs for repairs and related spare parts are much more difficult to predict. And although all cost components tend to increase as the turbine gets older, costs for repair and spare parts are particularly influenced by turbine age, starting low and increasing over time.

Due to the relative infancy of the wind energy industry, there are only a few turbines that have reached

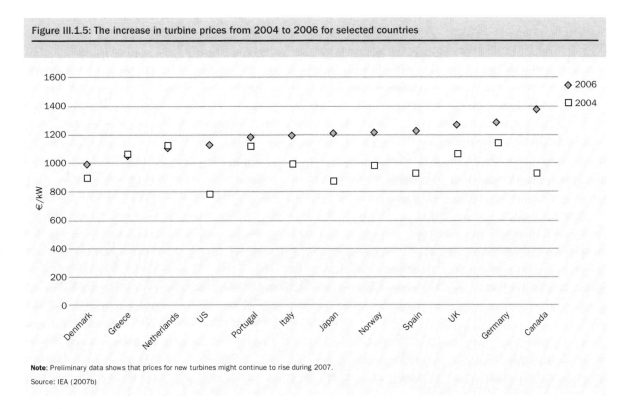

Figure III.1.5: The increase in turbine prices from 2004 to 2006 for selected countries

Note: Preliminary data shows that prices for new turbines might continue to rise during 2007.

Source: IEA (2007b)

their life expectancy of 20 years. These turbines are much smaller than those currently available on the market. Estimates of O&M costs are still highly unpredictable, especially around the end of a turbine's lifetime; nevertheless a certain amount of experience can be drawn from existing older turbines.

Based on experiences in Germany, Spain, the UK and Denmark, O&M costs are generally estimated to be around 1.2 to 1.5 euro cents (c€) per kWh of wind power produced over the total lifetime of a turbine. Spanish data indicates that less than 60 per cent of this amount goes strictly to the O&M of the turbine and installations, with the rest equally distributed between labour costs and spare parts. The remaining 40 per cent is split equally between insurance, land rental[2] and overheads.

Figure III.1.6 shows how total O&M costs for the period between 1997 and 2001 were split into six different categories, based on German data from DEWI.

Expenses pertaining to buying power from the grid and land rental (as in Spain) are included in the O&M costs calculated for Germany. For the first two years of its lifetime, a turbine is usually covered by the manufacturer's warranty, so in the German study O&M costs made up a small percentage (2–3 per cent) of total investment costs for these two years, corresponding to approximately 0.3–0.4c€/kWh. After six years, the total O&M costs increased, constituting slightly less than 5 per cent of total investment costs, which is equivalent to around 0.6–0.7c€/kWh. These figures are fairly similar to the O&M costs calculated for newer Danish turbines (see below).

Figure III.1.7 shows the total O&M costs resulting from a Danish study and how these are distributed between the different O&M categories, depending on the type, size and age of the turbine. For a three-year-old 600 kW machine, which was fairly well represented in the study,[3] approximately 35 per cent of total O&M costs

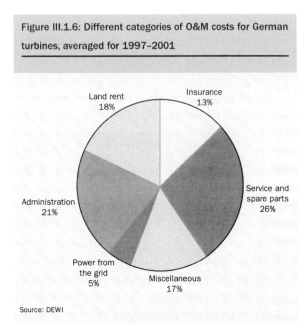

Figure III.1.6: Different categories of O&M costs for German turbines, averaged for 1997–2001

Source: DEWI

covered insurance, 28 per cent regular servicing, 11 per cent administration, 12 per cent repairs and spare parts, and 14 per cent other purposes. In general, the study revealed that expenses for insurance, regular servicing and administration were fairly stable over time, while the costs for repairs and spare parts fluctuated considerably. In most cases, other costs were of minor importance.

Figure III.1.7 also shows the trend towards lower O&M costs for new and larger machines. So for a three-year-old turbine, the O&M costs decreased from around 3.5c€/kWh for the old 55 kW turbines to less than 1c€/kWh for the newer 600 kW machines. The figures for the 150 kW turbines are similar to the O&M costs identified in the three countries mentioned above. Moreover, Figure III.1.7 shows clearly that O&M costs increase with the age of the turbine.

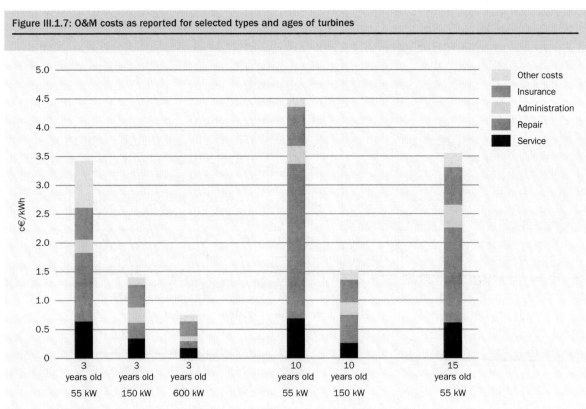

Figure III.1.7: O&M costs as reported for selected types and ages of turbines

Source: Jensen et al. (2002)

With regard to the future development of O&M costs, care must be taken in interpreting the results of Figure III.1.7. First, as wind turbines exhibit economies of scale in terms of declining investment costs per kW with increasing turbine capacity, similar economies of scale may exist for O&M costs. This means that a decrease in O&M costs will be related, to a certain extent, to turbine up-scaling. And second, the newer and larger turbines are better aligned with dimensioning criteria than older models, implying reduced lifetime O&M requirements. However, this may also have the adverse effect that these newer turbines will not stand up as effectively to unexpected events.

The Cost of Energy Generated by Wind Power

The total cost per kWh produced (unit cost) is calculated by discounting and levelising investment and O&M costs over the lifetime of the turbine and then dividing them by the annual electricity production. The unit cost of generation is thus calculated as an average cost over the turbine's lifetime. In reality, actual costs will be lower than the calculated average at the beginning of the turbine's life, due to low O&M costs, and will increase over the period of turbine use.

The turbine's power production is the single most important factor for the cost per unit of power generated. The profitability of a turbine depends largely on whether it is sited at a good wind location. In this section, the cost of energy produced by wind power will be calculated according to a number of basic assumptions. Due to the importance of the turbine's power production, the sensitivity analysis will be applied to this parameter. Other assumptions include the following:

- Calculations relate to new land-based, medium-sized turbines (1.5–2 MW) that could be erected today.

- Investment costs reflect the range given in Chapter III.2 – that is, a cost of €1100–1400/kW, with an average of €1225/kW. These costs are based on data from IEA and stated in 2006 prices.
- O&M costs are assumed to be 1.45c€/kWh as an average over the lifetime of the turbine.
- The lifetime of the turbine is set at 20 years, in accordance with most technical design criteria.
- The discount rate is assumed to range from 5 to 10 per cent per annum. In the basic calculations, a discount rate of 7.5 per cent per annum is used, although a sensitivity analysis of the importance of this interest range is also performed.
- Economic analyses are carried out on a simple national economic basis. Taxes, depreciation and risk premiums are not taken into account and all calculations are based on fixed 2006 prices.

The calculated costs per kWh of wind-generated power, as a function of the wind regime at the chosen sites, are shown in Figure III.1.8. As illustrated, the costs range from approximately 7–10c€/kWh at sites with low average wind speeds to approximately 5–6.5c€/kWh at windy coastal sites, with an average of approximately 7c€/kWh at a wind site with average wind speeds.

In Europe, the good coastal positions are located mainly in the UK, Ireland, France, Denmark and Norway. Medium wind areas are mostly found inland in mid and southern Europe – in Germany, France, Spain, The Netherlands and Italy – and also in northern Europe – in Sweden, Finland and Denmark. In many cases, local conditions significantly influence the average wind speeds at a specific site, so significant fluctuations in the wind regime are to be expected even for neighbouring areas.

Approximately 75–80 per cent of total power production costs for a wind turbine are related to capital costs – that is, the costs of the turbine, foundations, electrical equipment and grid connection. Thus a wind turbine is capital-intensive compared with

Figure III.1.8: Calculated costs per kWh of wind-generated power as a function of the wind regime at the chosen site (number of full load hours)

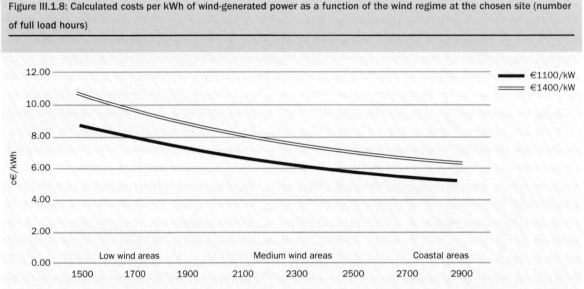

Note: In this figure, the number of full load hours is used to represent the wind regime. Full load hours are calculated as the turbine's average annual production divided by its rated power. The higher the number of full load hours, the higher the wind turbine's production at the chosen site.

Source: Risø DTU

conventional fossil fuel-fired technologies, such as natural gas power plants, where as much as 40–60 per cent of total costs are related to fuel and O&M costs. For this reason, the costs of capital (discount or interest rate) are an important factor for the cost of wind-generated power, a factor which varies considerably between the EU member countries.

In Figure III.1.9, the costs per kWh of wind-produced power are shown as a function of the wind regime and the discount rate (which varies between 5 and 10 per cent per annum).

As illustrated in Figure III.1.9, the costs range between around 6 and 8c€/kWh at medium wind positions, indicating that a doubling of the interest rate induces an increase in production costs of 2c€/kWh. In low wind areas, the costs are significantly higher, at around 8–11c€/kWh, while the production costs range between 5 and 7c€/kWh in coastal areas.

Development of the Cost of Wind-Generated Power

The rapid European and global development of wind power capacity has had a strong influence on the cost of wind power over the last 20 years. To illustrate the trend towards lower production costs of wind-generated power, a case that shows the production costs for different sizes and models of turbines is presented in Figure III.1.10.

Figure III.1.10 shows the calculated unit cost for different sizes of turbine, based on the same assumptions used in the previous section: a 20-year lifetime is assumed for all turbines in the analysis and a real discount rate of 7.5 per cent per annum is used. All costs are converted into constant 2006 prices. Turbine electricity production is estimated for two wind regimes – a coastal and an inland medium wind position.

Figure III.1.9: The costs of wind-produced power as a function of wind speed (number of full load hours) and discount rate; the installed cost of wind turbines is assumed to be €1225/kW

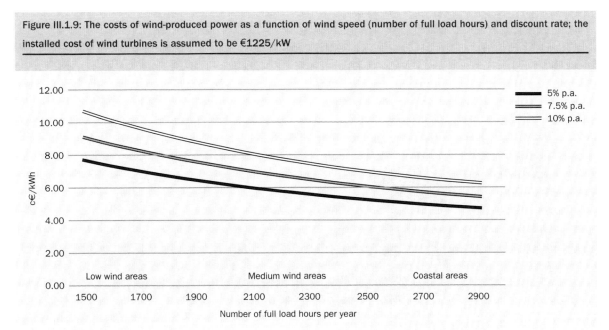

Source: Risø DTU

The starting point for the analysis is the 95 kW machine, which was installed mainly in Denmark during the mid-1980s. This is followed by successively newer turbines (150 kW and 225 kW), ending with the 2000 kW turbine, which was typically installed from around 2003 onwards. It should be noted that wind turbine manufacturers generally expect the production cost of wind power to decline by 3–5 per cent for each new turbine generation they add to their product portfolio. The calculations are performed for the total lifetime (20 years) of the turbines; calculations for the old turbines are based on track records of more than 15 years (average figures), while newer turbines may have a track record of only a few years, so the newer the turbine, the less accurate the calculations.

The economic consequences of the trend towards larger turbines and improved cost-effectiveness are clearly shown in Figure III.1.10. For a coastal position, for example, the average cost has decreased from around 9.2c€/kWh for the 95 kW turbine (mainly installed in the mid-1980s) to around 5.3c€/kWh for a fairly new 2000 kW machine, an improvement of more than 40 per cent over 20 years (constant 2006 prices).

Future Evolution of the Costs of Wind-Generated Power

In this section, the future development of the economics of wind power is illustrated by the use of the experience curve methodology. The experience curve approach was developed in the 1970s by the Boston Consulting Group; it relates the cumulative quantitative development of a product to the development of the specific costs (Johnson, 1984). Thus, if the cumulative sale of a product doubles, the estimated learning rate gives the achieved reduction in specific product costs.

The experience curve is not a forecasting tool based on estimated relationships. It merely shows the development as it would be if existing trends continue.

Figure III.1.10: Total wind energy costs per unit of electricity produced, by turbine size (c€/kWh, constant 2006 prices)

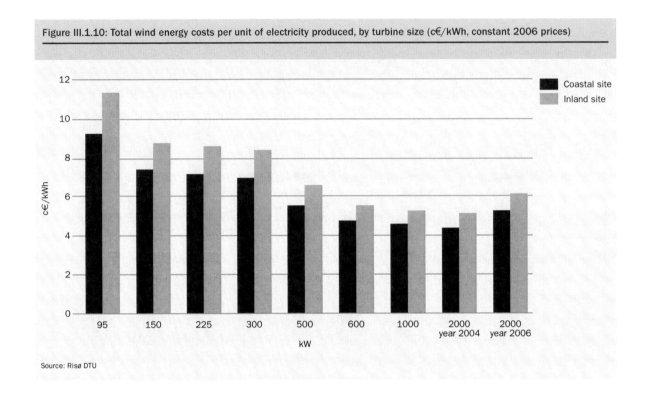

Source: Risø DTU

It converts the effect of mass production into an effect upon production costs, without taking other causal relationships into account. Thus changes in market development and/or technological breakthroughs within the field may change the picture considerably, as would fluctuations in commodity prices such as those for steel and copper.

Different experience curves have been estimated[4] for a number of projects. Unfortunately, different specifications were used, which means that not all of these projects can be directly compared. To obtain the full value of the experiences gained, the reduction in price of the turbine (€/kW) should be taken into account, as well as improvements in the efficiency of the turbine's production (which requires the use of an energy specification (€/kWh); see Neij et al., 2003). Thus, using the specific costs of energy as a basis (costs per kWh produced), the estimated progress ratios range from 0.83 to 0.91, corresponding to

learning rates of 0.17 to 0.09. So when the total installed capacity of wind power doubles, the costs per kWh produced for new turbines goes down by between 9 and 17 per cent. In this way, both the efficiency improvements and embodied and disembodied cost reductions are taken into account in the analysis.

Wind power capacity has developed very rapidly in recent years, on average by 25 to 30 per cent per year over the last ten years. At present, the total wind power capacity doubles approximately every three to four years. Figure III.1.11 shows the consequences for wind power production costs, based on the following assumptions:

• The present price-relation should be retained until 2010. The reason why no price reductions are foreseen in this period is due to a persistently high demand for new wind turbine capacity and sub-supplier constraints in the delivery of turbine components.

Figure III.1.11: Using experience curves to illustrate the future development of wind turbine economics until 2015

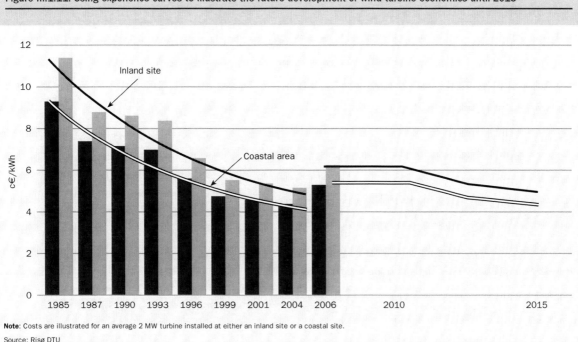

Note: Costs are illustrated for an average 2 MW turbine installed at either an inland site or a coastal site.

Source: Risø DTU

- From 2010 until 2015, a learning rate of 10 per cent is assumed, implying that each time the total installed capacity doubles, the costs per kWh of wind generated power decrease by 10 per cent.
- The growth rate of installed capacity is assumed to double cumulative installations every three years.

The curve illustrates cost development in Denmark, which is a fairly cheap wind power country. Thus the starting point for the development is a cost of wind power of around 6.1c€/kWh for an average 2 MW turbine sited at a medium wind regime area (average wind speed of 6.3 m/s at a hub height of 50 m). The development for a coastal position is also shown.

At present, the production costs for a 2 MW wind turbine installed in an area with a medium wind speed (inland position) are around 6.1c€/kWh of wind-produced power. If sited in a coastal location, the current costs are around 5.3c€/kWh. If a doubling time of total installed capacity of three years is assumed, in 2015 the cost interval would be approximately 4.3 to 5.0c€/kWh for a coastal and inland site respectively. A doubling time of five years would imply a cost interval, in 2015, of 4.8 to 5.5c€/kWh. As mentioned, Denmark is a fairly cheap wind power country; for more expensive countries, the cost of wind power produced would increase by 1–2c€/kWh.

III.2   OFFSHORE DEVELOPMENTS

Development and Investment Costs of Offshore Wind Power

Offshore wind only accounts for a small amount of the total installed wind power capacity in the world – approximately 1 per cent. The development of offshore wind has mainly been in northern European countries, around the North Sea and the Baltic Sea, where about 20 projects have been implemented. At the end of 2007, almost 1100 MW of capacity was located offshore.

Five countries have operating offshore wind farms: Denmark, Ireland, The Netherlands, Sweden and the UK, as shown in Table III.2.1. In 2007, the Swedish offshore wind farm Lillgrunden, with a rated capacity of 110 MW, was installed. Most of the capacity has been installed in relatively shallow waters (under 20 m deep) no more than 20 km from the coast in order to minimise the extra costs of foundations and sea cables.

Offshore wind is still around 50 per cent more expensive than onshore wind. However, due to the expected benefits of more wind and the lower visual impact of the larger turbines, several countries now have very ambitious goals concerning offshore wind.

The total capacity is still limited, but growth rates are high. Offshore wind farms are installed in large units – often 100–200 MW – and two new installed wind farms per year will result in future growth rates of between 20 and 40 per cent. Presently, higher costs and temporary capacity problems in the manufacturing stages, as well as difficulties with the availability of installation vessels, cause some delays, but even so, several projects in the UK and Denmark will be finished within the next three years, as can be seen in Tables III.2.2–III.2.6.

Offshore costs depend largely on weather and wave conditions, water depth and distance from the coast. The most detailed cost information on recent offshore installations comes from the UK, where 90 MW in

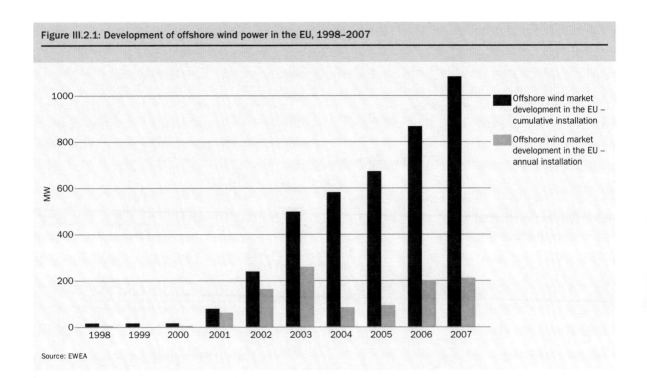

Figure III.2.1: Development of offshore wind power in the EU, 1998–2007

Source: EWEA

Figure III.2.2: Total offshore wind power installed by the end of 2007

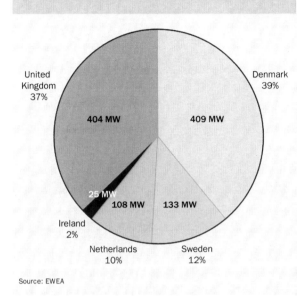

Source: EWEA

Investment costs per MW range from a low of €1.2 million/MW (Middelgrunden) to €2.7 million/MW (Robin Rigg) (Figure III.2.3).

The higher offshore capital costs are due to the larger structures and the complex logistics of installing the towers. The costs of offshore foundations, construction, installations and grid connection are significantly higher than for onshore. For example, offshore turbines are generally 20 per cent more expensive and towers and foundations cost more than 2.5 times the price of those for a similar onshore project.

In general, the costs of offshore capacity have increased in recent years, as is the case for land-based turbines, and these increases are only partly reflected in the costs shown in Figure III.2.3. As a result, the average costs of future offshore farms are expected to be higher. On average, investment costs for a new offshore wind farm are expected be in the range of €2.0–2.2 million/MW for a near-shore, shallow-water facility.

To illustrate the economics of offshore wind turbines in more detail, the two largest Danish offshore wind farms can be taken as examples. The Horns Rev project, located approximately 15 km off the west coast of Jutland (west of Esbjerg), was finished in 2002. It is equipped with 80 machines of 2 MW, with a total capacity of 160 MW. The Nysted offshore wind farm is located south of the isle of Lolland. It consists of 72 turbines of 2.3 MW and has a total capacity of 165 MW. Both wind farms have their own on-site transformer stations,

2006 and 100 MW in 2007 were added, and from Sweden with the installation of Lillgrunden in 2007.

Table III.2.7 gives information on some of the recently established offshore wind farms. As shown, the chosen turbine size for offshore wind farms ranges from 2 to 3.6 MW, with the newer wind farms being equipped with the larger turbines. The size of the wind farms also vary substantially, from the fairly small Samsø wind farm of 23 MW to Robin Rigg, the world's largest offshore wind farm, with a rated capacity of 180 MW.

Table III.2.1: Installed offshore capacity in offshore wind countries

Country	MW installed in 2006	Accumulated MW, end 2006	MW installed in 2007	Accumulated MW, end 2007
Denmark	0	409	0	409
Ireland	0	25	0	25
The Netherlands	108	108	0	108
Sweden	0	23	110	133
UK	90	304	100	404
Total global	198	869	210	1079

Source: EWEA

Table III.2.2: Operating and planned offshore wind farms in the UK

Project	Location	Region	Capacity (MW)	No of turbines	Water depth (m)	Distance to shore (km)	Online
In operation							
Barrow	7 km from Walney Island	Off England	90	30	>15	7	2006
Beatrice	Beatrice Oilfield, Moray Firth	Off Scotland	10	2	>40	unknown	2007
Blyth Offshore	1 km from Blyth Harbour	Off England	3.8	2	6	1	2000
Burbo Bank	5.2 km from Crosby	Off England	90	25	10	5.2	2007
Inner Dowsing	5.2 km from Ingoldmells	Off England	90	30	10	5.2	2008
Kentish Flats	8.5 km from Whitstable	Off England	90	30	5	8.5	2005
Lynn	5.2 km from Skegness	Off England	97	30	10	5.2	2008
North Hoyle	7.5 km from Prestatyn and Rhyl	Off Wales/England	60	30	5–12	7.5	2003
Scroby Sands	3 km NE of Great Yarmouth	Off England	60	30	2–10	3	2004
			403.8				
Under construction							
Greater Gabbard phase 1	Off Felixstowe/Clacton-on-Sea	Off England	300	-	-	-	2010
Greater Gabbard phase 2	Off Felixstowe/Clacton-on-Sea	Off England	200	-	-	-	2011
Ormonde	Off Walney Island	Off England	150	30	20	11	2010
Rhyl Flats	8 km from Abergele	Off Wales	90	25	8	8	2009
Solway Firth/ Robin Rigg A	9.5 km from Maryport/8.5 km off Rock Cliffe	Off England/ Scotland	90	30	>5	9.5	2010
Solway Firth/ Robin Rigg B	9.5 km from Maryport/8.5 km off Rock Cliffe	Off England/ Scotland	90	30	>5	9.5	2010
Thanet	Foreness Point, Margate	Off England	300	100	20–25	7–8.5	2010

Source: EWEA

Table III.2.3: Operating and planned offshore wind farms in The Netherlands

Project	Location	Capacity (MW)	No of turbines	Water depth (m)	Distance to shore (km)	Online
In operation						
Offshore Wind Farm Egmond aan Zee (OWEZ)	Egmond aan Zee	108	36	17–23	8–12	2006
Lely	Medemblik, Ijsselmeer (inland lake)	2	4	7.5	0.75	1994
Irene Vorrink (Dronten)	Dronten, Ijsselmeer (inland freshwater lake), to the outside of the dyke	16.8	28	2	0.03	1996
Princess Amalia	Ijmuiden	120	60	19–24	> 23	2008

Source: EWEA

Table III.2.4: Operating and planned offshore wind farms in Denmark

Project	Location	Capacity (MW)	No of turbines	Water depth (m)	Distance to shore (km)	Online
In operation						
Vindeby	Blæsenborg Odde, NW off Vindeby, Lolland	4.95	11	2.5–5	2.5	1991
TunøKnob	off Aarhus, Kattegat Sea	5	10	0.8–4	6	1995
Middelgrunden	Oresund, east of Copenhagen harbour	40	20	5–10	2–3	2001
Horns Rev I	Blåvandshuk, Baltic Sea	160	80	6–14	14–17	2002
Nysted Havmøllepark	Rødsand, Lolland	165.6	72	6–10	6–10	2003
Samsø	Paludans Flak, South of Samsø	23	4	11–18	3.5	2003
Frederikshavn	Frederikshavn Harbour	10.6	4	3	0.8	2003
Rønland I	Lim fjord, off Rønland peninsula, in the Nissum Bredning, off NW Jutland	17.2	8	3		2003
		426.35				
Under construction						
Avedøre	Off Avedøre	7.2	2	2	0.025	2009
Frederikshavn (test site)	Frederikshavn Harbour	12	2	15–20	4.5	2010
Rødsand 2	Off Rødsand, Lolland	200	89	5–15	23	2010
Sprøgø	North of Sprøgø	21	7	6–16	0.5	2009
Horns Rev II	Blåvandshuk, Baltic Sea (10 km west of Horns Rev)	209	91	9–17	30	2010

Source: EWEA

Table III.2.5: Operating and planned offshore wind farms in Sweden

Project	Location	Capacity (MW)	No of turbines	Water depth (m)	Distance to shore (km)	Online
In operation						
Bockstigen	Gotland	2.8	5	6–8	3	1998
Utgrunden I	Kalmarsund	10.5	7	4–10	7	2001
Yttre Stengrund	Kalmarsund	10.0	5	8–12	4	2002
Lillgrund	Malmö	110.0	48	2.5–9	10	2007
		133.25				
Under construction						
Gässlingegrund	Vänern	30	10	4–10	4	2009

Source: EWEA

Table III.2.6: Operating offshore wind farms in Ireland

Project	Location	Capacity (MW)	No of turbines	Water depth (m)	Distance to shore (km)	Online
Arklow Bank	Off Arklow, Co Wicklow	25.2	7	15	10	2004

Source: EWEA

Table III.2.7: Key information on recent offshore wind farms

	In operation	No of turbines	Turbine size (MW)	Total capacity (MW)	Investment costs (€ million)
Middelgrunden (DK)	2001	20	2	40	47
Horns Rev I (DK)	2002	80	2	160	272
Samsø (DK)	2003	10	2.3	23	30
North Hoyle (UK)	2003	30	2	60	121
Nysted (DK)	2004	72	2.3	165	248
Scroby Sands (UK)	2004	30	2	60	121
Kentish Flats (UK)	2005	30	3	90	159
Barrows (UK)	2006	30	3	90	–
Burbo Bank (UK)	2007	24	3.6	90	181
Lillgrunden (S)	2007	48	2.3	110	197
Robin Rigg (UK)		60	3	180	492

Note: Robin Rigg is under construction.

Source: Risø DTU

Figure III.2.3: Investments in offshore wind farms, €million/MW (current prices)

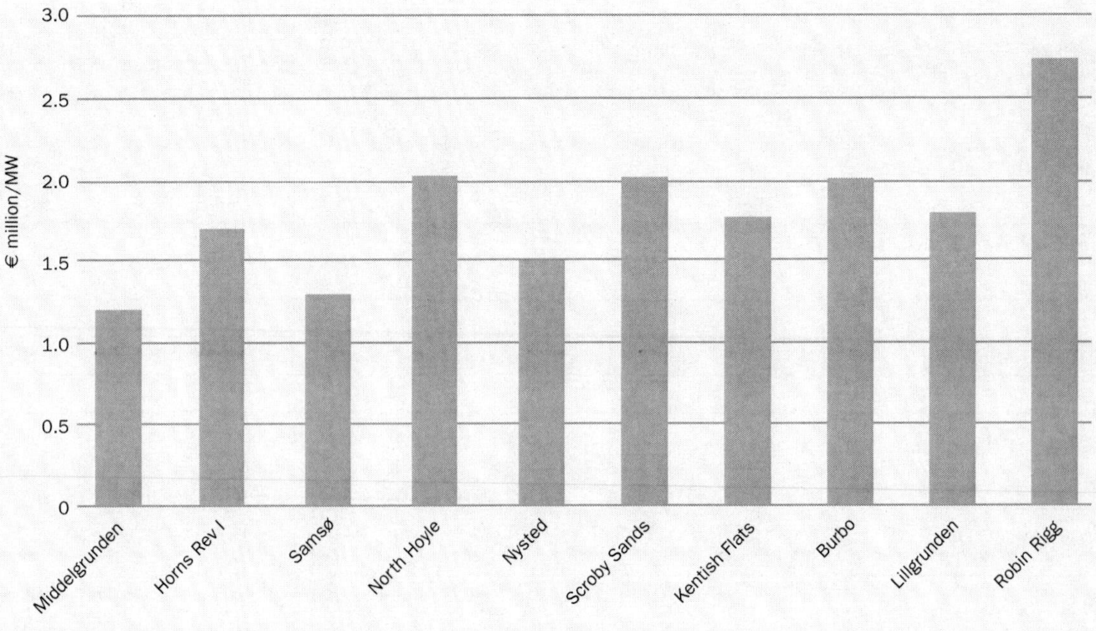

Source: Risø DTU

Table III.2.8: Average investment costs per MW related to offshore wind farms in Horns Rev and Nysted

	Investments (€1000 /MW)	Share (%)
Turbines ex-works, including transport and erection	815	49
Transformer station and main cable to coast	270	16
Internal grid between turbines	85	5
Foundations	350	21
Design and project management	100	6
Environmental analysis	50	3
Miscellaneous	10	<1
Total	1680	~100

Note: Exchange rate EUR1 = DKK7.45.

Source: Risø DTU

which are connected to the high voltage grid at the coast via transmission cables. The farms are operated from onshore control stations, so staff are not required at the sites. The average investment costs related to these two farms are shown in Table III.2.8.

In Denmark, all of the cost components above are covered by the investors, except for the costs of the transformer station and the main transmission cable to the coast, which are covered by TSOs in the respective areas. The total costs of each of the two offshore farms are around €260 million.

In comparison to land-based turbines, the main differences in the cost structure are related to two issues:

1. Foundations are considerably more expensive for offshore turbines. The costs depend on both the sea depth and the type of foundation being built (at Horns Rev monopiles were used, while the turbines at Nysted are erected on concrete gravity foundations). For a conventional turbine situated on land, the foundations' share of the total cost is normally around 5–9 per cent, while the average of the two projects mentioned above is 21 per cent (see Table III.2.8), and thus considerably more expensive. However, since considerable experience will be gained through these two wind farms, a further optimisation of foundations can be expected in future projects.

2. Transformer stations and sea transmission cables increase costs. Connections between turbines

and the centrally located transformer station, and from there to the coast, generate additional costs. For the Horns Rev and Nysted wind farms, the average cost share for the transformer station and sea transmission cables is 21 per cent (see Table III.2.8), of which a small proportion (5 per cent) goes on the internal grid between turbines.

Finally, a number of environmental analyses, including an environmental impact investigation (EIA) and graphic visualising of the wind farms, as well as additional research and development, were carried out. The average cost share for these analyses accounted for approximately 6 per cent of total costs, but part of these costs are related to the pilot character of these projects and are not expected to be repeated for future offshore wind farm installations.

The Cost of Energy Generated by Offshore Wind Power

Although the costs are higher for offshore wind farms, they are somewhat offset by a higher total electricity production from the turbines, due to higher offshore wind speeds. An on-land installation normally has around 2000–2300 full load hours per year, while for a typical offshore installation this figure reaches more

Table III.2.9: Assumptions used for economic calculations

	In operation	Capacity (MW)	€million/MW	Full load hours per year
Middelgrunden	2001	40	1.2	2500
Horns Rev I	2002	160	1.7	4200
Samsø	2003	23	1.3	3100
North Hoyle	2003	60	2.0	3600
Nysted	2004	165	1.5	3700
Scroby Sands	2004	60	2.0	3500
Kentish Flats	2005	90	1.8	3100
Burbo	2007	90	2.0	3550
Lillgrunden	2007	110	1.8	3000
Robin Rigg		180	2.7	3600

Note: Robin Rigg is under construction.

Source: Risø DTU

than 3000 full load hours per year. The investment and production assumptions used to calculate the costs per kWh are given in Table III.2.9.

In addition, the following economic assumptions are made:

- Over the lifetime of the wind farm, annual O&M costs are assumed to be €16/MWh, except for Middelgrunden, where these costs based on existing accounts are assumed to be €12/MWh for the entire lifetime.
- The number of full load hours is given for a normal wind year and corrected for shadow effects in the farm, as well as unavailability and losses in transmission to the coast.
- The balancing of the power production from the turbines is normally the responsibility of the farm owner. According to previous Danish experience, balancing requires an equivalent cost of around €3/MWh.[5] However, balancing costs are also uncertain and may differ substantially between countries.
- The economic analyses are carried out on a simple national economic basis, using a discount rate of 7.5 per cent per annum, over the assumed lifetime of 20 years. Taxes, depreciation, profit and risk premiums are not taken into account.

Figure III.2.4 shows the total calculated costs per MWh for the wind farms listed in Table III.2.9.

It can be seen that total production costs differ significantly between the illustrated wind farms, with Horns Rev, Samsø and Nysted being among the cheapest and Robin Rigg in the UK the most expensive. Differences can be related partly to the depth of the sea and distance to the shore and partly to increased investment costs in recent years. O&M costs are assumed to be at the same level for all wind farms (except Middelgrunden) and are subject to considerable uncertainty.

Costs are calculated on a simple national economic basis, and are not those of a private investor. Private investors have higher financial costs and require a risk premium and profit. So the amount a private investor would add on top of the simple costs would depend, to a large extent, on the perceived technological and political risks of establishing the offshore farm and on the competition between manufacturers and developers.

Development of the Cost of Offshore Wind Power up to 2015

Until 2004, the cost of wind turbines generally followed the development of a medium-term cost

Figure III.2.4: Calculated production cost for selected offshore wind farms, including balancing costs (2006 prices)

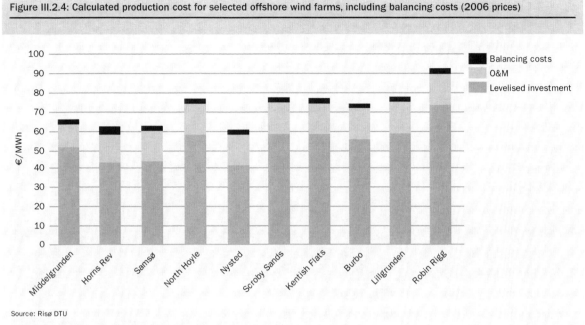

Source: Risø DTU

reduction curve (learning curve), showing a learning rate of approximately 10 per cent – in other words, each time wind power capacity doubled, the cost went down by approximately 10 per cent per MW installed. This decreasing cost trend changed in 2004–2006, when the price of wind power in general increased by approximately 20–25 per cent. This was caused mainly by the increasing costs of materials and a strong demand for wind capacity, which implied the scarcity of wind power manufacturing capacity and sub-supplier capacity for manufacturing turbine components.

A similar price increase can be observed for offshore wind power, although the fairly small number of finished projects, as well as a large spread in investment costs, make it difficult to identify the price level for offshore turbines accurately. On average, the expected investment costs for a new offshore wind farm are currently in the range of €2.0–2.2 million/MW.

In the following section, the medium-term cost development of offshore wind power is estimated using the learning curve methodology. However, it should be noted that considerable uncertainty is related to the use of learning curves, even for the medium term, and results should be used with caution.

The medium-term cost predictions for offshore wind power are shown in Table III.2.10 under the following conditions:

- The existing manufacturing capacity constraints for wind turbines will continue until 2010. Although there will be a gradual expansion of industrial capacity for wind power, a prolonged increase in demand will continue to strain the manufacturing capacity. Increasing competition among wind turbine manufacturers and sub-suppliers, resulting in unit reduction costs in the industry, will not occur before 2011.

- The total capacity development of wind power is assumed to be the main driving factor for the cost

Table III.2.10: Estimates for cost development of offshore wind turbines until 2015 (constant 2006 euros)

	Investment costs, €million/MW			O&M	Capacity factor
	Min	Average	Max	€/MWh	%
2006	1.8	2.1	2.4	16	37.5
2015	1.55	1.81	2.06	13	37.5

development of offshore turbines, since most of the turbine costs are related to the general wind power industry development. The growth rate of installed capacity is assumed to double cumulative installations every three years.

- For the period between 1985 and 2004, a learning rate of approximately 10 per cent was estimated (Neij et al., 2003). In 2011, this learning rate is again expected to be achieved by the industry and remain until at least 2015.

Given these assumptions, minimum, average and maximum cost scenarios are reported in Table III.2.10.

As shown in Table III.2.10, the average cost of offshore wind capacity is expected to decrease from €2.1 million/MW in 2006 to €1.81 million/MW in 2015, or by approximately 15 per cent. There will still be a considerable spread of costs, from €1.55 million/MW to €2.06 million/MW. A capacity factor of a constant 37.5 per cent (corresponding to an approximate number of full load hours of 3300) is expected for the whole period. This covers increased production from newer and larger turbines, moderated by sites with lower wind regimes and a greater distance to shore, which increase losses in transmission of power.

A study carried out in the UK (IEA, 2006) has estimated the future costs of offshore wind generation and the potential for cost reductions. The cost of raw materials, especially steel, which accounts for about 90 per cent of the turbine, was identified as the primary cost driver. The report emphasised that major savings can be achieved if turbines are made of lighter, more reliable materials and if major components are developed to be more fatigue-resistant. A model based on 2006 costs predicted that costs would rise from approximately £1.6 million/MW to approximately £1.75 million/MW (€2.37 to 2.6 million/MW) in 2011 before falling by around 20 per cent of the total cost by 2020.

III.3 PROJECT FINANCING

Over the last couple of decades, the vast majority of commercial wind farms have been funded through project finance. Project finance is essentially a project loan, backed by the cash flow of the specific project. The predictable nature of cash flows from a wind farm means they are highly suited to this type of investment mechanism.

Recently, as an increasing number of large companies have become involved in the sector, there has been a move towards balance sheet funding, mainly for construction. This means that the owner of the project provides all the necessary financing for the project, and the project's assets and liabilities are all directly accounted for at company level. At a later date, these larger companies will sometimes group multiple balance sheet projects in a single portfolio and arrange for a loan to cover the entire portfolio, as it is easier to raise a loan for the portfolio than for each individual project.

The structured finance markets (such as bond markets) in Europe and North America have also been used, but to a more limited extent than traditional project finance transactions. Such deals are like a loan transaction insomuch as they provide the project with an investment, in return for capital repayment and interest. However, the way in which transactions are set up is quite different to a traditional loan. Different types of funding for renewable energy have emerged in recent years in the structured debt market, which has significantly increased the liquidity in the sector.

Typical structures and transaction terms are discussed in more detail in this chapter.

Traditional Methods

WHAT IS PROJECT FINANCE?

Project finance is the term used to describe a structure in which the only security for a loan is the project itself. In other words, the owner of the project company is not personally, or corporately, liable for the loan. In a project finance deal, no guarantee is given that the loan will be repaid; however, if the loan is not repaid, the investor can seize the project and run or sell it in order to extract cash.

This process as rather like a giant property mortgage, since if a homeowner does not repay the mortgage on time, the house may be repossessed and sold by the lender. Therefore, the financing of a project requires careful consideration of all the different aspects, as well as the associated legal and commercial arrangements. Before investment, any project finance lender will want to know if there is any risk that repayment will not be made over the loan term.

DEAL STRUCTURE

A typical simple project finance deal will be arranged through a special purpose vehicle (SPV) company. The SPV is called 'Wind Farm Ltd' in Figure III.3.1. This would be a separate legal entity which may be owned by one company, consisting of several separate entities or a joint venture.

One bank may act alone if the project is very small, but will usually arrange a lending syndicate – this means that a group of banks will join together to provide the finance, usually with one bank as the 'lead arranger' of the deal. This is shown in Figure III.3.1, where Bank A syndicates the loan to Banks B, C and D.

A considerable amount of work is carried out before the loan is agreed, to check that the project is well planned and that it can actually make the necessary repayments by the required date. This process is called 'due diligence', and there is usually separate commercial, technical and legal due diligence carried out on behalf of the bank. The investors will make careful consideration of technical, financial and political risks, as well as considering how investment in a project fits in with the bank's own investment strategy.

Figure III.3.1: Typical wind farm finance structure

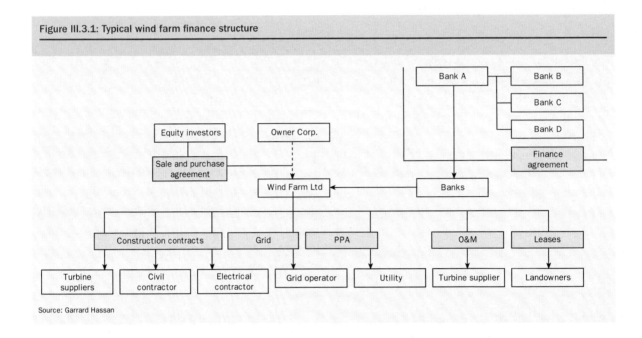

Source: Garrard Hassan

TYPICAL DEAL PARAMETERS

Generally, a bank will not lend 100 per cent of the project value and will expect to see a cash contribution from the borrower – this is usually referred to as 'equity'. It is typical to see 25–30 per cent equity and 70–75 per cent loan (money provided by the bank as their investment). Occasionally, a loan of 80 per cent is possible.

The size of the loan depends on the expected project revenue, although it is typical for investors to take a cautious approach and to assume that the long-term income will be lower than assumed for normal operation. This ensures that the loan does not immediately run into problems in a year with poor wind conditions or other technical problems, and also takes into account the uncertainty associated with income prediction.

Typically, a bank will base the financial model on the 'exceedance cases' provided within the energy assessment for the project. The mean estimated production of the project (P50) may be used to decide on the size of the loan, or in some cases a value lower than the mean

(for example P75 or P90). This depends on the level of additional cash cushioning that is available to cover costs and production variation over and above the money that is needed to make the debt payments. This is called the debt service cover ratio (DSCR) and is the ratio of cash available at the payment date to the debt service costs at that date. For example, if €1.4 million is available to make a debt payment (repayment and interest) of €1 million, the DSCR is 1.4:1.

The energy assumptions used for the financial model and associated DSCR are always a matter of negotiation with the bank as part of the loan agreement. Some banks will take a very cautious approach to the assumed energy production, with a low DCSR, and some will assume a more uncertain energy case, but with a high DSCR and sufficient cash cushioning to cover potential production variation.

The loan is often divided into two parts: a construction loan and a term loan. The construction loan provides funds for the construction of the project and becomes a term loan after completion. At the 'conversion' from a construction to a term loan, the terms and conditions associated with the loan change, as does the pricing of the debt. The term loan is usually less expensive than the construction loan, as the risks are lower during operation.

Typically, the length of a loan is between 10 and 15 years, but loan terms have become longer as banks have become more experienced in the wind industry.

The interest rate is often 1–1.5 per cent above the base rate at which the bank borrows their own funds (referred to as the interbank offer rate). In addition, banks usually charge a loan set-up fee of around 1 per cent of the loan cost, and they can make extra money by offering administrative and account services associated with the loan. Products to fix interest rates or foreign exchange rates are often sold to the project owner.

It is also typical for investors to have a series of requirements over the loan period; these are referred to as 'financial covenants'. These requirements are often the result of the due diligence and are listed within the 'financing agreement'. Typical covenants include the regular provision of information about operational and financial reporting, insurance coverage, and management of project bank accounts.

EXPERIENCE

In the last two decades, no wind industry project has ever had to be repossessed, although industry and project events have triggered some restructuring to adjust financing in difficult circumstances. The project finance mechanism has therefore served the industry and the banking community well. A decade ago, developers might have struggled to find a bank ready to loan to a project, whereas today banks often pursue developers to solicit their loan requirements. Clearly, this has improved the deals available to wind farm owners.

THE US

The description above covers most of the project-financed loans arranged outside the US. Inside the US

there are some very particular structures that are rather more complicated, as the US market is driven by tax considerations.

The renewable energy incentive in the US is the Production Tax Credit (PTC), and hence tax is of primary rather than secondary importance. Since this publication focuses on the European market, it is not appropriate to describe the US approach in any detail, but basically it includes another layer of ownership – the tax investors – who own the vast majority of the project, but only for a limited period of time (say ten years), during which they can extract the tax advantage. After that period there is an ownership 'flip', and the project usually returns to its original owner. Many sophisticated tax structures have been developed for this purpose, and this characteristic has had a major effect on the way in which the US industry has developed. The owners of US wind farms tend to be large companies with a heavy tax burden. Another group of passive tax investors has also been created that does not exist in the wind industry outside the US.

Recent Developments

STRUCTURED FINANCE

The last five years has seen the emergence of a number of new forms of transaction for wind financing, including public and private bond or share issues. Much of the interest in such structures has come from renewable energy funds, long-term investors, such as pension funds, and even high net worth individuals seeking efficient investment vehicles. The principle behind a structured finance product is similar to that of a loan, being the investment of cash in return for interest payment; however, the structures are generally more varied than project finance loans. As a result, there have been a number of relatively short-term investments offered in the market, which have been useful products for project owners considering project refinancing after a few years of operation. Structured finance investors have had a considerable appetite for cross-border deals and have had a significant effect on liquidity for wind (and other renewable energy) projects.

BALANCE SHEET FINANCING

The wind industry is becoming a utility industry in which the major utilities are increasingly playing a big role. As a result, while there are still many small projects being developed and financed, an increasing number are being built 'on balance sheet' (in other words with the utility's cash). Such an approach removes the need for a construction loan and the financing consists of a term loan only.

PORTFOLIO FINANCING

The arrival of balance sheet financing by the utilities naturally creates 'portfolio financing', in which banks are asked to finance a portfolio of wind farms rather than a single one. These farms are often operational and so data is available to allow for a far more accurate

projection of production. The portfolio will usually include a range of projects separated by significant physical distances, with a range of turbine types. The use of different turbine types reduces the risk of widespread, or at least simultaneous, design faults, and the geographical spread 'evens out the wind'. It is possible to undertake a rigorous estimation of the way in which the geographical spread reduces fluctuations (Marco et al., 2007). Figure III.3.2 shows the averaging effect of a portfolio of eight wind power plants in three countries.

Finally, if the wind farms are in different countries, then the portfolio also reduces regulatory risks.

The risk associated with such portfolio financing is significantly lower than that of financing a single wind farm before construction and attracts more favourable terms. As a result, the interest in such financing is growing. Portfolio financing can be adopted even after the initial financing has been in place for some time. It is now quite common to see an owner collecting together a number of individually financed projects and refinancing them as a portfolio.

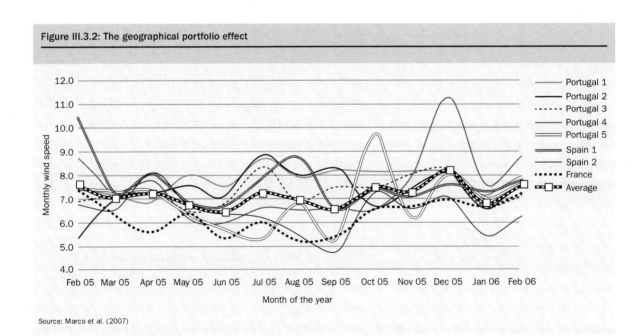

Figure III.3.2: The geographical portfolio effect

Source: Marco et al. (2007)

TECHNOLOGY RISK

The present 'sellers' market', characterised by the shortage in supply of wind turbines, has introduced a number of new turbine manufacturers, many of which are not financially strong and none of which has a substantial track record. Therefore, technology risk remains a concern for the banks, and the old-fashioned way of mitigating these risks, through extended warranties, is resisted forcefully by both new and experienced manufacturers. So technology risk has increased recently, rather than diminishing over time. However, some banks still show significant interest in lending to projects that use technology with relatively little operational experience.

OFFSHORE WIND

Offshore wind farms are now more common in Europe. The first few projects were financed in the way described above – by large companies with substantial financial clout, using their own funds. The initial involvement of banks was in the portfolio financing of a collection of assets, one of which was an offshore wind farm. Banks were concerned about the additional risks associated with an offshore development, and this approach allowed the risks to be diluted somewhat.

Although there are still relatively few offshore wind farms, banks are clearly interested in both term loans, associated with the operational phase of offshore wind farms, and the provision of construction finance. This clearly demonstrates the banks' appetite for wind energy lending. It is too early to define typical offshore financing, but it is likely to be more expensive than that for the equivalent onshore farm, at least until the banks gain greater confidence in the technology. The risk of poor availability as a result of poor accessibility is a particular concern.

BIG PROJECTS

Banks like big projects. The cost of the banks' own efforts and due diligence does not change significantly with the project (loan) size, so big projects are more attractive to them than smaller ones. Wind projects are only now starting to be big enough to interest some banks, so as project size increases, the banking community available to support the projects will grow. Furthermore, increasing project size brings more substantial sponsors, which is also reassuring for banks.

CONCLUSIONS

The nature of wind energy deals is changing. Although many small, privately owned projects remain, there has been a substantial shift towards bigger, utility-owned projects. This change brings new money to the industry, reduces dependence on banks for initial funding and brings strong sponsors.

Projects are growing and large-scale offshore activity is increasing. Since banks favour larger projects, this is a very positive change. If the general economic picture deteriorates, this may give rise to certain misgivings concerning project finance, in comparison to the last few years, but political and environmental support for renewable energy means that the funding of wind energy remains a very attractive proposition. Obtaining financing for the large-scale expansion of the industry will not be a problem.

III.4 PRICES AND SUPPORT MECHANISMS

Introduction: Types of RES-E Support Mechanism

When clustering the different types of support mechanisms available to electricity from renewable energy sources (RES-E), a fundamental distinction can be made between direct and indirect policy instruments. Direct policy measures aim to stimulate the installation of RES-E technologies immediately, whereas indirect instruments focus on improving long-term framework conditions. Besides regulatory instruments, voluntary approaches for the promotion of RES-E technologies also exist, mainly based on consumers' willingness to pay premium rates for green electricity. Further important classification criteria are whether policy instruments address price or quantity, and whether they support investments or generation.

Table III.4.1 provides a classification of the existing promotional strategies for renewables; there follows an explanation of the terminology used.

REGULATORY PRICE-DRIVEN STRATEGIES

Generators of RES-E receive financial support in terms of a subsidy per kW of capacity installed, or a payment per kWh produced and sold. The major strategies are:

- **investment-focused strategies**: financial support is given by investment subsidies, soft loans or tax credits (usually per unit of generating capacity); and
- **generation-based strategies**: financial support is a fixed regulated feed-in tariff (FIT) or a fixed premium (in addition to the electricity price) that a governmental institution, utility or supplier is legally obligated to pay for renewable electricity from eligible generators.

The difference between fixed FITs and premiums is as follows: for fixed FITs, the total feed-in price is fixed; for premium systems, the amount to be added to the electricity price is fixed. For the renewable plant owner, the total price received per kWh, in the premium scheme (electricity price plus the premium), is less predictable than under a feed-in tariff, since this depends on a volatile electricity price.

In principle, a mechanism based on a fixed premium/ environmental bonus that reflects the external costs of conventional power generation could establish fair trade, fair competition and a level playing field between RES and conventional power sources in a competitive electricity market. From a market

Table III.4.1: Types of RES-E support mechanism

	Direct		Indirect
	Price-driven	**Quantity-driven**	**Indirect**
Regulatory			
Investment-focused	• Investment incentives • Tax credits • Low interest/Soft loans	• Tendering system for investment grant	• Environmental taxes • Simplification of authorisation procedures
Generation-based	• (Fixed) feed-in tariffs • Fixed premium system	• Tendering system for long-term contracts Tradable Green Certificate system	• Connection charges, balancing costs
Voluntary			
Investment-focused	• Shareholder programmes • Contribution programmes		• Voluntary agreements
Generation-based	• Green tariffs		

Source: Ragwitz et al. (2007)

development perspective, the advantage of such a scheme is that it allows renewables to penetrate the market quickly if their production costs drop below the electricity price plus premium. If the premium is set at the 'right' level (theoretically at a level equal to the external costs of conventional power), it allows renewables to compete with conventional sources without the need for governments to set 'artificial' quotas. Together with taxing conventional power sources in accordance with their environmental impact, well-designed fixed premium systems are theoretically the most effective way of internalising external costs.

In practice, however, basing the mechanism on the environmental benefits of renewables is challenging. Ambitious studies, such as the European Commission's ExternE project, which investigates the external costs of power generation, have been conducted in both Europe and America; these suggest that establishing exact costs is a complex matter. In reality, fixed premiums for wind power and other renewable energy technologies, such as the Spanish model, are based on estimated production costs and are fixed based on the electricity price, rather than on the environmental benefits of RES.

REGULATORY QUANTITY-DRIVEN STRATEGIES

The desired level of RES generation or market penetration – a quota or a Renewable Portfolio Standard – is defined by governments. The most important points are:

- **Tendering or bidding systems**: calls for tender are launched for defined amounts of capacity. Competition between bidders results in contract winners that receive a guaranteed tariff for a specified period of time.
- **Tradable certificate systems**: these systems are better known in Europe as Tradable Green Certificate (TCG) systems and in the US and Japan as Renewable Portfolio Standards (RPSs). In such systems, the generators (producers), wholesalers, distribution

companies or retailers (depending on who is involved in the electricity supply chain) are obliged to supply or purchase a certain percentage of electricity from RES. At the date of settlement, they have to submit the required number of certificates to demonstrate compliance. Those involved may obtain certificates:

- from their own renewable electricity generation;
- by purchasing renewable electricity and associated certificates from another generator; and/or
- by purchasing certificates without purchasing the actual power from a generator or broker, that is to say purchasing certificates that have been traded independently of the power itself.

The price of the certificates is determined, in principle, according to the market for these certificates (for example NordPool).

VOLUNTARY APPROACHES

This type of strategy is mainly based on the willingness of consumers to pay premium rates for renewable energy, due to concerns over global warming, for example. There are two main categories:

1. **investment-focused**: the most important are shareholder programmes, donation projects and ethical input; and
2. **generation-based**: green electricity tariffs, with and without labelling.

INDIRECT STRATEGIES

Aside from strategies which directly address the promotion of one (or more) specific renewable electricity technologies, there are other strategies that may have an indirect impact on the dissemination of renewables. The most important are:

- eco-taxes on electricity produced with non-renewable sources;

- taxes/permits on CO_2 emissions; and
- the removal of subsidies previously given to fossil and nuclear generation.

There are two options for the promotion of renewable electricity via energy or environmental taxes:

1. exemption from taxes (such as energy and sulphur taxes); and
2. if there is no exemption for RES, taxes can be (partially or wholly) refunded.

Both measures make RES more competitive in the market and are applicable for both established and new plants.

Indirect strategies also include the institutional promotion of the deployment of RES plants, such as site planning and easy connection to the grid, and the operational conditions of feeding electricity into the system. First, siting and planning requirements can reduce the potential opposition to RES-E plants if they address issues of concern, such as noise and visual or environmental impacts. Laws can be used to, for example, set aside specific locations for development and/or omit areas that are particularly open to environmental damage or injury to birds.

Second, complementary measures also concern the standardisation of economic and technical connection conditions. Interconnection requirements are often unnecessarily onerous and inconsistent and can lead to high transaction costs for project developers, particularly if they need to hire technical and legal experts. Safety requirements are essential, particularly in the case of interconnection in weak parts of the grid. However, unclear criteria on interconnections can potentially lead to higher prices for access to the grid and use of transmission lines, or even unreasonable rejections of transmission access. Therefore, it is recommended that authorities clarify the safety requirements and the rules on the burden of additional expenses.

Finally, rules must be established to govern the responsibility for physical balancing associated with some technologies' variable production, in particular for wind power.

COMPARISON OF PRICE-DRIVEN VERSUS QUANTITY-DRIVEN INSTRUMENTS

In the following section, an assessment of direct promotional strategies is carried out by focusing on the comparison between price-driven (for example FITs, investment incentives and tax credits) and quantity-driven (for example Tradable Green Certificate (TGC)-based quotas and tendering systems) strategies. The different instruments can be described as follows:

- **Feed-in tariffs (FITs) are generation-based, price-driven incentives**. The price per unit of electricity that a utility, supplier or grid operator is legally obliged to pay for electricity from RES-E producers is determined by this system. Thus a federal (or provincial) government regulates the tariff rate. It usually takes the form of either a fixed price to be paid for RES-E production or an additional premium on top of the electricity market price paid to RES-E producers. Besides the level of the tariff, its guaranteed duration represents an important parameter for an appraisal of the actual financial incentive. FITs allow technology-specific promotion, as well as an acknowledgement of future cost reductions by applying dynamic decreasing tariffs.
- **Quota obligations based on Tradable Green Certificates (TGCs) are generation-based, quantity-driven instruments**. The government defines targets for RES-E deployment and requires a particular party in the electricity supply chain (for example the generator, wholesaler or consumer) to fulfil certain obligations. Once defined, a parallel market for renewable energy certificates is established and their price is set following demand and supply conditions (forced by the obligation). Hence for RES-E producers, financial support may arise from selling certificates, in addition to the

revenues from selling electricity on the power market. Technology-specific promotion in TGC systems, is also possible in principle. However, market separation for different technologies would lead to much smaller and less liquid markets. One solution could be to weight certificates from different technologies, but the key dilemma is how to find weights that are correct or at least widely accepted as fair.

- **Tendering systems are quantity-driven mechanisms**. Financial support can either be investment-focused or generation-based. In the first case, a fixed amount of capacity to be installed is announced and contracts are given following a predefined bidding process, which offers winners a set of favourable investment conditions, including investment grants per kW installed. The generation-based tendering systems work in a similar way; but instead of providing up-front support, they offer support in the form of a 'bid price' per kWh for a guaranteed duration.

- **Investment incentives are price-driven instruments** that establish an incentive for the development of RES-E projects as a percentage of total costs, or as a predefined amount of money per kW installed. The level of these incentives is usually technology-specific.

- **Production tax incentives are also price-driven, generation-based mechanisms** that work through payment exemptions from the electricity taxes applied to all producers. Hence this type of instrument differs from premium feed-in tariffs only in terms of the cash flow for RES-E producers; it represents a negative cost instead of additional revenue.

Overview of the Different RES-E Support Schemes in EU-27 Countries

Figure III.4.1 shows the evolution of the different RES-E support instruments from 1997 to 2007 in each of the EU-27 Member States. Some countries already have more than ten years' experience with RES-E support schemes.

Initially, in the old EU-15, only 8 out of the 15 Member States avoided a major policy shift between 1997 and 2005. The current discussion within EU Member States focuses on the comparison between two opposing systems – the FIT system and quota regulation in combination with a TGC market. The latter has recently replaced existing policy instruments in some European countries, such as Belgium, Italy, Sweden, the UK and Poland. Although these new systems were not introduced until after 2002, the announced policy changes caused investment instabilities prior to this date. Other policy instruments, such as tender schemes, are no longer used as the main policy scheme in any European country. However, there are instruments, such as production tax incentives and investment incentives, that are frequently used as supplementary instruments; only Finland and Malta use them as their main support scheme.

Table III.4.2 gives a detailed overview of the main support schemes for wind energy in the EU-27 Member States.

In Table III.4.3, a more detailed overview is provided on implemented RES-E support schemes in the EU-27 Member States in 2007, detailing countries, strategies and the technologies addressed. In the EU-27, FITs serve as the main policy instrument.

For a detailed overview of the EU Member States' support schemes, please refer to Appendix I.

Evaluation of the Different RES-E Support Schemes (Effectiveness and Economic Efficiency)

In reviewing and evaluating the different RES-E support schemes described above, the key question is whether each of these policy instruments has been a success. In order to assess the success of the

Figure III.4.1: Evolution of the main policy support schemes in EU-27 Member States

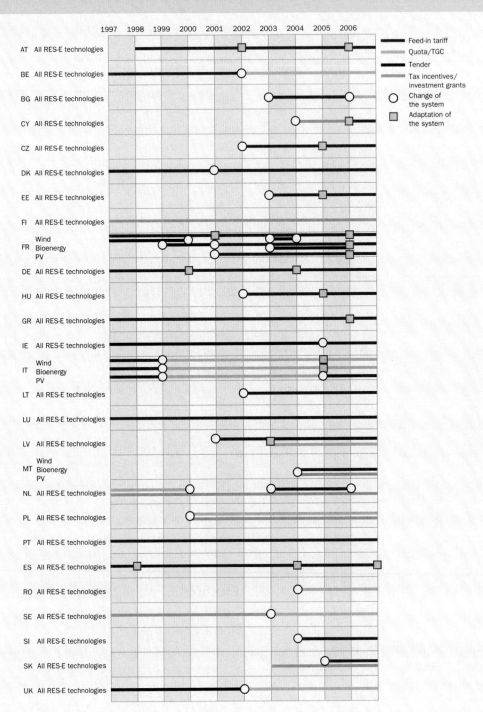

Source: Ragwitz et al. (2007)

Table III.4.2: Overview of the main RES-E support schemes for wind energy in the EU-27 Member States as implemented in 2007

Country	Main support instrument for wind	Settings of the main support instrument for wind in detail
Austria	FIT	New fixed feed-in tariff valid for new RES-E plants permitted in 2006 and/or 2007: fixed FIT for years 1–9 (€76.5/MWh for 2006 as a starting year; €75.5/MWh for 2007). Years 10 and 11 at 75% and year 12 at 50%.
Belgium	Quota obligation system with TGC, combined with minimum price for wind	Flanders, Wallonia and Brussels have introduced a quota obligation system (based on TGCs). The minimum price for wind onshore (set by the federal government) is €80/MWh in Flanders, €65/MWh in Wallonia and €50/MWh in Brussels. Wind offshore is supported at the federal level, with a minimum price of €90/MWh (the first 216 MW installed at €107/MWh minimum).
Bulgaria	Mandatory purchase price	Mandatory purchase prices (set by the State Energy Regulation Commission): new wind installations after 1 January 2006 (duration 12 years each): (i) effective operation >2250h/a: €79.8/MWh; (ii) effective operation <2250h/a: €89.5/MWh.
Cyprus	FIT	Fixed feed-in tariff since 2005: in the first 5 years €92/MWh, based on mean values of wind speeds; in the next 10 years €48–92/MWh according to annual wind operation hours (<1750–2000h/a: €85–92/MWh; 2000–2550h/a: €63–85/MWh; 2550–3300h/a: €48–63/MWh).
Czech Republic	Choice between FIT and Premium Tariff	Fixed feed-in tariff: €88–114/MWh in 2007 (duration: equal to the lifetime); Premium Tariff: €70–96/MWh in 2007 (duration: newly set every year).
Denmark	Market price and premium for wind onshore; tendering system for wind offshore	**Wind onshore**: Market price plus premium of €13/MWh (20 years); additionally, balancing costs are refunded at €3/MWh, leading to a total tariff of approximately €57/MWh. **Wind offshore**: €66–70/MWh (i.e. market price plus a premium of €13/MWh); a tendering system is applied for future offshore wind parks, balancing costs are borne by the owners.
Estonia	FIT	Fixed feed-in tariff for all RES: €52/MWh (from 2003 to present); current support mechanisms will be terminated in 2015.
Finland	Tax exemptions and investment subsidies	Mix of tax exemptions (refund) and investment subsidies: tax refund of €6.9/MWh for wind (€4.2/MWh for other RES-E). Investment subsidies up to 40% for wind (up to 30% for other RES-E).
France	FIT	**Wind onshore**: €82/MWh for 10 years; €28–82/MWh for the following 5 years (depending on the local wind conditions). **Wind offshore**: €130/MWh for 10 years; €30–130/MWh for the following 10 years (depending on the local wind conditions).
Germany	FIT	**Wind onshore (20 years in total)**: €83.6/MWh for at least 5 years; €52.8/MWh for further 15 years (annual reduction of 2% is taken into account). **Wind offshore (20 years in total)**: €91/MWh for at least 12 years; €61.9/MWh for further 8 years (annual reduction of 2% taken into account).
Greece	FIT	**Wind onshore**: €73/MWh (mainland); €84.6/MWh (autonomous islands). **Wind offshore**: €90/MWh (mainland); €90/MWh (autonomous islands); feed-in tariffs guaranteed for 12 years (possible extension up to 20 years).
Hungary	FIT	Fixed feed-in tariff (since 2006): €95/MWh; duration: according to the lifetime of technology.
Ireland	FIT	Fixed feed-in tariff (since 2006); guaranteed for 15 years: Wind >5 MW: €57/MWh; Wind <5 MW: €59/MWh.
Italy	Quota obligation system with TGCs	Obligation (based on TGCs) on electricity producers and importers. Certificates are issued for RES-E capacity during the first 12 years of operation, except biomass, which receives certificates for 100% of electricity production for first 8 years and 60% for next 4 years. In 2005 the average certificate price was €109/MWh.

continued

Table III.4.2: (continued)

Country	Main support instrument for wind	Settings of the main support instrument for wind in detail
Latvia	Main policy support instrument currently under development	Frequent policy changes and short duration of guaranteed feed-in tariffs (phased out in 2003) result in high investment uncertainty. Main policy currently under development.
Lithuania	FIT	Fixed feed-in tariff (since 2002): €63.7/MWh, guaranteed for 10 years.
Luxembourg	FIT	Fixed feed-in tariff: (i) <0.5 MW: €77.6/MWh; (ii) >0.5 MW: max €77.6/MWh (i.e. decreasing for higher capacities); guaranteed for 10 years.
Malta	No support instrument yet	Very little attention to RES-E (including wind) support so far. A low VAT rate is in place.
Netherlands	Premium Tariff (€0/MWh since August 2006)	Premium feed-in tariffs guaranteed for 10 years were in place from July 2003. For each MWh RES-E generated, producers receive a green certificate. The certificate is then delivered to the feed-in tariff administrator to redeem tariff. Government put all premium RES-E support at zero for new installations from August 2006 as it believed target could be met with existing applicants.
Poland	Quota obligation system; TGCs introduced end 2005 plus renewables are exempted from excise tax	Obligation on electricity suppliers with RES-E targets specified from 2005 to 2010. Poland has a RES-E and primary energy target of 7.5% by 2010. RES-E share in 2005 was 2.6% of gross electricity consumption.
Portugal	FIT	Fixed feed-in tariff (average value 2006): €74/MWh, guaranteed for 15 years.
Romania	Quota obligation system with TGCs	Obligation on electricity suppliers with targets specified from 2005 (0.7% RES-E) to 2010 (8.3% RES-E). Minimum and maximum certificate prices are defined annually by Romanian Energy Regulatory Authority. Non-compliant suppliers pay maximum price (i.e. €63/MWh for 2005–2007; €84/MWh for 2008–2012).
Slovakia	FIT	Fixed feed-in tariff (since 2005): €55–72/MWh; FITs for wind are set that way so that a rate of return on the investment is 12 years when drawing a commercial loan.
Slovenia	Choice between FIT and premium tariff	Fixed feed-in tariff: (i) <1 MW: €61/MWh; (ii) >1 MW: €59/MWh. Premium tariff: (i) <1 MW: €27/MWh; (ii) >1 MW: €25/MWh. Fixed feed-in tariff and premium tariff guaranteed for 5 years, then reduced by 5%. After ten years reduced by 10% (compared to original level).
Spain	Choice between FIT and premium tariff	Fixed feed-in tariff: (i) <5 MW: €68.9/MWh; (ii) >5 MW: €68.9/MWh; Premium tariff: (i) <5 MW: €38.3/MWh; (ii) >5 MW: €38.3/MWh. Duration: no limit, but fixed tariffs are reduced after either 15, 20 or 25 years, depending on technology.
Sweden	Quota obligation system with TGCs	Obligation (based on TGCs) on electricity consumers. Obligation level of 51% RES-E defined to 2010. Non-compliance leads to a penalty, which is fixed at 150% of the average certificate price in a year (average certificate price was €69/MWh in 2007).
UK	Quota obligation system with TGCs	Obligation (based on TGCs) on electricity suppliers. Obligation target increases to 2015 (15.4% RES-E; 5.5% in 2005) and guaranteed to stay at least at that level until 2027. Electricity companies which do not comply with the obligation have to pay a buy-out penalty (€65.3/MWh in 2005). Tax exemption for electricity generated from RES is available.

Sources: Auer (2008); Ragwitz et al. (2007)

Table III.4.3: Overview of the main RES-E support schemes in the EU-27 Member States as implemented in 2007

Country	Main electricity support schemes	Comments
Austria	FITs combined with regional investment incentives	Until December 2004, FITs were guaranteed for 13 years. In November 2005 it was announced that, from 2006 onwards, full FITs would be available for 10 years, with 75% available in year 11 and 50% in year 12. New FIT levels are announced annually and support is granted on a first-come, first-served basis. From May 2006 there has been a smaller government budget for RES-E support. At present, a new amendment is tabled, which suggests extending the duration of FIT fuel-independent technologies to 13 years (now 10 years) and fuel-dependent technologies to 15 years (now 10 years).
Belgium	Quota obligation system/ TGC combined with minimum prices for electricity from RES	The federal government has set minimum prices for electricity from RES. Flanders and Wallonia have introduced a quota obligation system (based on TGCs) with the obligation on electricity suppliers. In all three of the regions, including Brussels, a separate market for green certificates has been created. Offshore wind is supported at the federal level.
Bulgaria	Mandatory purchase of renewable electricity by electricity suppliers for minimum prices (essentially FITs) plus tax incentives	The relatively low level of incentives makes the penetration of renewables particularly difficult, since the current commodity prices for electricity are still relatively low. A green certificate system to support renewable electricity developments has been proposed, for implementation in 2012, to replace the mandatory purchase price. Bulgaria recently agreed upon an indicative target for renewable electricity with the European Commission, which is expected to provide a good incentive for further promotion of renewable support schemes.
Cyprus	FITs (since 2006), supported by investment grant scheme for the promotion of RES	An Enhanced Grant Scheme was introduced in January 2006, in the form of government grants worth 30–55% of investment, to provide financial incentives for all renewable energy. FITs with long-term contracts (15 years) were also introduced in 2006.
Czech Republic	FITs (since 2002), supported by investment grants	Relatively high FITs, with a lifetime guarantee of support. Producers can choose fixed FITs or a premium tariff (green bonus). For biomass co-generation, only green bonus applies. FIT levels are announced annually, but are increased by at least 2% each year.
Denmark	Premium FIT for onshore wind, tender scheme for offshore wind and fixed FITs for others	Duration of support varies from 10 to 20 years, depending on the technology and scheme applied. The tariff level is generally rather low compared to the formerly high FITs. A net metering approach is taken for photovoltaics.
Estonia	FIT system	FITs paid for 7–12 years, but not beyond 2015. Single FIT level for all RES-E technologies. Relatively low FITs make new renewable investments very difficult.
Finland	Energy tax exemption combined with investment incentives	Tax refund and investment incentives of up to 40% for wind and up to 30% for electricity generation from other RES.
France	FITs plus tenders for large projects	For power plants <12 MW, FITs are guaranteed for 15 or 20 years (offshore wind, hydro and PV). From July 2005, FIT for wind is reserved for new installations within special wind energy development zones. For power plants >12 MW (except wind) a tendering scheme is in place.
Germany	FITs	FITs are guaranteed for 20 years (Renewable Energy Act) and soft loans are also available.
Greece	FITs combined with investment incentives	FITs are guaranteed for 12 years, with the possibility of extension up to 20 years. Investment incentives up to 40%.
Hungary	FIT (since January 2003, amended 2005), combined with purchase obligation and grants	Fixed FITs recently increased and differentiated by RES-E technology. There is no time limit for support defined by law, so in theory guaranteed for the lifetime of the installation. Plans to develop TGC system; when this comes into effect, the FIT system will cease to exist.
Ireland	FIT scheme replaced tendering scheme in 2006	New premium FITs for biomass, hydropower and wind started in 2006. Tariffs guaranteed to supplier for up to 15 years. Purchase price of electricity from the generator is negotiated between generators and suppliers. However, support may not extend beyond 2024, so guaranteed premium FIT payments should start no later than 2009.

continued

Table III.4.3: (continued)

Country	Main electricity support schemes	Comments
Italy	Quota obligation system with TGCs; fixed FIT for PV	Obligation (based on TGCs) on electricity producers and importers. Certificates are issued for RES-E capacity during the first 12 years of operation, except for biomass, which receives certificates for 100% of electricity production for the first 8 years and 60% for the next 4 years. Separate fixed FIT for PV, differentiated by size, and building integrated. Guaranteed for 20 years. Increases annually in line with retail price index.
Latvia	Main policy under development; quota obligation system (since 2002) without TGCs, combined with FITs (phased out in 2003)	Frequent policy changes and short duration of guaranteed FITs result in high investment uncertainty. Main policy currently under development. Quota system (without TGCs) typically defines small RES-E amounts to be installed. High FIT scheme for wind and small hydropower plants (less than 2 MW) was phased out as from January 2003.
Lithuania	FITs combined with purchase obligation	Relatively high fixed FITs for hydro (<10 MW), wind and biomass, guaranteed for 10 years. Closure of Ignalina nuclear plant, which currently supplies the majority of electricity in Lithuania, will strongly affect electricity prices and thus the competitive position of renewables, as well as renewable support. Good conditions for grid connections. Investment programmes limited to companies registered in Lithuania. Plans exist to introduce a TGC system after 2010.
Luxembourg	FITs	FITs guaranteed for 10 years (20 years for PV). Also investment incentives available.
Malta	Low VAT rate and very low FIT for solar	Very little attention to RES support so far. Very low FIT for PV is a transitional measure.
Netherlands	FITs (tariff zero from August 2006)	Premium FITs guaranteed for 10 years have been in place since July 2003. For each MWh RES-E generated, producers receive a green certificate from the issuing body (CERTIQ). Certificate is then delivered to FIT administrator (ENERQ) to redeem tariff. Government put all premium RES-E support at zero for new installations from August 2006, as it believed target could be met with existing applicants. Premium for biogas (<2 MWe) immediately reinstated. New support policy under development. Fiscal incentives for investments in RES are available. Energy tax exemption for electricity from RES ceased 1 January 2005.
Poland	Quota obligation system; TGCs introduced from end 2005, plus renewables are exempted from the (small) excise tax	Obligation on electricity suppliers with targets specified from 2005 to 2010. Penalties for non-compliance were defined in 2004, but were not properly enforced until end of 2005. It has been indicated that from 2006 onwards the penalty will be enforced.
Portugal	FITs combined with investment incentives	Fixed FITs guaranteed for 15 years. Level dependent on time of electricity generation (peak/off peak), RES-E technology and resource. Is corrected monthly for inflation. Investment incentives up to 40%.
Romania	Quota obligation with TGCs; subsidy fund (since 2004)	Obligation on electricity suppliers, with targets specified from 2005 to 2010. Minimum and maximum certificate prices are defined annually by Romanian Energy Regulatory Authority. Non-compliant suppliers pay maximum price. Romania recently agreed on an indicative target for renewable electricity with the European Commission, which is expected to provide a good incentive for further promotion of renewable support schemes.
Slovak Republic	Programme supporting RES and energy efficiency, including FITs and tax incentives	Fixed FIT for RES-E was introduced in 2005. Prices set so that a rate of return on the investment is 12 years when drawing a commercial loan. Low support, lack of funding and lack of longer-term certainty in the past have made investors very reluctant.

continued

Table III.4.3: (continued)

Country	Main electricity support schemes	Comments
Slovenia	FITs, CO$_2$ taxation and public funds for environmental investments	Renewable electricity producers choose between fixed FITs and premium FITs. Tariff levels defined annually by Slovenian Government (but have not changed since 2004). Tariff guaranteed for 5 years, then reduced by 5%. After 10 years, reduced by 10% (compared to original level). Relatively stable tariffs combined with long-term guaranteed contracts makes system quite attractive to investors.
Spain	FITs	Electricity producers can choose a fixed FIT or a premium on top of the conventional electricity price. No time limit, but fixed tariffs are reduced after either 15, 20 or 25 years depending on technology. System very transparent. Soft loans, tax incentives and regional investment incentives are available.
Sweden	Quota obligation system with TGCs	Obligation (based on TGCs) on electricity consumers. Obligation level defined to 2010. Non-compliance leads to a penalty, which is fixed at 150% of the average certificate price in a year. Investment incentive and a small environmental bonus available for wind energy.
UK	Quota obligation system with TGCs	Obligation (based on TGCs) on electricity suppliers. Obligation target increases to 2015 and guaranteed to stay at that level (as a minimum) until 2027. Electricity companies that do not comply with the obligation have to pay a buy-out penalty. Buy-out fund is recycled back to suppliers in proportion to the number of TGCs they hold. The UK is currently considering differentiating certificates by RES-E technology. Tax exemption for electricity generated from RES is available (Levy Exemption Certificates, which give exemption from the Climate Change Levy).

Source: Ragwitz et al. (2007)

different policy instruments, the most important criteria are:

- **Effectiveness**: Did the RES-E support programmes lead to a significant increase in deployment of capacities from RES-E in relation to the additional potential? The effectiveness indicator measures the relationship of the new generated electricity within a certain time period to the potential of the technologies.
- **Economic efficiency**: What was the absolute support level compared to the actual generation costs of RES-E generators, and what was the trend in support over time? How is the net support level of RES-E generation consistent with the corresponding effectiveness indicator?

Other important performance criteria are the credibility for investors and the reduction of costs over time. However, effectiveness and economic efficiency

are the two most important criteria – these are discussed in detail in the following sections.

EFFECTIVENESS OF POLICY INSTRUMENTS

When analysing the effectiveness of RES-E support instruments, the quantities installed are of particular interest. In order to be able to compare the performance between the different countries, the figures are related to the size of the population. Here we look at all new RES-E in total, as well as wind and PV in detail.

Figure III.4.2 depicts the effectiveness of total RES-E policy support for the period 1998 to 2005, measured in yearly additional electricity generation in comparison to the remaining additional available potential for each EU-27 Member State. The calculations refer to following principle:

$$E_n^i = \frac{G_n^i - G_{n-1}^i}{ADDPOT_n^i} = \frac{G_n^i - G_{n-1}^i}{POT_{2020}^i - G_{n-1}^i}$$

Figure III.4.2: Policy effectiveness of total RES-E support for 1998–2005 measured in annual additional electricity generation in comparison to the remaining additional available potential for each EU-27 Member State

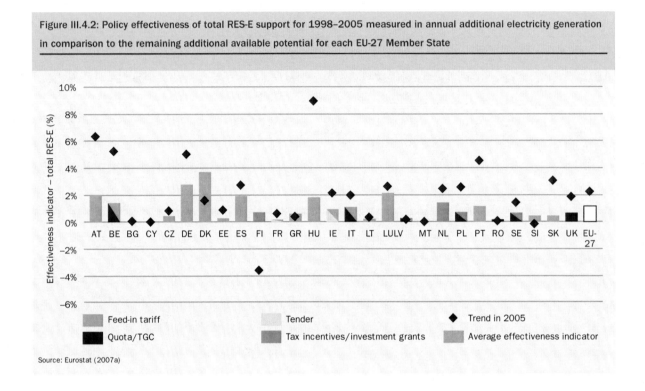

Source: Eurostat (2007a)

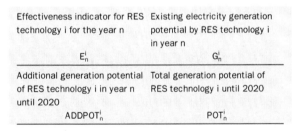

Effectiveness indicator for RES technology i for the year n	Existing electricity generation potential by RES technology i in year n
E_n^i	G_n^i
Additional generation potential of RES technology i in year n until 2020	Total generation potential of RES technology i until 2020
$ADDPOT_n^i$	POT_n^i

It is clearly indicated in Figure III.4.2 that countries with FITs as a support scheme achieved higher effectiveness compared to countries with a quota/TGC system or other incentives. Denmark achieved the highest effectiveness of all the Member States, but it is important to remember that very few new generation plants have been installed in recent years. Conversely, in Germany and Portugal there has been a significant increase in new installations recently. Among the new Member States, Hungary and Poland have implemented the most efficient strategies in order to promote 'new' renewable energy sources.

ECONOMIC EFFICIENCY

Next we compare the economic efficiency of the support programmes described above. In this context, three aspects are of interest:

1. absolute support levels;
2. total costs to society; and
3. dynamics of the technology.

Here, as an indicator, the support levels are compared specifically for wind power in the EU-27 Member States.

Figure III.4.3 shows that the support level and generation costs are almost equal. Countries with relatively high average generation costs frequently show a higher support level, but a clear deviation from this rule can be found in the three quota systems in Belgium, Italy and the UK, for which the support is presently significantly higher than the generation costs.

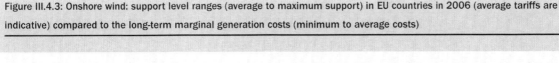

Figure III.4.3: Onshore wind: support level ranges (average to maximum support) in EU countries in 2006 (average tariffs are indicative) compared to the long-term marginal generation costs (minimum to average costs)

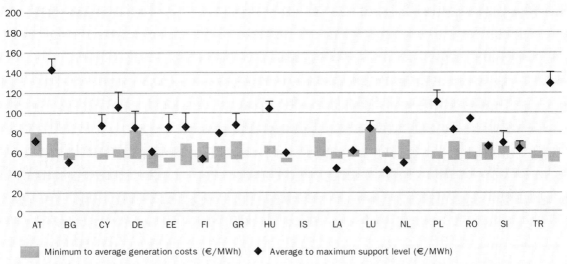

Minimum to average generation costs (€/MWh)   ◆ Average to maximum support level (€/MWh)

Note: Support level is normalised to 15 years.

Source: Adapted from Ragwitz et al. (2007)

The reasons for the higher support level, expressed by the current green certificate prices, may differ; but the main reasons are risk premiums, immature TGC markets and inadequate validity times of certificates (Italy and Belgium).

For Finland, the level of support for onshore wind is too low to initiate any steady growth in capacity. In the case of Spain and Germany, the support level indicated in Figure III.4.3 appears to be above the average level of generation costs. However, the potential with fairly low average generation costs has already been exploited in these countries, due to recent market growth. Therefore, a level of support that is moderately higher than average costs seems to be reasonable. In an assessment over time, the potential technology learning effects should also be taken into account in the support scheme.

Figure III.4.4 illustrates a comparative overview of the ranges of TGC prices and FITs in selected EU-27 countries. With the exception of Sweden, TGC prices

are much higher than those for guaranteed FITs, which also explains the high level of support in these countries, as shown in Figure III.4.4.

Policy Recommendations for the Design Criteria of RES-E Support Instruments

CONSIDERATION OF A DYNAMIC PORTFOLIO OF RES-E SUPPORT SCHEMES

Regardless of whether a national or international support system is concerned, a single instrument is not usually enough to stimulate the long-term growth of RES-E. Since, in general, a broad portfolio of RES technologies should be supported, the mix of instruments selected should be adjusted accordingly. Whereas investment grants are normally suitable for supporting immature technologies, FITs are appropriate for the interim stage of the market introduction of a

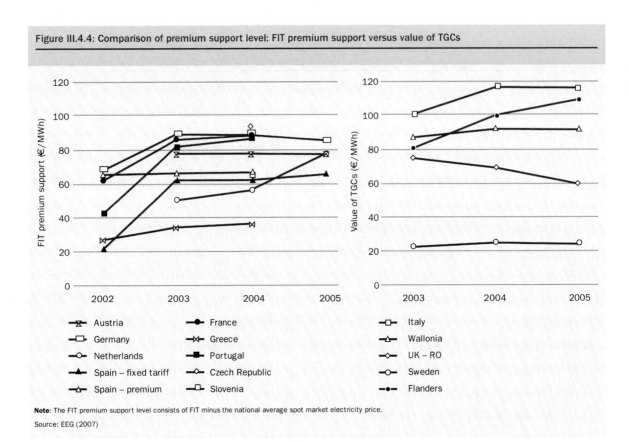

Figure III.4.4: Comparison of premium support level: FIT premium support versus value of TGCs

Legend:
- Austria
- Germany
- Netherlands
- Spain – fixed tariff
- Spain – premium
- France
- Greece
- Portugal
- Czech Republic
- Slovenia
- Italy
- Wallonia
- UK – RO
- Sweden
- Flanders

Note: The FIT premium support level consists of FIT minus the national average spot market electricity price.

Source: EEG (2007)

technology. A premium, or a quota obligation based on TGC, is likely to be the most relevant choice when:

- markets and technologies are sufficiently mature;
- the market size is large enough to guarantee competition among the market actors; and
- competition on the conventional power market is guaranteed.

Such a mix of instruments can then be supplemented by tender procedures, which are very efficient, for example, in the case of large-scale projects such as offshore wind.

STRIVING FOR OPTIMAL RES-E INSTRUMENT DESIGN IN TERMS OF EFFECTIVENESS AND EFFICIENCY

Most instruments still have significant potential for improvement, even after the minimum design criteria

described above have been met. A few examples of such optimisation options are as follows:

- In a feed-in system, a stepped design can clearly increase the economic efficiency of the instrument, especially in countries where the productivity of a technology varies significantly between different technology bands.
- In quota systems based on TGCs, the technology or band specification that is currently being tested in Italy (based on technology-specific certification periods) and in Belgium (based on technology-specific certificate values) may be a relevant option for increasing both the instrument's effectiveness and its efficiency. However, such technology specification should not be carried out by setting technology-specific quotas and separating the TGC market, as this would negatively influence market liquidity. Furthermore, the risk premium might go down if minimum tariffs were to be introduced in a quota system.

III.5   WIND POWER ON THE SPOT MARKET

Introduction

In a number of countries, wind power has an increasing share of total power production. This applies particularly to countries such as Denmark, Spain and Germany, where the shares of wind in terms of total power supply are currently 21 per cent, 12 per cent and 7 per cent respectively. In these cases, wind power is becoming an important player in the power market, and such high shares can significantly influence prices.

Different power market designs have a significant influence on the integration of wind power. In the following section, short descriptions of the most important market designs within the increasingly liberalised European power industry are presented, along with more detailed descriptions of spot and balancing markets. Finally, the impacts of Danish wind power on the Scandinavian power exchange, NordPool's Elspot, which comprises Denmark, Norway, Sweden and Finland, are discussed in more detail.

Power Markets

As part of the gradual liberalisation of the EU electricity industry, power markets are increasingly organised in a similar way, where a number of closely related services are provided. This applies to a number of liberalised power markets, including those of the Nordic countries, Germany, France and The Netherlands. Common to all these markets is the existence of five types of power market:

- **Bilateral electricity trade or OTC (over the counter) trading:** Trading takes place bilaterally outside the power exchange, and prices and amounts are not made public.
- **The day-ahead market (spot market):** A physical market where prices and amounts are based on supply and demand. Resulting prices and the overall amounts traded are made public. The spot market is a day-ahead market where bidding closes

at noon for deliveries from midnight and 24 hours ahead.
- **The intraday market:** Quite a long time period remains between close of bidding on the day-ahead market and the regulating power market (below). The intraday market is therefore introduced as an 'in-between market', where participants in the day-ahead market can trade bilaterally. Usually, the product traded is the one-hour-long power contract. Prices are published and based on supply and demand.
- **The regulating power market (RPM):** A real-time market covering operation within the hour. The main function of the RPM is to provide power regulation to counteract imbalances related to day-ahead operations planned. Transmission system operators (TSOs) alone make up the demand side of this market, and approved participants on the supply side include both electricity producers and consumers.
- **The balancing market:** This market is linked to the RPM and handles participant imbalances recorded during the previous 24-hour period of operation. The TSO alone acts on the supply side to settle imbalances. Participants with imbalances on the spot market are price takers on the RPM/balance market.

The day-ahead and regulating markets are particularly important for the development and integration of wind power in the power systems. The Nordic power exchange, NordPool, is described in more detail in the following section as an example of these power markets.

THE NORDIC POWER MARKET: NordPool SPOT MARKET

The NordPool spot market (Elspot) is a day-ahead market, where the price of power is determined by supply and demand. Power producers and consumers submit their bids to the market 12 to 36 hours in advance of delivery, stating the quantities of electricity supplied or demanded and the corresponding price. Then, for each hour, the price that clears the market (balancing supply with demand) is determined on the NordPool power exchange.

In principle, all power producers and consumers can trade on the exchange, but in reality, only big consumers (distribution and trading companies and large industries) and generators act on the market, while the smaller companies form trading cooperatives (as is the case for wind turbines), or engage with larger traders to act on their behalf. Approximately 45 per cent of total electricity production in the Nordic countries is traded on the spot market. The remaining share is sold through long-term, bilateral contracts, but the spot price has a considerable impact on prices agreed in such contracts. In Denmark, the share sold at the spot market is as high as 80 per cent.

Figure III.5.1 shows a typical example of an annual supply and demand curve for the Nordic power system. As shown, the bids from nuclear and wind power enter the supply curve at the lowest level, due to their low marginal costs, followed by combined heat and power plants, while condensing plants are those with the highest marginal costs of power production. Note that hydropower is not identified on the figure, since bids from hydro tend to be strategic and depend on precipitation and the level of water in reservoirs.

In general, the demand for power is highly inelastic (meaning that demand remains almost unchanged in spite of a change in the power price), with mainly Norwegian and Swedish electro-boilers and power-intensive industry contributing to the very limited price elasticity.

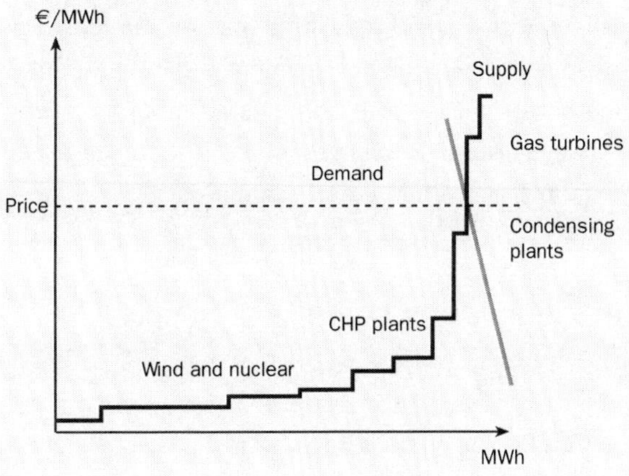

If power can flow freely in the Nordic area – that is to say, transmission lines are not congested – then there will only be one market price. But if the required power trade cannot be handled physically, due to transmission constraints, the market is split into a number of sub-markets, defined by the pricing areas. For example, Denmark splits into two pricing areas (Jutland/Funen and Zealand). Thus, if more power is produced in the Jutland/Funen area than consumption and transmission capacity can cover, this area would constitute a sub-market, where supply and demand would balance out at a lower price than in the rest of the NordPool area.

THE NORDIC POWER MARKET: THE REGULATING MARKET

Imbalances in the physical trade on the spot market must be levelled out in order to maintain the balance between production and consumption, and to maintain power grid stability. Totalling the deviations from bid volumes on the spot market yields a net imbalance for that hour in the system as a whole. If the grid is congested, the market breaks up into area markets, and equilibrium must be established in each area. The main tool for correcting such imbalances, which provides the necessary physical trade and accounting in the liberalised Nordic electricity system, is the regulating market.

Figure III.5.1: Supply and demand curve for the NordPool power exchange

Source: Risø DTU

The regulating power market and the balancing market may be regarded as one entity, where the TSO acts as an important intermediary or facilitator between the supply and demand of regulating power. The TSO is the body responsible for securing the system functioning in a region. Within its region, the TSO controls and manages the grid, and to this end, the combined regulating power and balancing market is an important tool for managing the balance and stability of the grid (Nordel, 2002). The basic principle for settling imbalances is that participants causing or contributing to the imbalance will pay their share of the costs for re-establishing balance. Since September 2002, the settling of imbalances among Nordic countries has been done based on common rules. However, the settling of imbalances within a region differs from country to country. Work is being done in Nordel to analyse the options for harmonising these rules in the Nordic countries.

If the vendors' offers or buyers' bids on the spot market are not fulfilled, the regulating market comes into force. This is especially important for wind electricity producers. Producers on the regulating market have to deliver their offers one or two hours before the hour of delivery, and power production must be available within 15 minutes of notice being given. For these reasons, only fast-response power producers will normally be able to deliver regulating power.

It is normally only possible to predict the supply of wind power with a certain degree of accuracy 12–36 hours in advance. Consequently, it may be necessary to pay a premium for the difference between the volume offered to the spot market and the volume delivered. Figure III.5.2 shows how the regulatory market functions in two situations: a general deficit on the market (left part of the figure) and a general surplus on the market (right part of figure).

If the market tends towards a deficit of power, and if power production from wind power plants is lower than offered, other producers will have to adjust regulation (up) in order to maintain the power balance. In this case, the wind producer will be penalised and get a lower price for their electricity production than the spot market price. The further off-track the wind producer is, the higher the expected penalty. The difference between the regulatory curves and the stipulated spot market price in Figure III.5.2 illustrates this. If wind power production is higher than the amount offered, wind power plants effectively help to eliminate market deficit and therefore receive the spot price for the full production without paying a penalty.

If the market tends towards an excess of power, and if power production from the wind power plant is higher than offered, other producers will have to adjust regulation (down) in order to maintain the power balance.

Figure III.5.2: The functioning of the regulatory market

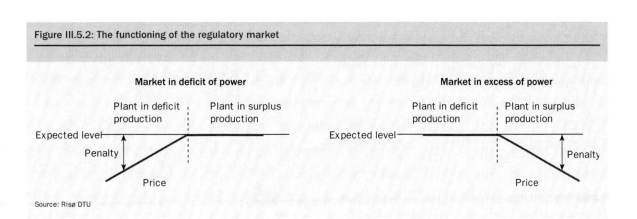

Source: Risø DTU

In this case, the wind producer will be penalised and get a lower price for their electricity production than the spot market price. Again, the further off-track the wind producer, the higher the expected premium. However, if wind power production is lower than the bid, then wind power plants help to eliminate surplus on the market, and therefore receive the spot price for the full amount of production without paying a penalty.

Until the end of 2002, each country participating in the NordPool market had its own regulatory market. In Denmark, balancing was handled by agreements with the largest power producers, supplemented by the possibility of TSOs buying balancing power from abroad if domestic producers were too expensive or unable to produce the required volumes of regulatory power. A common Nordic regulatory market was established at the beginning of 2003, and both Danish areas participate in this market.

In Norway, Sweden and Finland, all suppliers on the regulating market receive the marginal price for power regulation at the specific hour. In Denmark, market suppliers get the price of their bid to the regulation market. If there is no transmission congestion, the regulation price is the same in all areas. If bottlenecks occur in one or more areas, bids from these areas on the regulating market are not taken into account when forming the regulation price for the rest of the system, and the regulation price within the area will differ from the system regulation price.

In Norway, only one regulation price is defined, and this is used both for sale and purchase, at the hour when settling the imbalances of individual participants. This implies that participants helping to eliminate imbalances are rewarded even if they do not fulfil their actual bid. Thus if the market is in deficit of power and a wind turbine produces more than its bid, then the surplus production is paid a regulation premium corresponding to the penalty for those plants in deficit.

The Impact of Wind Power on the Power Market – Illustrated by the Case of Denmark

Denmark has a total capacity of a little more than 3200 MW of wind power – approximately 2800 MW from land turbines and 400MW offshore. In 2007, around 21 per cent of domestic power consumption was supplied by wind power, which makes Denmark the leading country in terms of wind power penetration (followed by Spain, where the share of wind as a total of electricity consumption is 12 per cent).

Figure III.5.3 shows wind power's average monthly coverage of power consumption in Denmark. Normally, the highest wind-generated production is from January to March. However, as 2006 was a bad wind year in Denmark, this was not the case. The contribution during the summer is normally at a fairly low level.

Considerable hourly variations are found in wind power production for Western Denmark, as illustrated in Figure III.5.4. January 2007 was a tremendously

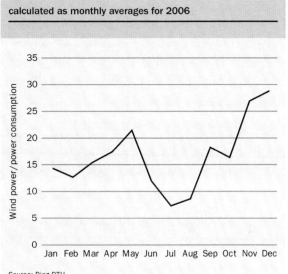

Figure III.5.3: The share of wind power in power consumption calculated as monthly averages for 2006

Source: Risø DTU

Figure III.5.4: Wind power as a percentage of domestic power consumption in January 2007 (hourly basis)

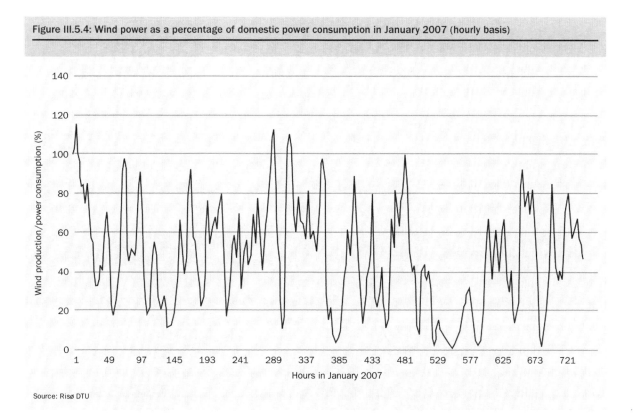

Source: Risø DTU

good wind month, with an average supply of 44 per cent of power consumption in Western Denmark, and, as shown, wind-generated power exceeded power consumption on several occasions. Nevertheless, there were also periods with low or no wind in January. In such cases, wind power can significantly influence price determination on the power market. This will be discussed in more detail in the following section.

HOW DOES WIND POWER INFLUENCE THE POWER PRICE AT THE SPOT MARKET?

Wind power is expected to influence prices on the power market in two ways:

1. Wind power normally has a low marginal cost (zero fuel costs) and therefore enters near the bottom of the supply curve. This shifts the supply curve to the right (see Figure III.5.5), resulting in a lower power price, depending on the price elasticity of the power demand. In general, the price of power is expected to be lower during periods of high wind than in periods of low wind.

2. As mentioned above, there may be congestions in power transmission, especially during periods with high wind power generation. Thus, if the available transmission capacity cannot cope with the required power export, the supply area is separated from the rest of the power market and constitutes its own pricing area. With an excess supply of power in this area, conventional power plants have to reduce their production, since it is generally not possible to limit the power production of wind. In most cases, this will lead to a lower power price in this sub-market.

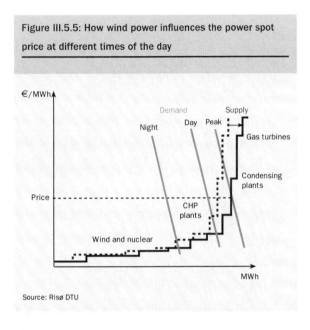

Figure III.5.5: How wind power influences the power spot price at different times of the day

Source: Risø DTU

The way in which wind power influences the power spot price, due to its low marginal cost, is shown in Figure III.5.5. When wind power supply increases, it shifts the power supply curve to the right. At a given demand, this implies a lower spot price on the power market, as shown. However, the impact of wind power depends on the time of day. If there is plenty of wind power at midday, during peak power demand, most of the available generation will be used. This implies that we are at the steep part of the supply curve (see Figure III.5.5) and, consequently, wind power will have a strong impact, reducing the spot power price significantly. But if there is plenty of wind-produced electricity during the night, when power demand is low and most power is produced on base load plants, we are at the flat part of the supply curve and consequently the impact of wind power on the spot price is low.

The congestion problem arises because Denmark, especially the Western Region, has a very high share of wind power, and in cases of high wind power production, transmission lines are often fully utilised.

In Figure III.5.6, this congestion problem is illustrated for January 2007, when the share of wind-generated electricity in relation to total power consumption for West Denmark was more than 100 per cent at certain periods (Figure III.5.6 left part). This means that during these periods, wind power supplied more than all the power consumed in that area. If the prioritised production from small, decentralised CHP plants is added on top of wind power production, there are several periods with a significant excess supply of power, part of which may be exported. However, when transmission lines are fully utilised, there is a congestion problem. In that case, equilibrium between demand and supply needs to be reached within the specific power area, requiring conventional producers to reduce their production, if possible. The consequences for the spot power price are shown in the right graph of Figure III.5.6. By comparing the two graphs, it can be clearly seen that there is a close relationship between wind power in the system and changes in the spot price for this area.

The consequences of the two issues mentioned above for the West Denmark power supply area are discussed below. It should be mentioned that similar studies are available for Germany and Spain, which show almost identical results.

IMPACTS OF WIND POWER ON SPOT PRICES

The analysis here entails the impacts of wind power on power spot prices being quantified using structural analyses. A reference is fixed, corresponding to a situation with zero contribution from wind power in the power system. A number of levels with increasing contributions from wind power are then identified and, relating to the reference, the effect of wind power's power production is calculated. This is illustrated in the left-hand graph in Figure III.5.7, where the shaded area between the two curves approximates the value of wind power in terms of lower spot power prices.

In the right-hand graph in Figure III.5.7, more detail is shown with figures from the West Denmark area. Five levels of wind power production and the corresponding

Figure III.5.6: Left – wind power as percentage of power consumption in Western Denmark; Right – spot prices for the same area and time period

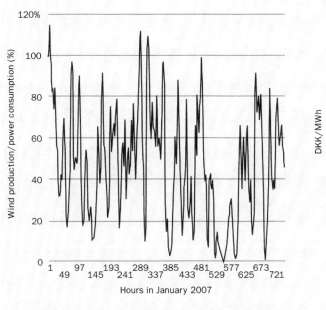

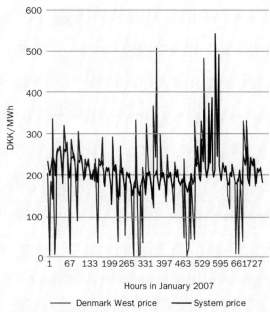

Source: Risø DTU

power prices are depicted for each hour of the day during December 2005. The reference is given by the '0–150 MW' curve, which thus approximates those hours of the month when the wind was not blowing. Therefore, this graph should approximate the prices for an average day in December 2005, in a situation with zero contribution from wind power. The other curves show increasing levels of wind power production: the 150–500 MW curve shows a situation with low wind, increasing to storm in the >1500 MW curve. As shown, the higher the wind power production, the lower the spot power price is in this area. At very high levels of wind power production, the power price is reduced significantly during the day, but only falls slightly during the night. Thus there is a significant impact on the power price, which might increase in the long term if even larger shares of wind power are fed into the system.

Figure III.5.7 relates to December 2005, but similar figures are found for most other periods during 2004 and 2005, especially in autumn and winter, owing to the high wind power production in these periods.

Of course, 'noise' in the estimations does exist, implying 'overlap' between curves for the single categories of wind power. Thus, a high amount of wind power does not always imply a lower spot price than that with low wind power production, indicating that a significant statistical uncertainty exists. Of course, factors other than wind power production influence prices on the spot market. But the close correlation between wind power and spot prices is clearly verified by a regression analysis carried out using the West Denmark data for 2005, where a significant relationship is found between power prices, wind power production and power consumption.

Figure III.5.7: The impact of wind power on the spot power price in the West Denmark power system in December 2005

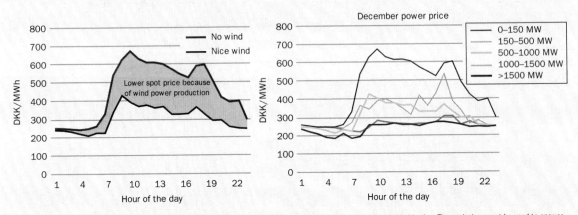

Note: The calculation only shows how the production contribution from wind power influences power prices when the wind is blowing. The analysis cannot be used to answer the question 'What would the power price have been if wind power was not part of the energy system?'.

Source: Risø DTU

When wind power reduces the spot power price, it has a significant influence on the price of power for consumers. When the spot price is lowered, this is beneficial to all power consumers, since the reduction in price applies to all electricity traded – not only to electricity generated by wind power.

Figure III.5.8 shows the amount saved by power consumers in Western and Eastern Denmark due to wind

Figure III.5.8: Annual percentage and absolute savings by power consumers in Western and Eastern Denmark in 2004–2007 due to wind power depressing the spot market electricity price

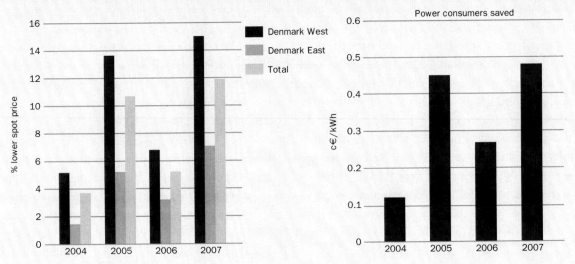

Source: Risø DTU

power's contribution to the system. Two calculations were performed: one using the lowest level of wind power generation as the reference ('0–150 MW'), in other words assuming that the power price would have followed this level if there was no contribution from wind power in the system, and the other more conservative, utilising a reference of above 500 MW. For each hour, the difference between this reference level and the levels with higher production of wind power is calculated. Summing the calculated amounts for all hours of the year gives the total benefit for power consumers of wind power lowering spot prices of electricity.

Figure III.5.8 shows how much higher the consumer price would have been (excluding transmission tariffs, taxes and VAT) if wind power had not contributed to power production.

In general in 2004–2007, the cost of power to the consumer (excluding transmission and distribution tariffs, taxes, and VAT) would have been approximately 4–12 per cent higher in Denmark if wind power had not contributed to power production. Wind power's strongest impact is estimated to have been for Western Denmark, due to the high penetration of wind power in this area. In 2007, this adds up to approximately 0.5c€/kWh saved by power consumers as a result of wind power lowering electricity prices, compared to the support given to wind power as FITs of approximately 0.7c€/kWh. Thus, although the expenses of wind power are still greater than the financial benefits for power consumers, a significant reduction of net expenses is certainly achieved due to lower spot prices.

Finally, though having a smaller impact, wind power clearly reduces power prices even within the large Nordic power system. Thus although wind power in the Nordic countries is mainly established in Denmark, all Nordic power consumers benefit financially due to the presence of Danish wind power on the market.

III.6 WIND POWER COMPARED TO CONVENTIONAL POWER GENERATION

In this chapter, the cost of conventionally generated power is compared with the cost of wind-generated power. To obtain a comparable picture, calculations for conventional technologies are prepared utilising the Recabs model, which was developed in the IEA Implementing Agreement on Renewable Energy Technology Deployment (IEA, 2008). The cost of conventional electricity production in general is determined by four components:

1. fuel cost;
2. cost of CO_2 emissions (as given by the European Trading System for CO_2, ETS);
3. operation and maintenance (O&M) costs; and
4. capital cost, including planning and site work.

Fuel prices are given by the international markets and, in the reference case, are assumed to develop according to the IEA's *World Energy Outlook 2007* (IEA, 2007c), which rather conservatively assumes a crude oil price of US$63/barrel in 2007, gradually declining to $59/barrel in 2010 (constant terms). Oil prices reached a high of $147/barrel in July 2008. As is normally observed, natural gas prices are assumed to follow the crude oil price (basic assumptions on other fuel prices: coal €1.6/GJ and natural gas €6.05/GJ). As mentioned, the price of CO_2 is determined by the EU ETS market; at present the price of CO_2 is around €25/t.

Here, calculations are carried out for two state-of-the-art conventional plants: a coal-fired power plant and a combined cycle natural gas combined heat and power plant, based on the following assumptions:

- Plants are commercially available for commissioning by 2010.
- Costs are levelised using a 7.5 per cent real discount rate and a 40-year lifetime (national assumptions on plant lifetime might be shorter, but calculations were adjusted to 40 years).
- The load factor is 75 per cent.
- Calculations are carried out in constant 2006-€.

When conventional power is replaced by wind-generated electricity, the costs avoided depend on the degree to which wind power substitutes for each of the four components. It is generally accepted that implementing wind power avoids the full fuel and CO_2 costs, as well as a considerable portion of the O&M costs of the displaced conventional power plant. The level of avoided capital costs depends on the extent to which wind power capacity can displace investments in new conventional power plants, and thus is directly tied to how wind power plants are integrated into the power system.

Studies of the Nordic power market, NordPool, show that the cost of integrating variable wind power is, on average, approximately 0.3–0.4c€/kWh of wind power generated at the present level of wind power capacity (mainly Denmark) and with the existing transmission and market conditions. These costs are completely in line with experiences in other countries. Integration costs are expected to increase with higher levels of wind power penetration.

Figure III.6.1 shows the results of the reference case, assuming the two conventional power plants are coming on-stream in 2010. As mentioned, figures for the conventional plants are calculated using the Recabs model, while the costs for wind power are taken from Chapter III.1.

As shown in the reference case, the cost of power generated at conventional power plants is lower than the cost of wind-generated power under the given assumptions of lower fuel prices. Wind-generated power at a European inland site is approximately 33–34 per cent more expensive than natural gas- and coal-generated power.

This case is based on the *World Energy Outlook* assumptions on fuel prices, including a crude oil price of $59/barrel in 2010. At present (September 2008), the crude oil price is $120/barrel. Although this oil price is combined with a lower exchange rate for US dollar, the present price of oil is significantly higher

Figure III.6.1: Costs of generated power comparing conventional plants to wind power, 2010 (constant 2006-€)

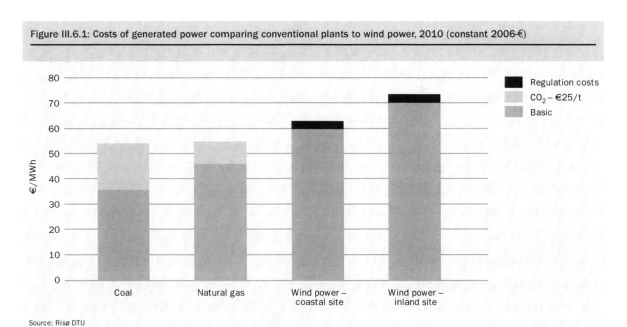

Source: Risø DTU

Figure III.6.2: Sensitivity analysis of costs of generated power comparing conventional plants to wind power, assuming increasing fossil fuel and CO_2 prices, 2010 (constant 2006-€)

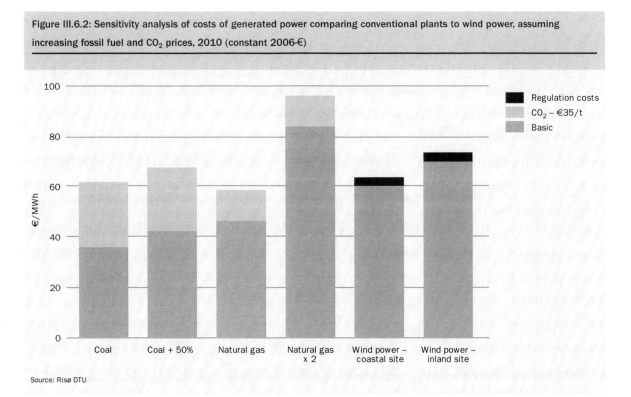

Source: Risø DTU

than the forecast IEA oil price for 2010. Therefore, a sensitivity analysis is carried through and results are shown in Figure III.6.2.

In Figure III.6.2, the natural gas price is assumed to double compared to the reference, equivalent to an oil price of $118/barrel in 2010, the coal price to increase by 50 per cent and the price of CO_2 to increase to €35/t from €25/t in 2008. As shown in Figure III.6.2, the competitiveness of wind-generated power increases significantly: costs at the inland site become lower than generation costs for the natural gas plant and only around 10 per cent more expensive than the coal-fired plant. At coastal sites, wind power produces the cheapest electricity.

Finally, as discussed by Awerbuch (2003a), the uncertainties related to future fossil fuel prices mentioned above imply a considerable risk for future generation costs of conventional plants. Conversely, the costs per kWh generated by wind power are almost constant over the lifetime of the turbine, following its installation. Thus, although wind power might currently be more expensive per kWh, it can account for a significant share in a utilities' portfolio of power plants, since it hedges against unexpected rises in prices of fossil fuels in the future. The consistent nature of wind power costs justifies a relatively higher cost compared to the uncertain risky future costs of conventional power.

III.7 EMPLOYMENT

Employment in the Wind Energy Sector

WIND ENERGY EMPLOYMENT IN EUROPE

Wind energy companies in the EU currently employ around 108,600 people.[6] For the purposes of this chapter, direct jobs relate to employment in wind turbine manufacturing companies and with sub-contractors whose main activity is supplying wind turbine components. Also included are wind energy promoters, utilities selling electricity from wind energy, and major R&D, engineering and specialised wind energy services. Any companies producing components, providing services, or sporadically working in wind-related activities are deemed to provide indirect employment.

The addition of indirect employment affects results significantly. The European Commission, in its EC Impact Assessment on the Renewable Energy Roadmap (EC, 2006), found that 150,000 jobs were linked to wind energy. The European Renewable Energy Council (EREC, 2007) report foresees a workforce of 184,000 people in 2010, but the installed capacity for that year has probably been underestimated. Therefore, the figure for total direct and indirect jobs is estimated at approximately 154,000 jobs.

These two figures of 108,600 direct and 154,000 total jobs can be compared with the results obtained by EWEA in its previous survey for *Wind Energy – The Facts* (EWEA, 2003) of 46,000 and 72,275 workers respectively. The growth experienced between 2003 and 2007 (236 per cent) is consistent with the evolution of the installed capacity in Europe (276 per cent – EWEA, 2008b) during the same period and with the fact that most of the largest wind energy companies are European.

A significant proportion of the direct wind energy employment (around 75 per cent) is in three countries, Denmark, Germany and Spain, whose combined installed capacity adds up to 70 per cent of the total in the EU. Nevertheless, the sector is less concentrated now than it was in 2003, when these three countries accounted for 89 per cent of the employment and 84 per cent of the EU installed capacity. This is due to the opening of manufacturing and operation centres in emerging markets and to the local nature of many wind-related activities, such as promotion, O&M, engineering and legal services.

Germany (BMU, 2006 and 2008) is the country where most wind-related jobs have been created, with around 38,000 directly attributable to wind energy companies[7] and a slightly higher amount from indirect effects. According to the German Federal Ministry of the Environment, in 2007 over 80 per cent of the value chain in the German wind energy sector was exported.

Table III.7.1: Direct employment from wind energy companies in selected European countries

Country	No of direct jobs
Austria	700
Belgium	2000
Bulgaria	100
Czech Republic	100
Denmark	23,500
Finland	800
France	7000
Germany	38,000
Greece	1800
Hungary	100
Ireland	1500
Italy	2500
The Netherlands	2000
Poland	800
Portugal	800
Spain	20,500
Sweden	2000
UK	4000
Rest of EU	400
TOTAL	**108,600**

Sources: Own estimates, based on EWEA (2008a); ADEME (2008); AEE (2007); DWIA (2008); BMU (2008)

In Spain (AEE, 2007), direct employment is 20,500 people. When indirect jobs are taken into account, the figure goes up to 37,730. According to the AEE, 30 per cent of the jobs are in manufacturing companies; 34 per cent in installation, O&M and repair companies, 27 per cent in promotion and engineering companies, and 9 per cent in other branches.

Denmark (DWIA, 2008) has around 23,500 employees in wind turbine and blade manufacturing and major sub-component corporations.[8]

The launch of new wind energy markets has fostered the creation of employment in other EU countries. Factors such as market size, proximity to one of the three traditional leaders, national regulation and labour costs determine the industry structure, but the effect is always positive.

France (2454 MW installed, 888 MW added in 2007 and an estimated figure of 7000 wind energy jobs), for instance, shows a wealth of small developers, consultants, and engineering and legal service companies. All the large wind energy manufacturers and developers and some utilities have opened up a branch in this country. France also counts on several wind turbine and component manufacturers producing in its territory.

In the UK, the importance of offshore wind energy and small-scale wind turbines is reflected by the existence of many job-creating businesses in this area. This country also has some of the most prestigious wind energy engineering and consultancy companies. The British Wind Energy Association is conducting a study of present and future wind energy employment; preliminary results point to the existence of around 4000 to 4500 direct jobs.

Another example is Portugal, where the growth of the market initially relied on imported wind turbines. From 2009 onwards, two new factories will be opened, adding around 2000 new jobs to the 800 that already exist.

Some other EU Member States, such as Italy, Greece, Belgium, The Netherlands, Ireland and Sweden, are also in the 1500 to 2500 band. The situation in the new Member States is diverse, with Poland in a leading position. Wind energy employment will probably rise significantly in the next three to five years, boosted by a combination of market attractiveness, a highly skilled labour force and lower production costs.

In terms of gender, the survey conducted by EWEA shows that males make up 78 per cent of the workforce. In the EU labour market as a whole, the figure is 55.7 per cent. Such a bias reflects the traditional predominance of men in production chains, construction work and engineering.

By type of company, wind turbine and component manufacturers account for most of the jobs (59 per cent). Within these categories, companies tend to be bigger and thus employ more people.

Wind energy figures can be measured against the statistics provided by Eurostat (2007). The energy sector employs 2.69 million people, accounting for 1.4 per cent of total EU employment. Approximately half this

Figure III.7.1: Direct employment by type of company, according to EWEA survey

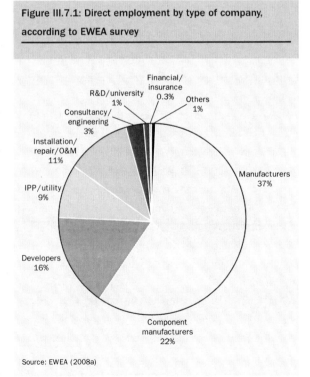

Source: EWEA (2008a)

amount is active in the production of electricity, gas, steam and hot water. Employment from the wind energy sector would then make up around 7.3 per cent of that amount; and it should be noted that wind energy currently meets 3.7 per cent of EU electricity demand. Although the lack of specific data for electricity production prevents us from making more accurate comparisons, this shows that wind energy is more labour intensive than the other electricity generating technologies. This conclusion is consistent with earlier research.

Finally, there is a well-documented trend of energy employment decline in Europe, particularly marked in the coal sector. For instance, British coal production and employment have dropped significantly, from 229,000 workers in 1981 to 5500 in 2006. In Germany, it is estimated that jobs in the sector will drop from 265,000 in 1991 to less than 80,000 in 2020. In EU countries, more than 150,000 utility and gas industry jobs disappeared in the second half of the 1990s and it is estimated that another 200,000 jobs will be lost during the first half of the 21st century (UNEP, ILO and ITUC, 2007). The outcomes set out in the previous paragraphs demonstrate that job losses in the European energy sector are independent of renewable energy deployment and that the renewable energy sector is, in fact, helping to mitigate these negative effects in the power sector.

JOB PROFILES OF THE WIND ENERGY INDUSTRY

The lack of any official classification of wind energy companies makes it difficult to categorise wind energy jobs. However, Table III.7.2 summarises the main profiles required by wind energy industries, according to the nature of their core business.

THE SHORTAGE OF WORKERS

In the last two to three years, wind energy companies have repeatedly reported a serious shortage of workers, especially within certain fields. This scarcity coincides with a general expansion of the European economy, where growth rates have been among the fastest since the end of the Second World War. An analysis of Eurostat (2008b) statistics proves that job vacancies have been difficult to cover in all sectors. The rotation of workers is high, both for skilled and non-skilled workers.

In the case of wind energy, the general pressure provoked by strong economic growth is complemented by the extraordinary performance of the sector since the end of the 1990s. In the 2000–2007 period, wind energy installations in the EU increased by 339 per cent (EWEA, 2008b). This has prompted an increase in job offers in all the sub-sectors, especially in manufacturing, maintenance and development activities.

Generally speaking, the shortage is more acute for positions that require a high degree of experience and responsibility:

- From a manufacturer's point of view, two major bottlenecks arise: one relates to engineers dealing with R&D, product design and the manufacturing processes; the other to O&M and site management activities (technical staff).
- In turn, wind energy promoters lack project managers; the professionals responsible for getting the permits in the country where a wind farm is going to be installed. These positions require a combination of specific knowledge of the country and wind energy expertise, which is difficult to gain in a short period of time.
- Other profiles, such as financiers or sales managers, can sometimes be hard to find, but generally this is less of a problem for wind energy companies, possibly because the necessary qualifications are more general.
- The picture for the R&D institutes is not clear: of the two consulted, one reported no problems, while the other complained that it was impossible to hire experienced researchers. It is worth noting that the remuneration offered by R&D centres, especially if

Table III.7.2: Typical wind energy job profiles demanded by different types of industry

Company type	Field of activity	Main job profiles
Wind energy manufacturers	Wind turbine producers, including manufacturers of major sub-components and assembly factories.	• Highly qualified chemical, electrical, mechanical and materials engineers dealing with R&D issues, product design, management and quality control of production process. • Semi-skilled and non-skilled workers for the production chains. • Health and safety experts. • Technical staff for the O&M and repair of wind turbines. • Other supporting staff (including administrative, sales managers, marketing and accounting).
Developers	Manage all the tasks related to the development of wind farms (planning, permits, construction and so on).	• Project managers (engineers and economists) to coordinate the process. • Environmental engineers and other specialists to analyse the environmental impacts of wind farms. • Programmers and meteorologists for wind energy forecasts and prediction models. • Lawyers and economists to deal with the legal and financial aspects of project development. • Other supporting staff (including administrative, sales managers, marketing and accounting).
Construction, repair and O&M	Construction of the wind farm, regular inspection and repair activities.[9]	• Technical staff for the O&M and repair of wind turbines. • Electrical and civil engineers for the coordination of construction works. • Health and safety experts. • Specialists in the transport of heavy goods. • Electricians. • Technical staff specialised in wind turbine installation, including activities in cranes, fitters and nacelles. • Semi-skilled and non-skilled workers for the construction process. • Other supporting staff (including administrative, sales managers and accounting).
Independent power producers, utilities	Operation of the wind farm and sale of the electricity produced.	• Electrical, environmental and civil engineers for the management of plants. • Technical staff for the O&M of plants, if this task is not sub-contracted. • Health and safety experts. • Financiers, sales and marketing staff to deal with the sale of electricity. • Other supporting staff (including administrative and accounting).
Consultancies, legal entities, engineering, financial institutions, insurers, R&D centres and others	Diverse specialised activities linked to the wind energy business.	• Programmers and meteorologists for the analysis of wind regimes and output forecasts. • Engineers specialised in aerodynamics, computational fluid dynamics and other R&D areas. • Environmental engineers. • Energy policy experts. • Experts in social surveys, training and communication. • Financiers and economists. • Lawyers specialised in energy and environmental matters. • Marketing personnel and event organisers.

they are governmental or university-related, is below the levels offered by private companies.

The quality of the university system does not seem to be at the root of the problem, although recently graduated students often need an additional specialisation that is given by the wind company itself. The general view is that the number of engineers graduating from European universities on an annual basis does not meet the needs of modern economies, which rely heavily on manufacturing and technological sectors.

In contrast, there seems to be a gap in the secondary level of education, where the range and quality of courses dealing with wind-related activities (mainly O&M, health and safety, logistics, and site management) are inadequate. Policies aimed at improving the educational programmes at pre-university level – dissemination campaigns, measures to encourage worker mobility and vocational training for the unemployed – can help overcome the bottleneck, and at the same time ease the transition of staff moving from declining sectors.

Employment Prediction and Methodology

METHODOLOGICAL APPROACHES TO EMPLOYMENT QUANTIFICATION

The quantification of wind energy employment is a difficult task for several reasons. First, it encompasses many company profiles, such as equipment manufacturing, electricity generation, consulting services, finance and insurance, which belong to different economic sectors. Second, we cannot rely on any existing statistics to estimate wind energy figures, as they do not distinguish between electricity and equipment manufacturing branches. And finally, the structure of the sector changes rapidly and historical data cannot be easily updated to reflect the current situation.

For these reasons, measurement initiatives must rely on a number of methodologies, which can largely be grouped under two headings:

1. data collection based on surveys and complemented by other written evidence; and
2. data collection based on estimated relationships between sectors, vectors of activity and input/output tables.

Surveys

Surveys are the best way to collect information on direct employment, especially when additional aspects – gender issues, employment profiles, length of contracts and other qualitative information – need to be incorporated. Surveys have significant limitations, notably the correct identification of the units that need to be studied and the low percentage of responses (see, for example, Rubio and Varas, 1999; Schuman and Stanley, 1996; Weisberg et al., 1996). When these problems arise, results need to be extrapolated and completed by other means.

Estimated Relationships

Estimated relationships, including input/output tables, can be used to estimate both direct and indirect employment impacts. These models require some initial information, collected by means of a questionnaire and/or expert interviews, but then work on the basis of technical coefficients (Leontief, 1986; Kulisic et al., 2007). The advantages of estimated models are based on the fact that they reflect net economic changes in the sector that is being studied, other related economic sectors and the whole of the economic system. These models also constitute the basis for the formulation of forecasts. The disadvantages relate to the cost of carrying out such studies and the need to obtain an appropriate model. In addition, they do not provide any details at sub-sector level and do not capture gender-related, qualification and shortage issues.

In the last six or seven years, coinciding with the boom of the wind energy sector, several studies have been conducted on the related employment repercussions. A list of the most relevant works can be found in Appendix J. A careful revision of their methodology shows that many of them are, in reality, meta-analyses (that is to say, a critical re-examination and comparison of earlier works), while research based on questionnaires and/or input/output tables is less common. Denmark, Germany and Spain, being the three world leaders in wind energy production and installation, have produced solid studies (AEE, 2007; DWIA, 2008; Lehr et al., 2008; BMU, 2008), but employment in the other EU markets remains largely unknown. In particular, there is a lack of information on some key features affecting the wind energy labour market, such as the profiles that are currently in demand, shortages and gender issues. These issues can best be dealt with through ad hoc questionnaires sent to wind energy companies.

EWEA SURVEY ON DIRECT EMPLOYMENT

As a response to the gaps mentioned above, EWEA has sought to quantify the number of people directly employed by the wind energy sector in Europe by means of a questionnaire. As explained in the previous section on wind energy employment in Europe (page 251), direct jobs relate to employment within wind turbine manufacturing companies and sub-contractors whose main activity is the supply of wind turbine components. Also taken into account are wind energy promoters, utilities selling electricity from wind energy, and major R&D, engineering and specialised wind energy services. Any other company producing components, providing services, or sporadically working in wind-related activities is deemed as providing indirect employment.

The analysts have attempted to minimise the main disadvantages linked to this type of methodology. Consequently, the questionnaire was drafted after careful analysis of previous research in this field, notably the questionnaires that had been used in the German, Danish and Spanish studies, and following a discussion with the researchers responsible for these. A draft was sent to a reduced number of respondents, who then commented on any difficulties in understanding the questions and using the Excel spreadsheet, the length of the questionnaire, and some other aspects. The document was modified accordingly.

The final version of the questionnaire was dispatched by email on 19 February 2008 to around 1100 organisations in 30 countries (the 27 EU Member States plus Croatia, Norway and Turkey). It reached all EWEA members and the members of the EU-27 national wind energy associations. The questionnaire was also distributed among participants of the last two European Wind Energy Conferences (EWEC 2006 and 2007). These included:

- wind turbine and component manufacturers;
- developers;
- independent power producers and utilities;
- installation, repair and O&M companies;

- consultancies;
- engineering and legal services;
- R&D centres;
- laboratories and universities;
- financial institutions and insurers;
- wind energy agencies and associations; and
- other interest groups directly involved in wind energy matters.

The document was translated into five EU languages (English, French, German, Spanish and Portuguese), and a number of national wind energy associations decided to write the introductory letter in their own language. A reminder was sent out on 11 March, followed up by telephone calls during April, May, June, July and August.

The questionnaire consisted of 11 questions, divided into three blocks:

1. The first four questions collected information on the profile of the company, its field of activity and the year in which it started operating in the wind energy sector.
2. The next three questions aimed to obtain relevant employment figures. The questionnaire requested both the total number of employees and the number of employees in the wind energy sector, and gave some indication about how to calculate the second figure when a worker was not devoted to wind-related activities full time. The figures were divided up by country, since some companies are transnational, and by sex. It would have been interesting to classify this data by age and level of qualification, but the draft sent to a sample of respondents showed us that this level of detail would be very difficult to obtain and that it would have had a negative impact on the number of replies.
3. The final four questions addressed the issue of labour force scarcity in the wind energy sector, and aimed to obtain information on the profiles that are in short supply and the prospects of wind energy companies in terms of future employment levels and profiles.

Questions 9 and 10 were more speculative, since it is difficult to quantify the exact employment demands in the next five years, but they gave an order of magnitude that could then be compared with the quantitative approaches of input/output tables used by other researchers.

The questionnaire was complemented by in-depth interviews with a selection of stakeholders that suitably reflected the main wind energy sub-sectors and EU countries. The interviews were carried out by phone, by email or face-to-face. They were aimed at verifying the data obtained from the questionnaires and at addressing some of the topics that could not be dealt with, notably a more thorough explanation of the job profiles demanded by the industry and the scarcity problem.

By the end of August 2008, 324 valid questionnaires had been received, implying a response rate of around 30 per cent. When looking at the responses, it is clear that it was mostly the largest turbine and component manufacturers, as well as the major utilities, that answered the questionnaire. The replies therefore do not provide an accurate representation of the industry as a whole.

The figures are good for this type of survey, but supplementary sources need to be used to fill in the gaps and validate results. This has been done in several ways:

- The use of thematic surveys and input/output analysis carried out in Denmark, France, Germany and Spain. The last two countries base their numbers on questionnaires very similar to the ones used by EWEA, an exhaustive analysis of the governmental registers for tax-related purposes, and the application of national input/output tables and other technical coefficients to estimate the indirect effects. The Danish Wind Energy Association collects information about employment from all its members on an annual basis and then predicts indirect and induced jobs through technical coefficients and

Table III.7.3: EWEA survey results[10]	
Country	**No of direct jobs**
Austria	270
Belgium	1161
Bulgaria	91
Cyprus	1
Czech Republic	52
Denmark	9875
Estonia	5
Finland	194
France	2076
Germany	17,246
Greece	812
Hungary	11
Ireland	870
Italy	1048
Latvia	6
Lithuania	6
The Netherlands	824
Poland	312
Portugal	425
Romania	27
Slovakia	22
Slovenia	4
Spain	10,986
Sweden	1234
UK	2753
Rest of Europe	70
TOTAL	**50,380**

Source: EWEA (2008a)

multipliers. The French Environment and Energy Management Agency (ADEME) bases its estimates on net production/employment ratios (imports have been disregarded).

- The review of the annual reports/websites of the main wind energy companies, notably the large wind energy manufacturers, component manufacturers, wind energy developers and utilities. As these

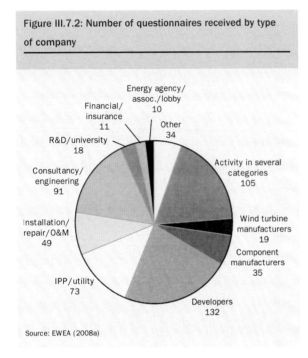

Source: EWEA (2008a)

companies are active in the stock market, they publish some information on their activities and structure that can be used to estimate wind energy figures.

- The registers and the expertise gained by the national wind energy associations. France, the UK and Portugal are currently carrying out thematic studies covering, among other things, employment issues. Their preliminary conclusions have been incorporated into this publication. In other cases, experts from the national associations and governments have been contacted.

Additionally, EWEA is engaged in an in-depth examination of the factors that are behind the repeatedly reported shortage of workers in the wind energy sector and the profiles that are particularly difficult to find. This has been done through in-depth interviews (conducted face-to-face, by email and by phone) with the human resources managers of a selection of wind

energy companies from different branches and geographical areas. The results were compared with those of the answers to questions 7–10 of the general questionnaire.

Part III Notes

1   'Ex-works' means that site work, foundation and grid connection costs are not included. Ex-works costs include the turbine as provided by the manufacturer, which includes the turbine, blades, tower and transport to the site.

2   In Spain land rental is counted as an O&M cost.

3   The number of observations was generally between 25 and 60.

4   See, for instance, Neij (1997), Neij et al. (2003) or Milborrow (2003).

5   This is in line with observed costs in other countries.

6   For more information on wind energy employment, see EWEA (2008c).

7   The 2006 BMU study found that 43 per cent of gross wind energy jobs (63,900) were direct; the rest, which also included O&M, were indirect. In 2008, the BMU published new data (84,300 jobs), but this does not distinguish between direct and indirect jobs. For the purposes of this publication, we have made the split based on the assumption that the earlier ratio still pertains (43 per cent direct and 57 per cent indirect).

8   Of those, 13,000 come from pure wind turbine and blade manufacturing companies. The remaining 4000 are attributed to major sub-suppliers. Most of these produce for more than one sector. In this publication, such companies are included within the category of 'direct employment' when at least 50 per cent of their turnover comes from sales to wind turbine manufacturers or operators. In addition, the questionnaire that was used as the basis for obtaining the statistics asked about 'jobs that can be attributed to wind-related activities', thus eliminating staff that are devoted to other activities.

9   The Windskill Project (www.windskill.eu/) funded by the European Commission offers a good summary of the profiles that are required in this area.

10   In a few cases, the questionnaires were filled in by the researchers themselves. This occurred when the figures were communicated through a phone call or by email, or when the information needed was available in an annual report or some other publicly available company document.

INDUSTRY AND MARKETS

Acknowledgements

Part IV was compiled by Angelika Pullen of the Global Wind Energy Council (GWEC); Keith Hays of Emerging Energy Research; Gesine Knolle of EWEA.

We would like to thank all the peer reviewers for their valuable advice and for the tremendous effort that they put into the revision of Part IV.

Part IV was carefully reviewed by the following experts:

Ana Estanquiero	INETI
Birger T. Madsen	BTM
Peter Madigan	BWEA
Frans van Hulle	EWEA
Peter Raftery	Airtricity
Isabel Blanco	UAH
Alberto Ceña	Aeeolica
Paz Nachón López	Acciona
Anna Pasławska	Polish Wind Energy Association
Charles Dugué	ENR
Guillaume DUCLOS	ENR

PART IV   INTRODUCTION

Europe is steadily moving away from conventional power sources and towards renewable energy technologies, a trend driven by the world's most significant piece of renewable energy legislation to date: the 2001 EU Renewable Electricity Directive. The move towards renewables has been further underpinned by the European Council's 2007 decision to establish a binding target of 20 per cent of EU energy from renewable sources by 2020.

Thermal power generation now stands at about 430 GW and, combined with large hydro and nuclear, has traditionally served as the backbone of Europe's power production. In recent years, however, renewable energy, and specifically wind energy, have become mainstream sources of power. Between 2000 and 2007, wind energy capacity in Europe more than quadrupled, from 13 GW to 57 GW. The 2007 total of 57 GW represents over 5 per cent of all power generation capacity in Europe and 30 per cent of all new power capacity installed since 2000. The bulk of European wind energy capacity has been concentrated in three countries, Germany, Spain and Denmark, which are now home to 72 per cent of all capacity in Europe.

Wind's spectacular European growth has attracted a broad range of players across the value chain, from local engineering firms to global vertically integrated utilities. There is strong competition, with about a dozen key suppliers vying for market share. However, the leading suppliers consolidated their dominant position in 2003–2004. Recent supply chain pressure has been a key competitive driver in wind turbine supply, and the relationships between turbine manufacturers and their component suppliers have become increasingly crucial. As more and more players look to develop, own or operate wind farms, wind power markets now include dozens of multinational players, illustrating the industry's increase in size and its geographic expansion. Between 2005 and 2006, the industry also saw more participation by utilities.

If wind energy investment has been tremendous in the past, there is no sign that the speed of development will decrease. Europe's top 15 wind utilities and independent power producers (IPPs) have announced that pipelines of over 18 GW will be installed between 2007 and 2010, translating into well over €25 billion worth of investments in wind plants. Overall, the European wind market is expected to grow at a rate of over 7–9 GW every year until 2010.

The growth of the European wind energy sector has also recently been mirrored in other continents, most notably in China, India and the US. In 2007, over 11 GW of new wind capacity was installed outside Europe, bringing the global total up to 94 GW. In terms of economic value, the global wind market was worth about €25 billion in 2007 in terms of new generating equipment.

Both in Europe and further afield, however, wind energy expansion is facing a number of barriers, both administrative and in terms of grid access. These barriers are created when administrative or financial procedures are opaque or inconsistent. They can also occur in relation to grid connections, and often pose serious obstacles to investment in wind energy, as well as preventing it from achieving competitiveness with other power-generating technologies.

IV.1 WIND IN THE EUROPEAN POWER MARKET

Renewable Energy Policies in Europe

Europe's electricity market is made up of rigid structures that do not take the environmental advantages of wind energy into account. New entrants face a number of barriers: they have to compete with conventional plants that were built decades ago and which are operated and maintained by government funds through former state-owned utilities in a monopoly market. In addition, incumbent electricity players tend to be powerful vertically integrated companies. New technologies experience obstacles when entering the market and often struggle to gain grid access and obtain fair and transparent connection costs.

The EU has acknowledged these problems, and set up a specific legal framework for renewable energies, including wind, which seeks to overcome such barriers.

The first step in this direction was the European Commission's (EC) 1997 White Paper on renewable sources of energy, which set a target for 40,000 MW of wind power to be installed in the EU by 2010. In the event, this target was reached in 2005, five years ahead of schedule. Part of the White Paper target was to increase electricity production from renewable energy sources by 338 TWh between 1995 and 2010.

In 2001, the EU passed what was until recently the world's most significant piece of legislation for electricity produced by renewable energies, including wind: EC Directive 2001/77/EC on the promotion of electricity produced from renewable energy sources in the internal electricity market. This directive has been tremendously successful in promoting renewables, particularly wind energy, and it is the key factor explaining the success of renewables worldwide. Its purpose was 'to promote an increase in the contribution of renewable energy sources to electricity production in the internal market for electricity and to create a basis for a future Community framework thereof'.

Thanks to this directive, Europe has become the world leader in renewable energy technology.

The strong development of wind power can continue in the coming years, as long as the clear commitment of the European Union and its Member States to wind power development continues to strengthen, backed up by effective policies.

Therefore, the adoption of the Renewable Energy Directive in December 2008 represents an historical moment for the further development of wind power in Europe. The directive is a breakthrough piece of legislation that will enable wind power and other renewables to push past barriers and confirms Europe as the leader of the energy revolution the world needs. Under the terms of the directive, for the first time each Member State has a legally binding renewables target for 2020 and by June 2010 each Member State will have drawn up a National Action Plan (NAP) detailing plans to meet their 2020 targets.

Key aspects of the directive are:

1. **Legally binding national targets and indicative trajectory**: the 20% overall EU renewables target is broken down into differentiated, legally binding national targets. The Member States are given an 'indicative trajectory' to follow in the run-up to 2020. By 2011–12, they should be 20 per cent of the way towards the target; by 2013–14, 30 per cent; by 2015–16, 45 per cent and by 2017–18, 65 per cent – all compared to 2005. In terms of electricity consumption, renewables should provide about 35 per cent of the EU's power by 2020. By 2020, wind energy is set to contribute the most – nearly 35 per cent of all the power coming from renewables.

2. **National Action Plans (NAPs)**: the directive legally obliges each EU Member State to ensure that its 2020 target is met and to outline the 'appropriate measures' it will take do so, by drafting a National Renewable Energy Action Plan (NAP) to be submitted by 30 June 2010 to the European Commission. The NAPs will set out how each EU country is to meet its overall national target, including elements such as sectoral targets for shares of renewable energy for transport, electricity and heating/

cooling and how they will tackle administrative and grid barriers. The NAPs will have to follow a binding template to be provided by the European Commission in June 2009; if the Commission considers an NAP to be inadequate, it will consider initiating infringement proceedings against that particular Member State. If they fall significantly short of their interim trajectory over any two-year period, Member States will have to submit an amended NAP stating how they will make up for the shortfall.

3. **Priority access to the electricity grid**: the directive requires that EU countries take 'the appropriate steps to develop transmission and distribution grid infrastructure, intelligent networks, storage facilities and the electricity system ... to accommodate the further development' of renewable electricity, as well as 'appropriate steps' to accelerate authorisation procedures for grid infrastructure and to coordinate approval of grid infrastructure with administrative and planning procedures.

Assuming that the reliability and safety of the grid is maintained, EU countries must ensure that transmission system operators and distribution system operators guarantee the transmission and distribution of renewable electricity and provide for either priority access or guaranteed access to the grid system. According to the agreement, 'priority access' to the grid provides an assurance given to connected generators of renewable electricity that they will be able to sell and transmit their electricity in accordance with connection rules at all times, whenever the source is available.

When the renewable electricity is integrated into the spot market, 'guaranteed access' ensures that all electricity sold and supported gets access to the grid, allowing the use of a maximum of renewable electricity from installations connected to the grid.

Furthermore, priority during dispatch (which was also the case in the 2001 directive) is a requirement for renewables, and EU countries must now also ensure that appropriate grid and market related operational measures are taken in order to minimise the curtailment of renewable electricity.

Europe can go a long way towards an energy mix that is superior to the business-as-usual scenario, offering greater energy independence, lower energy costs, reduced fuel price risk, improved competitiveness and more technology exports. Over the coming years, wind energy will play a major role in reaching this superior energy mix.

The EU Energy Mix

While thermal generation totalling over 430 GW, combined with large hydro and nuclear, has long served as the backbone of Europe's power production, Europe is steadily making the transition away from conventional power sources and towards renewable energy technologies. Between 2000 and 2007, the total EU power capacity increased by 200 GW to reach 775 GW by the end of 2007. The most notable change in the capacity is the near doubling of gas capacity to 164 GW, but wind energy also more than quadrupled, from 13 GW to 57 GW.

The addition of ten new Member States in May 2004 put another 112 GW into the EU generation mix, including 80 GW of coal, 12 GW of large hydro, 12 GW of natural gas, 6.5 GW of nuclear and 186 MW of wind power (see Figure IV.1.1).

Changes in net installed capacity for the various electricity generating technologies are shown in Figure IV.6.2. The figures include the EU-10 (ten new Member States) from 2005 and EU-12 (the EU-10 plus Romania and Bulgaria) from 2007. The growth of natural gas and wind power has taken place at the expense of fuel oil, coal and nuclear power. In 2007, 21.2 GW of new capacity was installed in the EU-27, of which 10.7 GW was gas (50 per cent) and 8.6 GW was wind power (40 per cent).

Gas and wind power also lead in terms of new capacity if decommissioning of old capacity is taken into account. Net installation of power capacity in the EU totalled 98 GW between 2000 and 2007. Gas and wind power accounted for 77 GW and 47 GW respectively

Figure IV.1.1: Installed power capacity, EU, 2000-2007 (in MW)

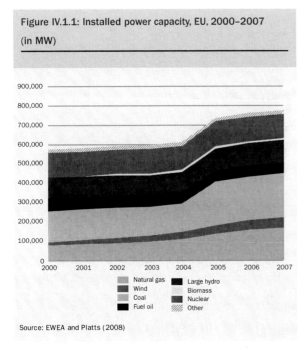

Source: EWEA and Platts (2008)

while more oil (−14 GW net), coal (−11 GW net) and nuclear (−6 GW net) have been removed than installed since 2000.

The share of EU capacity covered by natural gas has more than doubled since 1995 to reach 21 per cent. Coal's share is unchanged, while oil, large hydro and nuclear have all decreased in their share. Wind energy's share has increased from 0 per cent in 1995 to 7 per cent in 2007 (Figure IV.1.3).

The obstacles hindering more combined cycle gas turbine plant installations, including the rising costs of gas and volatile supply security from Russia and the Middle East, are most acutely felt in highly import-dependent countries such as Italy, The Netherlands and Portugal.

Nuclear accounts for around 17 per cent of the total installed capacity in Europe. While nuclear power emits only low amounts of carbon, safety concerns and costs remain key obstacles. Most of Europe stopped adding nuclear generation capacity in the 1980s, and several countries face major decommissioning programmes over the next ten years and looming capacity gaps to fill. In Germany alone, over

Figure IV.1.2: Net increase/decrease in power capacity, EU, 2000-2007 (in MW)

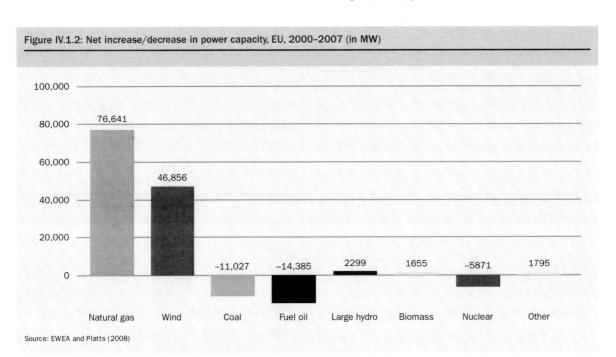

Source: EWEA and Platts (2008)

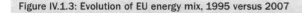

Figure IV.1.3: Evolution of EU energy mix, 1995 versus 2007

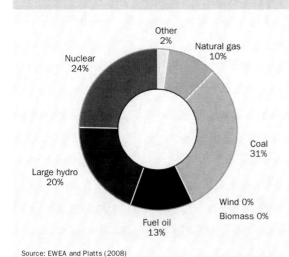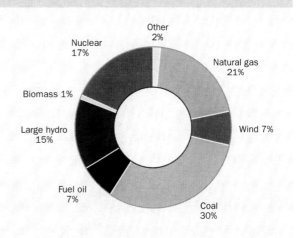

Source: EWEA and Platts (2008)

20 GW of nuclear capacity stands to be decommissioned by 2020, while France's 63 GW installed base will also require modernisation. At present, there is just one nuclear reactor currently under construction in the EU, in Finland, and this will add less than 5 GW to the country's capacity in the medium term.

Against this backdrop of rising costs and emissions for heavily thermal-dependant Europe, with its significant resistance to new nuclear construction, renewable energy technologies have been able to flourish in the past ten years. Europe's renewables targets, and the need to fill the generation gap, have resulted in a mix of support mechanisms for key technologies, including wind energy, biomass, solar, small hydro, ocean energy and geothermal. These generation options have resulted in a race to position these technologies as cost-competitive options for national energy mixes, with wind clearly in the lead.

The European Commission expects a 76 per cent decline in EU oil production between 2000 and 2030. Gas production will fall by 59 per cent and coal by 41 per cent. By 2030, the EU will be importing 95 per cent of its oil, 84 per cent of its gas and 63 per cent of its coal.

Wind in the EU's Energy Mix

With an impressive compound annual growth rate of over 20 per cent in MW installed between 2000 and 2007, and now accounting for over 5 per cent of total generation, wind energy has clearly established itself as a relevant power source. In 2007, 40 per cent of all new generating capacity installed in the EU was wind power. Shifting trends in generation mix planning, brought on by the challenges of supply security, climate change and cost competitiveness, are increasing support for wind as a mainstream generation technology able to meet a substantial share of Europe's electricity demand. Based on their existing generation mix, European countries will move at different speeds to incorporate wind into their energy portfolios; however, the changing political will and the improving performance of wind power underline its increasing competitiveness.

Of the main RES technologies, wind is the most rapidly deployable, clean and affordable, which explains why Europe is choosing this technology to help reach its 20 per cent renewable energy target by 2020. Wind

Figure IV.1.4: Sixteen years of global wind energy development, 1991–2006, compared to the first 16 years of nuclear development

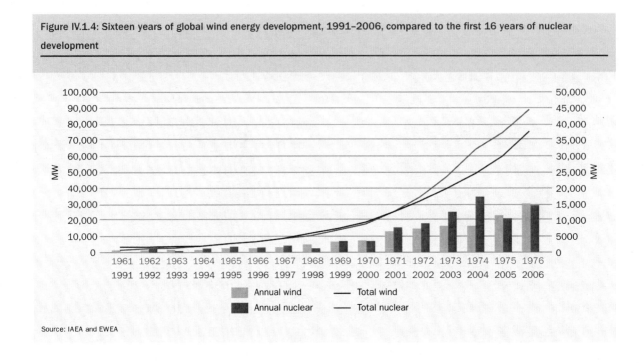

Source: IAEA and EWEA

has already made solid steps forward, penetrating national transmission systems by as much as 10 per cent in several markets and as much as 21 per cent in Denmark. Key features of wind's role in the national energy mix of European countries include its increasing weight relative to competing technologies, the level of penetration it has reached in specific markets and the speed at which it has been deployed. Wind's production variability will have an impact on a grid control area's generation mix, though transmission operators are increasingly capable of managing higher penetration levels as long as they maintain a flexible balance in the portfolio with dispatchable generation plant.

Wind power has experienced dramatic growth over recent years, and now represents over 10 per cent of the total installed power capacity, and more than 5 per cent of national electricity demand, in five European markets, Germany, Spain, Denmark, Portugal and Ireland, surpassing 10 per cent of the electricity

Figure IV.1.5: New power capacity, EU, 2000–2007

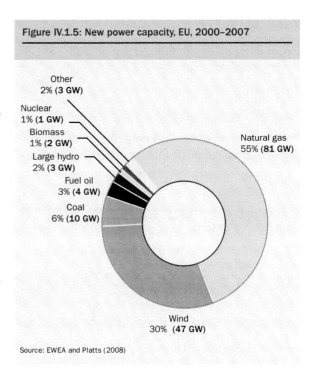

Source: EWEA and Platts (2008)

Figure IV.1.6: New power capacity, EU, 2000–2007 (in MW)

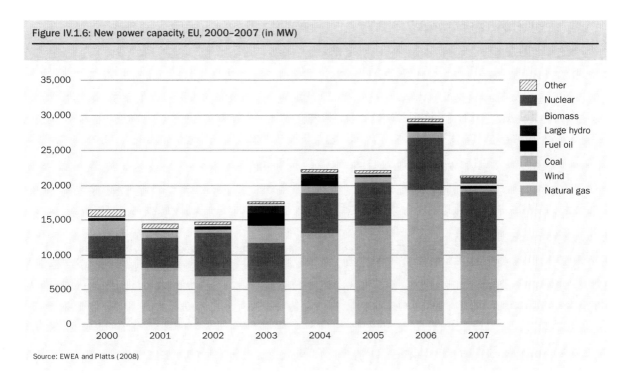

Source: EWEA and Platts (2008)

demand in both Spain and Denmark. As the industry continues to work with grid planners, utilities and developers to accommodate the variable nature of wind power generation, it is expected that the threshold for wind power penetration in several markets will increase. This will be particularly crucial for tapping offshore potential, as new transmission lines will be required for wind to see a greater surge in large-scale capacity additions.

Wind power has developed similarly to other power sources. Figure IV.1.4 shows the global development of wind energy (1991–2006) compared with nuclear power (1961–1976).

Wind energy increased its share of total capacity in the EU to 7 per cent in 2007, and its impact on new generation capacity has been noticeable. Thirty per cent of all power capacity installed between 2000 and 2007 was wind power, making it the second largest contributor to new EU capacity over the last eight years after natural gas (55 per cent). In 2007, no other electricity generating technology increased more than wind power in the EU. Six per cent of all new capacity over the eight-year period was coal, 3 per cent fuel oil and 2 per cent large hydro, with nuclear and biomass coming in at 1 per cent each (Figures IV.1.5 and IV.1.6).

IV.2 EUROPEAN MARKET OVERVIEW

The Current Status of the EU Wind Energy Market

Wind has become an integral part of the generation mix of markets like Germany, Spain and Denmark, alongside conventional power sources. Nonetheless, it continues to face the double challenge of competing against other technologies while proving itself as a sound energy choice for large power producers seeking to enlarge and diversify their portfolios.

By the end of 2007, there was 56,535 MW of wind power capacity installed in the EU-27, of which 55,860 MW was in the EU-15. In EWEA's previous scenario, drawn up in October 2003, we expected 54,350 MW to be installed in the EU-15 by the end of 2007. Thus the total capacity was underestimated by 1510 MW over the five-year period. In 2003, EWEA expected total annual installations in 2007 to be 6600 MW, whereas the actual market was significantly higher, at 8291 MW in the EU-15 (8554 MW in the EU-27).

In the EU, installed wind power capacity has increased by an average of 25 per cent annually over the past 11 years, from 4753 MW in 1997 to 56,535 MW in 2007. In terms of annual installations, the EU market for wind turbines has grown by 19 per cent annually, from 1277 MW in 1997 to 8554 MW in 2007.

In 2007, Spain (3522 MW) was by far the largest market for wind turbines, followed by Germany (1667 MW), France (888 MW) and Italy (603 MW). Eight countries – Germany, Spain, Denmark, Italy, France, the UK, Portugal and The Netherlands – now have more than 1000 MW installed. Germany, Spain and Denmark – the three pioneering countries of wind power – are home to 72 per cent of installed wind power capacity. That share is expected to decrease to 62 per cent of installed capacity in 2010.

Germany and Spain continue to attract the majority of investments. In 2007, these two countries represented 61 per cent of the EU market. However, there is a healthy trend towards less reliance on Germany and Spain, although this trend was broken in 2007 due to unprecedented Spanish growth. In 2000, 468 MW of European wind power capacity was installed outside Germany, Spain and Denmark; in 2007, the figure was 3362 MW.

Excluding Germany, Spain and Denmark, there has been an almost fivefold increase in the annual market in the past five years, confirming that a second wave of European countries are investing in wind power,

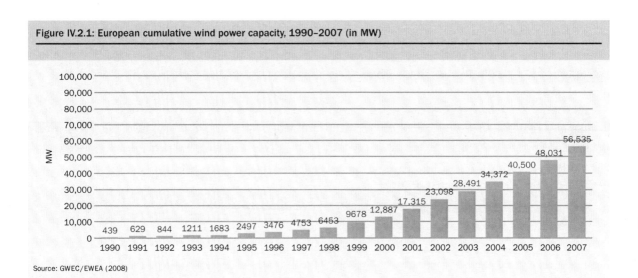

Figure IV.2.1: European cumulative wind power capacity, 1990–2007 (in MW)

Source: GWEC/EWEA (2008)

Figure IV.2.2: European annual wind power capacity, 1991–2007 (in MW)

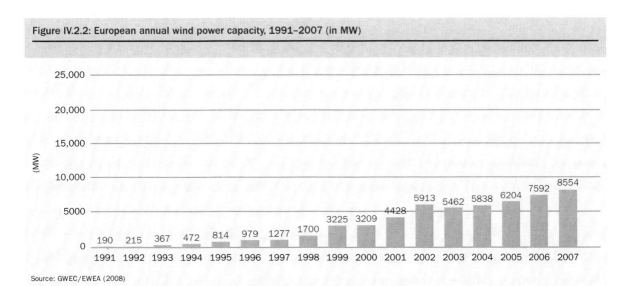

Source: GWEC/EWEA (2008)

Figure IV.2.3: 2007 Member State market shares of new capacity

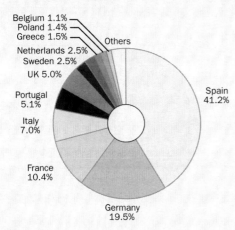

Others

Czech Republic	0.7%	Cyprus	0.0%
Ireland	0.7%	Denmark	0.0%
Estonia	0.3%	Hungary	0.0%
Finland	0.3%	Latvia	0.0%
Bulgaria	0.4%	Luxembourg	0.0%
Austria	0.2%	Malta	0.0%
Lithuania	0.1%	Slovakia	0.0%
Romania	0.1%	Slovenia	0.0%

Source: EWEA (2008a)

Figure IV.2.4: End 2007 Member State market shares of total capacity

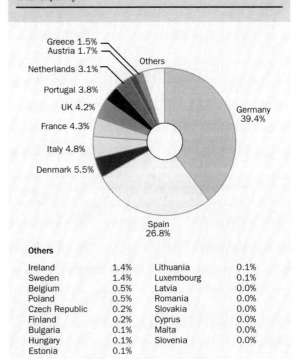

Others

Ireland	1.4%	Lithuania	0.1%
Sweden	1.4%	Luxembourg	0.1%
Belgium	0.5%	Latvia	0.0%
Poland	0.5%	Romania	0.0%
Czech Republic	0.2%	Slovakia	0.0%
Finland	0.2%	Cyprus	0.0%
Bulgaria	0.1%	Malta	0.0%
Hungary	0.1%	Slovenia	0.0%
Estonia	0.1%		

Source: EWEA (2008a)

Table IV.2.1: Cumulative installations of wind power in the EU and predictions for 2010 (in MW)

Country	2000	2001	2002	2003	2004	2005	2006	2007	2010
Austria	77	94	140	415	606	819	965	982	1200
Belgium	13	32	35	68	96	167	194	287	800
Bulgaria					10	10	36	70	200
Cyprus			0	0	0	0	0	0	0
Czech Republic			3	9	17	28	54	116	250
Denmark	2417	2489	2889	3116	3118	3128	3136	3125	4150
Estonia			2	2	6	32	32	58	150
Finland	39	39	43	52	82	82	86	110	220
France	66	93	148	257	390	757	1567	2454	5300
Germany	6113	8754	11,994	14,609	16,629	18,415	20,622	22,247	25,624
Greece	189	272	297	383	473	573	746	871	1500
Hungary			3	3	3	17	61	65	150
Ireland	118	124	137	190	339	496	746	805	1326
Italy	427	682	788	905	1266	1718	2123	2726	4500
Latvia			24	27	27	27	27	27	100
Lithuania			0	0	6	6	48	50	100
Luxembourg	10	15	17	22	35	35	35	35	50
Malta			0	0	0	0	0	0	0
Netherlands	446	486	693	910	1079	1219	1558	1746	3000
Poland			27	63	63	83	153	276	1000
Portugal	100	131	195	296	522	1022	1716	2150	3500
Romania			1	1	1	2	3	8	50
Slovakia			0	3	5	5	5	5	25
Slovenia			0	0	0	0	0	0	25
Spain	2235	3337	4825	6203	8264	10,028	11,623	15,145	20,000
Sweden	231	293	345	399	442	510	571	788	1665
UK	406	474	552	667	904	1332	1962	2389	5115
EU	**12,887**	**17,315**	**23,098**	**28,491**	**34,372**	**40,500**	**48,031**	**56,535**	**80,000**

Source: EWEA (2008a)

partly as a result of the EU Renewable Electricity Directive of 2001.

The total wind power capacity installed at the end of 2007 will produce 3.7 per cent of the EU-27 electricity demand in a normal wind year. Wind power in Denmark covers more than 20 per cent of its total electricity consumption, by far the largest share of any country in the world. Five EU countries – Denmark, Spain, Portugal, Ireland and Germany – have more than 5 per cent of their electricity demand met by wind energy (Figure IV.2.6).[1]

By the end of 2007, 116 kW of wind energy capacity was installed for every 1000 people in the EU (Figure IV.2.7). Denmark tops the list, with 579 kW/1000

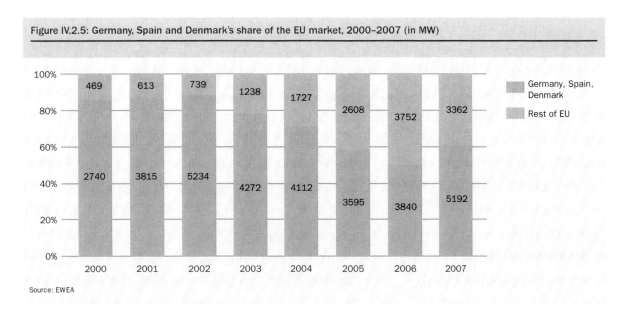

Figure IV.2.5: Germany, Spain and Denmark's share of the EU market, 2000–2007 (in MW)

Source: EWEA

people, followed by Spain (367 kW) and Germany (270 kW). If all EU countries had installed the same amount of wind power capacity per population as Spain, the total installed capacity in the EU would be 180,000 MW, equal to EWEA's 2020 target. If all EU countries had the same amount of capacity per capita as Denmark, total EU installations would be 282,000 MW, slightly less than the EWEA 2030 target.

There is 12.2 MW of wind power capacity installed per 1000 km² of land area in the EU. Not surprisingly, being a small country, wind power density is highest in Denmark, but Germany comes a close second. It is interesting that Spain's wind power density is less than half that of Germany, indicating a large remaining potential, at least from a visual perspective. The Netherlands, Portugal and Luxembourg are also above the EU average.

Many geographically large Member States, such as France, Sweden, Finland, Poland and Italy, still have very low density compared with the first-mover countries. If Sweden had the same wind power density as Germany, there would be more than 28 GW of wind power capacity installed there (0.8 GW was operating by the end of 2007) and France would have more than

34 GW of wind power capacity (compared to 2.5 GW in December 2007).

Tiered Growth Led by Germany, Spain, the UK, Italy and France

Europe's onshore wind power sector can be divided into three main market types – consolidating, scaling and growth markets – reflecting each market's maturity as a development environment and the intensity of competition within the country:

- **Growth markets** include those with a nascent wind energy sector, where the gradual creation and implementation of a stable regulatory framework is expected to facilitate sporadic project activation. The total wind power installed in growth markets remains under 1100 MW, accounting for less than 3 per cent of the region's total installed power capacity.
- **Scaling markets** are characterised by strong remaining resources coupled with stable regulatory frameworks that will facilitate project development. As such, these markets will experience high project volume in the near term, and are expected to be Europe's main driver of growth, accounting for an

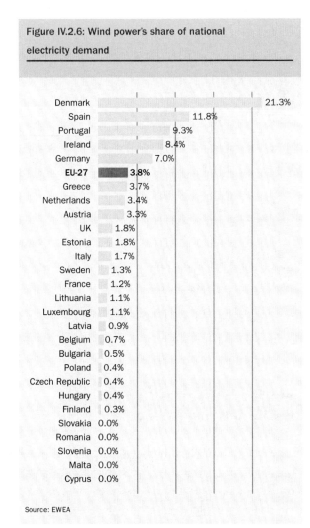

Figure IV.2.6: Wind power's share of national electricity demand

Denmark	21.3%
Spain	11.8%
Portugal	9.3%
Ireland	8.4%
Germany	7.0%
EU-27	**3.8%**
Greece	3.7%
Netherlands	3.4%
Austria	3.3%
UK	1.8%
Estonia	1.8%
Italy	1.7%
Sweden	1.3%
France	1.2%
Lithuania	1.1%
Luxembourg	1.1%
Latvia	0.9%
Belgium	0.7%
Bulgaria	0.5%
Poland	0.4%
Czech Republic	0.4%
Hungary	0.4%
Finland	0.3%
Slovakia	0.0%
Romania	0.0%
Slovenia	0.0%
Malta	0.0%
Cyprus	0.0%

Source: EWEA

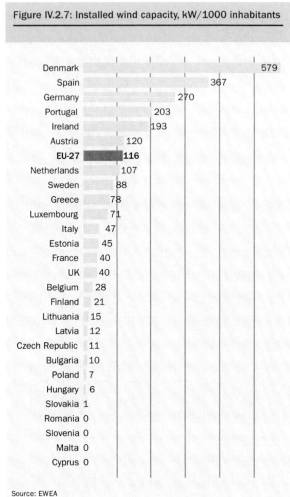

Figure IV.2.7: Installed wind capacity, kW/1000 inhabitants

Denmark	579
Spain	367
Germany	270
Portugal	203
Ireland	193
Austria	120
EU-27	**116**
Netherlands	107
Sweden	88
Greece	78
Luxembourg	71
Italy	47
Estonia	45
France	40
UK	40
Belgium	28
Finland	21
Lithuania	15
Latvia	12
Czech Republic	11
Bulgaria	10
Poland	7
Hungary	6
Slovakia	1
Romania	0
Slovenia	0
Malta	0
Cyprus	0

Source: EWEA

increasing share of the region's total installed energy capacity. With large wind resources and market environments propitious to investments in wind, scaling markets are now experiencing an influx of foreign players that is leading to heightened competition for projects.

- **Consolidating markets** have reached a good level of maturity, have high penetration levels (>10 per cent of national installed generation capacity) and have limited greenfield opportunities available. Denmark and Germany are the archetypal consolidating markets, where wind has become a mainstream source

of energy and the best sites have already been tapped out. As such, the majority of projects coming online in the next few years have already been fully permitted or are now in the final permitting stages.

At national level, these market groupings have several implications in terms of the expected amount of MW needed to tap remaining potential, as well as the intensity of the competition among players to tap undeveloped sites. Consolidating markets are most likely to turn in relatively flat or declining MW additions in the short term until the markets begin going

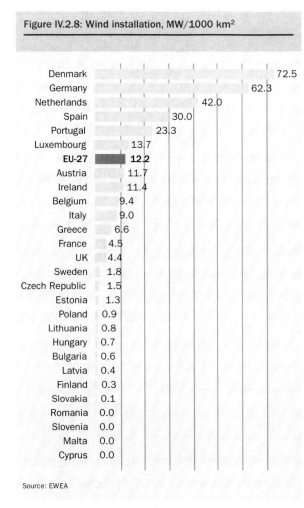

Figure IV.2.8: Wind installation, MW/1000 km²

Source: EWEA

offshore or facilitating widespread repowering (replacing older turbines). Germany is gearing up for its first offshore project, while Denmark and The Netherlands have already launched theirs.

Scaling markets, led by Spain, the UK, Italy, France and Portugal, will probably host the bulk of Europe's onshore growth in the near term. These countries' wind potential, relatively untapped pool of sites and supportive renewables policies will create the most competitive markets for project permits and asset ownership. These markets will be followed by smaller growing markets in Eastern Europe that still have to develop a steady flow of projects to support the industry.

Growth Potential in Emerging Markets

Entry barriers to emerging markets remain high due to unstable regulatory regimes, siting issues and/or grid-related barriers such as the lack of infrastructure at windy sites.

After Poland, Hungary is likely to be one of the most active Eastern European wind markets. However, although the combination of a high incentive scheme with a relatively straightforward permitting process facilitates project development, grid capacity constraints will cap market growth at 330 MW by 2010.

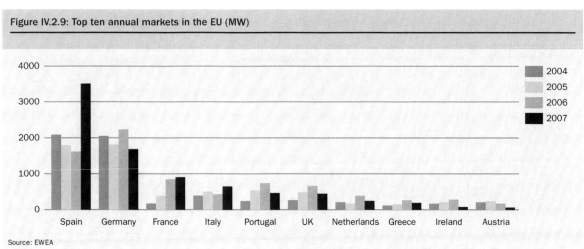

Figure IV.2.9: Top ten annual markets in the EU (MW)

Source: EWEA

- 56 GW installed capacity, including 1.08 GW offshore
- Annual installations of 8.5 GW, including 0.2 GW offshore
- Electricity production of 119 TWh, including 4 TWh offshore
- Meeting 3.7% of total EU electricity demand.
- 40.3% of the annual new electricity generating capacity
- 55% of annual net increase in installed electricity generating capacity
- 7.3% of total installed electricity generating capacity
- Providing power equivalent to the needs of 30 million average EU households (15% of EU households)
- Annual avoided fuel costs of €3.9 billion
- Annual investments in wind turbines of €11.3 billion
- Total lifetime avoided fuel costs of wind power capacity installed in 2007 of €16 billion (assuming fuel prices equivalent to $90/barrel of oil)
- Total lifetime avoided CO_2 cost of wind power capacity installed in 2007 of €6.6 billion (assuming CO_2 price of €25/t)
- European manufacturers have a 75% share of the global market for wind turbines (2006)[2]

Turkey, as the largest emerging market in terms of both size and population, is driven by large local industrial groups, with the number of projects installed per year as well as the average project size expected to increase over time. Although the country's incentive scheme recently became more stable, thereby lowering investment risks, the authorisation process is rather congested due to growing site speculation and remains difficult to navigate. As such, it is likely that Turkish independent power producers (IPPs) will continue to dominate the wind market.

In both Estonia and the Czech Republic, there are interesting opportunities for growth, given that wind investment risks are moderate despite the existing regulatory frameworks, site approval processes and competitive environments. However, project size in these markets will remain constrained by space and resource availability.

Bulgaria, Lithuania and Romania lead this group, because each has recently adopted a support system, with Lithuania and Bulgaria offering feed-in tariffs and

Figure IV.2.10: Wind power onshore market maturity

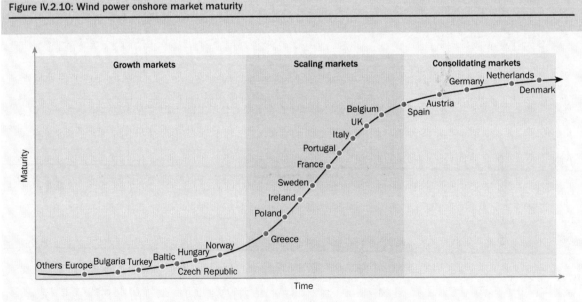

Source: Emerging energy research

Romania a green certificate scheme. At the moment, the authorisation processes in these markets are relatively expedient, although delays and congestion are anticipated as project queues get longer, taxing inexperienced permitting authorities.

The European Offshore Market

With a total of 1080 MW by the end of 2007, offshore accounted for 1.9 per cent of installed EU capacity and 3.5 per cent of the electricity production from wind power in the EU (Figure IV.2.11). The market is still below its 2003 level and development has been slower than previously anticipated.

Since 2003, the only country to consistently activate at least one offshore project per year has been the UK, when the 60 MW North Hoyle wind farm was commissioned. Denmark, Europe's earliest offshore pioneer, has not added any new projects since the 17 MW Ronland wind plant was commissioned in 2004, while Germany's first offshore wind turbine, a N90/2500, was installed in March 2006 in Rostock

international port. The Netherlands, Sweden and Ireland are the only other European markets with operational offshore projects.

With 409 MW, Denmark now has the most offshore wind capacity, but the UK (404 MW) is a very close second, having installed 100 MW in 2007. Sweden installed 110 MW in 2007. The Netherlands and Ireland also have operating offshore wind farms.

The main barriers to European offshore wind project development include:
- lengthy permitting processes that need to consider key issues such as defence, shipping, fishing and tourism;
- technical issues relating to the construction of the wind farms, including transport of turbines, turbine supports depending on the type of seabed, meteorological restrictions on building wind farms (often limited to a few months of the year in northern Europe) and connecting the wind farm to the mainland;
- incentive schemes that are not in line with the existing risks and/or costs associated with offshore investments, making it difficult to finance projects;

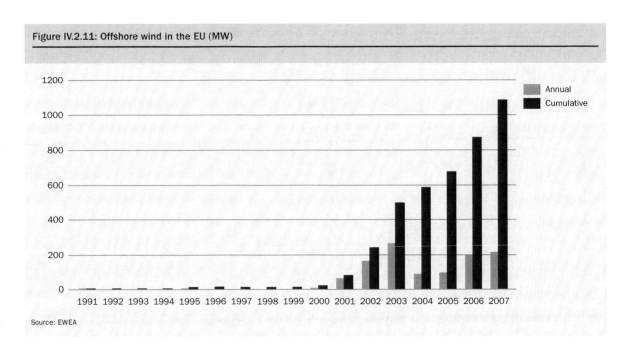

Figure IV.2.11: Offshore wind in the EU (MW)

Source: EWEA

Table IV.2.2: European offshore capacity, end 2007

Country	Total installed by end 2007 (MW)	Installed in 2007 (MW)
Denmark	409.15	0
UK	404	100
Sweden	133.25	110
Netherlands	108	0
Ireland	25.2	0
TOTAL	**1079.6**	**210**

Source: EWEA

- high costs associated with every stage of the project development, from development to construction, and turbine size;
- lack of turbine availability;
- lack of experience in offshore development;
- lack of smarter foundation types for deeper waters (>20 m); and
- an urgent need for more research and development (R&D) and demonstration projects.

Despite these issues, firms are moving offshore in some European markets, driven by high resources, limited onshore potential or even government pressure. As a result, it is estimated that the market will draw near to 10 per cent of total installed wind power within the next decade.

For the most part, utilities will be the main drivers of growth, as these firms can finance projects on their balance sheets, although some IPPs are looking at offshore installations in order to secure their position in the European wind energy market. In addition, more offshore opportunities will arise as developers tend to outsource these activities to firms with the technical know-how required for offshore project construction. In terms of turbine suppliers, Siemens and Vestas are currently the main two, although a handful of other firms are looking to challenge this duopoly with machines of 2.5 MW and 5 MW, and there are even larger models in the design phase.

Although Germany, the UK and Sweden are positioned to become the largest European offshore markets, other markets are now also ready to move offshore, exploiting existing resources. France's first offshore projects are due to be commissioned in 2008–2010, while in Spain several initiatives have been launched by key IPPs and utilities for large-scale projects, mainly in the Sea of Trafalgar and off the coast of Valencia. In Italy, Enel claims that it will develop a 150–200 MW wind farm in the Mediterranean.

Key Elements for Wind Energy Markets in Europe

Within the context of the Kyoto Protocol, Western European countries and a growing number of Eastern European markets have laid out renewables policies that are dependent on each market's energy balance, political will and level of liberalisation. Security of supply is also a key issue, as Europe works towards reducing its dependence on high-risk fossil fuel providers. An integral part of this policy is the regulatory mechanism that supports renewables, because it determines the economic model of renewables projects in most markets.

In addition to support mechanisms, renewable energy markets depend on resources, site approval, grid issues and the competitive environment. For the wind industry, these issues have been critical in defining both the market opportunity and its rules for participation. Economically viable tariffs and efficient and flexible permitting, combined with available grid capacity, are the key elements of a sound market.

With a policy and support mechanism in place, wind energy markets then depend on the resources available. Europe's highest wind speeds are seen offshore, where it is more expensive to install turbines, and on the coasts of countries in the northwest of the area. While significant untapped potential exists offshore, two onshore markets – Germany and The Netherlands – have begun to exhibit a degree of saturation as turbine procurement moves toward IEC Class III models with

Figure IV.2.12: Europe offshore market installed and targets by country

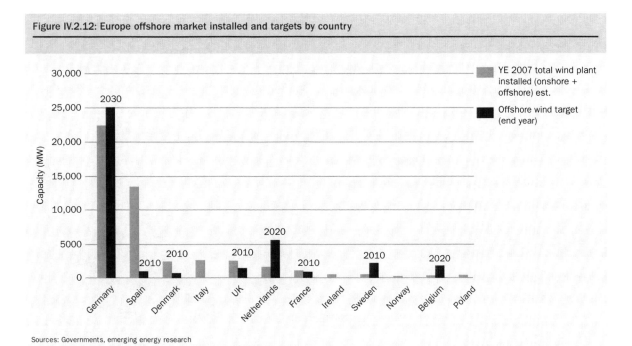

Sources: Governments, emerging energy research

Figure IV.2.13: European wind market environment and trends

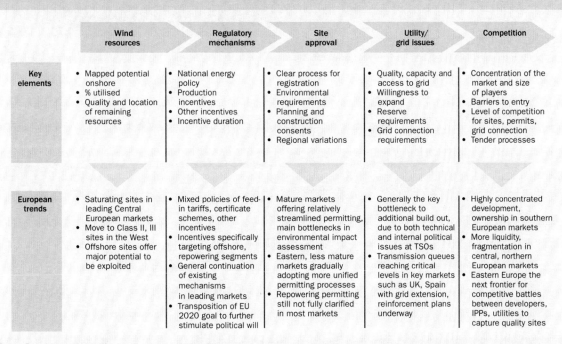

	Wind resources	Regulatory mechanisms	Site approval	Utility/ grid issues	Competition
Key elements	• Mapped potential onshore • % utilised • Quality and location of remaining resources	• National energy policy • Production incentives • Other incentives • Incentive duration	• Clear process for registration • Environmental requirements • Planning and construction consents • Regional variations	• Quality, capacity and access to grid • Willingness to expand • Reserve requirements • Grid connection requirements	• Concentration of the market and size of players • Barriers to entry • Level of competition for sites, permits, grid connection • Tender processes
European trends	• Saturating sites in leading Central European markets • Move to Class II, III sites in the West • Offshore sites offer major potential to be exploited	• Mixed policies of feed-in tariffs, certificate schemes, other incentives • Incentives specifically targeting offshore, repowering segments • General continuation of existing mechanisms in leading markets • Transposition of EU 2020 goal to further stimulate political will	• Mature markets offering relatively streamlined permitting, main bottlenecks in environmental impact assessment • Eastern, less mature markets gradually adopting more unified permitting processes • Repowering permitting still not fully clarified in most markets	• Generally the key bottleneck to additional build out, due to both technical and internal political issues at TSOs • Transmission queues reaching critical levels in key markets such as UK, Spain with grid extension, reinforcement plans underway	• Highly concentrated development, ownership in southern European markets • More liquidity, fragmentation in central, northern European markets • Eastern Europe the next frontier for competitive battles between developers, IPPs, utilities to capture quality sites

Source: Emerging energy research

larger rotors. Southern Europe, however, offers higher wind speeds.

The issue of grid access to the local distribution or transmission network has generally obstructed the development of wind power. Part of this challenge has stemmed from project permits greatly exceeding infrastructure capabilities, as is the case in Spain, the UK and other parts of Europe. At the same time, TSOs are in the process of understanding the technical requirements needed to integrate greater amounts of wind power. The application of new grid codes, improved fault ride-through and more accurate production forecasting are all contributing to resolving transmission challenges across Europe, although problems in connecting wind farm projects to the grid are likely to persist at the local level (this is even truer offshore, where new lines must be built to absorb GW-size offshore power stations).

Given sufficient transmission capacity, another key way of increasing installed wind power capacity is to facilitate the authorisation process for projects. Authorisation processes must not only be streamlined, but they must enable plans to be successfully realised, bolstered by flexibility and an infrastructure that allows new wind plants to be connected to the grid. Initiatives to unify overly bureaucratic permitting processes have been seen in the UK, Italy and Greece; though they have shown mixed results, these are key to shortening project execution time. Germany's reputation for efficiency in turning around applications in well under a year, and regional designations of wind development zones in markets such as Spain, have proven important ways of facilitating new projects.

IV.3   INDUSTRY ACTORS AND INVESTMENT TRENDS

Wind Turbine Manufacturing Trends

The global wind turbine market remains regionally segmented due to the wide variations in demand. With markets developing at different speeds and because there are different resource characteristics everywhere, market share for turbine supply has been characterised by national industrial champions, highly focused technology innovators and new start-ups licensing proven technology from other regions.

The industry is becoming ever more globalised. Europe's manufacturing pioneers have begun to penetrate North America and Asia. Wind turbine sales and supply chain strategies will take on a more international dimension as volumes increase.

As the region responsible for pioneering widespread, larger-scale uptake of wind power, Europe hosts the strongest competition for market share, with roughly a dozen suppliers. The European market has seen highly stable market share distribution with few major shifts since a round of consolidation among leading suppliers in 2003–2004. Between 2004 and 2007, three players held an average of over 15 per cent of the annual market share, followed by four players with a 5–10 per cent share each. During the same period, a handful of other players with a less than 5 per cent market share vied to establish themselves in the market for longer-term positioning. In Europe, manufacturers are primarily focusing on Class II machines of 2 MW and larger. There are several key players in this competitive region:

- Global leader **Vestas** averaged 30 per cent of annual MW added in Europe between 2001 and 2007 and is competing for offshore dominance with its 3 MW V90 turbine.
- **Enercon**, Vestas's chief European rival, has held steady at 20 per cent of MW supplied since 2001 with its bestselling 2 MW E-80 turbine.
- **Gamesa** has settled at 18–19 per cent of European MW added annually since 2005, based on steady sales of its 2 MW G80 turbine.

- **Siemens** has maintained 7–10 per cent of annual installations since acquiring Bonus in 2004, leveraging the success of the 2.3 MW turbine and 3.6 MW turbine offshore.
- **GE** has shifted much of its sales focus from Europe to its home US market with its 1.5 MW platform, and has seen its market share drop from 11 per cent in 2003 to 2 per cent in 2007, though it is seeking to regain its position with its 2.5 MW turbine.
- **Nordex** averaged 6 per cent of annual MW installed in Europe between 2001 and 2007, mainly due to steady sales in France and its home market Germany, followed by Portugal and other markets, where it mainly deploys its 2.3–2.5 MW N90 series.
- **Other suppliers** looking for market positioning include **Acciona Windpower**, **Alstom Ecotécnia** and **REpower**, with 5 per cent or less of the market.
- At the same time, **Pfleiderer** licensees **WinWinD** and **Multibrid** are continuing to scale up their pilot turbines, while **Fuhrländer** continued its lower volume deliveries in Germany.

PRODUCT PORTFOLIOS FIRMLY POSITIONED FOR MULTI-MEGAWATT SALES

Manufacturers are shifting their product strategies in order to address larger-scale project implementations, higher capacity turbines and lower wind speeds. Individual site characteristics will always determine which turbines enable buyers to hit their cost of energy targets; however, several trends can be seen in terms of the types of products suppliers are introducing to the market and what is in the development pipeline.

Manufacturers have taken significantly different approaches in terms of generator size and rotor size, based on the varying demands of their target markets. Most suppliers have at least one model available in three segments – 500 kW to 1 MW, 1 MW to 2 MW, and above 2 MW – with varying rotor sizes at 2 MW and above to meet Class II or I wind site conditions. However, some suppliers have created platforms using

Figure IV.3.1: European wind turbine market share, 2001–2007

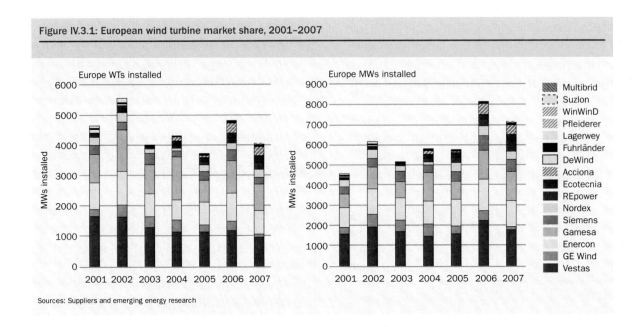

Sources: Suppliers and emerging energy research

one generator size with varying rotors (Vestas, GE, Gamesa and Enercon), while others have kept a tighter focus on rotor and generator size, with fewer variations on existing models (Nordex, Suzlon and Mitsubishi).

MANUFACTURERS TAILOR SALES STRATEGIES TO KEY CUSTOMERS

Soaring demand for wind turbines has led to bursting order books and a global shortage of key components. Demand has also driven manufacturers to redefine and sharpen their sales strategies over the past two years. Customer profiles have rapidly increased from 20 MW to 50 MW orders in selected countries and 1000 MW agreements spanning multiple regions. In turn, turbine suppliers have had to alter their sales approaches to maximise profitability, while positioning themselves for long-term market share.

Prior to 2005, turbine buyers could generally base their procurement decisions purely on site characteristics and price; however, the recent turbine shortage has made availability a key element in choosing suppliers. Buyers must now carefully balance their development plans with turbine availability and increasing

turbine cost in what is clearly a seller's market. This has led to multiple procurement strategies based on a developer's site portfolio, including multinational bulk orders, individual project contracts and a mix of

Figure IV.3.2: Interlinked factors determining turbine procurement

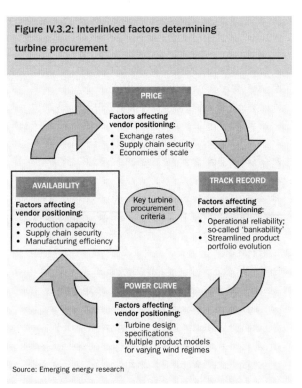

Source: Emerging energy research

both, with varying payment schedules based on the type of customer and of their position in the order book.

SUPPLY CHAIN KEY TO DELIVERY

Supply chain management is key to wind turbine supply. The relationships between manufacturers and their component suppliers have become increasingly crucial, and have come under increasing stress in the past three years as soaring demand has required faster ramp-up times, larger investments and greater agility to capture value in a rapidly growing sector. Supply chain issues have dictated delivery capabilities, product strategies and pricing for every turbine supplier. Manufacturers have sought to strike the most sustainable, competitive balance between a vertical integration of component supply and full component outsourcing to fit their turbine designs.

These procurement trends have given rise to unique market structures for each component segment, underlining the complexity of wind turbine design and manufacturing. Figure IV.3.3 illustrates the fact that the market for multiple segments, including blades, bearings and gearboxes, is highly concentrated and produces pinch points in the supply chain. These segments have high entry barriers based on size of investment and manufacturing ramp-up time. At the same time, controls, generators, castings and tower segments have lower entry barriers, with a larger number of players.

It is evident that with such uneven market structures across the supply chain, turbine manufacturers will see an opportunity to vertically integrate in order to reduce risk. In addition, this supply chain structure makes turbine shortages likely, as pinch points ripple through the market due to disparities in the availabilities of the different components. This means that in

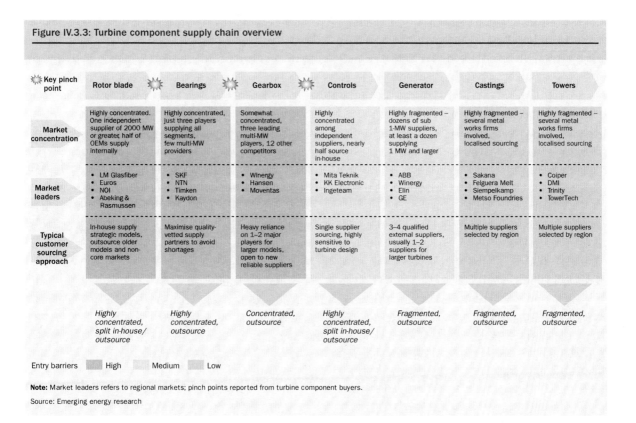

Figure IV.3.3: Turbine component supply chain overview

	Rotor blade	Bearings	Gearbox	Controls	Generator	Castings	Towers
Market concentration	Highly concentrated. One independent supplier of 2000 MW or greater, half of OEMs supply internally	Highly concentrated, just three players supplying all segments, few multi-MW providers	Somewhat concentrated, three leading multi-MW players, 12 other competitors	Highly concentrated among independent suppliers, nearly half source in-house	Highly fragmented – dozens of sub 1-MW suppliers, at least a dozen supplying 1 MW and larger	Highly fragmented – several metal works firms involved, localised sourcing	Highly fragmented – several metal works firms involved, localised sourcing
Market leaders	• LM Glasfiber • Euros • NOI • Abeking & Rasmussen	• SKF • NTN • Timken • Kaydon	• Winergy • Hansen • Moventas	• Mita Teknik • KK Electronic • Ingeteam	• ABB • Winergy • Elin • GE	• Sakana • Felguera Melt • Siempelkamp • Metso Foundries	• Coiper • DMI • Trinity • TowerTech
Typical customer sourcing approach	In-house supply strategic models, outsource older models and non-core markets	Maximise quality-vetted supply partners to avoid shortages	Heavy reliance on 1–2 major players for larger models, open to new reliable suppliers	Single supplier sourcing, highly sensitive to turbine design	3–4 qualified external suppliers, usually 1–2 suppliers for larger turbines	Multiple suppliers selected by region	Multiple suppliers selected by region
	Highly concentrated, split in-house/outsource	Highly concentrated, outsource	Concentrated, outsource	Highly concentrated, split in-house/outsource	Fragmented, outsource	Fragmented, outsource	Fragmented, outsource

Key pinch point

Entry barriers   High   Medium   Low

Note: Market leaders refers to regional markets; pinch points reported from turbine component buyers.

Source: Emerging energy research

today's seller's market, turbine assembly volume is dictated by the number of units that slip through the tightest pinch point. Generally, a proliferation in suppliers is anticipated throughout the supply chain, due to strong growth in the wind industry.

Value Chain Trends

Wind's spectacular growth as a vehicle for new generation capacity investment has attracted a broad range of players from across the industry value chain. From local, site-focused engineering firms to global, vertically integrated utilities, all have formed part of wind's European growth story. Since Europe's surge in 2005 to an annual market of over 6.5 GW of new capacity, the industry's value chain has become increasingly competitive, as a multitude of firms seek the most profitable balance between vertical integration and specialisation. The overall scaling up of the sector has meant that large-scale utilities have started to build sizeable project pipelines with long-term investment plans that indicate their commitment to adding wind to their generation portfolio, while at the same time a market remains for independent players able to contribute development skills, capital and asset management experience.

Europe's wind energy value chain is seeing dynamic shifts, as asset ownership is redistributed, growth is sought in maturing markets and players seek to maximise scale on an increasingly pan-European stage. While utilities build up GW-size portfolios, through their own strategy initiatives or government prompting, IPPs seek to compete for asset ownership in booming Western European markets, while development activity continues to shift towards new regions in the east. The proliferation of players looking to develop, own or operate wind farms has pushed competition to a new level, underlining the key elements of local market knowledge, technical expertise and financial capacity as crucial to positioning on the value chain.

Most of Europe's utilities have now taken position on the wind energy value chain as they comply with national renewables targets, while some have also taken the initiative of seeking international expansion with this newer generation technology. To maximise profitability, utilities have steadily migrated from risk-averse turnkey project acquisition to greater vertical integration, with in-house teams for development and operations and maintenance (O&M). Strategies devised by these players for meeting their objectives have largely depended on their experience in the sector as well as on their desire to expand geographically.

Utilities adopting a 'green' strategy are among the few European wind players that combine in-house experience and sufficient balance sheet to ramp up capacity at the pan-European level, whilst risking project development in less mature markets to sustain growth in the portfolio. These utilities have generally originated from countries with more pro-active renewables policies, including Spain and Denmark.

Another set of utilities have taken a more gradual approach to adding wind into their generation portfolios. These players, from Portugal, Italy and Germany, have moved into neighbouring markets and looked to build wind alongside thermal plants; they have made major acquisitions to support growth while following green utility strategies of building up internal teams for more vertical integration.

A third set of utilities, mostly working with conventional energies, has remained more domestically focused. Whether due to lower resource conditions, or a lack of scale to pursue larger opportunities, these players tend to focus on complying with national targets.

IPPs SEEK POSITIONS IN AN INCREASINGLY COMPETITIVE MARKET

Even before utilities began adopting wind energy, Europe's vertically integrated IPPs started aggressively exploiting wind turbine technology to improve their positioning. Led primarily by Spanish firms that were connected to large construction and industrial companies, the IPP model has evolved in various forms to represent a significant group of players in the value chain.

There are two main types of independent power producer in Europe: integrated ones, which have capabilities across the project development value chain and

exploit these for maximum control and returns on their project portfolio, and wind project buyers, which tend not to play a direct role in the development of wind plants in their portfolio, as these firms are often financial investors, rather than energy players. The number of these players that are active has continuously increased over the past three years, as utilities have sought acquisitions among this field of asset and pipeline-holding competitors, though those that are already a significant size may be positioned for long-term growth.

In terms of development, integrated IPPs are continuing to expand internationally, through greenfield project development and acquisitions, in order to compete with utilities. Players with strongholds in Spain,

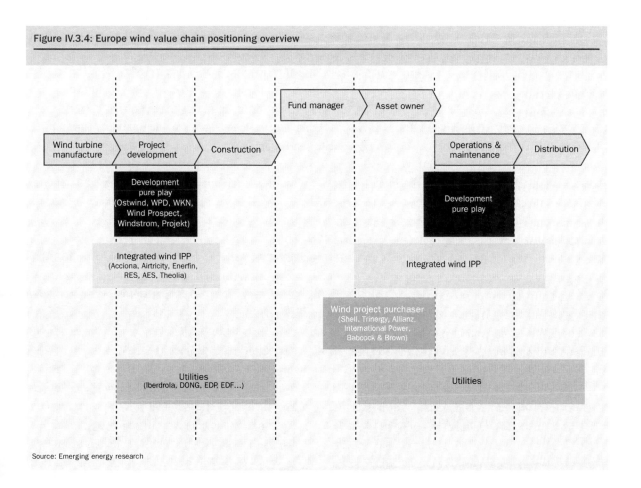

Figure IV.3.4: Europe wind value chain positioning overview

Source: Emerging energy research

France or Germany consistently look for growth in Eastern Europe, while some are also taking the plunge offshore. More risk-averse IPPs are seeing the number of quality projects available for acquisition in mature markets continue to dwindle.

As wind power owners, IPPs are facing stiffer competition from utilities as several project portfolios have been acquired in markets such as Spain, Germany, France and the UK. IPPs generally have higher capital costs than utilities, and those that can create assets organically through development on their own are generally better positioned to enlarge their portfolio.

As asset managers on the value chain, integrated wind IPPs and project purchasers are distinctly different, with integrated players increasingly focusing on O&M to maximise asset values. The boom in MW additions in the last three years means many turbines are coming out of their warranty periods, requiring IPPs to make key strategic decisions on how to manage their installations.

DEVELOPERS ADJUST STRATEGIES TO THE CHANGING ENVIRONMENT

European developers follow two distinct growth strategies: a develop-and-sell approach or a develop-and-own approach. With greenfield opportunities across consolidating and even scaling markets drying up, some pure play developers have transitioned into IPPs as a means of ensuring a steady revenue flow, often operating as pure plays in some markets and IPPs in others. However, a large number of traditional develop-and-sell players remain, and these are now focusing on capitalising on remaining opportunities in Europe's markets exhibiting wide levels of maturity

- While developers in high-growth markets such as the Balkans are focused on grabbing the best sites, in scaling markets like France, developers are looking to sell their projects. Greenfield project buyers tend to be wind operators with development capabilities, like IPPs and utilities, while new market

and/or industry entrants, such as financial investors, tend to buy turnkey.

- In consolidating markets such as Germany and Denmark, existing developers are mostly focused on realising and selling off the projects in their pipelines, while moving to offshore or new markets to ensure a steady revenue flow.
- In three to four years, developers will most likely see new opportunities emerge onshore as markets launch repowering schemes, which require operating wind parks to re-enter the permitting process to increase site output.

Key Player Positioning

Europe's shifting distribution of wind power asset ownership clearly illustrates the industry's scaling up and geographic expansion. From an industry concentrated in Denmark and Germany with single, farmer-owned turbines at the end of the 1990s, wind power ownership now includes dozens of multinational players that own several GWs of installed capacity. The European market is made up of five main ownership types:

1. **Utilities**: This group is made up of over 20 utilities, including pan-European, regional and local players that hold incumbent positions in electricity distribution and generation, and often transmission.
2. **Top IPPs**: Members of this group own over 300 MW each, mainly including vertically integrated players primarily working in Spain, Germany, France, the UK and Italy.
3. **Other Spanish IPPs** are all independent power producers with a presence in Spain, except for the top 20 IPPs with a presence in this market. Spain's heavy weighting in the European wind market, at 25 per cent of total installed capacity, represents a major ownership block for these players.
4. **German investors**: This block is composed of IPPs as well as institutional and private investors that own significant shares of Germany's total installed

capacity of over 22 GW, or 40 per cent of the European market.

5. **Other European investors/IPPs**: This group includes private and institutional investors and IPPs with a wind presence in European countries other than Spain and Germany and which are not among Europe's largest wind power operators.

Over the past five years, the most salient trend has been the increased participation of utilities in the industry. Utilities' share of the total wind power installed increased from 17 per cent in 2002 to 25 per cent in 2007. The biggest jump took place between 2005 and 2006, when the region's top wind utilities saw annual additions of well over 500 MW.

With consolidation in Europe's mature and scaling markets, it is anticipated that utilities and IPPs will have a bigger role in the future. Utility growth will be largely driven by pan-regional players realising their near-term projects, which currently range from 1000 to 4000 MW.

IPPs will also continue to increase their participation in wind, led by experienced vertically integrated players and larger investors able to develop internally or buy turnkey and leverage their strong financial capacity. At the same time, several of these firms may fall prey to expanding utilities, as seen in the past year, in which these firms' share fell to 9 per cent.

While German investors will continue to be the largest wind power ownership block in the next few years, their participation will diminish over time, as Germany's contribution to Europe's total wind power market decreases. On the other hand, Spanish IPPs are expected to decline in the near term despite the Spanish market's continued growth, as these smaller IPPs are either acquired by larger players or struggle to realise their modest pipelines amidst an increasingly competitive development environment.

Planned Future Investment

Capital-intensive construction of large wind capacity pipelines requires major investments by the utilities and IPPs planning to own assets. Sources of equity have taken a turn towards larger-scale, longer-term capital expenditure plans, with bond issues, IPOs and debt facilities proliferating among the top players. IPOs of utility renewable units have been consistently

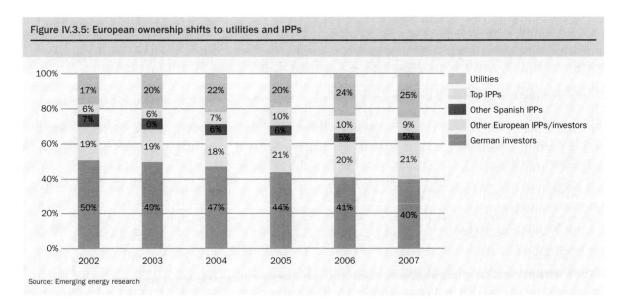

Figure IV.3.5: European ownership shifts to utilities and IPPs

Source: Emerging energy research

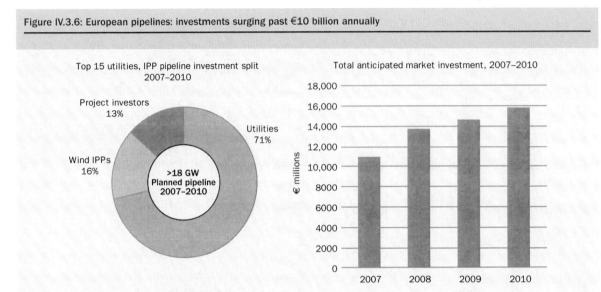

Figure IV.3.6: European pipelines: investments surging past €10 billion annually

Note: *Pipeline includes bottom-up announced, approved projects 2007–2008, plus estimates for meeting strategic goals through 2009–2010.

Sources: Utilities, IPPs, emerging energy research

oversubscribed in equity markets, while the overall volume of funds earmarked for wind capacity is reaching new heights.

For the 2007 to 2010 time period, Europe's top 15 utilities and IPPs (in terms of MW owned) declared project pipelines totalling over 18 GW, which translates into well over €25 billion in wind plant investments based on current cost estimates per MW installed. Overall, the European wind market is expected to grow at a rate of over 8–9 GW of annual installations up to 2010, which translates into yearly investments of between €10 billion and €16 billion. There will be several key investment trends in this period:

- Utilities are expecting to move into pole position as the clear leaders in wind capacity construction through three to five year CAPEX investment plans. For some players, these plans are worth over €6 billion and include offshore projects and expansion into Eastern Europe, combined with consolidation of their domestic market positions.
- The financial capacity of vertically integrated IPPs will be tested as they go head to head with utilities.

These players will reach deeper into the pockets of their parent companies to carry on accumulating assets in their target markets.

- Project buyers, or non-integrated IPPs, are likely to pick off individual turnkey opportunities with smaller investments plans of under €1 billion.

Industry Scales by Project and Turbine Size

The average size of wind projects is steadily increasing as wind becomes more integrated into the generation portfolios of leading utilities and IPPs that are looking to realise economies of scale. At the same time, project installation sizes are highly sensitive to local market and site restrictions, leading to wide discrepancies in average project sizes in the various European markets. There are several key trends for the different project segments:

- Projects of under 20 MW were the mainstay of European wind development, particularly in Germany, until 2004, when larger-scale markets like Spain, Portugal and the UK began growing in volume. These projects now represent less than 40 per cent of annual added capacity; however, this size of project remains important in tapping remaining market potential.

- Europe is seeing increasing saturation onshore in terms of mid-sized projects in the 20–50 MW range. While projects this size will maintain Europe's share of the global market, several coastal markets, led by the UK and Germany, are moving offshore for 100 MW and larger projects. Europe's steady growth onshore with average project sizes of 20–30 MW is likely to continue in the near term.
- At the same time, 50–99 MW projects have gradually increased their market share, reaching nearly 20 per cent of the market by 2007. Projects of this size often obtain the necessary permits more quickly from national governments.
- Based on installations in 2007, offshore projects look to be evolving towards the 60–200 MW range. 50–99 MW projects represent an intermediate phase of market development between pilot and large-scale developments, as observed in the UK, and will serve as a means for initiating larger installations in several markets.

As the global pioneer in wind turbine technology, Europe has seen a rapid increase in average turbine size, reaching multi-megawatt capacities. European markets have served as a base in establishing a track record for larger turbine deployment. However, suppliers look to continue maximising investments in work-horse product platforms of under 2 MW where possible. Key trends include:

- **Turbines of under 1 MW** are the most proven models of the majority of turbine suppliers, and installations peaked in 2004 with major deliveries in southern Europe. Since then, the introduction of more advanced systems with larger rotors has moved the industry towards higher-capacity machines. However, bigger is not always better, as O&M track records and performance are generally better understood with machines in the lower capacity segment.
- **1 to 1.49 MW** turbines carved out a niche in Europe with a few suppliers, for 3 per cent of annual

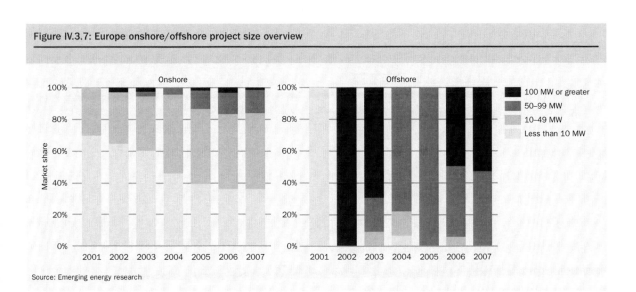

Figure IV.3.7: Europe onshore/offshore project size overview

Source: Emerging energy research

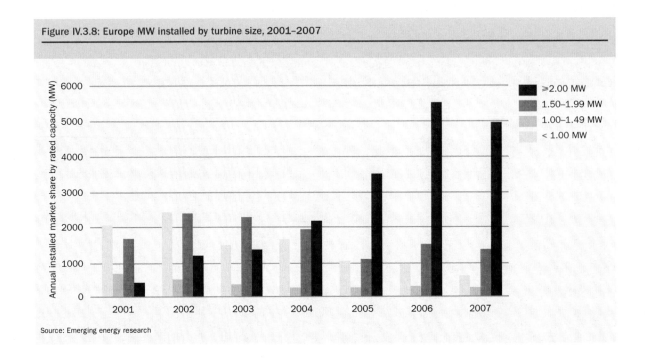

Figure IV.3.8: Europe MW installed by turbine size, 2001–2007

Source: Emerging energy research

European MW installations. However, the industry's trend towards multi-megawatt machines has reduced demand, as greater output at the same sites can be captured with 1.5 MW and larger models. The segment has now dropped to under 500 MW of annual installations, as suppliers seek to push their larger platforms.

- **1.50 to 1.99 MW** turbine installations peaked in 2002 and have generally levelled off at around 1500 MW of capacity installed in the past three years. This segment saw a major drop in demand, from over 30 per cent to less than 20 per cent of

installations, between 2004 and 2005, as leading suppliers and new entrants have pushed 2 MW and larger models into serial production.

- **2 MW and larger** turbines have become virtually standard in Europe since 2005, when this size machine jumped to over half of total installations in terms of megawatts. This surge continued in 2006, as the amount of MW installed in turbines this size pushed well past 5 GW. While 2007 saw the segment hit more severely by component shortages, Europe continues to rely on these larger turbines for the bulk of its installations.

IV.4 GLOBAL WIND ENERGY MARKETS

The Status of the Global Wind Energy Markets

In 2007, its best year yet, the wind industry installed close to 20,000 MW worldwide. This development was led by the US, China and Spain, and it brought global installed capacity to 93,864 MW. This is an increase of 31 per cent compared with the 2006 market, and represents an overall increase in global installed capacity of about 27 per cent.

The top five countries in terms of installed capacity are Germany (22.3 GW), the US (16.8 GW), Spain (15.1 GW), India (7.8 GW) and China (5.9 GW). In terms of economic value, the global wind market in 2007 was worth about €25 billion (US$37 billion) in new generating equipment, and attracted €34 billion (US$50.2 billion) in total investment.

Europe remains the leading market for wind energy, and new installations there represented 43 per cent of the global total in 2007.

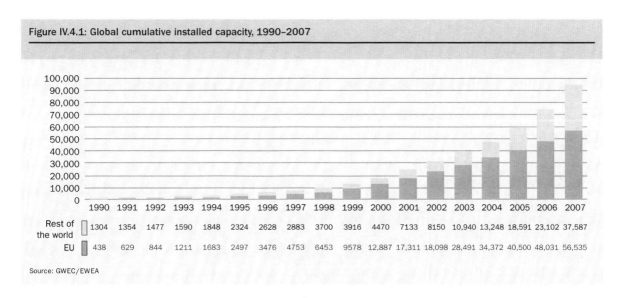

Figure IV.4.1: Global cumulative installed capacity, 1990–2007

	1990	1991	1992	1993	1994	1995	1996	1997	1998	1999	2000	2001	2002	2003	2004	2005	2006	2007
Rest of the world	1304	1354	1477	1590	1848	2324	2628	2883	3700	3916	4470	7133	8150	10,940	13,248	18,591	23,102	37,587
EU	438	629	844	1211	1683	2497	3476	4753	6453	9578	12,887	17,311	18,098	28,491	34,372	40,500	48,031	56,535

Source: GWEC/EWEA

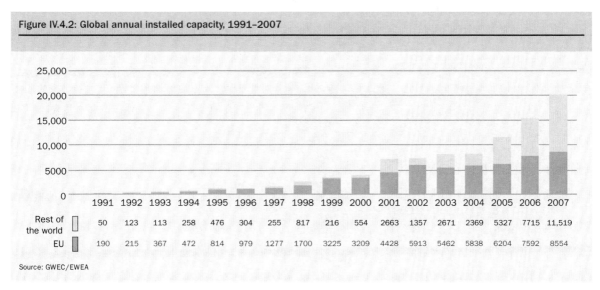

Figure IV.4.2: Global annual installed capacity, 1991–2007

	1991	1992	1993	1994	1995	1996	1997	1998	1999	2000	2001	2002	2003	2004	2005	2006	2007
Rest of the world	50	123	113	258	476	304	255	817	216	554	2663	1357	2671	2369	5327	7715	11,519
EU	190	215	367	472	814	979	1277	1700	3225	3209	4428	5913	5462	5838	6204	7592	8554

Source: GWEC/EWEA

Figure IV.4.3: Top ten installed capacity (end 2007)

	MW	%
Germany	22,247	23.7
US	16,818	17.9
Spain	15,145	16.1
India	7845	8.4
PR China	5906	6.3
Denmark	3125	3.3
Italy	2726	2.9
France	2454	2.6
UK	2389	2.5
Portugal	2150	2.3
Rest of world	13,060	13.9
Total top ten	80,805	86.1
Total	93,864	100.0

Source: GWEC

Figure IV.4.4: Top ten new capacity (end 2007)

	MW	%
US	5244	26.4
Spain	3522	17.7
PR China	3304	16.6
India	1575	7.9
Germany	1667	8.4
France	888	4.5
Italy	603	3.0
Portugal	434	2.2
UK	427	2.1
Canada	386	1.9
Rest of world	1815	9.1
Total top ten	18,050	90.9
Total	19,865	100.0

Source: GWEC

Figure IV.4.5: Annual installed capacity by region, 2003-2007

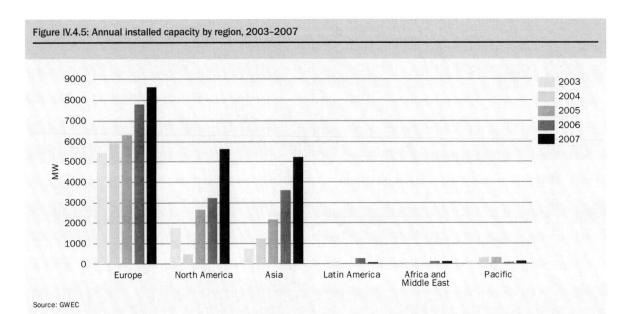

Source: GWEC

Global Markets Outside of Europe in 2007

NORTH AMERICA

The US

The US reported a record 5244 MW installed in 2007, more than double the 2006 figure, accounting for about 30 per cent of the country's new power-producing capacity. Overall, US wind power generating capacity grew by 45 per cent in 2007, with total installed capacity now standing at 16.8 GW.

Wind farms installed in the US by the end of 2007 will generate an estimated 48,000 GWh in 2008, just over 1 per cent of the country's electricity supply. The current US electricity mix consists of about 50 per cent coal, 20 per cent nuclear, 20 per cent natural gas, 6 per cent hydropower, with the rest generated from oil and non-hydro renewables, according to the US Energy Information Administration.

Most interesting, perhaps, is how quickly wind's share of current investment is growing: new wind projects account for about 30 per cent of the entire new power-producing capacity added in the US in 2007, establishing wind power as a mainstream option for new electricity generation.

In 2007, wind power production was extended to 34 US states, with Texas consolidating its lead and the Midwest and Northwest also setting a fast pace. The states with the most cumulative wind power capacity installed are Texas (4356 MW), California (2439 MW), Minnesota (1299 MW), Iowa (1273 MW) and Washington (1163 MW) (see Table IV.4.1).

Historically, the political framework conditions for wind power in the US have been very unstable. This sustained growth in wind power is the direct result of availability of the federal Production Tax Credit (PTC) over the past three years. The PTC is the only existing federal incentive in the US for wind power. It provides a 1.9 cent-per-kWh tax credit for electricity generated with wind turbines over the first ten years of a project's operations, and is a critical factor in financing new wind farms. In order to qualify, a wind farm must be completed and start generating power while the credit is in place. The energy sector is one of the most heavily subsidised in the US economy, and this incentive is needed to help level the playing field for wind and other renewable energy sources.

The PTC was set to expire in October 2008 but was extended for one more year, until the end of 2009. The tax credit is set at $2.1/kW at present value (January 2009). Previously when the credit was not extended, well before its expiry date installation growth rates fell

Table IV.4.1: Top ten states by megawatts of wind power generating capacity, as of 31 December 2007

State	Existing (MW)	Under construction (MW)	Share of total installations (existing) (%)	Rank (existing)
Texas	4356.35	1238.28	25.9	1
California	2438.83	45	14.5	2
Minnesota	1299.75	46.4	7.7	3
Iowa	1273.08	116.7	7.6	4
Washington	1163.18	126.2	6.9	5
Colorado	1066.75	0	6.3	6
Oregon	885.39	15	5.3	7
Illinois	699.36	108.3	4.2	8
Oklahoma	689	0	4.1	9
New Mexico	495.98	0	2.9	10

Source: AWEA

by 93 per cent (2000), 73 per cent (2002) and 77 per cent (2004).

The American Wind Energy Association's initial estimates indicate that another 5 GW of new wind capacity will be installed in 2008. However, the pace of growth in 2008 and beyond will largely depend not on turbine availability, but on the timing and duration of the potential extension of the PTC.

Canada

Canada's wind energy market experienced its second best year ever in 2007. A total of 386 MW of new wind energy capacity was installed in 2007, increasing Canada's total by 26 per cent. In 2007, Canada had 1856 MW of installed wind energy capacity.

Ten wind energy projects were commissioned in 2007 in five different Canadian provinces:

- Alberta led all provinces in 2007, installing three new projects totalling 139 MW. Alberta is now Canada's leading province for wind energy, with 524 MW.
- The largest wind energy project commissioned in Canada in 2007 was the 100.5 MW Anse-a-Valleau project in Quebec, the second project to be commissioned from Hydro-Quebec's earlier 1000 MW request for proposals.
- Two new projects, totalling 77.6 MW, were commissioned in Ontario. These two projects brought Ontario's total installed capacity to 491 MW.
- Three smaller projects, totalling 59 MW, were installed in Canada's smallest province, Prince Edward Island (P.E.I.). The installation of these facilities means that P.E.I. has now met its target to produce wind energy equivalent to 15 per cent of its total electricity demand three years ahead of schedule.
- One new 10 MW project was commissioned in Nova Scotia.

Turbines for these projects were provided by three manufacturers: Enercon (169 MW), Vestas (116.4 MW) and GE (100.5 MW).

Canada began 2008 with contracts signed ready for the installation of an additional 2800 MW of wind energy, most of which is to be installed by no later than 2010.

In addition, several new calls for tender for wind energy projects were launched in 2007 in Manitoba, Quebec, New Brunswick and Nova Scotia, and contracts have now been signed for 4700 MW of projects to be constructed in the period 2009–2016.

In addition, three tendering processes were issued in Quebec, Ontario and British Columbia in 2008, and contracts will be awarded in late 2008 or early 2009.

The Canadian Wind Energy Association forecasts that Canada will have 2600 MW of installed capacity by the end of 2008, an increase of around 800 MW on the previous year. Provincial government targets and objectives in Canada, if met, will give a minimum of 12,000 MW of installed wind energy in Canada by 2016.

ASIA

China

China added 3304 MW of wind energy capacity during 2007, representing market growth of 145 per cent over 2006, and now ranks fifth in total installed wind energy capacity worldwide, with 5906 MW at the end of 2007. However, experts estimate that this is just the beginning, and that the real growth in China is yet to come. Based on current growth rates, the Global Wind Energy Council (GWEC) forecasts a capacity of around 200 GW by 2020. The official government target of 30 GW by 2020 is likely to be met as early as 2012 or 2013.

The regions with the best wind regimes are located mainly along the southeast coast and in Inner Mongolia, Xinjiang, Gansu Province's Hexi Corridor, and some parts of Northeast China, Northwest China, Northern China and the Qinghai-Tibetan Plateau.

Satisfying rocketing electricity demand and reducing air pollution are the main driving forces behind the

Figure IV.4.6: Growth of the Chinese market, 1995–2007

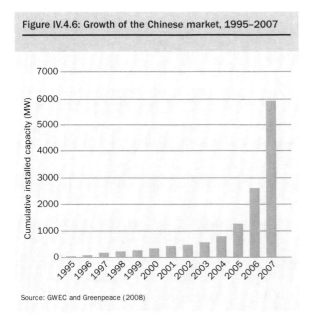

Source: GWEC and Greenpeace (2008)

Figure IV.4.7: Foreign and domestic players in the Chinese market (annual installed capacity)

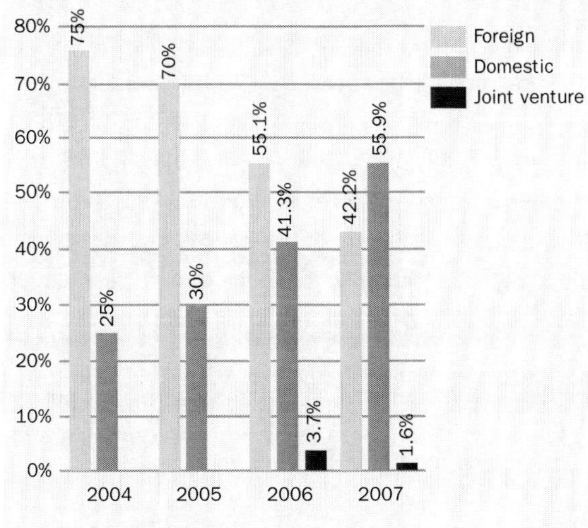

Source: Junfeng and Hu (2007)

addressed through the development of large-scale projects and boosting local manufacturing of wind turbines.

The Chinese Government estimates that the localisation of wind turbine manufacturing brings benefits to the local economy and helps keep costs down. Moreover, since most good wind sites are located in remote and poorer rural areas, wind farm construction benefits the local economy through the annual income tax paid to county governments, which represents a significant proportion of their budget. Other benefits include grid extension for rural electrification, and employment in wind farm construction and maintenance.

The wind manufacturing industry in China is booming. While in the past, imported wind turbines dominated the Chinese market, this is changing rapidly as the growing wind power market and the clear policy direction have encouraged domestic production, and most European manufacturers now produce in China.

The total manufacturing capacity is now about 5000 MW and is expected to reach 10–12 GW by 2010.

India

Wind energy is continuing to grow strongly in India, with over 1500 MW of new installed capacity in 2007, hitting 7845 MW in total. This represents year-on-year growth of 25 per cent.

The development of Indian wind power has so far been concentrated in a few regions, especially the southern state of Tamil Nadu, which accounts for more than half of all installations. This is beginning to change, with other states, including Maharashtra, Gujarat, Rajasthan and Karnataka, West Bengal, Madhya Pradesh and Andhra Pradesh starting to catch up. As a result, wind farms can be seen under construction right across the country, from the coastal plains to the hilly hinterland and sandy deserts.

development of wind energy in China. However, given the country's substantial coal resources and the still relatively low cost of coal-fired generation, reducing the cost of wind power is a crucial issue; this is being

The Indian Government envisages annual capacity addition of up to 2000 MW in the coming years. Official government estimations set the total wind energy potential in India at around 45 GW.

While there is no countrywide support for renewable energies, the Indian Ministry of New and Renewable Energy (MNRE) has issued guidelines to all state governments to create an attractive environment for the export, purchase and banking of electricity generated by wind power projects. State Electricity Regulatory Commissions (SERCs) were set up in most of the states in the country, with the mandate of promoting renewables, including wind energy, through preferential tariffs and a minimum obligation on distribution companies to source a certain share of electricity from renewable energy. Ten out of India's 29 states have set up renewable purchase obligations, requiring utilities to source up to 10 per cent of their power from renewable sources.

There are also a number of fiscal incentives for the wind energy sector established at national level, including:

- direct taxes – 80 per cent depreciation in the first year of installation of a project;
- a ten-year tax holiday;
- no income tax to be paid on power sales to utilities; and
- foreign direct investments are cleared very fast.

The Indian Government is considering accelerating depreciation and replacing the ten-year tax holiday with tradable tax credits or other instruments. While this would be an issue for established companies, new investors are less reliant on the tax holiday, since they often have little or no tax liability.

India has a solid domestic manufacturing base, including global player Suzlon, which accounts for over half of the market, and Vestas RRB. In addition, other international companies have set up production facilities in India, including Enercon, Vestas, REpower, Siemens and LM Glasfiber.

LATIN AMERICA

Brazil

Between 1999 and 2005, wind energy capacity in Brazil increased only by very small amounts, but in 2006, 208 MW were installed in one year, bringing the total to 237 MW. In 2007, only one wind farm came online: Eólica Millennium, a 10.2 MW project acquired by Pacific Hydro from local company Bioenergy. This brought the total to 247 MW.

The main obstacles to Brazilian wind power are significant import duties and taxes, which make projects less profitable unless complete local production and sourcing are established. Also, the country has prioritised the development of its biomass potential in the past few years. Wind power, however, is expected to grow substantially in the near future.

In 2002, the Brazilian Government passed a programme called the Programme of Incentives for Alternative Electricity Sources (PROINFA) to stimulate the development of biomass, wind and small hydro power generation. This law was revised in November 2003.

In the first stage (up to 2008, although the deadline has been extended until the end of 2008, and will possibly be extended into 2009), the programme guaranteed power sale contracts of projects with a total capacity of 3300 MW using these technologies, originally divided into three equal parts of 1100 MW per technology. Wind's share was later increased to 1400 MW. The Brazilian state-controlled electricity utility, Eletrobrás, will buy power produced by RES under power purchase agreements (PPAs) of 20 years at predetermined preferential prices.

Originally, a second stage of PROINFA was planned for when the 3300 MW objective had been met, with the aim of increasing the share of the three renewable sources to 10 per cent of annual electricity consumption within 20 years. Renewable energy generators would then have been required to issue a number of renewable energy certificates in proportion to the amount of clean energy produced.

However, despite the high expectations raised by the PROINFA programme, the scheme has to date failed to deliver the great number of wind projects the government had aimed for. As a result, the current government is showing little interest in taking PROINFA to its second stage, and is considering replacing it with an auction system. The Brazilian Wind Power Association (ABEEolica) is lobbying to proceed with PROINFA II while at the same time introducing an auction process.

The outlook for 2008 is quite optimistic: there are 14 wind energy plants financed by the PROINFA programme under construction, amounting to 107.3 MW of installed capacity. In addition, experts estimate that another 27 wind farms, representing 901.29 MW, could be added to the grid in 2009, provided that PROINFA is extended until the first semester of 2009.

Mexico

Despite the country's tremendous potential, the uptake of wind energy in Mexico has been slow, mainly due to the lack of government incentives for the use of renewable energy and the lack of a clear regulatory framework that would allow for private-sector participation in the development of wind facilities. At present, Mexico has a total installed capacity of 85 MW.

In 1984, the Confederation for Electricity (CFE – Comisión Federal de Electricidad) built the demonstration project La Venta I, with seven wind turbines and a total capacity of 1.6 MW, located south of the Isthmus of Tehuantepec, 30 km Northeast of Juchitán in the state of Oaxaca.

Another individual 600 kW plant was put into operation by CFE at the end of 1998, near Guerrero Negro in the federal state of Baja California Sur, operating in an isolated urban grid.

In October 2006, a bid for an 83.3 MW wind facility, La Venta II, and a demonstration project was granted to the Spanish consortium Iberdrola-Gamesa, for 98 turbines of 850 kW each. The Global Environment Facility (GEF) launched a programme to subsidise the cost per kWh of electricity produced at La Venta II in order to allow CFE to comply with its legal obligation to purchase power at the lowest cost. This programme is being implemented by the World Bank.

In terms of private-sector involvement, a number of companies have participated in wind energy development in Mexico, including major players such as Cisa-Gamesa, Demex, EDF-EN, Eoliatec, Fuerza Eólica, Iberdrola, Preneal and Unión Fenosa. The combined development portfolio in private wind energy facilities could reach 2600 MW in Oaxaca and 1000 MW in Baja California for the period from 2008 to 2010.

The monopoly of the state suppliers is the main obstacle to a more widespread use of renewable energy in Mexico. In addition, larger projects have failed to materialise due to the lack of favourable building and planning legislation, as well as the lack of experienced developers and officials. Moreover, strong pressure to provide electricity at very low prices has failed to make wind energy installations economically viable.

THE MIDDLE EAST AND AFRICA

Egypt

Egypt enjoys an excellent wind regime, particularly in the Suez Gulf, where average wind speeds reach over 10 m/s.

The Egyptian wind energy market increased from just 5 MW in 2001 to 310 MW at the end of 2007, with 80 MW of new capacity being added in 2007 to the Zafarana wind farm. Over 3000 MW are earmarked for wind power developments in the near future on the Gulf of Suez coast.

In April 2007, Egypt's Supreme Council of Energy announced an ambitious plan to generate 20 per cent of the country's electricity from renewable sources by 2020, including a 12 per cent contribution from wind energy, translating into 7200 MW of grid-connected wind farms. This plan will provide investor security and stimulate private investment in wind energy.

Moreover, a new draft energy act has recently been submitted to the Egyptian Parliament to encourage renewable energy deployment and private-sector involvement. In addition to guaranteeing third party access, power generation from renewable energy would enjoy priority grid access under this law.

With the Zafarana project, Egypt has moved on from limited experimental projects to large-scale grid-connected wind farms. Overall, 305 MW has been installed in different stages: 63 MW in 2001, 77 MW in 2003/2004, 85 MW in July 2006 and 80 MW in December 2007. The electricity production from the Zafarana farm is over 1000 GWh per year at an average capacity factor of 40.6 per cent. A further 240 MW extension of the wind farm is currently being put into place.

In addition to this, an area of 656 km² has been earmarked to host a 3000 MW wind farm at Gulf of El-Zayt on the Gulf of Suez coast. Studies are being conducted to assess the site potential to host large-scale grid-connected wind farms of 200 MW capacity (in cooperation with Germany), 220 MW (in cooperation with Japan) and 400 MW (a private-sector project).

Morocco

In April 2007, Morocco's new Amogdoul wind farm, situated on Cap Sim, 15 km south of Essaouira, started operations, thereby bringing the country's total installed capacity up to 124 MW. Other wind farms in Morocco include a 50 MW project in El Koudia El Baida (Tlat Taghramt, Province of Tetouan) installed in 2000, followed by a 3.5 MW project at the same site in 2001.

The annual electricity production from wind energy now stands at 450 GWh, accounting for around 2 per cent of Morocco's power consumption.

The Moroccan National Programme for Development of Renewable Energies and Energy Efficiency

(PNDEREE) is to raise the contribution of renewable energies to 20 per cent of national electricity consumption and 10 per cent of primary energy by 2012 (the figures are currently 7.9 per cent and 3.4 per cent respectively, including large hydropower installations).

With 3000 km of coastline and high average wind speeds (7.5–9.5 m/s in the south and 9.5–11 m/s in the north), wind power is one of the most promising sectors for renewable energy generation in Morocco. Taking into account this vast potential, the Moroccan Government decided to raise the wind energy capacity from the current 124 MW to 1000 MW by 2012. Between 2008 and 2010, the Moroccan Government is planning to add 600 MW of installed wind energy capacity near the towns of Tetouan, Tarfaya and Taza.

THE PACIFIC REGION

Australia

With some of the world's best wind resources, Australia is a prime market for wind energy. The growing industry can take advantage of a stable, growing economy, good access to grid infrastructure, and well-organised financial and legal services.

While Australia had an exceptionally weak year in 2007, with only 7 MW of new installations, the change in government at the end of the year encourages hope for a brighter future for wind energy. Within hours of being sworn in to office, the new Labour Prime Minister, Kevin Rudd, signed the ratification of the Kyoto Protocol, thereby dramatically changing Australia's commitment to reducing greenhouse gas emissions. This is likely to have positive long-term impacts for wind energy development in the country.

The total operating wind capacity at the end of 2007 was 824 MW. While there were only three new project commitments during 2007 – amounting to €440 million of investment – the 2008 outlook is rosier as a result of the growing political and public support. Significant amounts of wind capacity are moving through the project planning stages, with over 400 MW of projects receiving planning approval during 2007.

Nine projects (over 860 MW in total) were commissioned, although not yet operational, by December 2007, including three new projects totalling 290 MW of capacity.

The new government expanded Australia's national target of 2 per cent of electricity from renewable energy by 2020 to 20 per cent. To meet this target, around 10,000 MW of new renewable energy projects will be built over the next decade. The wind industry is poised to play a major role in meeting this demand.

New Zealand

New Zealand's wind energy industry is small, but it is growing steadily. Wind energy capacity almost doubled in 2007, increasing from 170.8 MW to 321.8 MW. New Zealand's exceptional wind resource means there is a high capacity factor by international standards. In 2006 the average capacity factor for New Zealand's wind farms was 41 per cent. The estimate for 2007 is 45 per cent, with turbines in some wind farms achieving up to 70 per cent capacity in the windier months.

New Zealand's wind industry does not receive direct financial support or subsidies from the government. Nonetheless, the development of a new wind farm near Wellington, West Wind, and ongoing investigations at other sites shows that with the right conditions, wind

energy is competitive with other forms of electricity generation.

In 2007 the government announced its target for New Zealand to generate 90 per cent of its electricity from renewable sources by 2025. New Zealand currently generates about 65 per cent of its electricity from renewable sources, primarily from hydro. To reach 90 per cent, renewable energy capacity needs to grow by about 200 MW each year.

Wind provides about 1.5 per cent of New Zealand's current electricity needs. With limited opportunities for the expansion of hydro and geothermal generation, the renewable energy target gives added impetus to New Zealand's wind industry. Wind energy's contribution is set to grow over the coming years, and developers are currently seeking consent to build projects with a combined capacity of more than 1800 MW.

IV.5   ADMINISTRATIVE AND GRID ACCESS BARRIERS: AN ANALYSIS OF EXISTING EU STUDIES IN THE FIELD

Barriers and EU Action

There are many barriers preventing electricity from renewable energy sources being integrated into the European electricity market. This chapter is written from a developer's point of view, and describes these barriers, taking four EU Member States as case studies. Barriers are related to issues such as obtaining building permits, spatial planning licences and grid access. There are often unclear, or unnecessarily complex, administrative and financial procedures.

Such problems are to be found in every Member State, but their impact on the deployment of renewable energy differs depending on the country. There are also grid connection obstacles, which can discourage investment in wind energy, as well as preventing it from achieving competitiveness with other power-generating technologies.

The European Commission has recognised the importance of the issue, and addresses administrative barriers in Article 6 of Directive 2001/77/EC on the promotion of electricity produced from renewable energy sources in the internal electricity market. It has also raised the issue more recently in its proposed Directive 2008/0016 (COD): energy and climate change: promotion of the use of energy from renewable sources (RES-E), and in its Communications 2004/366 and 2005/627.

In order to develop effective policy and regulatory improvements, EU studies in the form of research and innovation activities address and evaluate these obstacles. Under the Framework Programmes for Research and Technological Development (FP7), 60 projects have been carried out so far. The Intelligent Energy-Europe (IEE) programme accounts for 13 projects in which EU officials work together with experts from the renewable energy industry. The following brief account of barriers is based on two comprehensive studies conducted by the Intelligent Energy-Europe

programme: a) the Assessment and Optimisation of Renewable Support Schemes in the European Electricity Market – OPTRES[3] (2005–2006), is an analysis of the main barriers for the development of RES-E in the EU-25 and b) the Promotion and Growth of Renewable Energy Sources and Systems – PROGRESS[4] (2006–2008) focuses on authorisation and grid barriers.

ADMINISTRATIVE BARRIERS

Before building a wind power plant, the project developer needs to obtain permits from the local authorities and has to carry out an impact assessment of his project. This process is riddled with obstacles:

- there are a large number of authorities involved;
- a bad or total lack of coordination between authorities can result in project delays;
- lengthy waiting periods to obtain permits can result in rejection of the project;
- renewable energy sources are insufficiently taken into account in spatial planning;
- there are highly complex and non-transparent procedures for the whole licensing chain; and
- there is a low awareness of benefits of renewable energy sources within local and regional authorities.

GRID-RELATED BARRIERS

One of the most important elements for the success of wind energy projects is access and connection to the grids. Based on stakeholder consultation, OPTRES[5] and PROGRESS[6] identified several obstacles and other grid-related problems faced by project developers in the EU (see Part II on grid integration for more details):

- insufficient grid capacity available;
- grid connection procedure not fully transparent;
- objectiveness in the evaluation of applications not fully guaranteed;

- excessive grid connection costs in some cases;
- long lead times for grid connection authorisation; and
- grid extension problems

SOCIAL AND FINANCIAL BARRIERS

The OPTRES stakeholder consultation showed that social barriers fall into three categories:

1. opposition from local public (NIMBY);
2. lack of awareness of the benefits of renewable energy; and
3. invisibility of the full costs of electricity from non-renewable energy sources.

Financial barriers can be caused by the existing national frameworks and vary depending on the Member State's electricity system. The RESPOND (Renewable Electricity Supply interactions with conventional Power generations, Networks and Demand) project (2006–2009) claims that the growing amount of RES-E affects the electricity system and can only be efficiently integrated if it leads to economically efficient, market-based responses from different stakeholders. In practice, however, current electricity market regulation does not always give sufficient incentives to market participants for an optimal support of integration of RES-E.

The financial barriers that were identified during the OPTRES stakeholder consultation can be divided into two main categories:

1. lack of certainty among banks or investors; and
2. capital subsidies and cash flows that are hard to predict

Case Studies

POLAND

Poland is a relatively new market for wind energy. The installed capacity in 2005 was 73 MW; it reached 152 MW in 2006 and 276 MW in 2007.

The Polish Wind Energy Association (PWEA) identifies the main administrative barriers as:

- There is a lack of transparency in the decision-making process for authorising grid connection.
- Requirements for the environmental impact assessment, spatial planning permission and grid-connection processes are badly defined and vary throughout the country.
- The Natura 2000 certification process slows down the realisation of wind farm projects. Candidate areas for Natura 2000 status are added permanently to a list which is subject to unexpected changes. In these areas, development is generally not allowed and can thus lead to the failure of a project.

One of the main administrative barriers that developers are confronted with is the fact that neither the timeframe nor the costs involved in the building permit application process can be accurately estimated. This is due to the fact that the application process is structured heterogeneously and varies from project to project. The number of authorities involved in the building permit application at community and local level does not depend only on the capacity of the individual project but also on the possible impact it may have on the environment and whether it endangers any species protected within the Natura 2000 network. According to the PWEA, developers are not given specific guidelines for environmental impact assessments and are often required to submit additional, time-consuming information at a later stage, which often results in the general application procedure being delayed. In the worst-case scenario, this time delay and the opposition from environmental organisations or local residents can lead to the failure of a project. Depending on whether the project is perceived as having a large, small or non-existent impact, the developer contacts the local and/or community authority. The bigger the anticipated impact, the higher the number of authorities and organisations for nature conservation involved in the environmental study. In the case of a low expected impact, the developer may or may not be required to provide additional input.

Moreover, the procedures to be followed to obtain all the necessary permits in Poland are felt by stakeholders to be highly unclear and ill-adapted to the requirements of wind turbines. The investor is obliged to contact the bodies responsible for grid connection, spatial planning and environmental concerns individually, which may lead to confusion and delays as these bodies do not cooperate effectively with each other. Nevertheless, the lead time for the authorisation procedure is around two years, which is relatively short in comparison to other countries assessed in this chapter.

In terms of obtaining grid connection, the PWEA identifies four crucial barriers for wind energy projects in Poland:

1. the reserving of connection capacity, or 'the queue for the connection point';
2. the initial charge made by the system operator on the developer when obtaining connection offers;
3. the process's lack of transparency and the lack of published data; and
4. the limited validity of offers, which often expire between their being made and the start of the construction of the wind farm.

When applying for grid connection, the Polish distribution companies (DSOs) do not provide the developer with a specific deadline by which they will be granted grid access. This makes it uncertain as to when the wind plant will become operational. Moreover, the Polish grid has a limited capacity. Since developers do not receive information in terms of grid capacity available, and are unaware of their interconnection acquisition, many apply in advance for more capacity than actually needed in order to anticipate potential land gains for the project. This leads to the so-called 'queue for the connection point', in which the DSO treats all applications in the same way, without verifying

the feasibility of MW applied for, resulting in long time delays. Moreover, transmission and distribution operators can curtail production from wind, arguing that wind generation poses a threat to the security of the smooth functioning of the grid. The PROGRESS report confirms that, in Poland, more than 50 per cent of the planned projects encounter serious problems due to the constraints of the existing grid capacity. The same is valid regarding priority grid access. The PWEA claims that, despite the Polish law granting priority grid access to renewable energy, neither the transmission system operators (TSOs) nor the distributors commit to this piece of legislation. This is because the legislation is not well defined, which creates a loophole in the system: due to the numerous exceptions to the law, the TSO has the power to decide that including electricity from wind energy in the grid is not imperative. Furthermore, the situation in terms of transparency for connection costs seems rather controversial. The PWEA confirms that these costs differ widely between investors. In this context, in order to deploy wind energy successfully in Poland, it is necessary to establish effective central and local grid systems.

FRANCE

Over the last decades, France has invested massively in nuclear power and designed its grids for this purpose. Nevertheless, it is a strong emerging wind energy market. The installed capacity in 2005 was 757 MW and reached 1.6 GW in 2006, representing an annual increase of 112 per cent. By the end of 2007, the installed capacity reached an impressive 2,454 MW.

The main administrative barriers in France, according to the French Wind Energy Association (FEE), are:

- the frequent addition of new constraints to the environmental impact assessment studies; and
- the regular changes in legislation.

France has the second largest wind potential in Europe. Despite this strongly developing market, the deployment of the technology is often slowed down or even stopped.[7] In general, the highest barriers in France are seen to be administrative and legislative ones. According to OPTRES, this is because RES-E policies are not fully clear or consistent, and a large number of authorities are involved in granting the building permit.[8] In their assessment of administrative procedures in France, the Boston Consulting Group observes a 'vicious cycle' faced by the project developers, as failing to obtain one permit can result in the refusal of additional permits, which might lead to the failure of the planned project. The procedure takes place in three rounds, and involves 25 different offices, according to the French electricity board. Small-scale and large-scale project developers have to comply with different procedures: projects with a hub height below 50 m and those below 4.5 MW face a slightly simplified application process. In France, the time needed to get a building permit for a wind park is usually between one and two years, although the official length of time is given as five months. Project developers cannot undertake any actions against administrations which do not fulfil the legal terms. Although the procedures for the licensing chain are transparent in France, they tend to be lengthy and complex. Lead times for the authorisation procedure can also be lengthy.

On an environmental level, before being able to install a wind plant, the location has to be determined and approved after a thorough impact assessment. France has established local committees which give advice on the siting of every project that might affect the landscape, including wind farms. These committees work in cooperation with the army, civil aviation and the meteorology services. This impact assessment process, despite its lack of transparency and legal value, is often used by local authorities to reject projects without taking their benefits into account.

A discrepancy between the attitudes of national authorities and local/regional authorities towards RES-E projects can be observed in many Member States. OPTRES observes that environmental impact assessments currently only take into account the negative impacts of RES-E projects, without highlighting the positive points. Furthermore, OPTRES and the Boston Consulting Group have observed significant misinformation about legislative rules and how to apply them, as well as a lack of knowledge about the environmental, social and economic benefits of wind energy, especially at local authority level. In this context, wind energy project development is severely hindered, as the results are not only delays in the granting of building permits, but also significant and unnecessary increases in administration costs for the developers. The European Commission therefore urges the development of guidelines on the relationship with European environmental law.

In terms of priority grid access for wind energy, the grid operators' mistaken belief that wind energy is a potential threat to grid security has changed over the last few years. Nevertheless, many grid operators and power producers do not want to reduce the capacity of existing power plants in favour of energy produced by wind power plants, as it could pose a financial risk to the producer. In France, permits are granted based on grid studies, which last 6 to 12 months. In particular, the lack of connection capacity in large areas like the North, Picardie or on the Massif Central creates barriers. In these areas, the connection is in need of grid infrastructure development and better connection to the transmission grid. In this case, the grid studies requested by the grid operators are very complex and expensive, and the high initial costs can make projects less profitable, thus discouraging project developers and investors from installing new capacities.

The attribution of the costs of grid reinforcements is controversial. Since new financial rules were introduced, only a part of the connection cost (60 per cent) has to be covered by the project developer. The impossibility of dividing the cost of reinforcements between several producers is a real problem in all areas with a lack of capacity. Where small distribution networks are concerned, grid connection procedure is not fully transparent, meaning that the grid owner is sometimes reluctant to disclose information on available connection capacities and points.

SPAIN

In Spain, the installed capacity in 2005 was 10 GW and reached 11.6 GW in 2006, representing an annual increase of 16 per cent. This impressive total had reached 15,950 GW in July 2008. The main barriers are related to grid connections:

- Authorisation procedures are slow and sometimes non-transparent.
- Lack of coordination between the different levels of government involved and the lack of heterogeneity of the procedure to be followed (which mostly differs in each region). This can lead to conflicts over which level is responsible for what.
- Delays by the authorities can have a significant impact on the finance of the project.

Despite these barriers, wind energy deployment is successful in Spain.

In Spain, 25 different permits are needed from regional and national authorities, each of which requires a different set of documentation. According to the experiences of the Spanish stakeholders and OPTRES, the permitting process for small-scale projects is just as complex as for large-scale projects. Furthermore, there is no real difference between the processes for different RES-E technologies. In Spain the various administrative bodies are sometimes not well coordinated, thus causing authorisation application deadlines to be missed.

Regarding administrative licences in Spain, the administration often requests that project developers process the administrative licences for the wind farm and the connecting line together. This can be a single dossier or more, depending on whether the line is to be used by a single producer or by several. Negotiations with the owners of the land necessary to build a wind farm are usually quick, while negotiations with the owners of the land necessary to build a connecting line are more difficult. When no agreement can be reached with the landowners, it is possible to expropriate the land as long as the installations are declared to be 'useful to the public' by the authorities.

The authorisation procedures regarding connection to the grid and the environmental impact assessment of RES-E plants often overlap, causing confusion. In the past, conflicts between investors and environmental organisations in some regions, for example Cataluña, have impeded the development of the wind power sector. Moreover, the environmental assessment of the projects is a part of the administrative licensing process and is necessary in order to get the administrative licence necessary to build the installations. The administrative licence can be processed at the same time as the access and connection licences. In practice, the environmental assessment process takes about six or seven months.

The electricity grid needs reinforcement and investment, as it is of limited scope and cannot accommodate all approved RES-E projects. According to a project developer in Spain, the bottleneck in the development of wind power projects has changed. In the past, the bottleneck was the administrative licensing issue, while nowadays the bottleneck is the connection issue. According to the Spanish Wind Energy Association (AEE), if a solution is not found, the Spanish wind energy market could stop growing. As for France and Poland, OPTRES has found that it is often impossible for Spanish renewable energy project developers to know the available grid capacity; hence they cannot verify technical and cost data of the grid connection presented to them by the grid operator. In the different regions, there have been different ways of allocating the connection capacity to the different project developers, such as calls for tender. According to the wind power sector in Spain, the target of 20,155 MW installed wind power capacity is likely to be met. However, it is more uncertain whether this capacity is going to be built by 2010, as stated in the political target in the Plan de Energías Renovables (Renewable Energy Plan) published by the Spanish Institute for Energy Diversification and Saving in 2005. The reason behind a possible delay is not the lack of investment but the possible lack of capacity in the grid to transport the produced electric power. The construction of the required infrastructure takes a long time to complete.

In terms of priority grid access, the AEE confirms that all forms of renewable power generation are granted priority grid access under Spanish law. Furthermore, Spanish energy distributors are obliged to buy the energy surplus, which amounts to 20 per cent of the demand coverage. In practice, almost all wind generation, amounting to 95 per cent of the total production, disposes of priority access as long as wind producers offer power at zero price in the electricity market. In this way, guaranteed transmission and distribution of electricity produced from renewable energy sources are highly important, as they secure the purchase of the excess power production once connected to the grid.

Looking at grid connection costs, OPTRES observes that the Spanish system is significantly less strict against electricity generators selling the energy to distributors for a fixed price, so it applies lower penalties. During the OPTRES stakeholder consultation, some respondents questioned whether it is fair that the investor has to pay for all hardware and renewal costs, while ownership of the installations goes to the grid operator. In Spain, the RES-E developers can get an estimate of connection costs from the grid owners, as this is required by Spanish legislation. However, the estimate is often not detailed and comprehensive enough, and sometimes the costs are exaggerated. In many countries it is unclear how the connection costs should be shared between the grid operator and the RES-E developers. There is a high need for a legal framework with clear, objective and non-discriminatory rules for cost-sharing.

OFFSHORE: UK

Installed offshore capacity in Europe accounted for 886 MW by the end of 2006. The technology is developing fast – 1.08 GW were installed in the EU-27 in December 2007. However, offshore development is being slowed by the high level of financial and technical risk associated with the projects.

Offshore wind is a significant potential contributor to the 20 per cent target, but certain obstacles hamper its development:

- In most countries, the maritime policy framework is not adapted to electricity production at sea.
- There is often an absence of transparency in permitting and subsidising processes.
- There is a lack of strategic planning for offshore wind sites.
- There is a lack of coordination regarding offshore grid extension.
- A high level of environmental scrutiny and application of the precautionary principle do not take into account the environmental benefits of wind energy.

Moreover, the integration of offshore wind energy into the grid is strongly affected by the possibilities for trans-European power exchange. The main barriers concerning the connection of offshore wind farms to the national power systems are:

- transmission bottlenecks;
- offshore transmission infrastructure and grid access; and
- balancing.

The following case study of the UK illustrates on the one hand the barriers confronted by developers wishing to build an operational offshore wind farm, and on the other hand how successful offshore wind energy can be once administrative and grid barriers have been overcome.

In the UK, offshore wind is growing fast. In 2007, 404 MW of offshore wind capacity was in operation, and a further 460 MW was under construction. In addition, permits have been granted for over 2700 MW of new offshore capacity. Current programmes for offshore wind make a total over 8 GW of future installed capacity. Work is currently underway to build a new programme, which will aim to deliver up to a further 25 GW of installed capacity by 2020.

Recognising the huge potential for offshore wind electricity generation, the UK regulates the offshore energy installations through the Energy Act of 2004. This Act establishes renewable energy zones (REZs) adjacent to the UK's territorial waters – taking into consideration the rights accorded in the United Nations Convention on the Law of the Sea (1982) – and creating a comprehensive legal framework for offshore energy projects. The 2004 Energy Act facilitates the streamlining of the consent process within the REZs and inshore waters.

Barriers

Administrative Barriers

In the UK, six authorities are involved in the authorisation procedures for onshore and offshore renewable energy projects, depending on local circumstances. In addition to this, the offshore project developers can decide whether to submit the applications separately to the different authorities or to manage the process through the Offshore Renewables Consents Unit (ORCU) of the Department for Trade and Industry (DTI). The ORCU has established a so-called 'one stop shop', which provides developers with a single liaison point for questions regarding the administration of applications, clarifies issues and provides updates on the progress of all consent applications.

Legislation is currently being reviewed on marine consenting, and a new Infrastructure Planning Commission will become responsible for offshore wind consents. Marine planning and strategy frameworks are also being developed. A new system of offshore transmission regulation is currently being developed, where all connections of over 132 kV will require a transmission operator to be established via competitive tendering. The operator will be required to design, build and maintain the transmission system in return for a stream of revenue.

Key barriers identified in the UK are the rather slow approval rate of building applications and limited transmission capacity in parts of the country, in particular between the very windy Scottish locations and the southern part of the UK. At present, nearly 8 GW of capacity are held up in the onshore planning system, equivalent to nearly 6 per cent of potential UK electricity supply. A further 9 GW from offshore projects are awaiting decision or due to be submitted for consent. In 2006 it took local authorities an average of 16 months to decide on wind farm applications, even though the statutory time period for decisions is 16 weeks.

Grid Capacity

The grid connection application for offshore wind farms is given by the UK's National Grid:

- Grid reinforcements are necessary in order to facilitate the grid connection of offshore wind farms in the future; however, they require very long lead times.
- In wind transmission grid codes there is a trend towards active control of large wind farms within the boundaries of the legal frameworks. This contributes to grid stability, although some contractual issues are still unclear. The capabilities required from large wind farms should be harmonised with TSO-specific set points.
- Common offshore cables bundling several wind farms would be beneficial. Moreover, they can become initial nodes of an international offshore grid. Up to now no bundling has taken place.
- Grid access, energy pricing and balancing are inter-related. To increase the value of wind energy,

measures such as adapted demand control, backup generation and storage are needed. Furthermore, good short-term forecasting will increase the value of wind energy on the energy markets.

- In order to take advantage of the geographical aggregation of wind speed, the transmission of wind power must be possible over distances comparable to the extensions of meteorological systems. Strong trans-European networks are essential for this.

In conclusion, many things need to be done on a technical level in order to integrate large amounts of offshore wind power into the UK power system. The feasibility of integrating large amounts of offshore wind power is mainly a question of finance, and hence closely related to political decisions and the creation of a favourable framework.

Renewables Obligation Certificates

In the past, only a few offshore wind farms were installed, as they had to deal with low Renewables Obligation Certificate (ROC) prices and technical difficulties. As a result, the building of offshore wind farms in the rather harsh offshore environment around the UK was very difficult and came almost to a standstill two years ago. With increasing ROC prices and more suitable wind turbine technology, the interest in offshore wind power has now picked up again; however, due to the high worldwide demand for wind turbines, both onshore and offshore, it is rather difficult to find wind turbine suppliers that are interested in delivering wind turbines for large offshore wind projects in the UK.

It is expected that in 2009, 1.5 ROCs per MWh will be established for offshore wind. This increase will be beneficial; however, the increasing steel, aluminium and copper prices will also have an impact on wind turbine costs.

Conclusion

Offshore wind is expected to make a large impact on the UK's 2020 renewable energy targets, and a major expansion is planned. The transmission system will be crucial for the success of this expansion, and efficient and appropriate connections and access will be required.

Part IV Notes

[1] Source: Eurelectric and EWEA. The national wind power shares are calculated by taking the electricity that the capacity installed by the end of 2007 will produce in a normal wind year and dividing it by the actual 2006 electricity demand. The statistical methodology used differs from the methodology otherwise used throughout this chapter, which explains the difference of 0.1 per cent for the total EU share. The figures may differ from the shares reported by national wind energy associations, due to difference in methodology.

[2] See www.btm.dk/documents/pressrelease.pdf.

[3] The OPTRES full report is available at www.optres.fhg.de/.

[4] The PROGRESS full report is available at www.res-progress.eu/.

[5] see Coenraads et al. (2006).

[6] see Held (2008), p9.

[7] 'Administrative and grid barriers', EWEA internal paper, July 2007.

[8] Coenraads et al. (2006).

PART V

ENVIRONMENTAL ISSUES

Acknowledgements

Part V was compiled by Carmen Lago, Ana Prades, Yolanda Lechón and Christian Oltra of CIEMAT, Spain; Angelika Pullen of GWEC; Hans Auer of the Energy Economics Group, University of Vienna.

We would like to thank all the peer reviewers for their valuable advice and for the tremendous effort that they put into the revision of Part V.

Part V was carefully reviewed by the following experts:

Maarten Wolsink	University of Amsterdam, Amsterdam Study Centre for the Metropolitan Environment AME
Josep Prats	Ecotecnia, European Wind Energy Technology Platform
Manuela de Lucas	Estación Biológica de Doñana (CSIC)
Glória Rodrigues	European Wind Energy Association
Claus Huber	EGL
Daniel Mittler	Greenpeace
John Coequyt	Sierra Club
Yu Jie	Heinrich Boell Foundation, China
John Twidell	Editor of the international journal 'Wind Engineering', AMSET Center
Patrik Söderholm	Lulea University of Technology
António Sá da Costa	APREN
Paulis Barons	Latvian Wind Energy Association

PART V INTRODUCTION

The energy sector greatly contributes to climate change and atmospheric pollution. In the EU, 80 per cent of greenhouse gas emissions (GHGs) come from this sector (European Environment Agency, 2008). The 2008 European Directive promoting renewable energy sources recognises their contribution to climate change mitigation through the reduction of GHGs. Renewable energies are also much more sustainable than conventional power sources. In addition, they can help provide a more secure supply of energy, they can be competitive economically, and they can be both regional and local. Wind energy is playing an important role in helping nations reach Kyoto Protocol targets. The 97 GW of wind energy capacity installed at the end of 2007 will save 122 million tonnes of CO_2 every year (GWEC, 2008), helping to combat climate change.

Wind energy is a clean and environmentally friendly technology that produces electricity. Its renewable character and the fact it does not pollute during the operational phase makes it one of the most promising energy systems for reducing environmental problems at both global and local levels. However, wind energy, like any other industrial activity, may cause impacts on the environment which should be analysed and mitigated. The possible implications of wind energy development may be analysed from different perspectives and views. Accordingly, this part covers the following topics:

- environmental benefits and impacts;
- policy measures to combat climate change;
- externalities; and
- social acceptance and public opinion.

Environmental benefits of wind energy will be assessed in terms of the avoided environmental impacts compared to energy generation from other technologies. In order to compute these avoided environmental impacts, the life-cycle assessment (LCA) methodology has been used. LCA, described in the international standards series ISO 14040-44, accounts for the impacts from all the stages implied in the wind farm cycle. The analysis of the environmental impacts along the entire chain, from raw materials acquisition through production, use and disposal, provides a global picture determining where the most polluting stages of the cycle can be detected. The general categories of environmental impacts considered in LCA are resource use, human health and ecological consequences.

Focusing on the local level, the environmental impacts of wind energy are frequently site-specific and thus strongly dependent on the location selected for the wind farm installation.

Wind energy has a key role to play in combating climate change by reducing CO_2 emissions from power generation. The emergence of international carbon markets, which were spurred by the flexible mechanisms introduced by the Kyoto Protocol as well as various regional emissions trading schemes such as the European Union Emissions Trading Scheme (EU ETS), could eventually provide an additional incentive for the development and deployment of renewable energy technologies and specifically wind energy. Chapter V.3 pinpoints the potential of wind energy in reducing CO_2 emissions from the power sector, gives an overview of the development of international carbon markets, assesses the impact of Clean Development Mechanism (CDM) and Joint Implementation (JI) on wind energy, and outlines the path towards a post-2012 climate regime.

Wind energy is not only a favourable electricity generation technology that reduces emissions (of other pollutants as well as CO_2, SO_2 and NO_x), it also avoids significant amounts of external costs of conventional fossil fuel-based electricity generation. However, at present electricity markets do not include external effects and/or their costs. It is therefore important to identify the external effects of different electricity generation technologies and then to monetise the related external costs. Then it is possible to compare the external costs with the internal costs of electricity, and to compare competing energy systems, such as conventional electricity generation technologies and wind energy. Chapters V.4 and V.5 present the results of the empirical analyses of the avoided emissions and avoided external costs due to the replacement of conventional fossil fuel-based electricity generation by wind energy in each of the EU27 Member States (as well as at aggregated EU-27 level) for 2007 as well as for future projections of conventional electricity generation and wind deployment (EWEA scenarios) in 2020 and 2030.

Wind energy, being a clean and renewable energy, is traditionally linked to strong and stable public support. Experience in the implementation of wind projects in the EU shows that social acceptance is crucial for the successful development of wind energy. Understanding the divergence between strong levels of general support towards wind energy and local effects linked to specific wind developments has been a key challenge for researchers. Consequently, social research on wind energy has traditionally focused on two main areas: the assessment of the levels of public support for wind energy (by means of opinion polls) and the identification and understanding of the dimensions underlying the social aspects at the local level (by means of case studies), both onshore and offshore.

Chapter V.5, on the social acceptance of wind energy and wind farms, presents the key findings from the most recent research in this regard, in light of the latest and most comprehensive formulations to the concept of 'social acceptance' of energy innovations.

V.1 ENVIRONMENTAL BENEFITS

It is widely recognised that the energy sector has a negative influence on the environment. All the processes involved in the whole energy chain (raw materials procurement, conversion to electricity and electricity use) generate environmental burdens that affect the atmosphere, the water, the soil and living organisms. Environmental burdens can be defined as everything producing an impact on the public, the environment or ecosystems. The most important burdens derived from the production and uses of energy are:

- greenhouse gases;
- particles and other pollutants released into the atmosphere;
- liquid wastes discharges on water and/or soil; and
- solid wastes.

However, not all energy sources have the same negative environmental effects or natural resources depletion capability. Fossil fuel energies exhaust natural resources and are mostly responsible for environmental impacts. On the other hand, renewable energies in general, and wind energy in particular, produce significantly lower environmental impacts than conventional energies.

Ecosystems are extremely complex entities, including all living organisms in an area (biotic factors) together with its physical environment (abiotic factors). Thus the specific impact of a substance on the various components of the ecosystem is particularly difficult to assess, as all potential relationships should be addressed. This is the role of impact assessments: the identification and quantification of the effects produced by pollutants or burdens on different elements of the ecosystem. It is important because only those impacts that can be quantified can be compared and reduced.

Results from an environmental impact assessment could be used to reduce the environmental impacts in energy systems cycles. Also, those results should allow the design of more sustainable energy technologies, and provide clear and consistent data in order to define more environmentally respectful national and international policies. For all these reasons, the use of suitable methodologies capable of quantifying in a clear and comparable way the environmental impacts becomes essential.

This chapter describes the LCA methodology and, based on relevant European studies, shows the emissions and environmental impacts derived from electricity production from onshore and offshore wind farms throughout the whole life cycle. Also, the avoided emissions and environmental impacts achieved by wind electricity compared to the other fossil electricity generation technologies have been analysed.

The Concept of Life-Cycle Assessment

Life-cycle assessment (LCA) is an objective process to evaluate the environmental burdens associated with a product, process or activity by identifying energy and materials used and wastes released to the environment and to evaluate and implement opportunities to effect environmental improvements (ISO, 1999).

The assessment includes the entire life cycle of the product, process or activity, encompassing extracting and processing raw materials; manufacturing, transportation and distribution; use, reuse and maintenance; recycling; and final disposal (the so-called 'cradle to grave' concept).

According to the ISO 14040 and 14044 standards, an LCA is carried out in four phases:
1. goal and scope definition;
2. inventory analysis: compiling the relevant inputs and outputs of a product system;
3. impact assessment: evaluating the potential environmental impacts associated with those inputs and outputs; and
4. interpretation: the procedure to identify, qualify, check and evaluate the results of the inventory analysis and impact assessment phases in relation to the objectives of the study.

Figure V.1.1: Conceptual framework on LCA

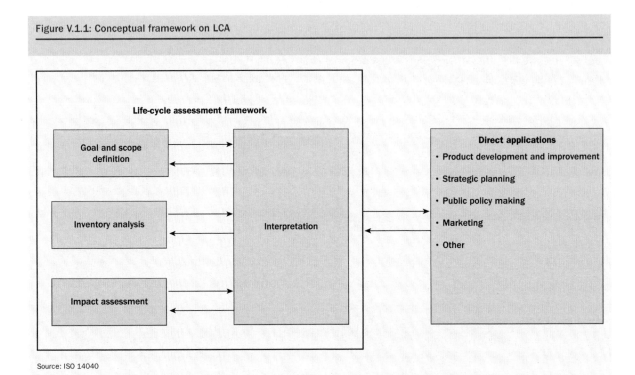

Source: ISO 14040

In the phase dealing with the goal and scope definition, the aim, the breadth and the depth of the study are established. The inventory analysis (also called life-cycle inventory – LCI), is the phase of LCA involving the compilation and quantification of inputs and outputs for a given product system throughout its life cycle. LCI establishes demarcation between what is included in the product system and what is excluded. In LCI, each product, material or service should be followed until it has been translated into elementary flows (emissions, natural resource extractions, land use and so on).

The third phase, life-cycle impact assessment, aims to understand and evaluate the magnitude and significance of the potential environmental impacts of a product system. This phase is further divided into four steps. The first two steps are termed classification and characterisation, and impact potentials are calculated based on the LCI results. The next steps are normalisation and weighting, but these are both voluntary according to the ISO standard. Normalisation provides a basis for comparing different types of environmental impact categories (all impacts get the same unit). Weighting implies assigning a weighting factor to each impact category depending on the relative importance.

The two first steps (classification and characterisation) are quantitative steps based on scientific knowledge of the relevant environmental processes, whereas normalisation and valuation are not technical, scientific or objective processes, but may be assisted by applying scientifically based analytical techniques.

Impact Categories

The impact categories (ICs) represent environmental issues of concern to which LCI results may be assigned. The ICs selected in each LCA study have to describe the impacts caused by the products being considered or the product system being analysed. The selection of the list of ICs has to fulfil several conditions (Lindfors et al., 1995):

- The overall recommendation regarding the choice of ICs is to include all the ICs for which international consensus have been reached.
- The list should not contain too many categories.
- Double counting should be avoided by choosing independent ICs.
- The characterisation methods of the different ICs should be available.

Some baseline examples considered in most of the LCA studies are illustrated in Table V.1.1.

As there is no international agreement on the different approaches regarding ICs, different methods are applied in current LCAs. Moreover, some studies do not analyse all the ICs described in the previous table, while others use more than the previous impact categories mentioned.

LCA in Wind Energy: Environmental Impacts through the Whole Chain

The LCA approach provides a conceptual framework for a detailed and comprehensive comparative evaluation of environmental impacts as important sustainability indicators.

Table V.1.1: Baseline examples

Impact category	Category indicator	Characterisation model	Characterisation factor
Abiotic depletion	Ultimate reserve, annual use	Guinee and Heijungs 95	ADP[9]
Climate change	Infrared radiative forcing	IPCC model[3]	GWP[10]
Stratospheric ozone depletion	Stratospheric ozone breakdown	WMO model[4]	ODP[11]
Human toxicity	PDI/ADI[1]	Multimedia model, e.g. EUSES[5], CalTox	HTP[12]
Ecotoxicity (aquatic, terrestrial, etc)	PEC/PNEC[2]	Multimedia model, e.g. EUSES, CalTox	AETP[13], TETP[14], etc
Photo-oxidant formation	Tropospheric ozone formation	UNECE[6] Trajectory model	POCP[15]
Acidification	Deposition critical load	RAINS[7]	AP[16]
Eutrophication	Nutrient enrichment	CARMEN[8]	EP[17]

Source: CIEMAT

[1] PDI/ADI	Predicted daily intake/Aceptable daily intake	
[2] PEC/PNEC	Predicted environmental concentrations/Predicted no-effects concentrations	
[3] IPCC	Intergovernmental Panel on Climate Change	
[4] WMO	World Meteorological Organization	
[5] EUSES	European Union System for the Evaluation of Substances	
[6] UNECE	United Nations Economic Commission For Europe	
[7] RAINS	Regional Acidification Information and Simulation	
[8] CARMEN	Cause Effect Relation Model to Support Environmental Negotiations	
[9] ADP	Abiotic depletion potential	
[10] GWP	Global warming potential	
[11] ODP	Ozone depletion potential	
[12] HTP	Human toxicity potential	
[13] AETP	Aquatic ecotoxicity potential	
[14] TETP	Terrestrial ecotoxicity potential	
[15] POCP	Photochemical ozone creation potential	
[16] AP	Acidification potential	
[17] EP	Eutrophication potential	

Recently, several LCAs have been conducted to evaluate the environmental impact of wind energy. Different studies may use different assumptions and methodologies, and this could produce important discrepancies in the results among them. However, the comparison with other sources of energy generation can provide a clear picture about the environmental comparative performance of wind energy.

An LCA considers not only the direct emissions from wind farm construction, operation and dismantling, but also the environmental burdens and resources requirement associated with the entire lifetime of all relevant upstream and downstream processes within the energy chain. Furthermore, an LCA permits quantifying the contribution of the different life stages of a wind farm to the priority environmental problems.

Wind energy LCAs are usually divided into five phases:

1. **Construction** comprises the raw material production (concrete, aluminium, steel, glass fibre and so on) needed to manufacture the tower, nacelle, hub, blades, foundations and grid connection cables.
2. **On-site erection and assembling** includes the work of erecting the wind turbine. This stage used to be included in the construction or transport phases.
3. **Transport** takes into account the transportation systems needed to provide the raw materials to produce the different components of the wind turbine, the transport of turbine components to the wind farm site and transport during operation.
4. **Operation** is related to the maintenance of the turbines, including oil changes, lubrication and transport for maintenance, usually by truck in an onshore scheme.
5. **Dismantling:** once the wind turbine is out of service, the work of dismantling the turbines and the transportation (by truck) from the erection area to the final disposal site; the current scenario includes recycling some components, depositing inert components in landfills and recovering other material such as lubricant oil.

ONSHORE

Vestas Wind Systems (Vestas, 2005 and 2006) conducted several LCAs of onshore and offshore wind farms based on both 2 MW and 3 MW turbines. The purpose of the LCAs was to establish a basis for assessment of environmental improvement possibilities for wind farms through their life cycles.

Within the framework of the EC project entitled 'Environmental and ecological life cycle inventories for present and future power systems in Europe' (ECLIPSE), several LCAs of different wind farm configurations were performed[1]. The technologies studied in ECLIPSE were chosen to be representative of the most widely used wind turbines. Nevertheless, a wide range of the existing technological choices were studied:

- four different sizes of wind turbines: 600 kW (used in turbulent wind conditions), 1500 kW, 2500 kW and 4500 kW (at the prototype stage);
- a configuration with a gearbox and a direct drive configuration, which might be developed in the offshore context;
- two different kinds of towers: tubular or lattice; and
- different choices of foundations, most specifically in the offshore context.

Within the EC project NEEDS (New energy externalities development for sustainability)[2], life-cycle inventories of offshore wind technology were developed along with several other electricity generating technologies. The wind LCA focused on the present and long-term technological evolution of offshore wind power plants. The reference technology for the present wind energy technology was 2 MW turbines with three-blade upwind pitch regulation, horizontal axis and monopile foundations. An 80-wind-turbine wind farm located 14 km off the coast was chosen as being representative of the contemporary European offshore wind farm.

In the framework of the EC project 'Cost Assessment for Sustainable Energy Systems' (CASES)[3], an estimation of the quantity of pollutants emitted at each production stage per unit of electricity for several electricity generation technologies, among them onshore and offshore wind farms, is performed.

Finally, the Ecoinvent v2.0 database[4] (Frischknecht et al., 2007) includes LCA data of several electricity generation technologies including an onshore wind farm using 800 kW turbines and an offshore wind farm using 2 MW turbines.

LCI Results: Onshore Wind Farms

Results extracted from the above-mentioned LCA studies for onshore wind farms regarding several of the most important emissions are shown in Figure V.1.2. Bars show the variability of the results when several wind farm configurations are considered in a study.

Carbon dioxide emissions vary from 5.6 to 9.6 g/kWh in the consulted references. Methane emissions

range from 11.6 to 15.4 mg/kWh. Nitrogen oxides emissions range from 20 to 38.6 mg/kWh. Non-methane volatile organic compounds (NMVOCs) are emitted in quantities that range from 2.2 to 8.5 mg/kWh, particulates range from 10.3 to 32.3 mg/kWh and, finally, sulphur dioxide emissions range from 22.5 to 41.4 mg/kWh. All of these quantities, with the only exception being particulates, are far below the emissions of conventional technologies such as natural gas (see Figure V.1.2).

Another main outcome of all the reviewed studies is that the construction phase is the main contributor to the emissions and hence the environmental impacts. As can be observed in Figure V.1.3, the construction phase causes about 80 per cent of the emissions. The operational stage, including the maintenance and replacement of materials, is responsible for 7–12 per cent of the emissions and the end-of-life stage of the wind farm is responsible for 3–14 per cent.

Regarding the construction stage, Figure V.1.4 shows the contribution of the different components. Important items in the environmental impacts of the

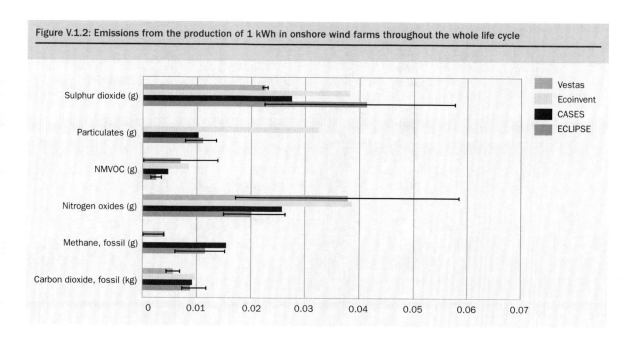

Figure V.1.2: Emissions from the production of 1 kWh in onshore wind farms throughout the whole life cycle

Figure V.1.3: Contribution of the different life-cycle phases to the relevant emissions

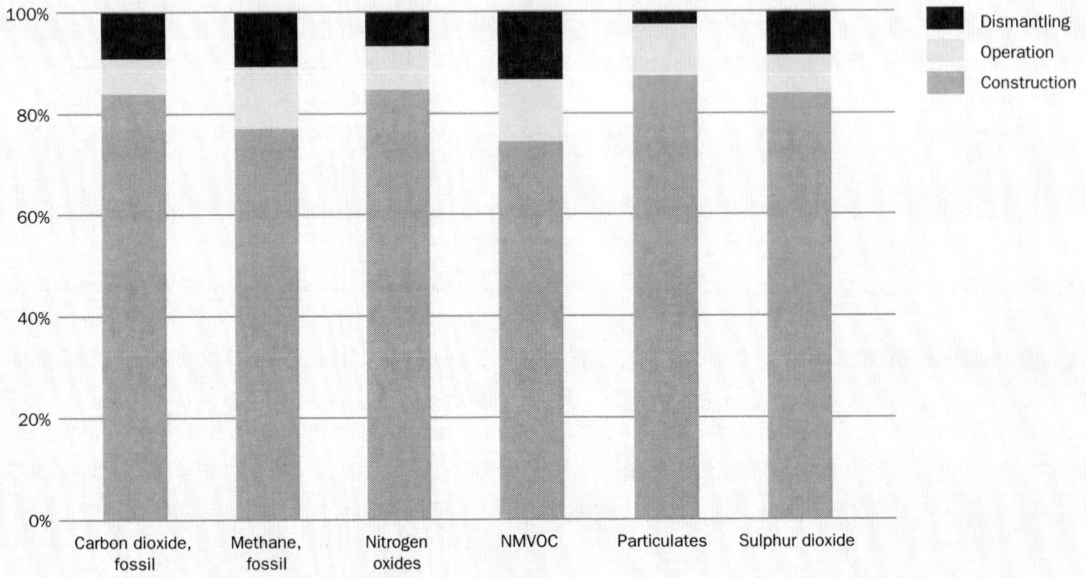

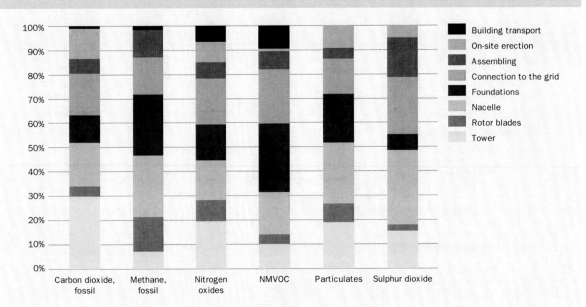

Source: Own elaboration using ECLIPSE results

Figure V.1.4: Contribution of the components of the construction phase to the different emissions

Source: Own elaboration based on ECLIPSE results

construction phase of an onshore wind farm are the tower and the nacelle but not the rotor blades. Foundations are another important source of emissions, and connection to the grid also contributes an important share. Emissions from transport activities during the construction phase are only relevant in the case of nitrogen oxides (NO_x) and NMVOC emissions.

LCA Results: Onshore Wind Farms

Results of LCAs have shown that wind farm construction is the most crucial phase because it generates the biggest environmental impacts. These impacts are due to the production of raw materials, mostly steel, concrete and aluminium, which are very intensive in energy consumption. The energy production phase from wind is clean because no emissions are released from the turbine. LCAs have also concluded that environmental impacts from the transportation and operation stages are not significant in comparison with the total impacts of the wind energy.

The contribution of the different stages to the ICs selected by the LCA of the Vestas V82 1.65 MW wind turbine is shown in Figure V.1.5.

In the Vestas study, the disposal scenario involves the dismantling and removal phases. Thus negative loads of recycling must be deducted, since some materials are returned to the technosphere. The disposal scenarios considered have great influence on the results.

This study evaluated the influence of small- and large-scale wind power plants on the environmental impacts, based on the V82 1.65 MW wind turbine. According to Figure V.1.6, a variation in the size of the wind power plant from 182 to 30 turbines did not

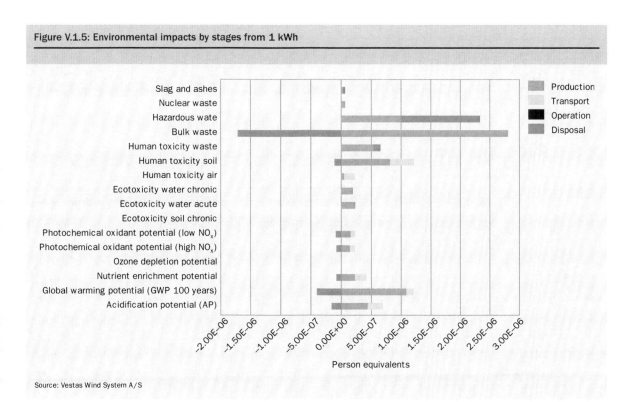

Figure V.1.5: Environmental impacts by stages from 1 kWh

Person equivalents

Source: Vestas Wind System A/S

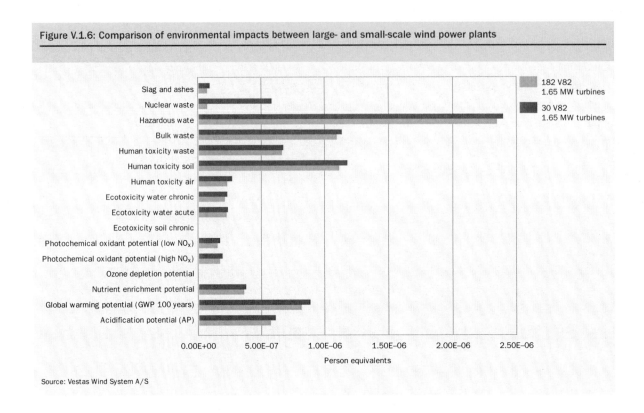

Figure V.1.6: Comparison of environmental impacts between large- and small-scale wind power plants

Source: Vestas Wind System A/S

produce significant changes in the environmental impacts.

OFFSHORE

LCI Results: Offshore Wind Farms

Results extracted from the reviewed LCA studies for offshore wind farms regarding several of the most relevant emissions are shown in Figure V.1.7. Bars show the variability of the results when several wind farm configurations are considered in a single study.

Carbon dioxide emissions vary from 6.4 to 12.3 g/kWh in the consulted references. Methane emissions range from 2.8 to 16.9 mg/kWh. Nitrogen oxides emissions range from 18 to 56.4 mg/kWh. NMVOCs are emitted in quantities that range from 1.7 to 11.4 mg/kWh, particulates range from 10.5 to 54.4 mg/kWh and, finally, sulphur dioxide emissions

range from 22.1 to 44.7 mg/kWh. All of these quantities are quite similar to those obtained for onshore wind farms, with the only exception being that particulates are far below the emissions of conventional technologies such as natural gas (see Figure V.1.7).

In Figure V.1.8, the contribution of different life-cycle phases to the emissions is depicted. In an offshore context, the contribution of the construction phase is even more important, accounting for around 85 per cent of the emissions and hence of the impacts.

Within the construction stage, Figure V.1.9 shows the contribution of the different components. Important items in the environmental impacts of the construction phase of an offshore wind farm are the nacelle and the foundations, followed by the tower. The rotor blades are not found to play an important part. Emissions from transport activities during construction phase are quite relevant in the case of NO_x and NMVOCs emissions.

Figure V.1.7: Emissions from the production of 1 kWh in offshore wind farms throughout the whole life cycle

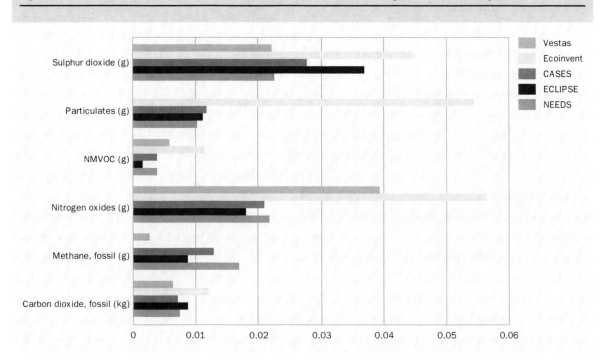

Figure V.1.8: Contribution of the different life-cycle phases of an offshore wind farm to the relevant emissions

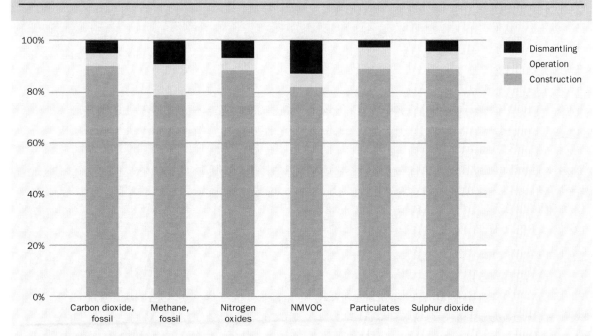

Source: Own elaboration using ECLIPSE results

Figure V.1.9: Contribution of the components of the construction phase to the different emissions

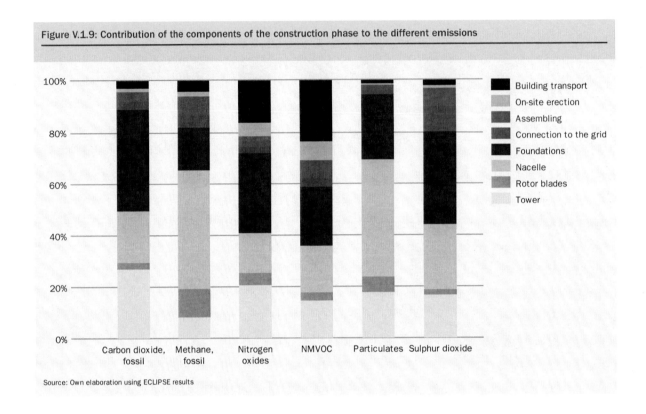

Source: Own elaboration using ECLIPSE results

LCA Results: Offshore Wind Farms

As far as offshore technology is concerned, Vestas Wind Systems A/S and Tech-wise A/S, on behalf of Elsam A/S, have developed a project titled 'LCA and Turbines'. The goal of the project was to create a life-cycle model for a large Vestas offshore turbine. Based on this offshore model, an analysis was carried out to identify the most significant environmental impacts of a turbine during its life cycle (Elsam-Vestas, 2004). Environmental impacts are shown in Figure V.1.10.

Results showed that the volume of waste is the largest normalised impact from a turbine. The bulk of waste is produced during the manufacturing phase, primarily from the steel production needed for the foundation and the tower.

The environmental impacts of the life phases and component systems are illustrated in Figure V.1.11. The largest environmental impacts are found in the

manufacturing phase. The disposal scenario also makes a very important contribution to the entire environmental impact. In the disposal scenario, about 90 per cent of the steel and iron could be recycled, while 95 per cent of the copper could be recycled. With less recycling, there is more waste. The other two life phases (operation and removal) do no contribute significantly to the environmental impacts.

The environmental impacts produced from the manufacturing phase by components shows that the foundation has the highest contribution to several impact categories. Tower and nacelle manufacturing also have a significant contribution. The impacts distribution is showed in Figure V.1.12.

A comparison between the onshore and offshore impact of the same wind turbine (a Vestas V90 3.0 MW) was carried out by Vestas (Vestas, 2005) (see Figure V.1.13). Results of this LCA show similar environmental profiles in both cases. Offshore wind turbines

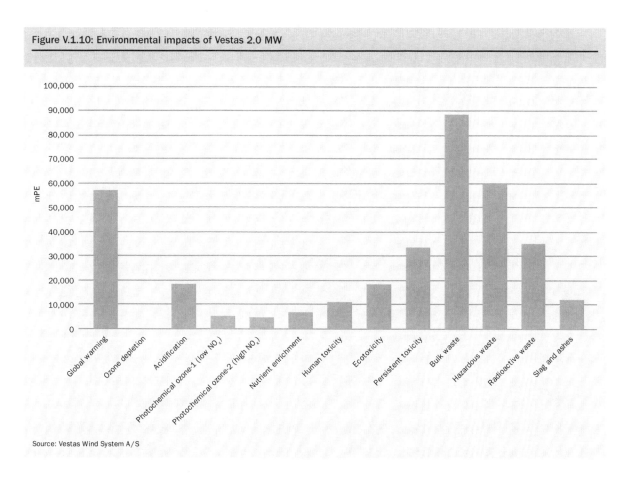

Figure V.1.10: Environmental impacts of Vestas 2.0 MW

Source: Vestas Wind System A/S

produce more electricity (11,300–14,800 MWh/ turbine) than onshore wind turbines (6900–9100 MWh/ turbine). However, offshore turbines are more resource demanding. Thus these two parameters are offset in some cases.

Energy Balance Analysis

The energy balance is an assessment of the relationship between the energy consumption of the product and the energy production throughout the lifetime. The energy balance analysis in the case of the Vestas V90 3.0 MW shows that, for an offshore wind turbine, 0.57 years (6.8 months) of expected average energy production are necessary to recover all the energy

consumed for manufacturing, operation, transport, dismantling and disposal.

As far as an onshore wind turbine is concerned, the energy balance is similar but shorter than the offshore one, with only 0.55 years (6.6 months) needed to recover the energy spent in all the phases of the life cycle. This difference is due to the larger grid transmission and steel consumption for the foundations in an offshore scheme.

The V80 2 MW turbines installed in Horns Rev only needed 0.26 years (3.1 months) to recover the energy spent in the offshore installation. The same turbines installed in the Tjaereborg onshore wind farm had an energy payback period of about 0.27 years (3.2 months).

Figure V.1.11: Contribution of environmental impacts by life-cycle stages: Vestas 2.0 MW

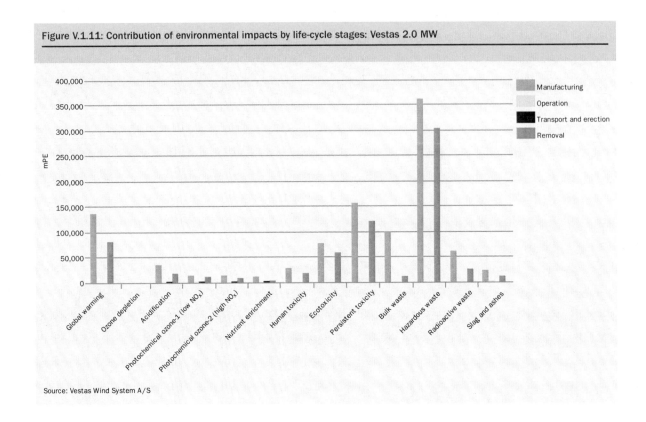

Source: Vestas Wind System A/S

Comparative Benefits with Conventional and Renewable Technologies Systems

Several studies have been conducted by different institutions and enterprises in order to quantify the environmental impacts of energy systems. The Vestas study[5] also analysed the environmental impacts produced by average European electricity in 1990, using data from the Danish method for environmental design of industrial products (EDIP) database, compared with the electricity generated by an offshore wind power plant and an onshore wind power plant. The reason for using data from 1990 is that the EDIP database did not include reliable updated data. The comparison shows that wind electricity has a much better environmental profile than the average Danish electricity for the year of the project. The impacts are considerably lower in the case of wind energy than European electricity in all the analysed impacts categories. However, the comparison is not quite fair, as the system limits of the two systems differ from each other (current data for wind turbines and 1990 data for European electricity). The comparison was made to see the order of magnitude (See Figure V.1.14).

Vattenfall Nordic Countries have carried out LCAs of its electricity generation systems. The results of the study showed that:

- Construction is the most polluting phase for technologies that do not require fuel, but instead use a renewable source of energy (hydro, wind and solar power).

- The operational phase dominates for all fuel-burning power plants, followed by fuel production.

- Wind energy generates low environmental impact in all the parameters analysed: CO_2, NO_x, SO_2 and

Figure V.1.12: Contribution of the components of the construction phase to the different impacts

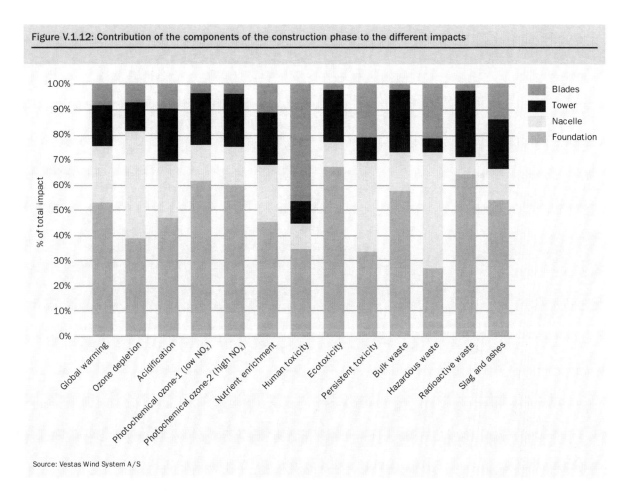

Source: Vestas Wind System A/S

particulate matter emissions and radioactive waste. Only the use of copper from mines presents a significant impact.

- The demolition/dismantling phase causes a comparatively low impact since, for example, metals and concrete can be recycled.

AVOIDED EMISSIONS

Environmental benefits of wind electricity can be assessed in terms of avoided emissions compared to other alternative electricity generation technologies.

LCI results for some relevant emissions from electricity production in a coal condensing power plant and

in a natural gas combined cycle power plant are shown in Figure V.1.15, compared with the results obtained for onshore and offshore wind energy.

As observed in Figure V.1.15, emissions produced in the life cycle of wind farms are well below those produced in competing electricity generation technologies such as coal and gas. The only exception is the emissions of particles in the natural gas combined cycle (NGCC), which are of the same order of those from wind farms in the whole life cycle.

Emissions avoided using wind farms to produce electricity instead of coal or natural gas power plants are quantified in Tables V.1.2 and V.1.3.

Figure V.1.13: Onshore/offshore comparison of environmental impacts

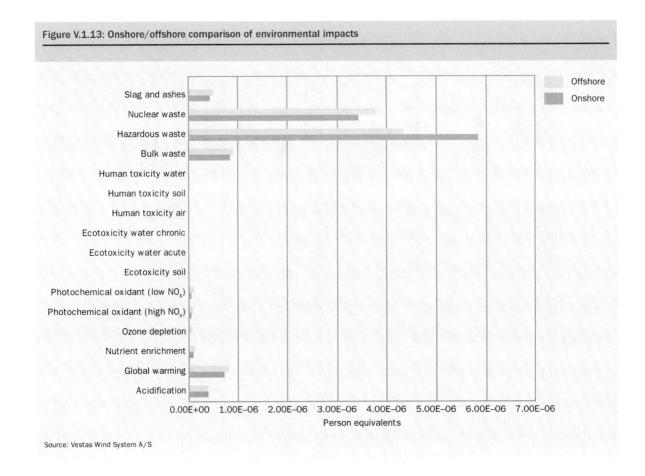

Source: Vestas Wind System A/S

Results show that as much as 828 g of CO_2 can be avoided per kWh produced by wind instead of coal, and 391 g of CO_2 per kWh in the case of natural gas. Quite significant nitrogen and sulphur oxides and NMVOC emission reductions can also be obtained by substituting coal or gas with wind energy.

As in the case of fossil energies, wind energy results show in general lower emissions of CO_2, methane, nitrogen and sulphur oxides, NMVOCs and particulates than other renewable sources. In this sense, it is possible to obtain avoided emissions, using wind (onshore and offshore) technologies in the power generation system.

Figure V.1.14: Onshore, offshore and electricity system comparison on environmental impacts

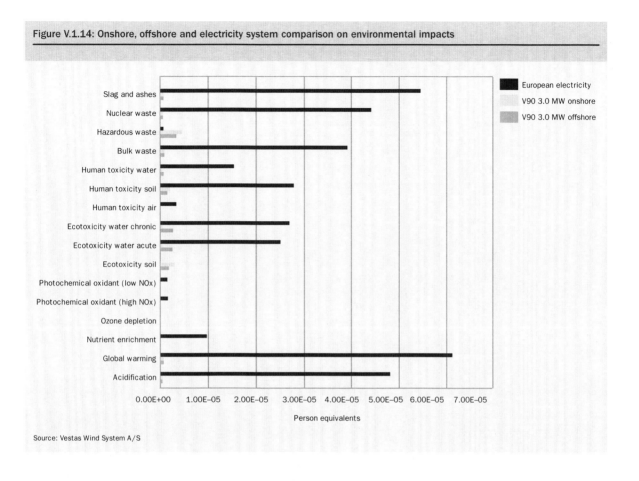

Source: Vestas Wind System A/S

Conclusions

LCA methodology provides an understandable and consistent tool to evaluate the environmental impact of the different phases of wind plant installations. LCA estimates the benefits of electricity from renewable energy sources compared to conventional technologies in a fully documented and transparent way.

The construction of the wind turbine is the most significant phase in terms of the environmental impacts produced by wind energy, both for offshore wind power plants and onshore wind power plants. Environmental

Figure V.1.15: Comparison of the emissions produced in the generation of 1 kWh in a coal and a natural gas combined cycle power plant and the emissions produced in an onshore and offshore wind farm

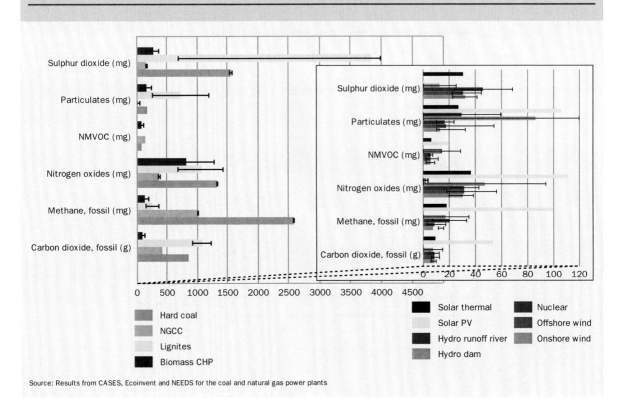

Source: Results from CASES, Ecoinvent and NEEDS for the coal and natural gas power plants

Table V.1.2: Emissions of relevant pollutants produced by wind electricity and coal and natural gas electricity in the whole life cycle, and benefits of wind versus coal and natural gas

	Emissions						Benefits		
	Onshore wind	Offshore wind	Average wind	Hard coal	Lignite	NGCC	vs. coal	vs. Lignite	vs. NGCC
Carbon dioxide, fossil (g)	8	8	8	836	1060	400	828	1051	391
Methane, fossil (mg)	8	8	8	2554	244	993	2546	236	984
Nitrogen oxides (mg)	31	31	31	1309	1041	353	1278	1010	322
NMVOC (mg)	6	5	6	71	8	129	65	3	123
Particulates (mg)	13	18	15	147	711	12	134	693	–6
Sulphur dioxide (mg)	32	31	32	1548	3808	149	1515	3777	118

Source: CIEMAT

Table V.1.3: Emissions and benefits of relevant pollutants produced by wind electricity and other renewable energies

	Emissions					Benefits			
	Average wind	Nuclear	Solar PV	Solar thermal	Biomass CHP	vs. Nuclear	vs. Solar PV	vs. Solar thermal	vs. Biomass CHP
Carbon dioxide, fossil (g)	8	8	53	9	83	0	45	1	75
Methane, fossil (mg)	8	20	100	18	119	12	92	10	111
Nitrogen oxides (mg)	31	32	112	37	814	1	81	6	784
NMVOC (mg)	6	6	20	6	66	0	14	1	60
Particulates (mg)	15	17	107	27	144	1	91	12	128
Sulphur dioxide (mg)	32	46	0	31	250	15	−31	−1	218

Source: CIEMAT

impacts generated in the transportation and operation phases cannot be considered significant in relation to the total environmental impacts of either offshore or onshore wind power plants. However, in offshore wind power plants, zinc is discharged from offshore cables during the operational stage.

The disposal scenario has great importance for the environmental profile of the electricity generated from wind power plants. Environmental impacts are directly dependent on the recycling level, with a higher amount of recycling resulting in a better environmental result.

The energy balance of wind energy is very positive. The energy consumed in the whole chain of wind plants is recovered in several average operational months. The comparison of wind energy with conventional technologies highlights the environmental advantages of wind energy. Quite significant emissions reductions can be obtained by producing electricity in wind farms instead of using conventional technologies such as coal and natural gas combined cycle power plants.

The significant benefits of wind energy should play an increasingly important role in deciding what kinds of new power plants will be built.

V.2 ENVIRONMENTAL IMPACTS

The energy supply is still dominated by fossil fuels, which contribute to the main environmental problems at the world level: climate change and air pollution. The use of renewable energies means lower greenhouse gas emissions and reduced air pollution, representing a key solution to reach a sustainable future.

Wind is clean, free, indigenous and inexhaustible. Wind turbines do not need any type of fuel, so there are no environmental risks or degradation from the exploration, extraction, transport, shipment, processing or disposal of fuel. Not only is generation produced with zero emissions of carbon dioxide (during the operational phase) but it also does not release toxic pollutants (for example mercury) or conventional air pollutants (for example smog-forming nitrogen dioxide and acid rain-forming sulphur dioxide). Furthermore, the adverse impacts caused by mountain-top mining and strip mining of coal, including acid mine drainage and land subsidence are avoided, and the negative effects of nuclear power, including radioactive waste disposal, security risks and nuclear proliferation risks, are not created. Finally, wind power can have a long-term positive impact on biodiversity by reducing the threat of climate change – the greatest threat to biodiversity.

At the same time, however, the construction and operation of both onshore and offshore wind turbines can result in negative local environmental impacts on birds and cetaceans, landscapes, sustainable land use (including protected areas), and the marine environment. The negative environmental impacts from wind energy installations are much lower in intensity than those produced by conventional energies, but they still have to assessed and mitigated when necessary.

EU Directive 85/337 defines environmental impact assessment (EIA) as the procedure which ensures that environmental consequences of projects are identified and assessed before authorisation is given. The main objective is to avoid or minimise negative effects from the beginning of a project rather than trying to counteract them later. Thus the best environmental policy consists of preventing pollution or nuisances at source so the environment is not damaged. The procedure requires the developer to compile an environmental statement (ES) describing the likely significant effects of the development on the environment and proposed mitigation measures. The ES must be circulated to statutory consultation bodies and made available to the public for comment. Its contents, together with any comments, must be taken into account by the competent authority (for example local planning authority) before it may grant consent.

A strategic environmental assessment (SEA) is the procedure used to evaluate the adverse impacts of any plans and programmes on the environment. National, regional and local governments must undertake SEAs of all wind energy plans and programmes that have the potential for significant environmental effects. Appropriate assessments (AAs) have to be carried out in accordance with the Habitats Directive to evaluate the effects on a Natura 2000 site. Where potential trans-boundary effects are foreseen, international cooperation with other governments should be sought. SEAs should be used to inform strategic site selection for renewable energy generation and identify the information requirements for individual EIAs.

Worldwide, biodiversity loss is in principle caused because of human activities on the environment (such as intensive production systems, construction and extractive industries), global climate change, invasions of alien species, pollution and over-exploitation of natural resources. In 2005 the transportation and energy (DG TREN) and environment (DG ENV) directorates at the European Commission created an ad hoc working group on wind energy and biodiversity. The group is composed of industry, governmental and non-governmental representatives. A draft guidance document is currently being debated and aims at facilitating the development of wind energy while preserving biodiversity.

Onshore

VISUAL IMPACT

The landscape is a very rich and complex concept. Defining landscape is not an easy task, as is made clear by the high number of definitions that exist. Landscape definitions can be found in different fields like art, geography, natural sciences, architecture or economics. According to the European Landscape Convention, landscape means an area, as perceived by people, whose character is the result of the action and interaction of natural and/or human factors. Landscapes are not static. The landscape is changing over time according to human and ecological development.

Landscape perceptions and visual impacts are key environmental issues in determining wind farm applications related to wind energy development as landscape and visual impacts are by nature subjective and changing over time and location.

Wind turbines are man-made vertical structures with rotating blades, and thus have the potential of attracting people's attention. Typically wind farms with several wind turbines spread on the territory may become dominant points on the landscape.

The characteristics of wind developments may cause landscape and visual effects. These characteristics include the turbines (size, height, number, material and colour), access and site tracks, substation buildings, compounds, grid connection, anemometer masts, and transmission lines. Another characteristic of wind farms is that they are not permanent, so the area where the wind farm has been located can return to its original condition after the decommissioning phase.

Landscape and visual assessment is carried out differently in different countries. However, within the EU, most wind farms are required to carry out an EIA. The EIA shall identify, describe and assess the direct and indirect effects of the project on the landscape.

Some of the techniques commonly used to inform the landscape and visual impact assessment are:

- zone of theoretical visibility (ZTV) maps define the areas from which a wind plant can be totally or partially seen as determined by topography; these areas represent the limits of visibility of the plant;
- photographs to record the baseline visual resource;
- diagrams to provide a technical indication of the scale, shape and positioning of the proposed wind development; and
- photomontages and video-montages to show the future picture with the wind farm installed.

Visual impact decreases with the distance. The ZTV zones can be defined as:

- Zone I – Visually dominant: the turbines are perceived as large scale and movement of blades is obvious. The immediate landscape is altered. Distance up to 2 km.
- Zone II – Visually intrusive: the turbines are important elements on the landscape and are clearly perceived. Blades movement is clearly visible and can attract the eye. Turbines not necessarily dominant points in the view. Distance between 1 and 4.5 km in good visibility conditions.
- Zone III – Noticeable: the turbines are clearly visible but not intrusive. The wind farm is noticeable as an element in the landscape. Movement of blades is visible in good visibility conditions but the turbines appear small in the overall view. Distance between 2 and 8 km depending on weather conditions.
- Zone IV – Element within distant landscape: the apparent size of the turbines is very small. Turbines are like any other element in the landscape. Movement of blades is generally indiscernible. Distance of over 7 km.

While visual impact is very specific to the site at a particular wind farm, several characteristics in the design and siting of wind farms have been identified to minimise their potential visual impact (Hecklau, 2005; Stanton, 2005; Tsoutsos et al., 2006):

- similar size and type of turbines on a wind farm or several adjacent wind farms;
- light grey, beige and white colours on turbines;
- three blades;
- blades rotating in the same direction;
- low number of large turbines is preferable to many smaller wind turbines; and
- flat landscapes fit well with turbine distribution in rows.

Mitigation measures to prevent and/or minimise visual impact from wind farms on landscape can be summarised as follows (Brusa and Lanfranconi, 2007):

- design of wind farm according to the peculiarities of the site and with sensitivity to the surrounding landscape;
- locate the wind farm at least a certain distance from dwellings;
- selection of wind turbine design (tower, colour) according to landscape characteristics;
- selection of neutral colour and anti-reflective paint for towers and blades;
- underground cables; and
- lights for low-altitude flight only for more exposed towers.

The effects of landscape and visual impact cannot be measured or calculated and mitigation measures are limited. However, experience gained recently suggests that opposition to wind farms is mainly encountered during the planning stage. After commissioning the acceptability is strong.

NOISE IMPACT

Noise from wind developments has been one of the most studied environmental impacts of this technology. Noise, compared to landscape and visual impacts, can be measured and predicted fairly easily.

Wind turbines produce two types of noise: mechanical noise from gearboxes and generators, and aerodynamic noise from blades. Modern wind turbines have virtually eliminated the mechanical noise through good insulation materials in the nacelle, so aerodynamic noise is the biggest contributor. The aerodynamic noise is produced by the rotation of the blades generating a broad-band swishing sound and it is a function of tip speed. Design of modern wind turbines has been optimised to reduce aerodynamic noise. This reduction can be obtained in two ways:

1. decreasing rotational speeds to under 65 m/s at the tip; and
2. using pitch control on upwind turbines, which permits the rotation of the blades along their long axis.

At any given location, the noise within or around a wind farm can vary considerably depending on a number of factors including the layout of the wind farm, the particular model of turbines installed, the topography or shape of the land, the speed and direction of the wind, and the background noise. The factors with the most influence on noise propagation are the distance between the observer and the source and the type of noise source.

The sound emissions of a wind turbine increase as the wind speed increases. However, the background noise will typically increase faster than the sound of the wind turbine, tending to mask the wind turbine noise in higher winds. Sound levels decrease as the distance from the wind turbines increases.

Noise levels can be measured and predicted, but public attitude towards noise depends heavily on perception. Sound emissions can be accurately measured using standardised acoustic equipment and methodologies (International Organization for Standardization – ISO Standards, International Electrotechnical Commission – IEC Standards, ETSU – Energy Technology Support Unit, UK Government and so on). Levels of sound are most commonly expressed in decibels (dB). The predictions of

Table V.2.1: Comparative noise for common activities

Source/activity	Indicative noise level (dB)
Threshold of hearing	0
Rural night-time background	20–40
Quiet bedroom	35
Wind farm at 350 m	35–45
Busy road at 5 km	35–45
Car at 65 km/h at 100 m	55
Busy general office	60
Conversation	60
Truck at 50 km/h at 100 m	65
City traffic	90
Pneumatic drill at 7 m	95
Jet aircraft at 250 m	105
Threshold of pain	140

Source: CIEMAT

sound levels in future wind farms are of the utmost importance in order to foresee the noise impact. Table V.2.1, based on data from the Scottish Government, compares noise generated by wind turbines with other everyday activities.

When there are people living near a wind farm, care must be taken to ensure that sound from wind turbines should be at a reasonable level in relation to the ambient sound level in the area. Rural areas are quieter than cities, so the background noise is usually lower. However, there are also noisy activities – agricultural, commercial, industrial and transportation. Wind farms are located in windy areas, where background noise is higher, and this background noise tends to mask the noise produced by the turbines. The final objective is to avoid annoyance or interference in the quality of life of the nearby residents.

Due to the wide variation in the levels of individual tolerance for noise, there is no completely satisfactory way to measure its subjective effects or the corresponding reactions of annoyance and dissatisfaction. The individual annoyance for noise is a very complex topic, but dose–response relationship studies have demonstrated a correlation between noise annoyance with visual interference and the presence of intrusive sound characteristics. In the same way, annoyance is higher in a rural area than in a suburban area and higher also in complex terrain (hilly or rocky) in comparison with a ground floor in a rural environment.

Low frequency noise (LFN), also known as infrasound, is used to describe sound energy in the region below about 200 Hz. LFN may cause distress and annoyance to sensitive people and has thus been widely analysed. The most important finding is that modern wind turbines with the rotor placed upwind produce very low levels of infrasound, typically below the threshold of perception. A survey of all known published measurement results of infrasound from wind turbines concludes that, with upwind turbines, infrasound can be neglected in evaluating environmental effects.

Experience acquired in developing wind farms suggests that noise from wind turbines is generally very low. The comparison between the number of noise complaints about wind farms and about other types of noise indicates that wind farm noise is a small-scale problem in absolute terms. Information from the US also suggests that complaints about noise from wind projects are rare and can usually be satisfactorily resolved.

LAND USE

National authorities consider the development of wind farms in their planning policies for wind energy projects. Decisions on siting should be made with consideration to other land users.

The administrative procedures needed to approve wind plants for each site have to be taken into account from the beginning of the project planning process. Regional and local land-use planners must decide whether a project is compatible with existing and planned adjacent uses, whether it will modify negatively the overall character of the surrounding area, whether it will disrupt established communities, and whether it will be integrated into the existing landscape. Developers, in the very early planning stage, should contact the most relevant authorities and stakeholders in the area: the Ministry of Defence, civil aviation authorities, radar and radio communication suppliers, the grid company, environmental protection authorities, the local population and relevant non-governmental associations, among others.

The authorities involved in reviewing and making land-use decisions on projects must coordinate and communicate with each other throughout the project. At the same time, local citizen participation as well as good communication with the main stakeholders (local authorities, developer, NGOs, landowners, etc.) would help to obtain a successful wind development.

Special attention must be paid to nature reserves, their surrounding zones and habitats of high value for nature conservation. There are additional obligations for assessment when Ramsar sites or Natura 2000 sites could be significantly affected by wind energy developments. The project or plan will only be approved if there is not an adverse effect on the integrity of the site. If it cannot be established that there will be no adverse effects, the project may only be carried out if there are no alternative solutions and if there are imperative reasons of public interest.

Recently, a new concern has been raised: wind farm installations over peatlands. Peatlands are natural carbon storage systems with a delicate equilibrium of waterlogging. According to the United Nations Environment Programme Division of Global Environment Facility Coordination (UNEP-GEF), peatlands cover only 3 per cent of the world's surface, but store the equivalent of 30 per cent of all global soil carbon, or the equivalent of 75 per cent of all atmospheric carbon. The impacts associated with drainage are carbon dioxide and methane emissions, erosion and mass movements, and dissolved organic carbon. The consequence

is the loss of the land's capability of acting as a carbon sink. Moreover, the EU Habitats Directive has designated several grassland formations as special areas of conservation. In these areas, Member States have the responsibility to apply the necessary conservation measures for the maintenance or restoration of the natural habitats and/or the populations of the species for which they are designated. The Scottish Government (Nayak et al., 2008) has recently developed an approach to calculate the impact of wind energy on organic soils. This method permits the calculation of potential carbon losses and savings of wind farms, taking into account peat removal, drainage, habitat improvement and site restoration. The method proposes to integrate the carbon losses by peatland use in the overall life-cycle assessment (LCA) of the wind farms, computing the global carbon saving by the use of wind energy and subtracting the carbon losses associated with wind farm installations. The study also provides some recommendations for improving carbon savings of wind farm developments:

- peat restoration as soon as possible after disturbance;
- employing submerged foundation in deeper area of peat;
- maintenance of excavated C-layer as intact as possible until restoration;
- good track design according to geomorphologic characteristics;
- improving habitats through drain blocking and re-wetting of areas; and
- using floating roads when peat is deeper than 1 m.

Another issue is the interaction between tourism and wind energy developments. Many tourist areas are located in beautiful and/or peaceful landscapes. Wind power plants could reduce the attractiveness of the natural scenery. The most recent study, carried out by the Scottish Government (Scottish Government, 2008), has analysed the impacts of wind farms on the tourism industry and reviewed 40 studies from Europe, the US and Australia. The conclusions from the review can be summarised as follows:

- The strongest opposition occurs at the planning stage.
- A significant number of people think there is a loss of scenic value when a wind farm is installed; however, to other people, wind farms enhance the beauty of the area.
- Over time, wind farms are better accepted.
- In general terms, there is no evidence to suggest a serious negative impact on tourism.
- A tourist impact statement is suggested as part of the planning procedure to decrease the impact on tourism, including analysis of tourist flows on roads and number of beds located in dwellings in the visual zone of the wind farm.

IMPACTS ON BIRDS

Wind farms, as vertical structures with mobile elements, may represent a risk to birds, both as residents and migratory birds. However, it is difficult to reach a clear conclusion about the impacts of wind energy on birds for several reasons:

- Impacts are very site-dependent (depending on landscape topography, wind farm layout, season, types of resident and migratory birds in the area, and so on).
- Impacts vary among the different bird species.

The types of risks that may affect birds are:

- collision with turbines (blades and towers) causing death or injury;
- habitat disturbance: the presence of wind turbines and maintenance work can displace birds from preferred habitats and the breeding success rate may be reduced;
- interference with birds' movements between feeding, wintering, breeding and moulting habitats, which could result in additional flights consuming more energy; and
- reduction or loss of available habitat.

The main factors which determine the mortality of birds by collision in wind farms are landscape topography, direction and strength of local winds, turbine design characteristics, and the specific spatial distribution of turbines on the location (de Lucas et al., 2007). Specific locations should be evaluated a priori when a wind farm is planned. Every new wind farm project must include a detailed study of the interaction between birds' behaviour, wind and topography at the precise location. This analysis should provide information to define the best design of the wind farm to minimise collision with the turbines. Raptors present a higher mortality rate due to their dependence on thermals to gain altitude, to move between locations and to forage. Some of them are long-lived species with low reproductive rates and thus more vulnerable to loss of individuals by collisions.

The mortality caused by wind farms is very dependent on the season, specific site (for example offshore, mountain ridge or migration route), species (large and medium versus small, and migratory versus resident) and type of bird activity (for example nocturnal migrations and movements from and to feeding areas).

Bird mortality seems to be a sporadic event, correlated with adverse weather or poor visibility conditions. Results from Altamont Pass and Tarifa on raptors showed some of the highest levels of mortality; however, the average numbers of fatalities were low in both places, ranging from 0.02 to 0.15 collisions/turbine. In Altamont Pass the overall collision rate was high due to the large number of small, fast, rotating turbines installed in the area. In Tarifa, the two main reasons for collisions were that the wind farms were installed in topographical bottlenecks, where large numbers of migrating and local birds fly at the same time through mountain passes, and the use of wind by soaring birds to gain lift over ridges. In Navarra, studies of almost 1000 wind turbines and including all types of birds showed a mortality rate of between 0.1 and 0.6 collisions per turbine and year. Raptors were the bird group more affected (78.2 per cent) during spring, followed by migrant passerines during post-breeding migration time (September/October).

At the global level, it can be accepted that many wind farms show low rates of mortality by collision (Drewitt and Langston, 2006). However, even these low collision mortality rates for threatened or vulnerable species could be significant and make it harder for a particular species to survive.

A comparative study of bird mortality by anthropogenic causes was carried out by Erickson et al. (2005). Table V.2.2 gives the distribution by human activities.

A more recent study stated current wind energy developments are only responsible for 0.003 per cent of bird mortalities caused by human activities.

Concerning habitat disturbance, the construction and operation of wind farms could potentially disturb birds and displace them from around the wind farm site. The first step in analysing this disturbance is to define the size of the potential disturbance zone. Wind turbines can trigger flight reactions on birds displacing them out of the wind farm area. Potential disturbance distances have been studied by several authors, giving an average of 300 m during the breeding season and 800 m at other seasons of the year. Approximately 2 per cent of all flights at hub height showed a sudden change of direction in the proximity of wind farm.

Table V.2.2: Anthropogenic bird mortality

Causes	Annual mortality estimate
Buildings/windows	550 million
Cats	100 million
High tension lines	130 million
Vehicles	80 million
Pesticides	67 million
Communication towers	4.5 million
Airplanes	25 thousand
Wind turbines	28.5 thousand

Source: Erickson et al. (2005)

An indirect negative impact of wind farms is a possible reduction in the available area for nesting and feeding by birds avoiding wind farm installations.

During construction, species can be displaced from their original habitat, but in most cases they return during the operational phase. However, exclusions may occur for other species during the breeding period.

Mitigation measures to minimise impacts vary by site and by species, but common findings in the literature are as follows:

- important zones of conservation and sensitivity areas must be avoided;
- sensitive habitats have to be protected by implementing appropriate working practices;
- an environmental monitoring programme before, during and after construction will provide the needed information to evaluate the impact on birds;
- adequate design of wind farms: siting turbines close together and grouping turbines to avoid an alignment perpendicular to main flight paths;
- provide corridors between clusters of wind turbines when necessary;
- increase the visibility of rotor blades;
- underground transmission cables installation, especially in sensitive areas, where possible;
- make overhead cables more visible using deflectors and avoiding use in areas of high bird concentrations, especially of species vulnerable to collision;
- implement habitat enhancement for species using the site;
- adequate environmental training for site personnel;
- presence of biologist or ecologist during construction in sensitive locations;
- relocation of conflictive turbines;
- stop operation during peak migration periods; and
- rotor speed reduction in critical periods.

ELECTROMAGNETIC INTERFERENCE

Electromagnetic interference (EMI) is any type of interference that can potentially disrupt, degrade or interfere with the effective performance of an electronic device. Modern society is dependent on the use of devices that utilise electromagnetic energy, such as power and communication networks, electrified railways, and computer networks. During the generation, transmission and utilisation of electromagnetic energy, the devices generate electromagnetic disturbance that can interfere with the normal operation of other systems.

Wind turbines can potentially disrupt electromagnetic signals used in telecommunications, navigation and radar services. The degree and nature of the interference will depend on:

- the location of the wind turbine between receiver and transmitter;
- characteristics of the rotor blades;
- characteristics of the receiver;
- signal frequency; and
- the radio wave propagation in the local atmosphere.

Interference can be produced by three elements of a wind turbine: the tower, rotating blades and generator. The tower and blades may obstruct, reflect or refract the electromagnetic waves. However, modern blades are typically made of synthetic materials which have a minimal impact on the transmission of electromagnetic radiation. The electrical system is not usually a problem for telecommunications, because interference can be eliminated with proper nacelle insulation and good maintenance.

Interference to mobile radio services is usually negligible. Interference to TV signals has been minimised with the substitution of metal blades with synthetic materials. However, when turbines are installed very close to dwellings, interference has been proven difficult to rule out.

The interference area may be calculated using the Fresnel zone. This area is around and between the transmitter and receiver and depends on transmission frequency, distance between them and local atmospheric conditions.

Technical mitigation measures for TV interference can be applied during the planning stage, siting the turbine away from the line-of-sight of the broadcaster transmitter. Once the wind farm is in operation, there are also a set of measures to mitigate the interference:

- installation of higher-quality or directional antenna;
- direct the antenna toward an alternative broadcast transmitter;
- installation of an amplifier;
- relocate the antenna;
- installation of satellite or cable TV; and
- construction of a new repeater station if the area affected is very wide.

There is common agreement that adequate design and location can prevent or correct any possible interference problems at relatively low cost using simple technical measures, such as the installation of additional transmitter masts. Interference on communication systems is considered to be negligible because it can be avoided by careful wind farm design.

CONSTRAINTS ON NATURAL RESERVES AREAS

There is a rough consensus about which are the most important environmental threats and what are their general influences on biological diversity. The continuous deterioration of natural habitats and the increasing number of wild species which are seriously threatened has prompted governments to protect the environment.

There are many types of protected areas at national and regional levels. At the EU level, the Birds Directive (1979) and the Habitats Directive (1992) are the base of the nature conservation policy.

The Birds Directive is one of the most important tools to protect all wild bird species naturally living in or migrating through the EU. The directive recognises that habitat loss and degradation are the most serious threats to the conservation of wild birds. The Birds Directive has identified 194 species and sub-species (listed in Annex I) as particularly threatened and in need of special conservation measurements.

The aim of the Habitats Directive is to promote the maintenance of biodiversity by preserving natural habitats and wild species. Annex I includes a list of 189 habitats and Annex II lists 788 species to be protected by means of a network of high-value sites. Each Member State has to define a national list of sites for evaluation in order to form a European network of Sites of Community Importance (SCIs). Once adopted, SCIs are designated by Member States as Special Areas of Conservation (SACs), and, along with Special Protection Areas (SPAs) classified under the EC Birds Directive, form a network of protected areas known as Natura 2000.

The development of wind farms in natural reserves should be assessed on site-specific and species-specific criteria to determine whether the adverse impacts are compatible with the values for which the area was designated.

Of special importance is the requirement of the Habitats Directive to include indicative 'sensitivity' maps of bird populations, habitats, flyways and migration bottlenecks as well as an assessment of the plan's probable effects on these in the SEA and AA procedures. These maps should provide enough information about feeding, breeding, moulting, resting, non-breeding and migration routes to guarantee biodiversity conservation.

Offshore

Offshore wind energy is a renewable technology capable of supplying significant amounts of energy in a sustainable way. According to EWEA estimates, between 20 GW and 40 GW of offshore wind energy capacity will be operating in the EU by 2020. This capacity could meet more than 4 per cent of EU electricity consumption. The total offshore installed capacity in Europe at the end of 2007 was almost 1100 MW, distributed in the coastal waters of Denmark, Ireland, The Netherlands, Sweden and the UK, representing almost 2 per cent of the total wind energy (56,536 MW) in the EU.

Offshore wind projects are more complex than onshore ones. Offshore developments include platforms, turbines, cables, substations, grids, interconnection and shipping, dredging and associated construction activity. The operation and maintenance activities include the transport of employees by ship and helicopter and occasional hardware retrofits.

From an ecological point of view, shallow waters are usually places with high ecological value and are important habitats for breeding, resting and migratory seabirds. Close participation and good communication between the countries involved in the new developments is essential to reduce environmental impacts from several wind farms in the same area.

Most of the experience gained in offshore wind energy comes from several years of monitoring three wind farms in Denmark (Middelgrunden, Horns Rev and Nysted) installed between 2001 and 2003. Valuable analysis has also been carried out by the Federal Environment Ministry (BMU) of Germany through technical, environmental and nature conservation research about offshore wind energy foundations.

VISUAL IMPACT

Offshore wind farms usually have more and bigger turbines than onshore developments. However, visual impact is lower due to the greater distance from the coastline. Nevertheless, the coastal landscape is often unique and provides some of the most valued landscapes, thus special attention could be required.

The visual impact of offshore wind farms can affect three components of the seascape:

1. an area of sea;
2. a length of coastline; and
3. an area of land.

Figure V.2.1: Components of seascape

Source: Wratten et al. (2005)

Offshore wind farms involve several elements which influence the character of the produced visual impact (Wratten et al., 2005):

- the site and size of wind farm area;
- the wind turbines: size, materials and colours;
- the layout and spacing of wind farms and associated structures;
- location, dimensions and form of ancillary onshore (substation, pylons, overhead lines, underground cables) and offshore structures (substation and anemometer masts);
- navigational visibility, markings and lights;
- the transportation and maintenance boats;
- the pier, slipway or port to be used by boats; and
- proposed road or track access, and access requirements to the coast.

Just as for onshore developments, ZTV maps, photomontages and video-montages are tools used to predict the potential effects of new offshore wind developments.

The potential offshore visibility depends on topography, vegetation cover and artificial structures existing on the landscapes. The visibility assessment of offshore developments includes the extent of visibility over the main marine, coastline and land activities (recreational activities, coastal populations and main road, rail and footpath). The effects of the curvature of the Earth and lighting conditions are relevant in the visibility of offshore wind farms. Rainy and cloudy days result in less visibility. Experience to date on Horns Rev proves that a wind farm is much less visible than the 'worst-case' clear photomontage assessment, due to prevailing weather conditions and distance.

The magnitude of change in the seascape with the construction of a new offshore wind farm is dependent of several parameters, such as distance, number of turbines, the proportion of the turbine that is visible, weather conditions and the navigational lighting of turbines. The distance between observer and wind farm usually has the strongest influence on the visual impact perception. Nevertheless, changes in lighting and weather conditions vary considerably the visual effects at the same distance.

The indicative thresholds established for highly sensitive seascapes during the DTI study on three SEA areas in the UK are shown in Table V.2.3.

More recently, research on visual assessment by Bishop and Miller (2005) found that distance and contrast are very good predictors of perceived impact. The study, based on North Hoyle wind farm 7 km off the coast of Wales, showed that in all atmospheres and lighting conditions (except a stormy sky), visual impacts decreased with distance. However, visual impact increased with increasing contrast. Further research is needed to analyse the dependence of visual effects on turbine numbers, orientation and distribution.

Table V.2.3: Thresholds for seascapes

Thresholds
<13 km possible major visual effects
13–24 km possible moderate visual effects
>24 km possible minor visual effects

Source: Wratten et al. (2005)

Cumulative effects may occur when several wind farms are built in the same area. The degree of cumulative impact is a product of the number of wind farms and the distance between them, the siting and design of the wind farms, the inter-relationship between their ZTVs, and the overall character of the seascape and its sensitivity to wind farms.

The Danish Energy Agency (DEA) has reported an absence of negative press during the development of Nysted and Horns Rev offshore wind farms. Opinion polls showed better acceptance levels for the projects in the post-construction phase.

NOISE IMPACT

Offshore wind farms are located far away from human populations, which are thus not affected by the noise generated by the turbines. However, marine animals could be affected by the underwater noise generated during the construction and operation of wind turbines. Any effects of the noise will depend on the sensitivity of the species present and their ability to adjust to it.

The procedures to calculate the acoustic noise from offshore wind turbines should include the following:

- wind turbine parameters: rated power, rotor diameter and so on;
- type of foundation, material, pile depth and so on;
- effective pile driving and/or vibration energy;
- period of construction phase and blow or vibrator frequency; and
- depth of water at the site.

Construction and Decommissioning Noise

Construction and decommissioning noise comes from machines and vessels, pile-driving, explosions and installation of wind turbines. Measurements carried out by the German Federal Ministry of the Environment on two platforms reached peak levels of 193 dB at 400 m from the pile (North Sea) and 196 dB at 300 m (Baltic Sea). Nedwell reports peaks up to 260 dB in foundation construction and 178 dB in cable lying at 100 m from the sound source (Gill, 2005). These high sound levels may cause permanent or temporary damage to the acoustic systems of animals in the vicinity of the construction site. However, there is not enough scientific knowledge to determine the maximum thresholds permitted for certain effects. Close collaboration between physicists, engineers and biologists is necessary to get relevant information and obtain standardisation of the measurement procedures in offshore developments.

The measurements from FINO-1 at 400 m from source revealed peaks of 180 dB. The measurements carried out during construction of North Hoyle wind farm in the UK indicate that:

- The peak noise of pile hammering at 5 m depth was 260 dB and at 10 m depth was 262 dB.
- There were no preferential directions for propagation of noise.
- The behaviour of marine mammals and fish could be influenced several kilometres away from the turbine.

Table V.2.4 shows the avoidance reaction expected to occur due to pile-driving during the North Hoyle wind farm construction.

The behaviour of marine organisms may be modified by the noise, resulting in an avoidance of the area during construction. The possible effects on sealife will depend on the sensitivity of the species present in

Table V.2.4: Calculated ranges for avoidance distance for different marine species

Species	Distance
Salmon	1400 m
Cod	5500 m
Dab	100 m
Bottlenose dolphin	4600 m
Harbour porpoise	1400 m
Harbour seal	2000 m

Source: Nedwell et al. (2004)

the area and will be reduced when the noise decreases at the end of the construction (or decommissioning) phase.

Different working groups are currently discussing mitigation measures to reduce damage to sealife:

- soft start in the ramp-up procedure, slowly increasing the energy of the emitted sound;
- using an air-bubble curtain around the pile, which could result in a decrease of 10–20 dB;
- mantling of the ramming pile with acoustically insulated material such as plastic could result in a decrease of 5–25 dB in source level;
- extending the duration of the impact during pile-driving could result in a decrease of 10–15 dB in source level; and
- using acoustic devices which emitted sounds to keep away mammals during ramp-up procedure; several pingers might be necessary at different distances from the sound source.

Operational Noise

In the operation phase, the sound generated in the gearbox and the generator is transmitted by the tower wall, resulting in sound propagation underwater. Measurements of the noise emitted into the air from wind turbines and transformers have shown a negligible contribution to the underwater noise level. The underwater noise from wind turbines is not higher than the ambient noise level in the frequency range above approximately 1 kHz, but it is higher below approximately 1 kHz. The noise may have an impact on the benthic fauna, fish and marine mammals in the vicinity of wind turbine foundations (Greenpeace, 2005).

Operational noise from single turbines of maximum rated power of 1.5 MW was measured in Utgruden, Sweden, at 110 m distance by Thomsen et al. (2006). At moderate wind speeds of 12 m/s, the 1/3 octave sound pressure levels were between 90 and 115 dB.

This anthropogenic noise may have both behavioural and physiological impacts on sealife. Impacts on behaviour include:

- attraction to or avoidance of the area;
- panic; and
- increases in the intensity of vocal communication.

Reports about noise impact on fish have shown a range of effects, from avoidance behaviour to physiological impacts. Changes in behaviour could make fish vacate feeding and spawning areas and migration routes. Studies of noise impact on invertebrates and planktonic organisms have a general consensus of very few effects, unless the organisms are very close to the powerful noise source. Measurements from one 1500 kW wind turbine carried out by the German Federal Ministry of the Environment has found that operational noise emissions do not damage the hearing systems of sealife. Concerning behaviour, the same study stated that it is not clear whether noise from turbines has an influence on marine animals.

Ships are involved in the construction of wind parks and also during the operation phase for maintenance of wind turbines and platforms. The noise from ships depends on ship size and speed, although there are variations between boats of similar classes. Ships of medium size range produce sounds with a frequency mainly between 20 Hz and 10 kHz and levels between 130 and 160 dB at 1 m.

Standardised approaches to obtain noise certificates, similar to those existing onshore, are necessary.

Electromagnetic Fields and Marine Organisms

The electricity produced by offshore wind turbines is transmitted by cables over long distances. The electric current generated produces magnetic fields. Studies of possible effects of artificial static magnetic fields have been carried out on various species under various experimental conditions. Artificial electromagnetic fields could interact with marine organisms to produce detectable changes. Usually, however, only very slight differences in control groups have been recorded.

The magnetic field may affect molluscs, crustaceans, fish and marine mammals that use the Earth's magnetic field for orientation during navigation. But it is still unknown whether the magnetic fields associated with wind turbines influence marine organisms.

Elasmobranches, one of the more electro-sensitive species, are attracted by electrical fields in the range of 0.005–1 μVcm[1] and avoid fields over 10 μVcm[1].

Electro-sensitive species could be attracted or repelled by the electrical fields generated by submarine cables. Special attention must be paid in areas of breeding, feeding or nursing because of the congregation or dispersion of sensitive individuals in the benthic community.

Experimental analysis on several benthic organisms exposed to static magnetic fields of 3.7 mT for several weeks have shown no differences in survival between experimental and control populations. Similarly, mussels living under these static magnetic field conditions for three months during the reproductive period do not present significant differences with the control group. The conclusions are that static magnetic fields of power cable transmissions don't seem to influence the orientation, movement or physiology of the tested benthic organisms.

The results from a study carried out at Nysted on the influence of electromagnetic fields on fish are not conclusive. Some impact on fish behaviour has been recorded, but it was not possible to establish any correlation. There is not enough knowledge about this topic and additional research is needed.

The magnetic fields of both types of cable (bipolar and concentric) used in marine wind farms are small or zero. The Greenpeace study mentioned earlier concludes that the electromagnetic fields of submarine cables have no significant impacts on the marine environment. Studies with a long-term perspective are necessary to confirm the negligible impact of electromagnetic fields of wind energy on marine ecosystems.

Impacts on Benthos

The benthos include the organisms that live on or in the sediment at the bottom of a sea, lake or deep river. The benthic community is complex and is composed of a wide range of plants, animals and bacteria from all levels of the food chain. It can be differentiated by habitat: *infauna* are animals and bacteria of any size that live in bottom sediments, such as worms and clams. They form their own community structures within the sediments, connected to the water by tubes and tunnels; *epifauna* are animals that live either attached to a hard surface (for example rocks or pilings) or move on the surface of the sediments. Epifauna include oysters, mussels, barnacles, snails, starfish, sponges and sea squirts.

These communities are highly dependent on some abiotic factors such as depth of water, temperature, turbidity and salt content. Fluctuations of any of those parameters result in changes in species composition and the numbers of individuals.

The introduction of hard bottom structures such as turbine foundations provides a new artificial substrate which helps to develop a new habitat for marine epifaunal organisms. These structures can attract specific benthos species, generating changes in the previous benthic associations by the colonization of these new substrates. The most susceptible groups are non-mobile (for example mussels, barnacles and sponges), hardly mobile species (snails, starfish) or sand-filtering species (oysters). Small fish species depredating over benthic animals and plants may also appear in the new area. Furthermore, larger benthic or pelagic fish as well as sea birds may be attracted from the surroundings areas. Therefore, the construction of offshore wind farms will modify the relationships of benthic communities, changing the existing biodiversity in the area and creating a new local ecosystem. (Köller et al., 2006).

The knowledge gained from Horns Rev monitoring shows that indigenous infauna habitats have been replaced by the epifauna community associated with hard bottom habitats with an estimated 60-fold increase in availability of food for fish and other organism in the wind farm area compared with the native infauna biomass. An increase of general biodiversity in the wind farm area and progress succession in the benthic community has been verified. The new hard bottom substrates have provided habitats as nursery grounds for larger and more mobile species like the edible crab *Cancer parugus*. The most noticeable news is the introduction of two new species: the ross worm *Sabellaria spinulosa* and the white weed *Sertularia cupressina* in the Horns Rev wind farm area, both considered as threatened or included on the Red List in the Wadden Sea area. The epifauna community in artificial underwater structures differs from the natural marine fauna in the vicinity of wind turbines not only in its species composition but also in the dynamics of its faunal succession.

The installation of steel structures in the western Baltic marine waters has also increased the diversity and abundance of benthic communities.

The construction work phase temporarily increased the water turbidity. This effect may have had a negative impact on vegetation, because of a decrease in the sunlight. However, this impact was transient so the habitat loss caused is expected to be negligible.

Impacts on Fish

The potential effects from offshore wind energy installations may be divided into:
- introduction of new artificial habitat;
- noise; and
- electromagnetic fields.

The construction phase probably disturbs many of the fish species. However, the underwater movements, noise and increased turbidity of the water associated with the works period disappear at the end of this stage.

The response from fish species to the introduction of wind turbine foundations is comparable with artificial reefs. Fish attraction behaviour to artificial reefs has been demonstrated in several European studies. It is expected that fish abundance and species diversity will be increased around the turbine foundations as the new habitat becomes more integrated with the marine environment.

The new artificial habitats created by the construction of Horns Rev and Nysted wind farms have had insignificant effects on fish. The species composition was similar inside and outside of the wind farm areas. Only sand eels show a different pattern, with the population increasing by about 300 per cent in the Horns Rev wind farm and decreasing by 20 per cent outside of it. More clear and definitive results will be obtained in the coming years, when the colonisation process becomes more mature.

Positive impacts from offshore wind energy are foreseen with the ban on fishing, especially demersal trawling, in the wind farm area resulting in more local fish. The increase of biomass in benthos communities as a result of the construction of new foundations would support this supposition.

The low frequency noise may be audible to many fish species. The frequency, intensity and duration of the noise will determine the grade of disturbance. Studies on goldfish, cod and Atlantic salmon have indicated that they can detect offshore turbines from 0.4 to 25 km at wind speeds of 8 to 13 m/s. The detection distance depends on the size and numbers of wind turbines, the hearing organs of the fish, the water depth and bottom substrate. The fish produce a variety of sound for communication that may be interfered with by the noise from turbines. This could decrease the effective range of communication by fish. However, the extent of this interference and its influence on the behaviour and fitness of fish is not known and additional studies are needed. There is no evidence that turbines damage the hearing of fish, even at low distances of a few metres. The avoidance distance is about 4 m, but only at high wind speeds of 13 m/s. The noise impact mainly masks communication and orientation signals, whereas it does not produce serious damage to hearing organs or strong avoidance reactions are produced.

Overall, the environmental monitoring in Horns Rev and Nysted shows that the effects of noise and vibrations from the wind farms on fish are negligible. However, the current knowledge about wind energy impacts on fish presents large uncertainties. Knowledge of the behavioural response of fish to noise and vibrations from offshore wind developments is still limited (Boesen and Kjaer, 2005; Thomsen et al., 2006). Future studies must gather better data on the nature of the acoustic field around wind turbines and the physiological and behavioural impacts on fish (Wahlberg and Westerberg, 2005; Thomsen et al., 2006).

Maintenance of wind farms needs more-or-less daily activity, with ships moving into the wind farm area. This associated noise should create more impacts than the operating turbines (Greenpeace, 2005).

Impacts on Marine Mammals

Offshore wind farms can negatively affect marine mammals during both construction and operation stages. The physical presence of turbines, the noise during construction, the underwater noise, and boat and helicopter traffic can disturb mammals, causing them to avoid wind farms.

Monitoring marine mammals living and moving below sea level is very difficult. Fortunately, the traditional visual surveys from ships and aircraft are being supplemented or replaced by new, more accurate technologies such as acoustic monitoring by stationary data loggers, remotely controlled video monitoring and tagging of individuals with satellite transmitters.

Mammals are very dependent on their hearing systems, which are used for several purposes: communication between individuals of the same species, orientation, finding prey and echolocation. The behavioural response by marine mammals to noise includes modification of normal behaviour, displacement from the noisy area, masking of other noises, and the impossibility of acoustically interpreting the environment. The consequences from this disturbance could cause problems of viability of individuals, increased vulnerability to disease, and increased potential for impacts due to cumulative effects from other impacts such as chemical pollution combined with stress induced by noise.

The noise measured by the German Federal Ministry of Environment doesn't seem to damage the hearing organs of marine animals, but it is not well known how it will affect their behaviour in the area surrounding the turbines. Although the sound level is moderate, it is permanent (until decommissioning), thus more research about its influence on marine animals behaviour is needed.

Horns Rev and Nysted wind farms in Denmark carried out a comprehensive environmental monitoring programme between 1999 and 2006, covering baseline analysis, construction and operation phases. The highlight of the study shows different reactions between seals and porpoises. Seals were only affected during the construction phase, due to the high sound levels in pile-driving operations. In the operation phase, it seems wind farms did not have any effect on seals. However, harbour porpoises' behaviour was dissimilar at the two offshore wind farms. In Horns Rev, the population decreased slightly during construction, but recovered to the baseline situation during operation. In Nysted, porpoise densities decreased significantly during construction and only after two years of operation did the population recover. The reason for this slow recovery is unknown.

Nysted wind farm is located 4 km away from the Rosland seal sanctuary. The presence of the wind farm had no measurable effects on the behaviour of seals on land.

The foundations of wind farms create new habitats, which are colonised by algae and benthic community. This further availability of food may attract new species of fish and subsequently mammals. This change could be neutral or even positive to mammals.

It is very difficult to assess the long-term impacts on reproduction and population status with the current state of knowledge. The possible behaviour modification of marine mammals due to the presence of wind turbines at sea is presumably a species-specific subject. Other factors which also require further research relate to oceanographic parameters (hydrography, bathymetry, salinity and so on) and the hearing systems of mammals.

Impacts on Sea Birds

The influence of offshore wind farms on birds can be summarised as follows:
- collision risk;
- short-term habitat loss during construction phase;
- long-term habitat loss due to disturbance from wind turbines installed and from ship traffic during maintenance;
- barriers to movement in migration routes; and
- disconnection of ecological units.

The methodology proposed by Fox et al. (2006) to support EIAs of the effects on birds of offshore wind farms reveals the great complexity of the analysis. The relationships between offshore wind farms and bird impacts must be analysed by gathering information about avoidance responses, energetic consequences of habitat modification and avoidance flight, and demographic sensitivity of key species.

Collisions have the most direct effect on bird populations. Collision rates for wintering waterfowl, gulls and passerines on coastal areas in northwest Europe range from 0.01 to 1.2 birds/turbine. No significant population decline has been detected. Direct observations from Blyth Harbour, UK, have demonstrated that collisions with rotor blades are rare events in this wind farm located within a Site of Special Scientific Interest and Special Protection Area, under the Birds Directive.

In poor visibility conditions, large numbers of terrestrial birds could collide with offshore wind farms, attracted by their illumination. However, this occurs only on a few nights. Passerines are the group mainly involved in these collisions. One of the most useful mitigation measures to avoid this type of impact is to replace the continuous light with an intermittent one.

Information about bird mortality at offshore wind farms is very scarce for two reasons: the difficulty of detecting collisions and the difficulty in recovering dead birds at sea. Further investigations on this topic are needed to get reliable knowledge.

There is a lack of good data on migration routes and flight behaviour of many of the relevant marine bird species. But this data is essential for assessing the potential impacts of collisions and barriers to movements. The large scale of proposed offshore wind farms and the expected cumulative effects increase the need to fill in these gaps.

The degree of disturbance differs between different species. The disturbance may be determined by several factors such as availability of appropriate habitats, especially roosting and feeding areas, time of year, flock size and the layout of the wind farms.

Disturbances during construction are produced by ships and/or helicopter activities and noise generated by ramming piles. After that, in the operation stage, disturbances by boat traffic still have an impact on birds.

The impacts of marine wind farms are higher on sea birds (resident, coastal and migrant) than on onshore birds. The reasons for this higher impact at offshore developments are related to the larger height of marine wind turbines, the larger size of wind farms and

Figure V.2.2: Flow chart of hazards factors to birds by offshore developments

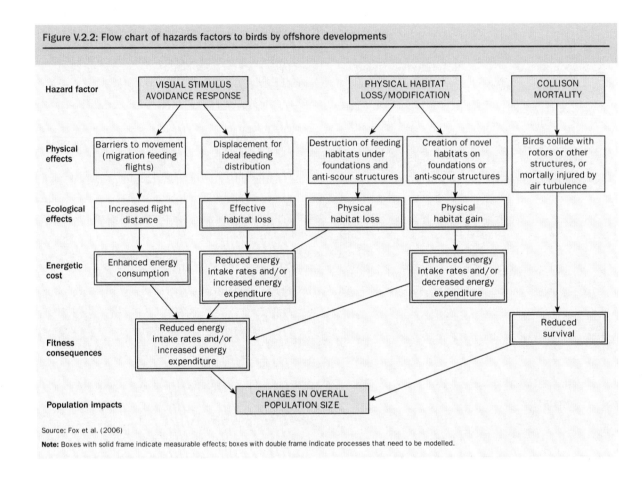

Source: Fox et al. (2006)

Note: Boxes with solid frame indicate measurable effects; boxes with double frame indicate processes that need to be modelled.

the higher abundance of large bird species, which are more sensitive to disturbance.

The most important findings after seven years of monitoring at Horns Rev and Nysted wind farms indicate negligible effects on overall bird populations. The majority of bird species showed avoidance of the wind farms. Although there was considerable movement of birds around wind farms, between 71 and 86 per cent of flocks avoided flying between the turbine rows of the wind farm. Changes in flying directions, for most of the species, were verified at 0.5 km from wind farms at night and at 1.5 km in the day. This avoidance represents an effective habitat loss, but the proportion of feeding area lost due to the presence of these two wind farms, in relation to the total feeding area, is relatively small and is considered of little biological importance.

Avoidance behaviour reduces the collision with turbines. The displacement of birds because of wind farm installations makes the collision risk at the two installations low. The predicted collision rates of common eiders at Nysted were around 0.02 per cent, which means 45 birds of a total of 235,000 passing each autumn in the area. Monitoring has also confirmed that waterbirds (mainly eider) reduce their flight altitude, to below rotor height, at the Nysted wind farm.

Avoidance observed in Nysted and Horns Rev affects flying, resting and foraging between turbines. New wind farm proposals in the same area have to be carefully analysed because they may cause important habitat loss for certain species.

EIAs on marine ecosystems must take into account the cumulative effects from all the wind farms in the surrounding area, including cable connections to the network on the mainland.

During the last several years, a lot of methodologies on collision risk models, baseline surveys using both ship and aerial techniques and post-construction monitoring have been developed. This data is needed to properly assess and predict the future impacts of proposed wind farms. Several sophisticated technologies, such as radar and infrared cameras, have helped to acquire a better understanding.

When there is not enough knowledge about specific species or taxonomic groups in unstudied habitats, the potential disturbance distances could be unknown. The most appropriate approach to define the disturbance distance may be to determine the bird numbers at different ranges of distances from wind farm, ensuring that all the affected area is covered in the study.

There is a common opinion on the need for more information about potential impacts of wind farms on birds. Further research is required on avian responses to wind farms, models to predict the future impacts of a new single wind farm installation and groups of wind farms on an area, the collection of information on bird movements to design marine sanctuaries, and data gathering standardisation methodologies.

Mitigation measures for onshore schemes are also applicable to offshore wind farms.

Ship Collisions

Ship collisions with the turbines are one of the potential risks associated with offshore wind energy development. Colliding with a wind turbine foundation could damage or possibly destroy a ship. The potential danger to the environment is the spillage of oil or chemicals from the ship into the water.

Evaluation of several collision scenarios between three different types of turbine foundations (monopile, jacket and tripod) and different ship types (single and double hull tankers, bulk carriers, and container ships) has been carried out in several locations in the North Sea and the Baltic Sea off Germany. The results have demonstrated two main results: the first is that monopile and jacket foundations are safer than tripod structures, and the second is related to the risk of collision, which can be reduced, but not totally avoided.

There are several safety approaches applicable to avoid or minimise this potential risk:

- redundant navigation and control systems such as radar and ships optimised to survive collisions;
- prohibition on navigation into the wind farm area for certain kinds of unsafe ships;
- introduction of traffic management systems;
- wind farm monitoring;
- availability of tug boats for emergencies; and
- crew training.

Radar and Radio Signals

The wind turbines may impact on aviation activity, both civil and military, due to interference with radars that manage aircraft operations. Radar is a system for detecting the presence or position or movement of objects by transmitting radio waves, which are reflected back to a receiver. The radio wave transmitted by radar can be interrupted by an object (also called a target), then part of the energy is reflected back (called echo or return) to a radio receiver located near the transmitter.

Wind turbines are vertical structures that can potentially interfere with certain electromagnetic transmissions. Mobile structures such as rotating blades may generate more interference on the radars than stationary structures. The effects depend on type of radar, specific characteristics of wind turbines and the distribution of wind turbines. Air traffic management is susceptible to being negatively affected by wind turbine installations. The systems managed by radars are air traffic control, military air defence and meteorological radars.

Table V.2.5 summarises the functions and the mitigation measures according to the different types of radar and wind turbine effects in the UK.

Table V.2.5: Effects and mitigation measures by radar types

Systems	Air traffic control		Meteorological control		Air defence	
Mission	Control of arrival, departure and transit in vicinity of airport and transit over the country		Weather forecasting; very important to aviation safety		Detect and identify aircraft approaching, leaving or flying over the territory of a country	
Types	Primary radar	Secondary surveillance radar	Weather radar	Wind profile radar	Ground based radars	Airborne radars
Wind turbines' effects	False radar responses or returns	Masking genuine aircraft returns; reflection from wind turbines could cause misidentification or mislocation of aircraft	Reflection	Reflection	Highly complex and not completely understood	Highly complex and not completely understood
Mitigation measures at the beginning of project planning	Ensuring location in area with low aircraft traffic; ensuring location not in line of sight of any aircraft radar	Avoiding close vicinity to radars; minimum safe distance between wind farms and these types of radars not defined	Avoiding wind farm installation at 10 km or less of radar facility		Minister of Defence of UK does not permit any wind farm located at less than 74 km from an air defence radar, unless developers can demonstrate no interferences with the defence radar	Moving the location of wind farm or adjusting the configuration of turbines to avoid interference; providing alternative site for the affected radar; contribute to investment in additional or improved radar system

Source: Based on DTI (2002)

The impacts associated with wind turbines are masking, returns/clutter and scattering.

MASKING

Radar systems work at high radio frequencies and therefore depend on a clear 'line of sight' to the target object for successful detection. When any structure or geographical feature is located between the radar and the target, it will cause a shadowing or masking effect. The interference varies according to turbine dimensions, type of radar and the aspect of the turbine relative to the radar. The masking of an aircraft can occur by reflecting or deflecting the returns when the aircraft is flying in the 'shadow' of wind turbines and thus it is not detected. Also the masking can occur when returns from the towers and blades of the wind turbines are so large that returns from aircraft are lost in

the 'clutter' (radar returns from targets considered irrelevant to the purpose of the radar).

RETURNS/CLUTTER

Radar returns may be received from any radar-reflective surface. In certain geographical areas, or under particular meteorological conditions, radar performance may be adversely affected by unwanted returns, which may mask those of interest. Such unwanted returns are known as radar clutter. Clutter is displayed to a controller as 'interference' and is of concern primarily to air surveillance and control systems – ASACS and aerodrome radar operators, because it occurs more often at lower altitudes.

The combination of blades from different turbines at a wind farm can give the appearance of a moving object, which could be considered as an unidentified aircraft requiring controllers to take action to avoid a crash with another aircraft.

SCATTERING, REFRACTION AND/OR FALSE RETURNS

Scattering occurs when the rotating wind turbine blades reflect or refract radar waves in the atmosphere. The source radar system or another system can absorb the waves and provide false information to that system. This effect is not well known, but it has been reported in Copenhagen airport as a result of the Middelgrunden offshore wind farm.

The possible effects are:

- multiple, false radar returns such as blade reflections are displayed to the radar operator as false radar contacts;
- radar returns from genuine aircraft are recorded but in an incorrect location; and
- garbling or loss of information.

Marine radars and communication and navigation systems may suffer interference from nearby wind farms. However, Howard and Brown (2004) stated that most of the effects of Hoyle offshore wind farm do not significantly compromise marine navigation or safety. Mitigation measures in open water include the definition of vessel routes distant from wind farms, while in restricted areas the boundaries of wind farms must be kept at appropriate distances from navigation routes or port approaches.

V.3 POLICY MEASURES TO COMBAT CLIMATE CHANGE

The Kyoto Protocol

The Kyoto Protocol is an international treaty subsidiary to the United Nations Framework Convention on Climate Change (UNFCCC/1992). Negotiations for the Kyoto Protocol were initiated at the first Conference of the Parties (COP 1) of the UNFCCC in Berlin in 1995, in recognition that the voluntary measures included in the UNFCCC were ineffective. The major feature of the Kyoto Protocol is that it sets 'quantified emission limitation or reduction obligations' (QUELROs) – binding targets – for 38 industrialised countries and the European Community (Annex B countries)[6] for reducing greenhouse gas (GHG) emissions by an aggregate 5.2 per cent against 1990 levels over the five-year period 2008–2012, the so-called 'first commitment period'.

The Protocol was agreed in December 1997 at the third Conference of the Parties (COP 3) in Kyoto, Japan. After COP 7, in Marrakech in late 2001, the Protocol was considered ready for ratification, and over the course of the next three years sufficient countries ratified it for it to enter into force on 16 February 2005. Industrialised countries that ratify the Protocol commit to reducing their emissions of carbon dioxide and a basket of five other GHGs according to the schedule of emissions targets laid out in Annex B to the Protocol.

The Parties to the UNFCCC agreed that the effort to combat climate change should be governed by a number of principles, including the principle of 'common but differentiated responsibilities', in recognition that:

- the largest share of historical emissions of greenhouse gases originated in developed countries;
- per capita emissions in developing countries are still very low compared with those in industrialised countries; and
- in accordance with the principle of equity, the share of global emissions originating in developing countries will need to grow in order for them to meet their social and development needs.

As a result of this, most provisions of the Kyoto Protocol apply to developed countries, listed in Annex I to the UNFCCC. China, India and other developing countries have not been given any emission reduction commitments, in recognition of the principles of common but differentiated responsibilities and equity enumerated above. However, it was agreed that developing countries share the common responsibility of all countries in reducing emissions.

OBJECTIVES AND COMMITMENTS

The overall objective of the international climate regime, articulated in Article 2 of the UNFCCC, is to achieve 'stabilisation of greenhouse gas concentrations in the atmosphere at a level that would prevent dangerous anthropogenic (human) interference with the climate system'.

The goal of the Kyoto Protocol is to reduce overall emissions of six GHGs – carbon dioxide, methane, nitrous oxide, sulphur hexafluoride, hydrofluorocarbons and perfluorocarbons – by an aggregate 5.2 per cent over the period 2008–2012, the first commitment period, which would then be followed by additional commitment periods with increasingly stringent emissions reduction obligations.

National emissions reduction obligations range from 8 per cent for the EU and some other countries to 7 per cent for the US, 6 per cent for Japan, 0 per cent for Russia, and permitted increases of 8 per cent for Australia and 10 per cent for Iceland. Emission figures exclude international aviation and shipping.

One of the most heavily contested issues in the Kyoto Protocol negotiations was the legally binding nature of the emissions reductions and the compliance regime. As the detailed architecture of the Protocol emerged in the period from 1997 to 2001, it became clear that for carbon markets to function effectively, the private sector needed to be able to 'bank' on the efficiency of the regime and the legality of the carbon credits.

What was agreed in the end was a compliance mechanism which held national governments accountable for their emissions reduction obligations, imposing a penalty of 30 per cent on countries for failing to meet their obligations by the end of the first commitment period (2012). Their obligation in the second commitment period, in addition to what was negotiated, would be increased by 1.3 tonnes for each tonne of shortfall in meeting their first commitment period obligation. Furthermore, if they were judged by the Protocol's Compliance Committee to be out compliance, then their right to use the flexible mechanisms would be suspended until they were brought back into compliance. Thus far, these legal arrangements have been sufficient to allow the functioning of the carbon markets, although the true test is yet to come.

STATUS OF RATIFICATION

As of May 2008, 182 parties have ratified the protocol. Out of these, 38 developed countries (plus the EU as a party in its own right) are required to reduce GHG emissions to the levels specified for each of them in the treaty. The protocol has been ratified by 145 developing countries, including Brazil, China and India. Their obligations focus on monitoring and reporting emissions, which also enable them to participate in the Clean Development Mechanism (CDM). To date, the US and Kazakhstan are the only signatory nations of the UNFCCC not to have ratified the protocol.

FLEXIBLE MECHANISMS

For combating climate change, it does not matter where emissions are reduced, as it is the overall global reduction that counts. As a result, the Kyoto Protocol has taken a strong market approach, recognising that it may be more cost-effective for industrialised (Annex I) parties to reduce emissions in other countries, whether also Annex I or developing. In order to achieve their targets set under the Kyoto Protocol, industria-

lised countries thus have the ability to apply three different mechanisms in which they can collaborate with other parties and thereby achieve an overall reduction in GHG emissions. These are:
1. Joint Implementation (JI);
2. the Clean Development Mechanism (CDM); and
3. emissions trading.

Joint Implementation

The Joint Implementation procedure is set out in Article 6 of the Kyoto Protocol. This stipulates that an Annex I country can invest in emissions reduction projects in any other Annex I country as an alternative to reducing emissions domestically. This allows countries to reduce emissions in the most economical way, and to apply the credit for those reductions towards their commitment goal. Most JI projects are expected to take place in so-called 'transition economies', as specified in Annex B of the Kyoto Protocol, mainly Russia, Ukraine and Central and East Europe (CEE) countries. Most of the CEE countries have since joined the EU or are in the process of doing so, thereby reducing the number of JI projects as the projects in these countries were brought under the European Union Emissions Trading Scheme (EU ETS) and its rules to avoid double counting. The JI development in Russia and Ukraine was relatively slow due to delays in developing the nations' domestic JI rules and procedures, although activity is now picking up.

The credits for JI emission reductions are awarded in the form of 'emission reduction units' (ERUs), with one ERU representing a reduction of one tonne of CO_2 equivalent. These ERUs come out of the host country's pool of assigned emissions credits, which ensures that the total amount of emissions credits among Annex I parties remains stable for the duration of the Kyoto Protocol's first commitment period.

ERUs will only be awarded for JI projects that produce emissions reductions that are 'additional to any that would otherwise occur' (the so-called 'additionality' requirement), which means that a project must prove that it would only be financially viable with the extra revenue of ERU credits. Moreover, Annex I parties may only rely on JI credits to meet their targets to the extent that they are 'supplemental to domestic actions'. The rationale behind these principles is to formally limit the use of the mechanism. However, since it is very hard to define which actions are 'supplemental' to what would have occurred domestically in any event, this clause is, sadly, largely meaningless in practice.

The Clean Development Mechanism

The Kyoto Protocol's Article 12 established the Clean Development Mechanism, whereby Annex I parties have the option to generate or purchase emissions reduction credits from projects undertaken by them in non-Annex I countries. In exchange, developing countries will have access to resources and technology to assist in development of their economies in a sustainable manner.

The credits earned from CDM projects are known as 'certified emissions reductions' (CERs). Like JI projects, CDM projects must meet the requirement of 'additionality', which means that only projects producing emissions reductions that are additional to any that would have occurred in the absence of the project will qualify for CERs. The CDM is supervised by an Executive Board, which is also responsible for issuance of the CERs. Other requirements, including compliance with the project and development criteria, the validation and project registration process, the monitoring requirements, and the verification and certification requirements, are done externally by a third party.

A wide variety of projects have been launched under the CDM, including renewable energy projects such as wind and hydroelectric; energy efficiency projects; fuel switching; capping landfill gases; better management of methane from animal waste; the control of coal mine methane; and controlling emissions of certain industrial gases, including HFCs and N_2O.

China has come to dominate the CDM market, and in 2007 expanded its market share of CDM transactions to 62 per cent. However, CDM projects have been registered in 45 countries and the UNFCCC points out that investment is now starting to flow into other parts of the world, such as Africa, Eastern Europe and Central Asia.

In 2007, the CDM accounted for transactions worth €12 billion (Point Carbon, 2008), mainly from private sector entities in the EU, EU governments and Japan.

The average issuance time for CDM projects is currently about one to two years from the moment that they enter the 'CDM pipeline', which counted over 3000 projects as of May 2008. Around 300 projects have received CERs to date, with over two-thirds of the issued CERs stemming from industrial gas projects, while energy efficiency and renewable energy projects seem to be taking longer to go through the approval process. However, there are now more than 100 approved methodologies and continuous improvement to the effective functioning of the Executive Board.

The rigorous CDM application procedure has been criticised for being too slow and cumbersome. The 'additionality' requirement has especially represented a stumbling block for some projects, since it is difficult to prove that a project would not be viable without the existence of CERs. The CDM also has the potential to create perverse incentives, in other words discouraging the implementation of rigorous national policies for fear of making the additionality argument more difficult.

There are many improvements yet to be made, and the additionality principle will be one of the many issues surrounding the flexible mechanisms to be discussed during the negotiations leading to a post-2012 climate agreement. The CDM, as a project-based market mechanism, is by definition going to be fundamentally limited in both scope and geographic application. A variety of options for sectoral approaches are under consideration for moving away from a project-based approach with its additionality requirements.

In addition, it has become very clear in retrospect that the large industrial gas projects which still count for a large share of the CERs on the market during this first period should in reality be dealt with legislatively rather than through the CDM.

Emissions Trading

Under the International Emissions Trading provisions, Annex I countries can trade so-called 'assigned amount units' (AAUs) among themselves, which are allocated to them at the beginning of each commitment period. The emissions trading scheme, which is established in Article 17 of the Kyoto Protocol, also foresees this to be 'supplemental to domestic actions' as a means of meeting the targets established for the Annex I parties. The total amount of allowable emissions for all Annex I countries (the 'cap') has been proposed under the Kyoto Protocol. The scheme then allocates an amount of these emissions as 'allowances' to each of the Annex I parties (the 'assigned amount'). The assigned amount for any Annex I country is based on its emissions reduction target specified under Annex B of the Kyoto Protocol. Those parties that reduce their emissions below the allowed level can then trade

some part of their surplus allowances (AAUs) to other Annex parties.

The 'transition economies', such as Russia, Ukraine and CEE countries, have a huge quantity of surplus AAUs in the first commitment period, which is largely as a result of the collapse of the Warsaw Pact economies in the early 1990s. As these surplus AAUs were not created from active emissions reductions, the EU and Japanese buyers have vowed not to purchase them from the region unless the AAU revenue is associated with some 'greening' activities. The problem is partly being solved through the introduction of a new mechanism called the Green Investment Scheme (GIS), in which the sales revenue from AAUs are channelled to projects with climate and/or environment benefits. The surplus AAUs are now beginning to enter the market, with some CEE countries taking a lead in establishing the scheme. Various estimates suggest the total amount of AAUs entering the market through the GIS could be very large – much larger than the World Bank estimate of demand of between 400 million and 2 billion AAUs in the market (World Bank, 2008). The exact figure of the supply is hard to predict, however, as the biggest reserve of surplus AAU is in Russia, whose participation in the GIS is not yet clear.

Box V.3.1: The EU Emissions Trading System

HOW THE EU ETS WORKS

In order to tackle climate change and help EU Member States achieve compliance with their commitments under the Kyoto Protocol, the EU decided to set up an Emissions Trading System (ETS). Directive 2003/87/EC of the European Parliament and of the Council of 13 October 2003 established the EU ETS. On 1 January 2005 the EU ETS commenced operation. The system is being implemented in two trading periods. The first trading period ran from 2005 to 2007. The second trading period is set

in parallel to the first commitment period of the Kyoto Protocol: it began on 1 January 2008 and runs until the end of 2012. A third trading period is expected to start in 2013 and to be implemented along with a reviewed ETS Directive.

The EU ETS is the first and largest international trading system for CO_2 emissions in the world (see following section, 'Carbon as a commodity'). Since January 2008 it has applied not only to the 27 EU Member States but also to the other three members of the European Economic Area – Norway, Iceland and Liechtenstein. It covers over 10,000 installations in the energy and industrial sectors, which are responsible for about 50 per cent of the EU's total CO_2 emissions and about 40 per cent of its total greenhouse gas emissions. The emission sources regulated under the system include combustion plants for power generation (capacities greater than 20 MW), oil refineries, coke ovens, iron and steel plants, and factories making cement, glass, lime, bricks, ceramics, pulp and paper. Discussions are under way on legislation to bring the aviation sector into the system from 2011 or 2012 (EC, 2006b).

The EU ETS is an emission allowance cap and trade system, that is to say it caps the overall level of emissions allowed but, within that limit, allows participants in the system to buy and sell allowances as they need. One allowance gives the holder the right to emit one tonne of CO_2.

Currently, for each trading period under the system, Member States draw up national allocation plans (NAPs), which determine their total level of ETS emissions and how many emission allowances each installation in their country receives. By allocating a limited number of allowances, below the current expected emissions level, Member States create scarcity in the market and generate a market value for the permits. A company that emits less than the level of their allowances can sell its surplus allowances. Those companies facing difficulties in keeping their emissions in line with their allowances have a choice between taking measures to reduce their own emissions or buying the extra allowances they need on the market.

The ETS Directive stipulates that at least 95 per cent of issued allowances should be given out for free by Member States for the period 2005–2007. For the second trading period (2008–2012), this value is 90 per cent. The remaining percentage can be charged for, for example in an auction. This process, referred to as 'allocation', has been carried out using what is known as a 'grandfathering' approach,

which is based on historical data (emissions or production levels).

The scheme is linked to the Kyoto Protocol's flexible mechanisms through Directive 2004/101/EC. According to the 'Linking Directive', in addition to domestic action, Member States may also purchase a certain amount of credits from Kyoto flexible mechanisms projects (CDM and JI) to cover their emissions in the same way as ETS allowances.

PERFORMANCE OF THE EU ETS (2005–2007)

The EU ETS has so far failed to achieve some of its main objectives, notably encouraging investment in clean technologies and the use of CO_2 emissions reduction certificates as a market signal to regulate greenhouse gas emissions (Carbon Trust, 2007; Open Europe, 2007). This is due to a combination of adverse incentives associated with the EU ETS design:

- political national influence on the allocation process and over-allocation of permits;
- counterproductive allocation methods; and
- limited scope of the system.

Political national influence on the allocation process and over-allocation of permits

As previously explained, in order to make sure that real trading emerges, Member States must make sure that the total amount of allowances issued to installations is less than the amount that would have been emitted under a business-as-usual scenario.

Under the current system, where a significant degree of freedom over the elaboration of the NAPs is retained by Member States, decisions concerning allocation hinge upon emission projections, national interests and business efforts to increase the number of allowances (del Río González, 2006; Kruger et al., 2007; Blanco and Rodrigues, 2008).

Actual verified emissions in 2005 showed allowances had exceeded emissions by about 80 million tonnes of CO_2, equivalent to 4 per cent of the EU's intended maximum level (Ellerman and Buchner, 2007). This happened because government allocation had been based on over-inflated projections of economic growth and participants had a strong incentive to overestimate their needs (ENDS Europe Report, 2007 – http://www.endseurope.com).

The publication of those figures provoked the collapse of the CO_2 prices to less than €10/t in spring 2006. By the end of 2006 and into early 2007, the price of allowances for the first phase of the EU ETS fell below €1/tCO_2 (€0.08/tCO_2 in September 2007) (www.pointcarbon.com). The over-allocation of permits and the consequent collapse of CO_2 prices have hampered any initiative of clean technology investment, as it is clear that most companies regulated by the EU ETS didn't need to make any significant change to their production processes to meet the target they had been assigned (Blanco and Rodrigues, 2008).

Counterproductive allocation methods

The first phase of the EU ETS has shown that free allocation based on absolute historical emissions (grandfathering) causes serious distortions in competition by favouring de facto fossil fuel generation (EWEA, 2007).

A controversial feature of the system has been the ability of the electric power sector to pass along the the marginal cost of freely allocated emissions to the price of electricity and to make substantial profits. This happens because in competitive markets the power generation sector sets prices relative to marginal costs of production. These marginal costs

include the opportunity costs of CO_2 allowances, even if allowances are received for free. As a consequence, fossil fuel power producers receive a higher price for each kWh they produce, even if the costs for emitting CO_2 only apply to a minor part of their merchandise. The effect is known as windfall profit.

In the first phase of the EU ETS, conventional power generators are believed to have made over €12.2 billion in windfall profits in the UK alone (Platts, 2008). There have been similar arguments over ETS windfall profits in other European countries, such as Germany and Spain (Platts, 2008). Carbon market experts see the situation as likely to arise again in the second trading period. According to a recent Point Carbon study of the UK, Germany, Spain, Italy and Poland, power companies could reap profits in excess of €71 billion over the next four years (Point Carbon, 2008).

Furthermore, as the economist Neuhoff remarks, any free allocation acts as a subsidy to the most polluting companies, which – in addition to not paying the environmental cost they entail – obtain substantial gains (Neuhoff et al., 2006). This is clearly in contradiction with the 'polluter pays principle' (established by Article 174 of the EC Treaty), which states that 'environmental damage should as a priority be rectified at source and that the polluter should pay'.

Grandfathering also penalises 'early action' and justifies 'non-action'. Since allowances are allocated as a function of emission levels, firms are clearly encouraged not to reduce their emissions, as this would result in fewer allowances in future phases (Neuhoff et al., 2006).

Limited scope of the ETS

About 55 per cent of the CO_2 emitted in the EU comes from sectors outside the EU ETS. In the same way, other more powerful greenhouse gases, such as nitrous oxide, sulphur hexafluoride and methane are excluded. The experience from recent years illustrates that it is in some of the sectors that have been left outside the ETS – notably transport – that the highest CO_2 emission growth rates have occurred (Eurostat, 2007).

PROPOSALS FOR THE POST-2012 PERIOD

As a preliminary step to design the third phase of the EU ETS (post-2012), the European Commission has embarked in a public consultation on what the new system should look like. The debate started in November 2006 with the publication of the Communication 'Building a global carbon market' (EC, 2006a). In the context of the European Climate Change Programme (ECCP), a Working Group on the review of the EU ETS was also set up to discuss the four categories of issues identified by the EC Communication (EC, 2006a):

1. scope of the scheme;
2. robust compliance and enforcement;
3. further harmonisation and increased predictability; and
4. participation of third countries.

As part of the Commission's climate change and energy package, and in the light of the European Council's 2020 commitments to reduce greenhouse gas emissions by 20 per cent compared to 1990 levels (30 per cent if other developed countries join the effort), a new proposal for reform of the EU ETS Directive was presented on 23 January 2008 (EC, 2008).

Although the proposal still needs to be approved by both the Council of the EU and the European Parliament, the main elements of the new system, which will enter into force in 2013 and run until 2020, seem to be the following:

- One EU-wide cap on the number of emission allowances. Allowances would be centrally allocated by the European Commission instead of through NAPs.
- Emissions from EU ETS installations would be capped at 21 per cent below 2005 levels by 2020 – thus a maximum of 1720 million allowances. The annual cap would fall linearly by 1.74 per cent annually as of 2013.
- 100 per cent auctioning for the power sector. For the other sectors covered by the ETS, a transitional system would be put in place, with free allocations being gradually phased out on an annual basis between 2013 and 2020.
- However, an exception could be made for installations in sectors judged to be 'at significant risk of carbon leakage', in other words relocation to third countries with less stringent climate protection laws. Sectors concerned by this measure are yet to be determined.
- At least 20 per cent of auction revenues would have to be ring-fenced to reduce emissions, to support climate adaptation and to fund renewable energy development.
- Extension of the system's scope to new sectors, including aluminium, ammonia and the petrochemicals sectors, as well as to two new gases, nitrous oxide and perfluorocarbons. Road transport and shipping would remain excluded, although the latter is likely to be included at a later stage.
- In the absence of an international climate agreement, the limit on the use of the CERs and ERUs is expected to be restricted to the unused portion of operators' phase two cap. This limit is to rise to 50 per cent of the reduction effort if a new international climate agreement is reached.

CARBON AS A COMMODITY

The Kyoto Protocol's efforts to mitigate climate change have resulted in an international carbon market that has grown tremendously since the entry into force of the Protocol in 2005. While previously, the relatively small market consisted mostly of pilot programmes operated either by the private sector or by international financial institutions such as the World Bank, it has experienced strong growth in the past two years, and was valued at €40 billion in 2007, 80 per cent more than the 2006 value. The total traded volume increased by 64 per cent from 1.6 MtCO$_2$ in 2006 to 2.7 Mt in 2007 (Figure V.3.1).

While the international carbon market has expanded to include a wide variety of project types and market participants, it has been dominated by two market-based mechanisms: the EU ETS and the CDM.

The EU ETS continues to be the largest carbon market, with a traded volume of 1.6 MtCO$_2$ and a value of €28 billion in 2007 (Point Carbon, 2008), which corresponds to nearly a doubling of both volume and value compared to the previous year. The EU ETS now contains more than 60 per cent of the physical global carbon market and 70 per cent of the financial market. The CDM market increased to 947 MtCO$_2$-equivalents and €12 billion in 2007. This is an increase of 68 per cent in volume terms and a staggering 200 per cent in value terms from 2006, and the CDM now constitutes 35 per cent of the physical market and 29 per cent of the financial market. The JI market, while still small, also finally started to take off in 2007, nearly doubling in volume to 38 MtCO$_2$ and more than tripling in value to €326 million.

However, experts predict that the potential for future market growth is much larger. Point Carbon forecasts 56 per cent market growth in 2008, increasing volumes to over 4 million tonnes of carbon, with a value of more than €60 billion, depending on prices. Current prices in the ETS hover around €25/tonne, and CDM prices range from anywhere between 9 and 17/tonne, depending on the type of project and its stage of development.

Providing that the price for carbon is high enough, the carbon market is a powerful tool for attracting investment, fostering cooperation between countries,

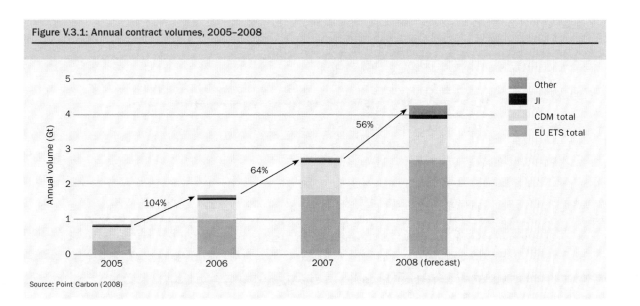

Figure V.3.1: Annual contract volumes, 2005–2008

Source: Point Carbon (2008)

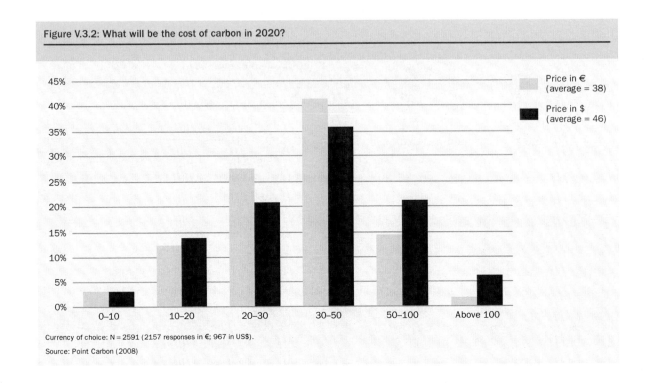

Figure V.3.2: What will be the cost of carbon in 2020?

Currency of choice: N = 2591 (2157 responses in €; 967 in US$).

Source: Point Carbon (2008)

companies and individuals, and stimulating innovation and carbon abatement worldwide. In theory, at least, the price of carbon should more or less directly reflect the rigorousness of the economy-wide caps of the Annex B countries. The reality is of course more complicated, since there is only one real 'compliance market' at present, which is the EU ETS, and the CDM and JI markets are in reality just getting started. It is also not clear what role Canada, Japan and Australia will play in the carbon market during the first commitment period; and of course, the original conception and design of the carbon market was predicated on the fact that the US would be a large buyer, which has not turned out to be the case, again, for the first commitment period. Governments negotiating the post-2012 climate agreement seem committed to 'building carbon markets' and/or 'keeping the CDM', but there is very little detail to go on at present. The UNFCCC negotiations in June 2008 produced little more than a shopping list of issues to be addressed in the further

development of carbon markets in general and the CDM in particular.

Point Carbon conducted a survey of carbon market practitioners at the end of 2007 and came up with the figures presented in Figure V.3.2. These give as good a prognostication as any as to the future price of carbon.

Wind Energy's Contribution to Emissions Reductions

EMISSIONS FROM THE POWER SECTOR

The power sector today accounts for 41 per cent of global CO_2 emissions (WEO, 2006), and continuing improvements of thermal power stations in terms of efficiency are offset by the strong growth in global power demand. The International Energy Agency (IEA, 2007a) estimates that by 2030, electricity production will account for over 17,000 $MtCO_2$, up from 10,500 $MtCO_2$ in 2004.

According to the IEA, electricity generation has had an average growth rate of 2.6 per cent since 1995 and is expected to continue growing at a rate of 2.1–3.3 per cent until 2030, which would result in a doubling of global electricity demand. The bulk of this growth is expected to occur in developing Asia, with India and China seeing the fastest growth in demand. World CO_2 emissions from power production are projected to increase by about 66 per cent over 2004–2030. China and India alone would account for 60 per cent of this increase.

These figures emphasise the strong responsibility and key role that the power sector has to play in reducing CO_2 emissions. According to the IEA, the power sector can be the most important contributor to global emission reductions, with potential CO_2 savings of 6–7 Gt by 2050 on the demand side and 14–18 Gt of CO_2 reductions on the supply side if the right policy choices are taken (IEA, 2008).

The carbon intensity of electricity production largely depends on a given country's generation mix. While inefficient coal steam turbines, which are still in use in many parts of the world, emit over 900 tCO_2/GWh (UNFCCC, 2006) and oil steam turbines around 800 tCO_2/GWh, modern combined cycle gas turbines only produce half these levels. China and India, which have a high share of coal in their power mix, see their electricity produced with over 900 tCO_2/GWh, while other countries, with a high share of renewable energy, such as Brazil, produce power with only 85 tCO_2/GWh. The global average for electricity production can be assumed to be at around 600 tCO_2/GWh, which is close to the OECD average.

WIND ENERGY'S POTENTIAL FOR EMISSIONS REDUCTIONS UP TO 2020

The Intergovernmental Panel on Climate Change (IPCC) released its Fourth Assessment Report in 2007. This left no doubt about climate change being real, serious and man-made. It warned that in order to avert the worst consequences of climate change, global emissions must peak and start to decline before the end of 2020. The potential of wind energy to curb global emissions within this timeframe is therefore key to the long-term sustainability of the power sector.

The benefit to be obtained from carbon dioxide reductions through wind energy again mainly depends on which other fuel, or combination of fuels, any increased wind power generation will replace, so this differs from country to country. For the purposes of this section, we assume a global average of 600 tCO_2/GWh.

Following the logic of the GWEC Wind Energy Scenarios (GWEC, 2008) presented in Part VI, global wind energy capacity could stand at more than 1000 GW by the end of 2020, producing 2,500,000 TWh annually. As a result, as much as 1500 $MtCO_2$ could be saved every year.

It is important to point out that modern wind energy technology has an extremely good energy balance. The CO_2 emissions related to the manufacture, installation and servicing over the average 20-year life cycle of a wind turbine are offset after a mere three to six months of operation, resulting in net CO_2 savings thereafter.

Wind Energy CDM Projects

The CDM has, to some extent, contributed to the deployment of wind energy globally. As of 1 September 2008, a total of 504 wind energy projects were in the 'CDM pipeline', totalling an installed capacity of 16,410 MW. This represents 13 per cent of the total number of projects introduced into the pipeline. Four million CERs have already been issued to wind projects, a number that will go up to 203 million by the end of the first commitment period in 2012.

The majority of these projects are located in China and India. In China, 90 per cent of wind energy projects have applied for CDM registration, and there are now 235 projects in the CDM pipeline, making up almost 11.93 GW of capacity. India now has 216 projects in the pipeline, totalling close to 4.24 GW.

The narrow focus of CDM-supported wind projects in a very few countries is unfortunate but is a reflection of the fact that while carbon finance is a *useful*, and in some cases *necessary* condition for the development of wind power, it is by no means *sufficient*. In the case of both India and China, carbon finance functions alongside a wide range of other measures necessary for countries to diversify and decarbonise their power supply sectors.

There are signs that some other countries may join the list of major host countries for wind power projects assisted by CDM carbon finance. However, it is clear that the ultimate responsibility for this lies with active government implementation of policies and measures

Table V.3.1: CDM projects in the pipeline

Type	Number	kCERs/yr	2012 kCERs	2020 kCERs
Afforestation	5	344	1864	7058
Agriculture	172	6570	43,494	77,570
Biogas	259	11,936	59,172	139,366
Biomass energy	582	33,850	184,661	427,249
Cement	38	6806	41,342	81,796
CO_2 capture	1	7	29	66
Coal bed/mine methane	55	23,597	121,634	301,687
Energy distribution	4	129	1053	1886
EE households	10	306	1504	3346
EE industry	168	6398	32,444	69,634
EE own generation	363	56,558	272,091	608,542
EE service	8	84	393	1034
EE supply side	36	8059	22,817	93,257
Fossil fuel switch	129	41,973	203,062	483,663
Fugitive	28	10,227	62,112	134,515
Geothermal	13	2457	13,775	32,443
HFCs	22	83,190	506,379	1,132,155
Hydro	1006	97,099	437,951	1,183,733
Landfill gas	290	45,793	253,685	559,304
N_2O	65	48,195	257,774	637,426
PFCs	8	1121	4785	11,806
Reforestation	22	1436	11,782	23,230
Solar	23	641	2816	7214
Tidal	1	315	1104	3631
Transport	7	711	3938	9451
Wind	504	40,801	203,081	496,050
Total	3819	528,602	2,744,744	6,527,113

Source: UNEP Risø DTU, CDM/JI Pipeline Analysis and Database, available at http://cdmpipeline.org/cdm-projects-type.htm

Table V.3.2: Wind CDM projects

JI projects in the pipeline (numbers, ERUs & issuance)	All JI projects		
Type	Projects	1000 ERUs	2012 kERUs
Afforestation	0	0	0
Agriculture	0	0	0
Biogas	3	351	1861
Biomass energy	22	1834	8960
Cement	1	306	1041
CO_2 capture	1	268	1071
Coal bed/mine methane	17	8758	43,790
Energy distribution	7	727	3636
EE households	0	0	0
EE industry	12	4870	22,807
EE own generation	1	1698	8491
EE service	0	0	0
EE supply side	15	3127	13,154
Fossil fuel switch	9	1965	9711
Fugitive	33	19,763	92,397
Geothermal	0	0	0
HFCs	3	1774	6579
Hydro	9	766	3295
Landfill gas	17	2436	11,758
N_2O	21	19,402	82,899
PFCs	1	233	1165
Reforestation	0	0	0
Solar	0	0	0
Tidal	0	0	0
Transport	0	0	0
Wind	18	1974	8610
Total	190	70,252	321,225

Source: http://cdmpipeline.org/cdm-projects-type.htm

to create the enabling environment within which carbon finance can play its role, in other words to be an important source to defray the marginal costs of wind power versus conventional fossil fuel plants. This is particularly the case in the absence of an economy-wide cap on carbon emissions.

While clarification and simplification of the carbon finance mechanisms can assist in the broadening and deepening participation of developing countries in the carbon finance market, the fundamental responsibility lies with the host governments, at least as far as wind power is concerned.

Wind Energy JI Projects

There are currently 18 wind energy projects in the JI pipeline (Table V.3.3), totalling an installed capacity of 961 MW. The biggest of these (300 MW) is located in Ukraine, in the autonomous Republic of Crimea. Other projects are based in Bulgaria, Poland, Lithuania and Estonia.

While the JI market is very small today in terms of traded volume, the mechanism could serve to incentivise large countries such as Russia and the Ukraine to tap into their important wind energy potential.

The Path to a Post-2012 Regime

The process to arrive at an international climate agreement for the period after 2012 has been long and arduous. As our understanding of the urgency of early action to avoid the worst dangers of climate change has increased, so has the political pressure on governments to conclude an effective agreement.

The Fourth Assessment Report of the IPCC, which shared the 2007 Nobel Peace Prize with former US Vice-President Al Gore, has promoted the powerful voice of the scientific community and led to a growing chorus of public support for this urgent call.

In addition, a number of independent studies, such as the report for the British Government by former World Bank Chief Economist Sir Nicholas Stern, have highlighted concerns that the economic and social costs associated with the increasing impacts of climate change will far outweigh the costs of effective mitigation of GHG emissions. In fact, the costs associated with mitigation of climate change seem relatively

Table V.3.3: JI projects in the pipeline

Type	Number	kERUs	2012 kERUs
Afforestation	0	0	0
Agriculture	0	0	0
Biogas	1	115	682
Biomass energy	16	1166	5618
Cement	1	306	1041
CO_2 capture	1	268	1071
Coal bed/mine methane	14	7418	37,088
Energy distribution	7	721	3401
EE households	0	0	0
EE industry	10	3207	15,252
EE own generation	1	1557	7787
EE service	0	0	0
EE supply side	13	2692	10,979
Fossil fuel switch	8	1912	9499
Fugitive	32	19,533	91,308
Geothermal	0	0	0
HFCs	2	1577	5789
Hydro	5	259	1325
Landfill gas	13	2088	10,226
N_2O	14	11,883	53,422
PFCs	1	233	1165
Reforestation	0	0	0
Solar	0	0	0
Tidal	0	0	0
Transport	0	0	0
Wind	10	1492	6314
Total	149	56,427	261,967

Source: UNEP Risø DTU, CDM/JI Pipeline Analysis and Database, available at http://cdmpipeline.org/cdm-projects-type.htm

small when viewed on a global basis over the next several decades, and in addition yield many potential economic, social and human health benefits.

However, despite the obvious conclusion that early action is required, questions of who does what, when and within what framework present political difficulties for government negotiators faced with the large task of, in effect, reshaping the global economy without

either a clear mandate as to how that should be achieved or unambiguous backing by all governments involved.

In the autumn of 2005, anticipation grew over a 'showdown' in Montreal over the future of the Kyoto Protocol, which had just (finally) entered into force in February of that year. The future of the global regime, which was in large part designed around a US-driven demand for legally binding emissions reductions obligations driving a global carbon market, and which was in fact designed to accommodate the US as a large buyer of credits, was seriously jeopardized by a change in Bush Administration policy in early 2001 and its subsequent argument that a global regime, particularly a *binding* global regime, was neither necessary nor desirable. However, the intervening years saw an uneasy but effective alliance between the EU and key developing countries to get the regime established, ratified and finally operational for the first commitment period, 2008–2012.

Article 3.9 of the Kyoto Protocol states that:

commitments for subsequent periods for Parties included in Annex I shall be established in amendments to Annex B to this Protocol. The Conference of the Parties serving as the meeting of the Parties to this Protocol shall initiate the consideration of such commitments at least seven years before the end of the first commitment period mentioned in Paragraph 7 above.

As a result, just as the Protocol was becoming operational, countries had to establish a process for negotiating targets for the second commitment period, which would start following the expiration of the first commitment period in 2012.

Most countries and many experts entirely discounted the possibility that the Kyoto Protocol signatories would in fact agree to move forward with these negotiations. The US in particular was hostile to any such negotiations, stating over and over again that the Kyoto Protocol was 'fatally flawed'. However, thanks to both the resolve of the majority of countries to move forward with global climate protection and the resolve of US civil society and business organisations, as well as the skilful leadership of the Canadian Presidency, in December 2005 the Montreal COP agreed to move forward on negotiations for a second commitment period as specified in the treaty. The US

delegation at first refused to participate in the talks, but at last came back to the table and agreed to proceed with negotiations. This major reversal marked the beginning of a new phase of the international climate negotiations.

Over the next two years, the negotiations proceeded on two parallel tracks: the Kyoto Protocol track mentioned above and the so-called 'dialogue' under the Convention, made up by a series of workshops covering a broad range of topics but with no formal relationship to the negotiations. In December 2007 at the 13th COP in Bali, countries achieved:

- an agreement to launch negotiations under the Convention to replace the 'dialogue';
- an agenda and process for conducting those negotiations; and
- an end date for the negotiations of COP 15 of December 2009 in Copenhagen.

The critical pieces of the negotiation process for the future regime will be conducted primarily under the following three processes:

1. The Ad Hoc Working Group on Long-Term Cooperative Action under the Convention (AWGLCAC – now shortened to AWG-LCA): this group was newly established at the COP in Bali, with the aim of creating a framework in which the US will negotiate until there is a new administration in place at the beginning of 2009; and in which China, India, Brazil and South Africa (and Japan to some extent) will negotiate until the US is fully engaged.
2. The Ad Hoc Working Group on Further Commitments for Annex I Parties under the Kyoto Protocol (AWG – now called the AWG-KP): this is the main ongoing working group of the Parties to the Kyoto Protocol to consider further commitments by Annex I Parties for the period beyond 2012. The aim is to ensure that no gap arises between the first and the second commitment periods. AWG-KP reports to the COP/MOP at each session on the progress of its work.

3. The Second Review of the Kyoto Protocol pursuant to its Article 9: the Article 9 review is where the formal consideration of the evaluation and potential improvements to the Kyoto Protocol takes place. This would be the major point of review for the entire climate regime if the US had ratified the Kyoto Protocol along with other countries, but in the current circumstances it has been difficult to get agreement on moving forward on this.

These bodies/processes will all feed into the formal COPs (at the end of each year), where all the final political decisions will be taken, either at the COP (Convention) or the COP/MOP (Kyoto Protocol) level, and this will be where all the pieces will have to be put together into a coherent whole.

For the wind sector, the outcomes of these negotiations are critical on a number of key points:

- the rigour of the emission reduction targets;
- the resultant future price of carbon;
- technology transfer agreements that actually work; and
- an expanded carbon market.

THE NEED FOR STRONG COMMITMENTS

The driver of the global climate regime must be rigorous, legally binding emission reduction targets for an increasing number of countries under the Kyoto Protocol or its successor agreement. Rigorous emission reduction targets for industrialised countries will send the most important political and market signal that governments are serious about creating a framework for moving towards a sustainable energy future. The indicative range of targets for Annex I countries agreed to by the Kyoto Protocol countries at Bali of CO_2 reductions of 25–40 per cent below 1990 levels by 2020 is a good starting point, although they would need to be closer to the upper end of that range to stay in line with the EU's stated policy objective of keeping global mean temperature rise to less than 2°C above pre-industrial levels.

CARBON PRICES

In addition to achieving climate protection goals, strong emission reduction targets are necessary to bolster the price of carbon on emerging carbon markets, and the regime needs to be broadened so that we move towards a single global carbon market, with the maximum amount of liquidity to achieve the maximum emission reductions at the least cost. While the EU ETS and the CDM are the two major segments of the market, and are growing enormously, they need to be broadened and deepened until they are truly global and the market is able to 'find' the right price for carbon. Achieving that objective must not be at the cost of the integrity of the system, and it will take significant experimentation and time; but it must be clear that that is the final objective, and that governments are agreed to sending the market a signal that the global economy needs to be largely decarbonised by 2050, and effectively completely decarbonised by the end of the century.

TECHNOLOGY TRANSFER

One of the fundamental building blocks of the UNFCCC when it was agreed in 1992 was the commitment by industrialised countries to provide for the development and transfer of climate-friendly technologies to developing countries. While a noble statement of intent at the time, when the world was contemplating how to spend the 'peace dividend' resulting from the end of the Cold War, reality has turned out somewhat differently. For the most part, governments do not own technology (other than military) and are therefore not in a position to 'transfer' it, even if they were in a financial position to do so, which most are not.

However, in the meantime, through economic globalisation, enormous quantities and varieties of technologies have been 'transferred' through direct and indirect foreign investment, world trade and a variety of means used by the private sector as the economy has become increasingly global. As a result, the abstract government and academic debate about technology transfer in its current form has at times become dated, as it no longer reflects the economic reality of today.

Having said that, the political and moral obligations on the part of industrialised countries to deliver on this promise do exist. Developing countries, rightly, are not slow to remind their industrialised country negotiating partners of that fact, nor that reaching some resolution of this subject will be a key part of a post-2012 agreement.

The development and dissemination of technology is a complex subject which varies widely from sector to sector and from country to country. In the first instance, it is useful to distinguish between three major categories in this discussion:

1. the dissemination of existing climate-friendly technology;
2. research and development and deployment of new technologies; and
3. the transfer, on a grant or concessional basis, of both mitigation and climate adaptation technologies to least developed countries and small island states.

Furthermore, it is necessary to define technology transfer activities under the UNFCCC framework and those which can be supported by public funds which would be established internationally as part of the post-2012 negotiation. From the wind industry perspective, category (1) above is most relevant, and the correct division of labours between government and the private sector in that area is of key importance. If these parameters were clear, it is possible that a useful role for the UN system on this subject might be devised.

EXPANDED CARBON MARKETS

The global climate regime can only be aided by the expansion and integration of the emerging carbon markets. Larger markets lead to more liquidity, which in turn results in more active markets and a greater likelihood of finding the 'right price' for carbon given the overall objectives to reduce emissions in the most cost-efficient fashion. However, markets are by definition imperfect, and require substantial and rigorous regulation to function effectively towards their stated goals.

In pursuit of the final objective of a global, seamless carbon market, there are a number of steps that can be taken. First and foremost, it is essential that the US, as the world's largest CO_2 polluter, joins the global carbon market, which was in fact designed largely at the instigation of the US and with the expectation that the US would be the major 'buyer' on the global market.

Second, the membership of Annex B needs to be expanded to include those countries which have recently joined the OECD and those whose economies have grown to reach or even exceed OECD or EU average income per capita. And third, there are many proposals under discussion for improving the scope and the effectiveness of the CDM in the period after 2012.

A SECTORAL APPROACH FOR THE POWER SECTOR

A sectoral approach has been proposed as one way to reform the CDM. A sectoral approach could also avoid the counterfactual and hypothetical questions of additionality at a project-by-project level. The concept was further developed into a broader discussion of using sectoral approaches to engage developing countries more fully in the post-2012 regime. The project-based approach does not satisfy the requirements for achieving rigorous measures to create the 'significant deviation' from baseline emissions growth in rapidly industrialising countries that models show are required to achieve an emissions pathway consistent with rigorous climate protection targets.

To ensure the maximum uptake of emissions-reducing technology for the power generation sector, the Global Wind Energy Council (GWEC) and others are exploring options for a voluntary electricity sector emissions reduction mechanism. The main characteristics of this proposal involve establishing a hypothetical baseline of future emissions in the electricity sector in an industrialising country, quantifying the effect of national policies and measures, and on that basis establishing a 'no regrets' target baseline for the entire electricity sector, which would usually mean a limitation in the growth of emissions in the sector. Reductions in emissions *below* that baseline in the electricity sector would then be eligible to be traded as credits on international carbon markets.

The advantages of this system over the current project-based CDM would be in terms of the simplicity and scope of its operation, encompassing both clean energy production as well as a built-in incentive for energy efficiency, while providing potentially very large sources of investment in the decarbonisation of the energy sector of a rapidly industrialising country. It would also be a good stepping stone between the current situations of non-Annex I countries and their eventual assumption of an economy-wide cap as the regime develops in the future.

V.4 EXTERNALITIES AND WIND COMPARED TO OTHER TECHOLOGIES

Introduction to Externalities

Analyses of the economics of wind energy have shown that it is increasingly competitive with conventional electricity generation technologies. However, in present market conditions the gap towards full competitiveness has to be covered by economic support instruments such as feed-in tariffs and tradable green certificates.

While wind energy and other renewable energy sources have environmental benefits compared with conventional electricity generation, these benefits may not be fully reflected in electricity market prices, despite a fledgling CO_2 Emission Trading Scheme. The question therefore is: 'Do present electricity market prices give an appropriate representation of the full costs to society of producing electricity?'. In other words, are externalities included in the price mechanisms?

The externalities of electricity generation deal with such questions in order to estimate the hidden benefit or damage of electricity generation not otherwise accounted for in the existing pricing system. The costs are real and 'external' because they are paid for by third parties and by future generations, and not directly by the generators or consumers. In order to establish a consistent and fair comparison of the different electricity generation technologies, all costs to society, both internal and external, need to be taken into account.

The following sections of Chapter V.4 explain the basic economic concept of external cost, the policy options to internalise external cost and the present knowledge of the external costs of different electricity generation technologies. Finally, empirical results on specific and total emissions, and on the external cost of fossil fuel-based electricity generation in the EU-27 are presented for the Member State level for 2005. Chapter V.5 continues with quantitative results on the environmental benefits of wind energy in terms of avoided emissions and external costs for different wind deployment scenarios in the EU-27 Member States up to 2020 and 2030.

The Economic Concept of External Effects

DEFINITION AND CLASSIFICATION

The different definitions and interpretations of external costs relate to the principles of welfare economics, which state that economic activities by any party or individual making use of scarce resources cannot be beneficial if they adversely affect the well-being of a third party or individual (see, for example, Jones, 2005).

From this, a generic definition of externalities is 'benefits and costs which arise when the social or economic activities of one group of people have an impact on another, and when the first group fails to fully account for their impacts' (European Commission, 1994).

By definition, externalities are not included in the market pricing calculations, and therefore it can be concluded that private calculations of benefits or costs may differ substantially from society's valuation if substantial external costs occur. Externalities can be classified according to their benefits or costs in two main categories:

1. **environmental and human health externalities:** these can additionally be classified as local, regional or global, referring to climate change caused by emissions of CO_2 or destruction of the ozone layer by emissions of CFCs or SF6; and

2. **non-environmental externalities:** hidden costs, such as those borne by taxpayers in the form of subsidies, research and development costs, or benefits like employment opportunities, although for the last it is debatable whether it constitutes an external benefit in the welfare economics sense.

If an external cost is recognised and charged to a producer, then it is said to have been 'internalised'.

IMPORTANCE OF EXTERNALITIES

By definition, markets do not include external effects or their costs. It is therefore important to identify the external effects of different energy systems and then to monetise the related external costs. It is then possible to compare the external costs with the internal costs of energy, and to compare competing energy systems, such as conventional electricity generation technologies and wind energy.

As markets do not intrinsically internalise external costs, internalisation has to be achieved by adequate policy measures, such as taxes or adjusted electricity rates. Before such measures can be taken, policymakers need to be informed about the existence and the extent of external costs of different energy systems. Analysing external costs is not an easy task. Science (to understand the nature of the impacts) and economics (to value the impacts) must work together to create analytical approaches and methodologies, producing results upon which policymakers can base their decisions for appropriate measures and policies.

Valuation procedures are needed, for example putting a value on a person becoming ill due to pollution, or on visual intrusion caused by a wind turbine, or on future climate change damage caused by a tonne of CO_2. Such evaluations of externalities have uncertainties due to assumptions, risks and moral dilemmas. This sometimes makes it difficult to fully implement the internalisation of externalities by policy measures and instruments (for example emission standards, tradable permits, subsidies, taxes, liability rules and voluntary schemes). Nevertheless, they offer a base for politicians to improve the allocation processes of the energy markets.

Subsequently, the question arises whether the internalisation of externalities in the pricing mechanism could impact on the competitive situation of different electricity generation technologies, fuels or energy sources. As Figure V.4.1 illustrates, a substantial difference in the external costs of two competing electricity generation technologies may result in a situation where the least-cost technology (where only internal costs are considered) may turn out to be the highest-cost solution to society if all costs (internal and external) are taken into account.

PRESENT STATE OF KNOWLEDGE

Serious study of external costs began in the late 1980s, when the first studies were published attempting to quantify and compare the external costs of electricity generation. The most important early studies are listed in the references. These studies seeded public interest in externalities, since they indicated that external costs could be of the same order of magnitude as the direct internal costs of generating electricity. Since that time more research and different approaches, better scientific information, and constant improvement of the analytical methodologies used have advanced the study of externalities, especially in Europe and the US.

This development has resulted in a convergence of methodologies, at least for calculating the external costs of fossil fuel-based electricity generation and

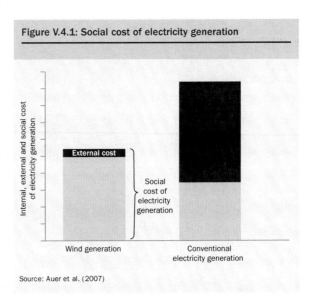

Figure V.4.1: Social cost of electricity generation

Source: Auer et al. (2007)

wind energy. Despite the uncertainties and debates about externalities, it can be stated that, with the exception of nuclear power and long-term impacts of GHGs on climate change, the results of the different research groups converge and can be used as a basis for developing policy measures aimed at a further internalisation of the different external costs of electricity generation.

Externalities of Different Types of Electricity Generation Technologies

PIONEERING STUDIES

The most noted project on determining the external cost of energy is the ExternE (Externalities of Energy) project, which attempted to develop a consistent methodology to assess the externalities of electricity generation technologies. Work and methodologies on the ExternE project are continuously updated (comprehensive details on ExternE are available at www.externe.info).

Prior to the ExternE project, studies were conducted in the late 1980s and beginning of the 1990s that gave an early insight into the importance of externalities for energy policy as a decision-making tool. An overview of the key aspects of these early studies is presented in Appendix I.

The ExternE methodology is a bottom-up approach which first characterises the stages of the fuel cycle of the electricity generation technology in question. Subsequently, the fuel chain burdens are identified. Burdens refer to anything that is, or could be, capable of causing an impact of whatever type. After having identified the burdens, an identification of the potential impacts is achieved, independent of their number, type or size. Every impact is then reported. This process just described for the fuel cycle is known as the 'accounting framework'. For the final analysis, the most significant impacts are selected and only their effects are calculated.

Afterwards, the 'impact pathway' approach developed by ExternE proceeds to establish the effects and spatial distribution of the burdens to see their final impact on health and the environment. Then, the 'economic valuation' assigns the respective costs of the damages induced by each given activity.

The methodology summarised above was implemented in the computer model EcoSense (also within the ExternE project). EcoSense is based on the impact pathway approach and is therefore widely used to assess environmental impacts and the resulting

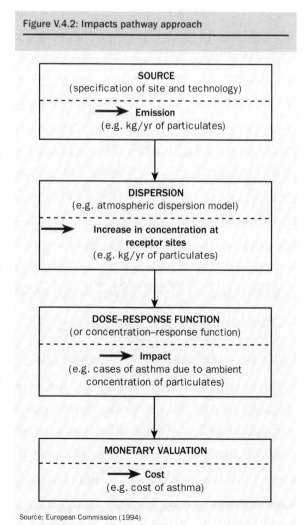

Figure V.4.2: Impacts pathway approach

SOURCE
(specification of site and technology)
— — — — — — — — — — — — — — — —
→ **Emission**
(e.g. kg/yr of particulates)

↓

DISPERSION
(e.g. atmospheric dispersion model)
— — — — — — — — — — — — — — — —
→ **Increase in concentration at receptor sites**
(e.g. kg/yr of particulates)

↓

DOSE–RESPONSE FUNCTION
(or concentration–response function)
— — — — — — — — — — — — — — — —
→ **Impact**
(e.g. cases of asthma due to ambient concentration of particulates)

↓

MONETARY VALUATION
— — — — — — — — — — — — — — — —
→ **Cost**
(e.g. cost of asthma)

Source: European Commission (1994)

external costs of electricity generation technologies. Moreover, EcoSense provides the relevant data and models required for an integrated impact assessment related to airborne pollutants.

The modelling approach of EcoSense is briefly summarised in 'Methodology for the calculation of external costs of different electricity generation technologies based on the EcoSense Model' below, where the different steps for the determination of empirical results of external costs of electricity generation in the EU-27 Member States are presented. It is important to note that the EcoSense model not only includes the external costs caused by conventional electricity generation in its own country but also models the pathway of emissions from conventional power plants to the different receptors (humans, animals, plants, crops, materials and so on) all over Europe (in other words including those located thousands of kilometres outside an EU Member State). The aspect that emissions from one country pass to other countries, and, especially for climate change, to the whole world is essential to derive robust results. The objective of the EcoSense model, however, is to model cross-border effects in Europe only, and not on a global scale.

Because air pollutants can damage a number of different receptors (humans, animals, plants and so on), the task of analysing the impacts of any given emission is complex. Moreover, the final values of external effects and external costs vary between different countries and regions, since specific peculiarities from every country have an influence on the results due to a different range of technologies, fuels and pollution abatement options as well as locations.

In general, the fossil fuel cycle of electricity generation demonstrates the highest values on external effects and external cost (coal, lignite, peat, oil and gas), of which gas is the least damaging. In the ExternE studies, nuclear and renewable energy show the lowest externalities or damages.

FUEL CYCLE OF ELECTRICITY GENERATION TECHNOLOGY

In almost all studies to date, the fossil fuel cycles of electricity generation are associated with higher external costs than nuclear and renewable energies. An exception are the studies undertaken by Hohmeyer (1988) and Ottinger et al. (1990), which also show significant external costs of nuclear energy:

- For the *fossil fuel cycles*, earlier studies derived the impacts of emissions from regional and national statistics as a base for the economic valuation of the damage (top-down approach). In contrast, the more recent studies made use of the damage function approach, in which emissions of a pollutant are site-specifically quantified and their dispersion in the environment modelled to quantify the impact through dose–response functions. Finally, a monetary value is assigned to the impact (bottom-up approach). The emissions, concentrations and impacts of earlier studies are greater than those for recent studies, leading also to diverse results. For instance, atmospheric sulphur oxides (SO_x), nitrogen oxides (NO_x), total suspended particles (TSPs), and carbon dioxide (CO_2) are greater in earlier studies, thereby, results for associated health effects are larger.

- In the case of *nuclear power*, the assessment of severe accidents is the major focus of the analyses. Factors contributing to result variation are risk perception, resource depletion, and public spending on research and development. Hohmeyer (1998) and Ottinger et al. (1990), in contrast to the other studies, used data from the Chernobyl accident as the basis for their external cost analysis from severe reactor accidents. Generally, all studies conclude that the issue of the public's perception of the risks of nuclear power remains unresolved. In conclusion, the weakest points of externality studies of electricity generation so far have been that in almost all studies it is assumed that (i) in the nuclear cycle waste and other hazardous impacts are well managed and (ii) the problem of accidents (for example severe core meltdown accidents with containment rupture) and their disastrous effects for society are not addressed accordingly and/or are completely neglected.

- For *renewable energies*, the external costs are usually lowest among all energy generation technologies. However, the use of hydro power can have significant external effects as it can impact high-value ecosystems and adjacent populations. External effects from wind energy, such as noise creation and visual impacts, can also be significant in certain areas (for a detailed discussion, see earlier chapters of this volume and 'Avoided emissions' below).

EMISSIONS OF FOSSIL FUEL-BASED ELECTRICITY GENERATION

The most important emissions concerning electricity generation are of CO_2, SO_2, NO_x and PM_{10} (particulate matter up to 10 micrometres in size). Emissions generally depend on the type of fuel used:

- CO_2 emissions are related to carbon content. There is no realistic opportunity of reducing such carbon dioxide emissions by using filters or scrubbers, although techniques such as burning fossil fuel with pure oxygen and capturing and storing the exhaust gas may reduce the carbon content of emissions. Carbon (dioxide) capture is the only possibility.
- For SO_2, the quantity of emissions per kWh electricity generated depends on the sulphur content of the input fuel. Furthermore, SO_2 emissions can be reduced by filtering the exhaust gases and converting SO_2 to gypsum or elementary sulphur. In general, the sulphur content of lignite is relatively high, fuel oil and hard coal have a medium sulphur content, and natural gas is nearly sulphur-free.
- In contrast, NO_x emissions are practically unrelated to input fuel. As NO_x gases are formed from the

nitrogen in air during combustion, their formation depends mainly upon the combustion temperature. Thus NO_x emissions can be reduced by choosing a favourable (low) combustion temperature or by denitrifying the exhaust gases (by wet scrubbing).

Benefits of Wind Energy under the Consideration of External Cost

AVOIDED EMISSIONS

In general, the benefits of wind energy are avoided emissions and avoided external costs as compared with conventional, mainly fossil fuel-based, electricity generation. Figure V.4.1 (comparison of social costs of different electricity generation technologies) indicates that a kWh of wind energy (as for renewable energy in general) presents a negligible external cost in comparison with fossil fuel-based power systems. This fact illustrates the social and environmental advantages of wind energy and other renewables over conventional energy systems. Consequently, it is desirable to increase wind energy and other renewables in the electricity supply systems.

In recent years, the implementation of a variety of different renewable promotion instruments in Europe has resulted in significant amounts of renewable electricity generation, particularly wind generation. Without this, the corresponding amount of electricity generation would have been from conventional power plants. This means that renewable electricity generation has already displaced conventional electricity generation technologies and, subsequently, avoided significant amounts of emissions. Therefore, the external costs of total electricity generation have decreased as compared with the situation without any renewable electricity generation.

In the empirical analyses of avoided emissions and external costs in the EU-27 Member States (see 'The

EcoSense computer model' below), country-specific results are presented according to the quantity of external costs that have been already avoided due to wind generation in the different EU-27 Member States. In Chapter V.5, the avoided emissions and avoided external costs for different scenarios of wind deployment in the electricity systems of the EU-27 Member States up to 2020 and 2030 are presented.

EXTERNALITIES OF WIND ENERGY

Although wind energy is a clean technology, mainly due to the avoidance of air-pollutant emissions, it is not totally free of impacts on the environment and human health. However, wind energy has very few environmental impacts in its operation. The most commonly discussed impacts on people are acoustic noise and visual intrusion. Visual intrusion of the turbines and ancillary systems in the landscape and noise are considered as amenity impacts of the technology. Other impacts include indirect pollution from the production of components and construction of the turbine; the collision of birds in flight with turbines and bird behavioural disturbance from blade avoidance; the impacts of wind turbine construction on terrestrial ecosystems; and accidents affecting workers in manufacturing, construction and operation. A comprehensive overview and discussion of these kinds of wind energy externalities is conducted in previous sections of this Part.

Methodology for the Calculation of External Costs of Different Electricity Generation Technologies Based on the EcoSense Model

THE ECOSENSE COMPUTER MODEL

To calculate the external costs of a given conventional power plant portfolio as well as the avoided external costs of wind energy, it is necessary to model the pathway of emissions from conventional power plants to the different receptors, such as humans, animals, plants, crops and materials, which may be located thousands of kilometres away. As air pollutants can damage a number of different receptors, the task of analysing the impacts of any given emission is fairly complex. To allow such complex analysis of external costs, a tool has been developed during the last ten years in a major coordinated EU research effort, the EcoSense Model. The basics of the model are explained below, as used in the calculations of the external costs of electricity generation in the EU-27 Member States in Chapter V.5.

EcoSense is a computer model for assessing environmental impacts and the resulting external costs of electricity generation systems. The model is based on the impact pathway approach of the ExternE project (see www.externe.info as well as 'Pioneering studies' above) and provides the relevant data and models required for an integrated impact assessment related to airborne pollutants (see also European Commission, 1994).

EcoSense provides the wind-rose trajectory model for modelling the atmospheric dispersion of emissions, including the formation of secondary air pollutants. For any given point source of emissions (for example a coal-fired power plant), the resulting changes in the concentration and deposition of primary and secondary pollutants can be estimated on a Europe-wide scale with the help of this model. Developed in the UK by the Harwell Laboratory, it covers a range of several thousand kilometres. The reference environment database, which is included in EcoSense, provides receptor-specific data as well as meteorological information based on the Eurogrid coordinate system.

The impact pathway approach can be divided into four analytical steps (see, for example, Mora and Hohmeyer, 2005):

1. *Calculation of emissions*: the first step is to calculate emissions of CO_2, SO_2 and NO_x per kWh from a specific power plant.

2. *Dispersion modelling*: then air-pollutant dispersion around the site of the specific plant is modelled. Based on meteorological data, changes in the concentration levels of the different pollutants can be calculated across Europe.

3. *Impact analysis*: based on data for different receptors in the areas with significant concentration changes, the impacts of the additional emissions on these receptors can be calculated on the basis of so-called dose–response functions. Important data on receptors included in the model database are, for example, population density and land-use patterns.

4. *Monetisation of costs*: the last step is to monetise the impacts per kWh caused by the specific power plant. In this stage, the calculated physical damage to a receptor is valued on a monetary scale, based on the best available approaches for each type of damage.

INPUT DATA TO THE MODEL

Because the EcoSense model requires a specified site as a starting point for its pollutant dispersion modelling, one typical electricity generation site has been chosen for each country to assess the impacts and to calculate the costs caused by emissions from fossil fuel-fired power plants which may be replaced by wind energy. The coordinates at each site are chosen in order to locate the reference plants centrally in the electricity generating activities of each country. Thus

it is assumed that the chosen site represents approximately the average location of electricity generating activities of each country that has been chosen.

To control for effects caused by this assumption and to prevent extreme data results, a sensitivity analysis was carried out by shifting the geographical location of the plant. This analysis showed a relatively high sensitivity of external costs to the location of the electricity generation facilities. This is due to the very heterogeneous distribution of the different receptors in different parts of a country.

In order to run the model, the capacity of the conventional power plant, its full load hours of operation and the volume stream of exhaust gas per hour are required. The assumptions made for the calculations are shown in Table V.4.1 for the different fossil fuels of conventional electricity generation.

For each country, calculations have been performed for a representative conventional power plant location based on the specific national emission data for each fuel and each pollutant.

The evaluation in the EcoSense Model includes damage from air-pollutant emissions like SO_2 and NO_x (PM_{10} is negligible compared to these) for the major receptors: humans, crops and materials. For each of the pollutants, high, medium and low specific external costs are derived per country.

The costs of the anthropogenic greenhouse effect resulting from CO_2 emissions are not modelled here in the EcoSense Model, but are based on estimates from Azar and Sterner (1996) and Watkiss et al. (2005).

Table V.4.1: Assumed data for the calculation of the reference fossil fuel-based power plant technology

Fuel type	Capacity (MW)	Full load hours per year	Volume stream per hour (m³)
Hard coal	400	5000	1,500,000
Lignite	800	7000	3,000,000
Fuel oil	200	2000	750,000
Natural gas derived gas	200	2000	750,000
Mixed firing not specified	400	5000	1,500,000

Source: Auer et al. (2007)

Three different levels of specific external costs are implemented (high, medium and low) for the empirical calculation of CO_2-related external costs in the different EU Member States.

Summing up: to calculate the external costs of conventional electricity generation technologies and, subsequently, the avoided external costs by the use of wind energy, the external costs resulting from air pollutants such as SO_2 and NO_x (calculated by EcoSense) have to be added to the external costs of the anthropogenic greenhouse effect resulting from CO_2 emissions (not calculated by EcoSense).

DETERMINATION OF AIR-POLLUTING CONVENTIONAL POWER PLANTS AND REPLACEABLE SEGMENT OF CONVENTIONAL ELECTRICITY GENERATION BY WIND ENERGY

This section identifies, as a first step, the different types of conventional power plant responsible for air pollution (and, subsequently, for external costs caused by CO_2, SO_2 and NO_x). Then a methodology is derived for the determination of the replaceable segment of conventional electricity generation technologies by wind energy. Finally, empirical data on annual generation and specific emissions is presented for the portfolio of air-polluting conventional power plants in several of the EU-27 Member States.

Determination of Air-Polluting Conventional Power Plants

In general, the load duration curve of different types of conventional power plants is split into three different segments: base load, intermediate load and peak load. Typical examples of conventional power plants operating in the different load duration segments are:

- base load: nuclear power plants, large run-of-river hydropower plants, lignite power plants, hard coal power plants;

- intermediate load: hard coal power plants, fuel oil-fired power plants, combined cycle gas turbine plants; and
- peak load: pumped-storage hydropower plants, open cycle gas turbine plants.

With respect to above-mentioned air-polluting emissions (CO_2, SO_2 and NO_x), several different kinds of fossil fuel conventional power plants are candidates for further investigation.

Replaceable Segment of Conventional Electricity Generation by Wind Energy

Due to its inherent variability, wind power can at present only replace specific segments of the load duration curve of conventional electricity generation. More precisely, wind energy can replace conventional power plants at the intermediate rather than base-load or peak-load segment.

Keeping this in mind, a reference system can be defined whereby wind energy can be expected to replace conventional power plants:

- First, neither nuclear nor large hydropower plants are replaceable by wind energy, as both almost exclusively operate in the base load segment.
- As pump hydro storage power plants are used to cover very short load peaks, they cannot be replaced by wind energy either, due to the latter's variable nature.
- This leaves conventional electricity generation from the fossil fuels: lignite, hard coal, fuel oil and gas.

However, this assumption can lead to an overestimation of the share of the replaced electricity generation by lignite, as this is predominantly used in the base load segment as well, and to an underestimation of substituted electricity from gas, which, due to the dynamic characteristics of gas-fired power plants, lends itself perfectly to balance fluctuations in wind generation. As the current mode of operation of conventional power plants, the rules of their dispatch based on the so-called 'merit order' and the dynamic behaviour of the different types of conventional power plants is all well known, it can be safely assumed that a replacement of intermediate load by wind energy is the best description of the complexities of actual operation in the 'real world'.

For detailed empirical analyses, the contributions of the different fossil fuel-based power plant technologies to the intermediate load segment need to be specified. As the best statistics in this context available, data for the power plant in Germany (VDEW, 2000) are used as the basis for such analysis:

- The load curves for one typical load day are derived for each relevant type of fuel and are taken as the basis for the calculation of shares of intermediate load.
- In general, the highest load variations during one day are displayed by fuel oil and gas; hard coal shows some variation, while electricity generation based on lignite is almost constant.

Although the characteristics of the load curves are based on the German electricity generation structure, conventional power plants have common fuel-specific technical and economic characteristics. Therefore, load curves are assumed to have similar day-to-day variations in several other European countries. Based on these considerations, Table V.4.2 sets out assumptions for the intermediate load shares, with the percentage figures being based on the total amount of electricity generated for each fuel.

The assumptions for intermediate load shares replaceable by wind energy will remain the same for 2020 and 2030 for the fuel types lignite (10 per cent), hard coal (30 per cent), mixed firing (50 per cent) and fuel oil (100 per cent). For natural/derived gas the replaceable share will decrease from 100 per cent (2007) to 92.5 per cent (2020) and 85 per cent (2030) in those EU Member States with no other flexible power plant technologies like pumped hydro storage power

Table V.4.2: Share of intermediate load of different types of fossil fuel power plant

Fuel type	Share of intermediate load (%)
Lignite	10
Hard coal	30
Mixed firing	50
Fuel oil	100
Natural/derived gas	100

Source: Auer et al. (2007)

plants. Due to significant shares of wind penetration in 2020 and 2030, a certain level of very flexible power plant types like combined cycle gas turbines (CCGTs) is absolutely necessary to balance the electricity systems (see, for example, Auer et al. 2007).

FOSSIL FUEL-BASED ELECTRICITY GENERATION AND ITS SPECIFIC EMISSIONS IN THE EU-27 IN 2007

The structure of total electricity generation (by fuel type) and the structure and fractions of fossil fuel-based electricity generation are presented at the EU-27 Member State level in Figures V.4.3 and V.4.4 respectively. The empirical data is mainly derived from the official 2006 Eurostat Statistics of the European Commission (*EU Energy in Figures – Pocket Book 2007/2008*) and updated for the reference year 2007 with the recent *Platts European Power Plant Data Base 2008*.

Figure V.4.3 indicates that the absolute levels and shares of input fuels for electricity generation vary greatly between the different EU-27 Member States. In particular, Figure V.4.3 indicates those fossil fuel-based electricity generation technologies that may well be replaced by wind energy and/or other renewable energy sources in the future. As already mentioned, neither base-load electricity generation technologies like nuclear and large hydro power nor peak-load technologies like pumped storage hydro

Figure V.4.3: Total electricity generation (by fuel type) in the different EU-27 Member States in 2007

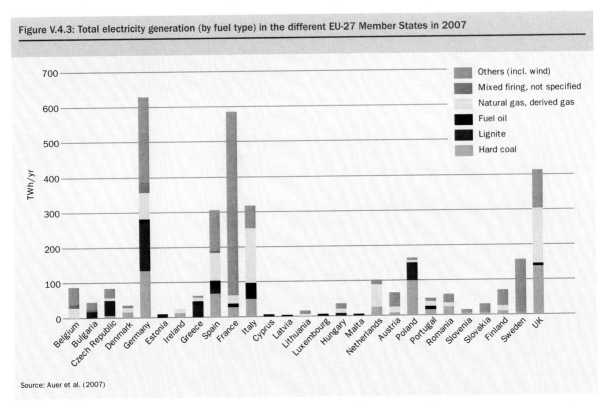

Source: Auer et al. (2007)

Figure V.4.4: Fossil fuel-based electricity generation in the different EU-27 Member States in 2007

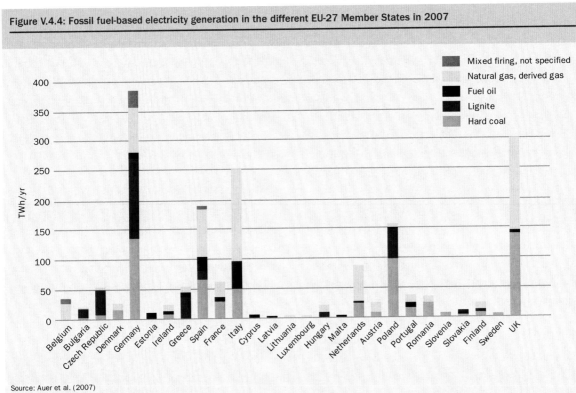

Source: Auer et al. (2007)

power plants are candidates to be replaced by wind energy. Also, several other kinds of renewable electricity generation technologies are incorporated into the category of 'others', since they are not of primary interest in further analyses when discussing fuel substitution by wind energy.

In Figure V.4.4, several fossil fuel-based electricity generation technologies at the EU-27 Member State level are presented in detail. The data presented here is the starting point for comprehensive in-depth analyses in subsequent sections.

Based on the portfolio of fossil fuel-based electricity generation in the different EU-27 Member States in 2007 presented in Figure V.4.4, the corresponding specific air-pollutant emissions (CO_2, SO_2, NO_x and PM_{10}) can be determined at the country level. The corresponding absolute air-pollutant emissions of fossil fuel-based electricity generation are derived from a variety

of different Member States' statistics as well as from official documents of the European Commission.

In this context it is important to note that the corresponding national studies and statistics take into account a variety of country-specific and technology-specific characteristics, for example the age structure of the different power plants (and as a consequence also indirectly the primary fuel efficiency) in each of the EU-27 Member States.

Figure V.4.5 finally presents the results on the average specific CO_2, SO_2 and NO_x emissions from fossil fuel-based electricity generation at the Member State's level for 2007. In principle, the average specific emissions are less in the 'old' EU-15 Member States (except Greece) than in the 'new' EU-12 Member States. Bulgaria, Estonia, Greece, Romania and Slovenia are among the most air-polluting countries within the EU-27, at least in terms of CO_2 emissions.

Figure V.4.5: Specific average emissions (CO_2, SO_2, NO_x) from fossil fuel-based electricity generation in the different EU-27 Member States in 2007

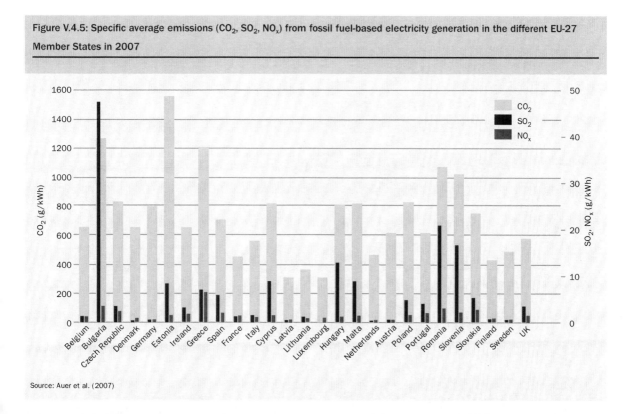

Source: Auer et al. (2007)

Figure V.4.5 indicates that the difference in specific CO_2 emissions from fossil fuel generation is more than a factor of four between various EU-27 Member States. This is related to differences in the fuel mix and because some Member States still have power plants with very low efficiencies. The distribution of SO_2 emissions per kWh is also very different, as shown in Figure V.4.5. This is related to the very heterogeneous sulphur content of fuel and the use of desulphurisation in only the most advanced Member States. Finally, NO_x emissions differ between the countries according to the combustion process used, the combustion temperature, which is not optimal in all the Member States, and the scrubbing technologies employed.

The picture presented in Figure V.4.5 will be further elaborated in subsequent sections when discussing the replaceable/avoidable as well as already avoided shares of fossil fuel-based electricity generation by wind energy (and other non-fossil fuel-based generation technologies) for 2007, 2020 and 2030.

V.5   ENVIRONMENTAL BENEFITS OF WIND ENERGY IN COMPARISON TO REMAINING ELECTRICITY GENERATION TECHNOLOGIES

Electricity Generation, Emissions and External Cost in the EU-27 Countries in 2007

AMOUNT OF FOSSIL FUEL-BASED ELECTRICITY GENERATION AVOIDABLE/REPLACEABLE BY WIND (AND OTHER RENEWABLE ELECTRICITY GENERATION TECHNOLOGIES) IN 2007

In general, the benefits of wind energy are the avoided emissions and external costs from fossil fuel-based electricity generation. The evaluation of external costs includes damage from:
- air-pollutant emissions;
- the anthropogenic greenhouse effect resulting from CO_2 and other emissions; and
- SO_2 and NO_x.

To analyse the environmental and health benefits of the use of wind energy, we need to know the specific emissions of fossil fuel-based electricity generation replaced thereby. These can be derived by dividing the absolute emissions produced by a type of fossil fuel in kilotonnes of CO_2 per year used for electricity generation in a country by the amount of electricity generated from this fuel in kWh per year.

In our model, wind energy only replaces intermediate load of conventional fossil fuel-based electricity generation. In general, the emissions avoided by wind energy depend on three factors:
1. the specific emissions from each type of fossil fuel-based electricity generation facility;
2. the fuel mix in each country; and
3. the percentage of each fuel replaced by wind energy.

Figure V.5.1 presents the absolute levels and shares of fossil fuel-based electricity generation replaceable/avoidable by wind energy (and other renewable electricity generation technologies) in each of the EU-27 Member States according to the individual replaceable shares of fossil fuels in the intermediate load segment comprehensively discussed in the previous chapter (Table V.4.2).

Derived from Figure V.5.1, Figure V.5.2 presents the total emissions (CO_2, SO_2, NO_x) replaceable/avoidable by wind (and other renewable electricity generation technologies) in the EU-27 Member States in 2007.

AMOUNT OF FOSSIL FUEL-BASED ELECTRICITY GENERATION REPLACED/AVOIDED BY WIND IN 2007

The previous section presents the empirical data on the replaceable/avoidable amount of fossil fuel-based electricity generation. The following section shows the amount of already replaced/avoided fossil fuel-based electricity generation by wind energy in the EU-27 Member States in 2007. The annual wind generation in each of the EU-27 Member States in 2007 has to be studied first (see Figure V.5.3).

Figure V.5.3 clearly indicates that already a significant number of EU Member States have implemented a considerable amount of wind energy in 2007. On top of the list are Germany and Spain (around 39 TWh per year) annual wind generation each); Denmark (8 TWh per year) and the UK (6.3 TWh per year) are next; and other EU Member States like Portugal, Italy, The Netherlands, and France are aiming at the 5 TWh per year benchmark of annual wind generation very fast. However, there still existed many EU-27 Member States with negligible wind penetration in 2007.

The total CO_2, SO_2 and NO_x emissions from fossil fuel-based electricity generation having been already avoided by wind energy in the different EU-27 Member States in 2007 are presented in Figure V.5.4.

Figure V.5.1: Fossil fuel-based electricity generation replaceable/avoidable by wind (and other renewable electricity generation technologies) in the EU-27 Member States in 2007

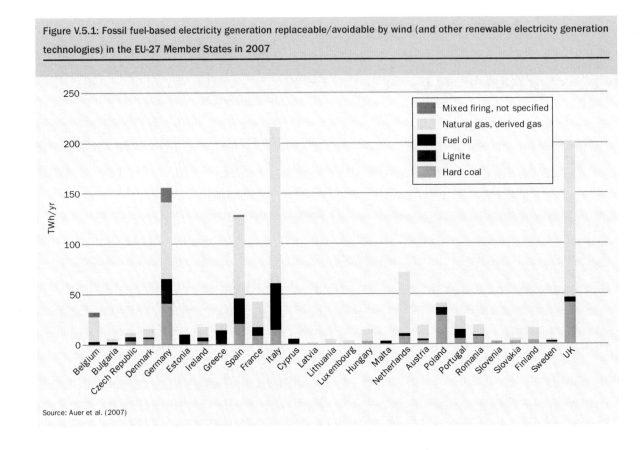

Source: Auer et al. (2007)

The 2007 results in Figure V.5.4 take into account the individual characteristics of conventional electricity generation at the country level (for example age structure and efficiency of the fossil fuel power plants) in terms of the specific average emissions (CO_2, SO_2, NO_x) from fuel-based electricity generation on the one hand (see Figure V.4.5), and annual wind generation in 2007, on the other hand (see Figure V.5.3). Not surprisingly, the total avoided emissions in 2007 perfectly correlate with annual wind generation in the different EU-27 Member States.

EXTERNAL COSTS OF FOSSIL FUEL-BASED ELECTRICITY GENERATION AND AVOIDED EXTERNAL COST BY WIND IN 2007

So far, empirical results have been presented for each of the EU-27 Member States on fossil fuel-based

electricity generation replaceable/avoidable by wind energy (and other renewable generation technologies) in each of the EU-27 Member States in 2007. The factors involved are:

- average specific emissions (CO_2, SO_2, NO_x) from fossil fuel-based electricity generation in each of the EU-27 Member States in 2007;
- total wind generation in each of the EU-27 Member States in 2007; and
- total emissions (CO_2, SO_2, NO_x) already avoided from fossil fuel-based electricity generation in each of the EU-27 Member States in 2007.

These analyses provide the basis for the final step to determine the external costs of fossil fuel-based electricity generation and the already avoided external costs from wind generation in the EU-27 Member States in 2007.

Figure V.5.2: Total emissions (CO_2, SO_2, NO_x) replaceable/avoidable by wind (and other renewable electricity generation technologies) in the EU-27 Member States in 2007

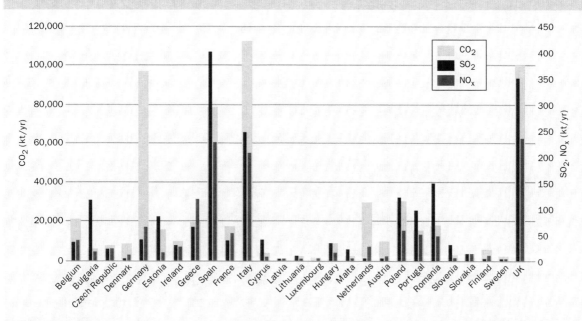

Source: Auer et al. (2007)

Figure V.5.3: Annual wind generation in the EU-27 Member States in 2007

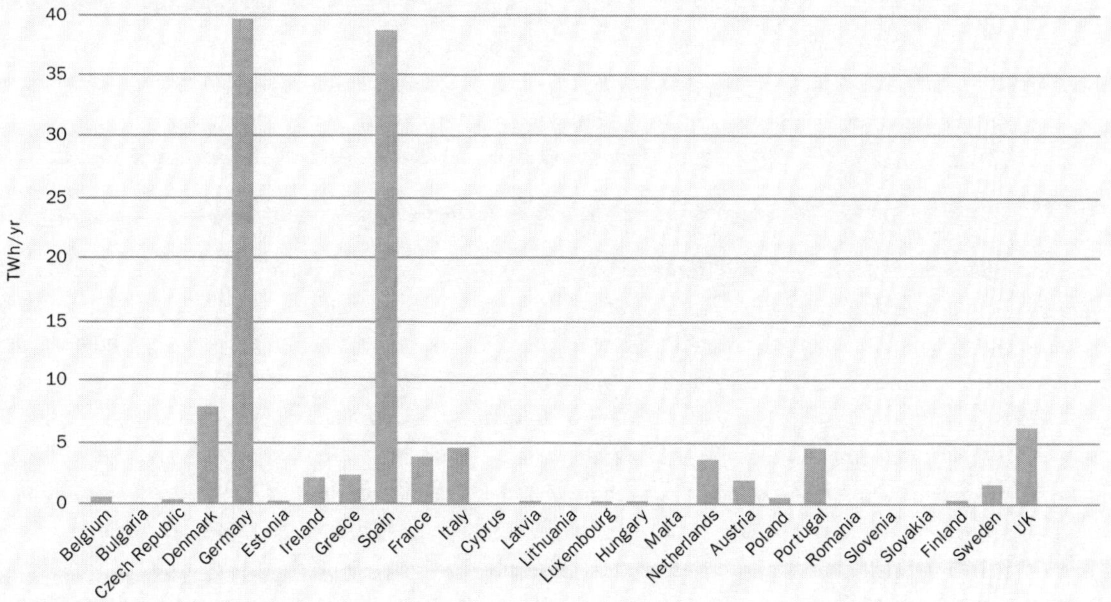

Source: Auer et al. (2007)

Figure V.5.4: Total emissions (CO_2, SO_2, NO_x) from fossil fuel-based electricity generation already avoided by wind energy in the EU-27 Member States in 2007

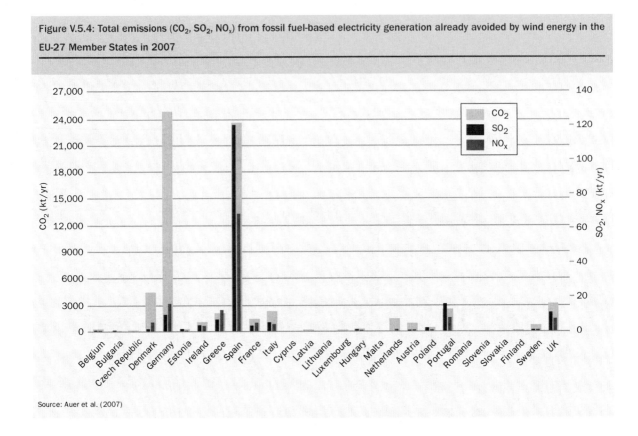

Source: Auer et al. (2007)

The external costs resulting from air pollutants such as SO_2 and NO_x (calculated by EcoSense; see page 370) have to be added to the external costs of the anthropogenic greenhouse effect resulting from CO_2 emissions (not calculated by EcoSense, but based on estimates by Azar and Sterner, 1996, and Watkiss et al., 2005).

Because air pollutants can damage a large number of different receptors, calculations of external costs will generally include a large number of types of damage, which tend to be restricted to the most important impacts to allow a calculation of external costs with a limited resource input. At present, EcoSense includes the following receptors: humans (health), crops, materials (in buildings and so on), forests and ecosystems, with monetary valuation only included for human health, crops and materials. For each of these a bandwidth (high, medium and low values) is determined.

There are two approaches to evaluating effects on human health: value of statistical life (VSL) and years of life lost (YOLL).

- The VSL approach measures a society's willingness to pay to avoid additional deaths.
- The YOLL approach takes human age into account. For each year of life lost approximately one-twentieth of the VSL value is used.

Unfortunately, outputs from the EcoSense Model used in this analysis do not provide a calculation based on the VSL approach. As pointed out above, VSL may lead to substantially larger external costs than the YOLL approach which is applied by the EcoSense Model. Results of former ExternE studies estimate external costs based on both approaches. These resulted in VSL results of approximately three times more than with YOLL. As the present version of

EcoSense does not calculate VSL values, the EcoSense results on human health effects based on the YOLL approach have been scaled. This has been done with a factor of one for low-damage cost estimates calculated for human health, a factor of two for medium cost estimates and a factor of three for high estimates.

Figure V.5.5 finally presents the results of the external costs of conventional fossil fuel-based electricity generation in each of the EU-27 Member States in 2007 (high/average/low values). Similar to the specific emissions of fossil fuel-based electricity generation presented in Figure V.4.5, there is a noticeable difference between external costs in different EU-27 Member States. Bulgaria, Romania and Slovenia are the Member States with the highest external costs of fossil fuel-based electricity generation (average values 20–25c€$_{2007}$/kWh), but Estonia and Greece also

reach nearly 20c€$_{2007}$/kWh (average values for external costs). Latvia, Lithuania, Luxembourg, Finland, Sweden and The Netherlands are characterised by external costs of fossil fuel-based electricity generation below 5c€$_{2007}$/kWh (average values for external costs).

By combining the avoidable external costs of fossil fuel-based electricity generation with the amount of electricity produced by wind energy, the total amount of already-avoided external costs can be calculated for 2007. Figure V.5.6 presents the corresponding results of already-avoided external costs by wind generation in each of the EU-27 Member States.

In 2007, in the EU-27 region around €$_{2007}$10.2 billion on external costs have been avoided by wind generation in total (summing up the average values in each of the EU-27 Member States shown in Figures V.5.6 and V.5.7).

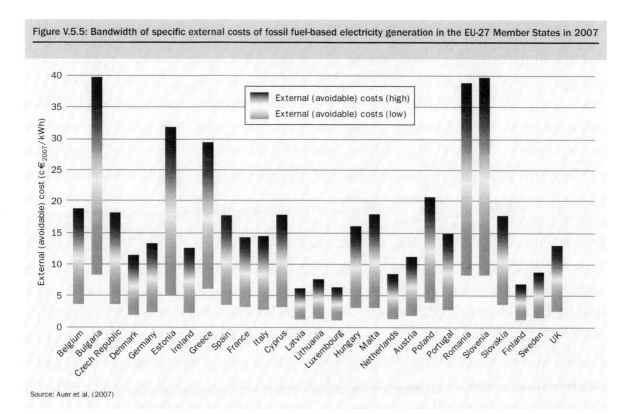

Figure V.5.5: Bandwidth of specific external costs of fossil fuel-based electricity generation in the EU-27 Member States in 2007

Source: Auer et al. (2007)

Figure V.5.6: Bandwidth of avoided external costs of fossil fuel-based electricity generation in the EU-27 Member States in 2007

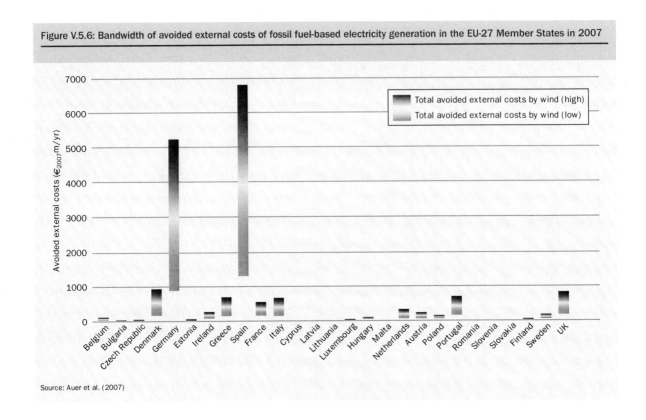

Source: Auer et al. (2007)

The following EU Member States are mainly responsible for the majority shares of this already impressive number: Spain (€$_{2007}$3.968 billion), Germany (€$_{2007}$3.027 billion), Denmark (€$_{2007}$0.518 billion), the UK (€$_{2007}$0.472 billion), Greece (€$_{2007}$0.400 billion), Portugal (€$_{2007}$0.388 billion) and Italy (€$_{2007}$0.377 billion).

Figure V.5.7 finally presents the absolute values of already avoided external costs by wind generation in each of the EU-27 Member States in 2007.

Avoided Emissions and External Cost for Different Wind Deployment Scenarios in the EU-27 Member States in 2020

The previous section presented 'real' life in 2007. In the following section different scenarios on the portfolio of electricity generation in the EU-27 Member States in 2020 are discussed and, subsequently, the same analyses are conducted in terms of:

- determination of the share of fossil fuel-based electricity generation and corresponding emissions in each of the EU-27 Member States;
- determination of the amount of fossil fuel-based electricity generation and corresponding emissions replaceable/avoidable by wind (and other renewable technologies) in each of the EU-27 Member States;
- determination of the replaced/avoided emissions by wind energy in the three EWEA wind generation scenarios; and
- determination of the external costs of fossil fuel-based electricity generation and, subsequently, avoided external costs by wind generation in the three EWEA wind generation scenarios.

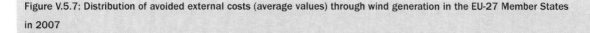

Figure V.5.7: Distribution of avoided external costs (average values) through wind generation in the EU-27 Member States in 2007

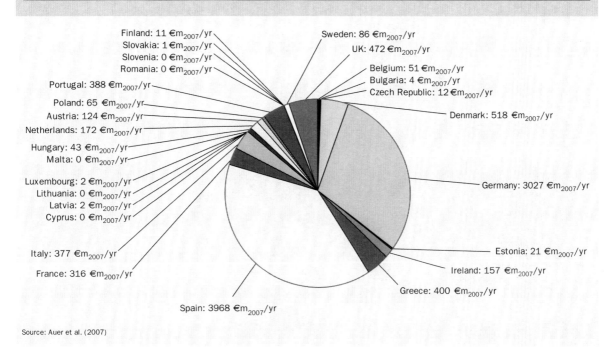

Source: Auer et al. (2007)

Before presenting the empirical results, there are at least the following three important points worth mentioning:

1. The business-as-usual scenarios in the portfolio of conventional electricity generation are based on the official documents of the European Commission (Eurelectric, 2006; Capros et al., 2008).

2. In general, the efficiency of new plants within each of the types of fossil fuel-based electricity generation technologies improves with time and, therefore, the specific emissions for 2020 decrease compared to 2007.

3. Due to expected electricity demand increase, the total amount of fossil fuel-based electricity generation in 2020 is supposed to be higher than in 2007 in almost all EU Member States (but specific emissions per power plant technology – see above – will be lower). However, the significant shares of wind

generation in the different EWEA scenarios are expected to be even greater (see subsequent sections for details).

FOSSIL FUEL-BASED ELECTRICITY GENERATION AND EMISSIONS IN 2020

Figure V.5.8 presents fossil fuel-based electricity generation in the EU-27 Member States in 2020 and Figure V.5.9 the corresponding specific average emissions (CO_2, SO_2, NO_x).

Figure V.5.10 presents the fossil fuel-based electricity generation replaceable/avoidable by wind (and other renewable electricity generation) in the EU27 Member States in 2020 and Figure V.5.11 the corresponding amount of total avoidable emissions (CO_2, SO_2, NO_x).

Figure V.5.8: Fossil fuel-based electricity generation in the EU-27 Member States in 2020

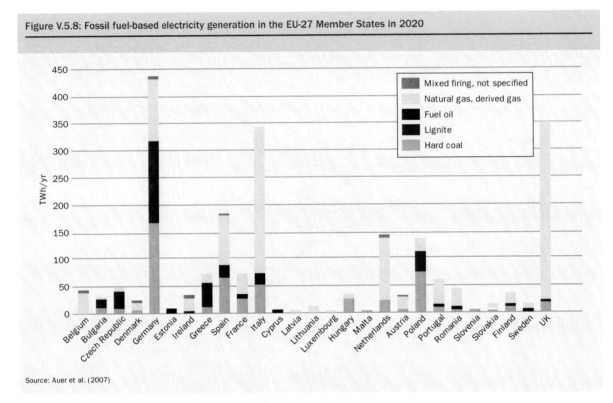

Source: Auer et al. (2007)

Figure V.5.9: Specific average emissions (CO_2, SO_2, NO_x) from fuel-based electricity generation in the EU-27 Member States in 2020

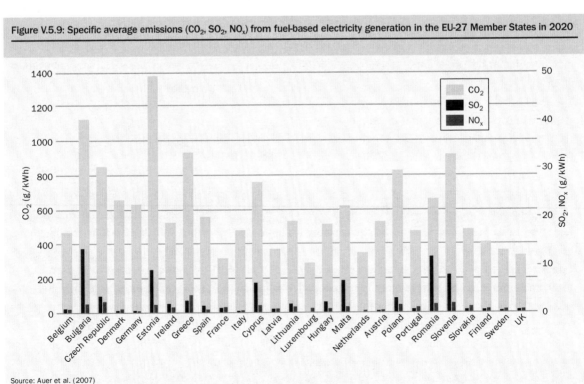

Source: Auer et al. (2007)

Figure V.5.10: Fossil fuel-based electricity generation replaceable/avoidable by wind (and other renewable electricity generation technologies) in the EU-27 Member States in 2020

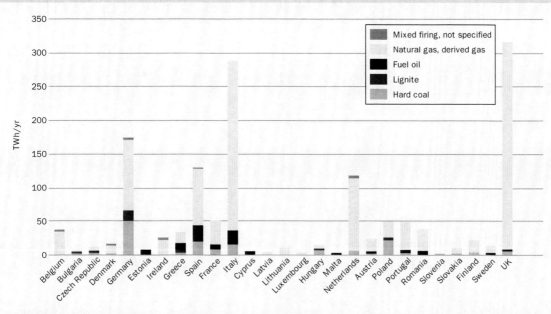

Source: Auer et al. (2007)

Figure V.5.11: Total emissions (CO_2, SO_2, NO_x) replaceable/avoidable by wind (and other renewable electricity generation technologies) in the EU-27 Member States in 2020

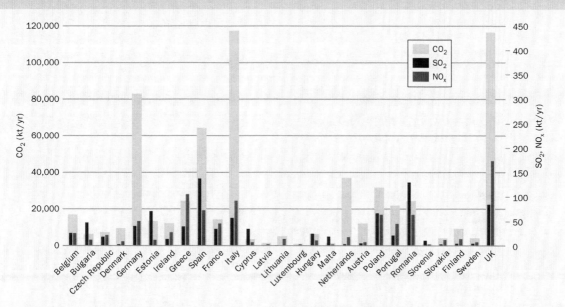

Source: Auer et al. (2007)

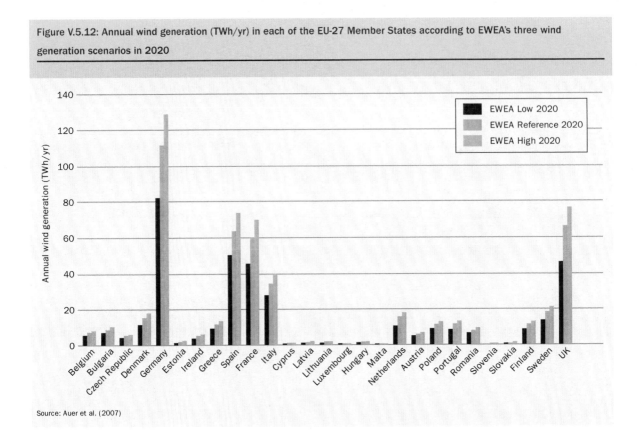

Figure V.5.12: Annual wind generation (TWh/yr) in each of the EU-27 Member States according to EWEA's three wind generation scenarios in 2020

Source: Auer et al. (2007)

BREAKDOWN OF EWEA'S WIND GENERATION SCENARIOS FOR 2020 (BY EU MEMBER STATE)

In order to be able to calculate the amount of replaced/avoided fossil fuel-based electricity generation by wind energy in the EU-27 Member States in 2020, wind penetration scenarios for 2020 are necessary. Figure V.5.12 presents EWEA's three wind generation scenarios for the EU-27 on a separate – EU Member State – level. The breakdown of EWEA's three wind generation scenarios at the EU Member State level is mainly based on comprehensive modelling and sensitivity analyses with the simulation software model GreenNet-Europe (see www.greennet-europe.org). GreenNet-Europe models show cost deployment of renewable electricity generation technologies (wind in particular) at EU Member State level up to 2020 and 2030, taking into account

several different country-specific potentials and cost of renewable (wind) generation, different renewable-promotion instruments, and a variety of other country-specific as well as general parameters and settings. The results of the breakdown of EWEA's three wind generation scenarios have also been cross-checked with other existing publications (for example, Resch et al., 2008, and Capros et al., 2008).

AVOIDED EMISSIONS (OF FOSSIL FUEL-BASED ELECTRICITY GENERATION) IN THE BREAKDOWN OF EWEA'S WIND GENERATION SCENARIOS FOR 2020

In Figures V.5.13–V.5.15, the total avoided emissions (CO_2, SO_2, NO_x) by wind generation are presented for EWEA's three wind generation scenarios in each of the EU-27 Member States in 2020.

Figure V.5.13: Total emissions (CO_2, SO_2, NO_x) from fossil fuel-based electricity generation avoided by wind energy according to EWEA's Reference Scenario in the EU-27 Member States in 2020

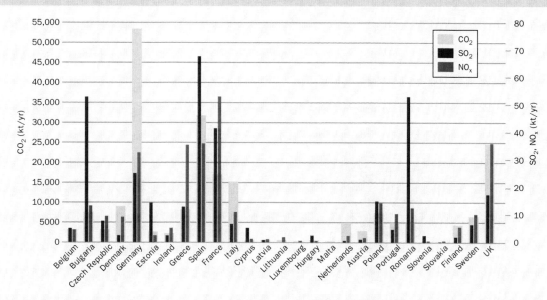

Source: Auer et al. (2007)

Figure V.5.14: Total emissions (CO_2, SO_2, NO_x) from fossil fuel-based electricity generation avoided by wind energy according to EWEA's High Scenario in the EU-27 Member States in 2020

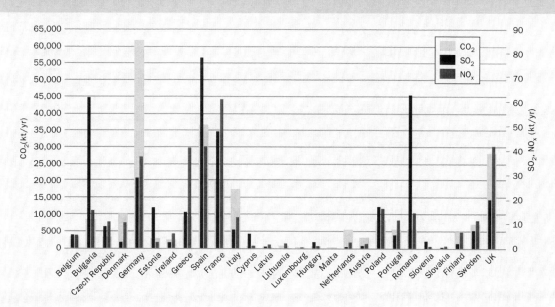

Source: Auer et al. (2007)

Figure V.5.15: Total emissions (CO_2, SO_2, NO_x) from fossil fuel-based electricity generation avoided by wind energy according to EWEA's Low Wind Scenario in the EU-27 Member States in 2020

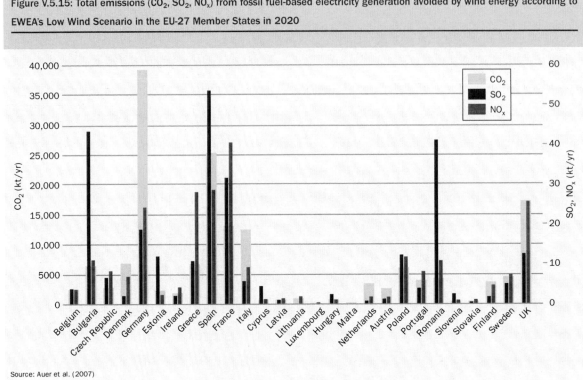

Source: Auer et al. (2007)

EXTERNAL COSTS OF FOSSIL FUEL-BASED ELECTRICITY GENERATION AND AVOIDED EXTERNAL COSTS IN THE BREAKDOWN OF EWEA'S WIND GENERATION SCENARIOS FOR 2020

Figure V.5.16 presents the results of the calculation of the external costs of conventional fossil fuel-based electricity generation in each of the EU-27 Member States in 2020 (high/average/low values), based on the same methodology used for 2007 (see Figure V.5.5). From this, we finally determine the avoided external costs of wind generation in 2020 in Figures 5.17–5.22 (according to EWEA's three wind generation scenarios).

It is important to note that the specific emissions of fossil fuel-based electricity generation technologies in 2020 are less than in 2007, and also the specific

external costs in 2020 (Figure V.5.16) are, on average, less than in 2007 (see Figure V.5.5). In general, the picture for 2020 is similar to 2007 – in other words there is still a noticeable difference between the different EU-27 Member States. Bulgaria, Slovenia and Estonia are those Member States with the highest external costs of fossil fuel-based electricity generation (average values around 20c€$_{2007}$/kWh), but Romania and Greece reach nearly 15c€$_{2007}$/kWh (average values for external costs). On the other hand, there are also a significant number of EU Member States with external costs below 5c€$_{2007}$/kWh (average values of external costs).

By combining the avoidable external costs of fossil fuel-based electricity generation (Figure V.5.16) with the amount of electricity produced by wind energy (Figure V.5.12), the total amount of avoided external costs can be calculated for 2020. Subsequent figures

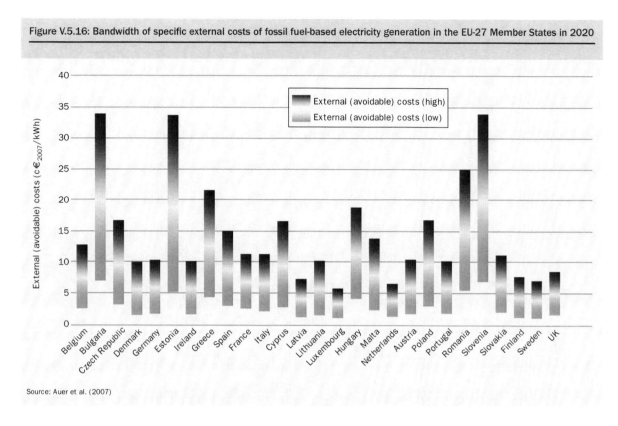

Figure V.5.16: Bandwidth of specific external costs of fossil fuel-based electricity generation in the EU-27 Member States in 2020

Source: Auer et al. (2007)

present the results of EWEA's three wind generation scenarios for each of the EU-27 Member States for 2020.

The corresponding total avoided external costs (using the values of the average specific external costs for each of the EU-27 Member States in Figure V.5.17) are presented in Figure V.5.18. At an aggregated EU-27 level, the total avoided external costs by wind generation in EWEA's reference scenario in 2020 is around €32 billion per year.

Avoided External Cost for Different Wind Deployment Scenarios in the EU-27 Member States in 2030

In this section, wind deployment scenarios for 2030 are addressed in the same way as for 2020. For clarity, empirical results are presented for EWEA's 2030 high

scenario only, since this is the most optimistic assumption at present.

Before presenting the avoided external costs due to wind generation in EWEA's 2030 high scenario, annual wind generation of EWEA's three wind generation scenarios at EU Member State level in 2030 is shown in Figure V.5.23.

Figures V.5.24 and V.5.25 show €69 billion per year (in total) of avoided external costs in the EU-27 Member States by wind generation in EWEA's high wind penetration scenario at aggregated EU-27 level.

Environmental Benefits of Wind Energy – Concluding Remarks

Empirical analyses in previous sections impressively demonstrate that there exist significant environmental benefits of wind generation compared to conventional electricity generation in the EU-27 Member States.

Figure V.5.17: Bandwidth of avoided external costs of fossil fuel-based electricity generation according to EWEA's Reference Scenario in the EU-27 Member States in 2020

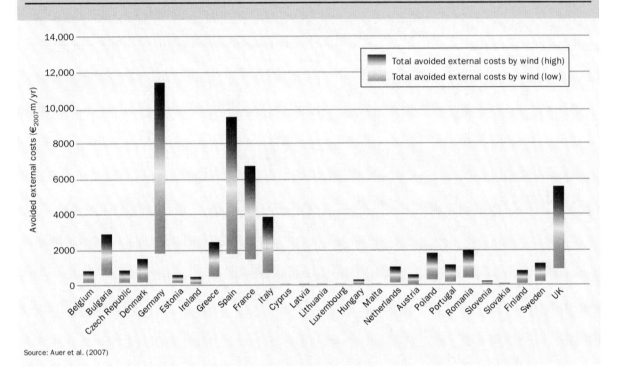

Source: Auer et al. (2007)

Figure V.5.18: Avoided external costs by wind generation according to EWEA's Reference Scenario in each of the EU-27 Member States in 2020 (a total of €32 billion per year)

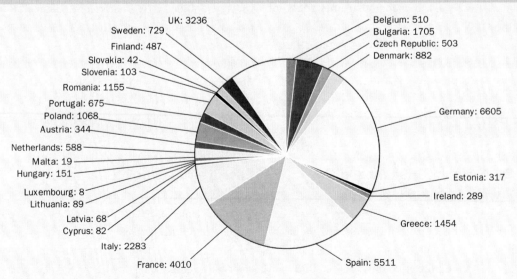

Source: Auer et al. (2007)

Figure V.5.19: Bandwidth of avoided external costs of fossil fuel-based electricity generation according to EWEA's High Scenario in the EU-27 Member States in 2020

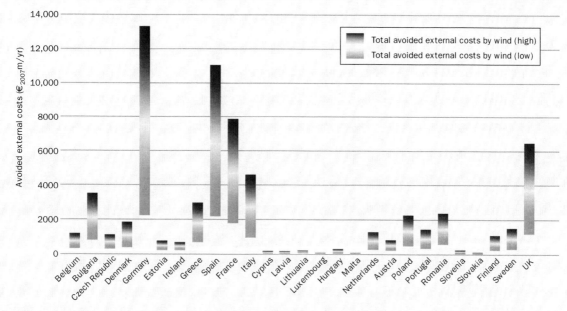

Source: Auer et al. (2007)

Figure V.5.20: Avoided external costs by wind generation according to EWEA's High Scenario in each of the EU-27 Member States in 2020 (a total of €39 billion per year)

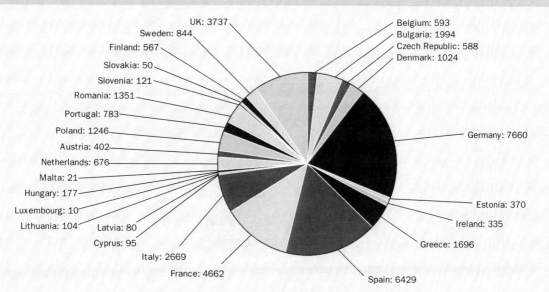

Source: Auer et al. (2007)

Figure V.5.21: Bandwidth of avoided external costs of fossil fuel-based electricity generation according to EWEA's Low Scenario in the EU-27 Member States in 2020

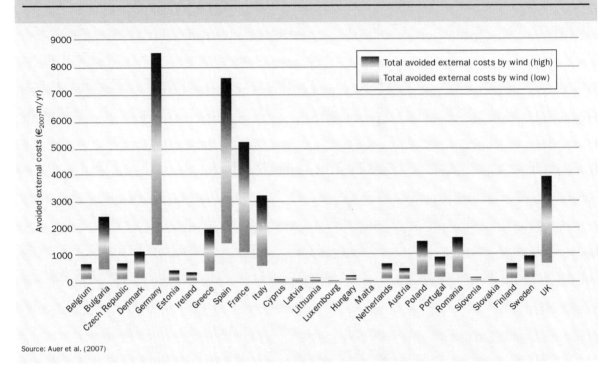

Source: Auer et al. (2007)

Figure V.5.22: Avoided external costs by wind generation according to EWEA's Low Scenario in each of the EU-27 Member States in 2020 (a total of €25 billion per year)

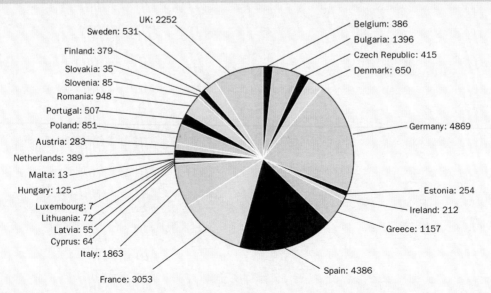

Source: Auer et al. (2007)

Figure V.5.23: Annual wind generation in each of the EU-27 Member States according to EWEA's three wind generation scenarios in 2030

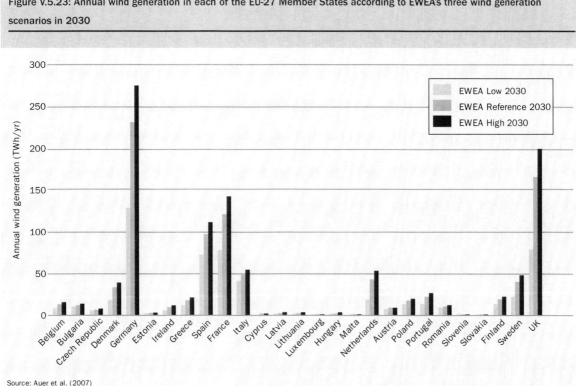

Source: Auer et al. (2007)

In 2007, total annual wind generation of 118.7 TWh per year at EU-27 Member State level has already avoided 70,412 kt per year of CO_2, 183.7 kt per year of SO_2 and 135.3 kt per year of NO_x (see Table V.5.1). The countries mainly contributing to these 2007 results in the EU Member States are Germany, Spain, Denmark and the UK.

In the next decade the share of wind generation will increase considerably in the European power plant mix. Moreover, mainly due to offshore wind deployment (but also a continual increase of onshore wind), the amount of fossil fuel-based electricity generation replaced by wind generation will increase further in all of the three wind deployment scenarios analysed (see Figure V.5.2).

At aggregated EU-27 Member State level, the following amounts of emissions and external cost are avoided due to wind generation in the different wind deployment scenarios:

- reference scenario (477.4 TWh per year) – avoided emissions CO_2: 217,236 kt per year, SO_2: 379.1 kt per year, NO_x: 313.5 kt per year; avoided external cost €$_{2007}$32,913 million;
- high scenario (554.0 TWh per year) – avoided emissions CO_2: 252,550 kt per year, SO_2: 442.0 kt per year, NO_x: 364.8 kt per year; avoided external cost: €$_{2007}$38,284 million;
- low scenario (360.3 TWh per year) – avoided emissions: CO_2: 165,365 kt per year, SO_2: 299.7 kt per year, NO_x: 240.8 kt per year; avoided external cost: €$_{2007}$25,237 million.

Finally, in the high scenario for 2030 (1103.8 TWh per year; see Table V.5.3), the estimates of the

Figure V.5.24: Bandwidth of avoided external costs of fossil fuel-based electricity generation according to EWEA's High Scenario in the EU-27 Member States in 2030

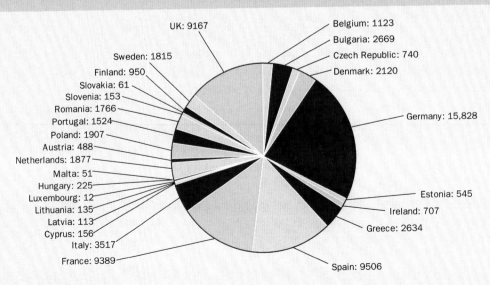

30,000
25,000
20,000
15,000
10,000
5000
0
Avoided external costs (€2007m/yr)
Total avoided external costs by wind (high)
Total avoided external costs by wind (low)
Belgium
Bulgaria
Czech Republic
Denmark
Germany
Estonia
Ireland
Greece
Spain
France
Italy
Cyprus
Latvia
Lithuania
Luxembourg
Hungary
Malta
Netherlands
Austria
Poland
Portugal
Romania
Slovenia
Slovakia
Finland
Sweden
UK

Source: Auer et al. (2007)

Figure V.5.25: Avoided external costs by wind generation according to EWEA's High Scenario in each of the EU-27 Member States in 2030 (a total of €69 billion per year)

UK: 9167
Belgium: 1123
Bulgaria: 2669
Czech Republic: 740
Denmark: 2120
Sweden: 1815
Finland: 950
Slovakia: 61
Slovenia: 153
Romania: 1766
Portugal: 1524
Poland: 1907
Austria: 488
Netherlands: 1877
Malta: 51
Hungary: 225
Luxembourg: 12
Lithuania: 135
Latvia: 113
Cyprus: 156
Italy: 3517
France: 9389
Germany: 15,828
Estonia: 545
Ireland: 707
Greece: 2634
Spain: 9506

Source: Auer et al. (2007)

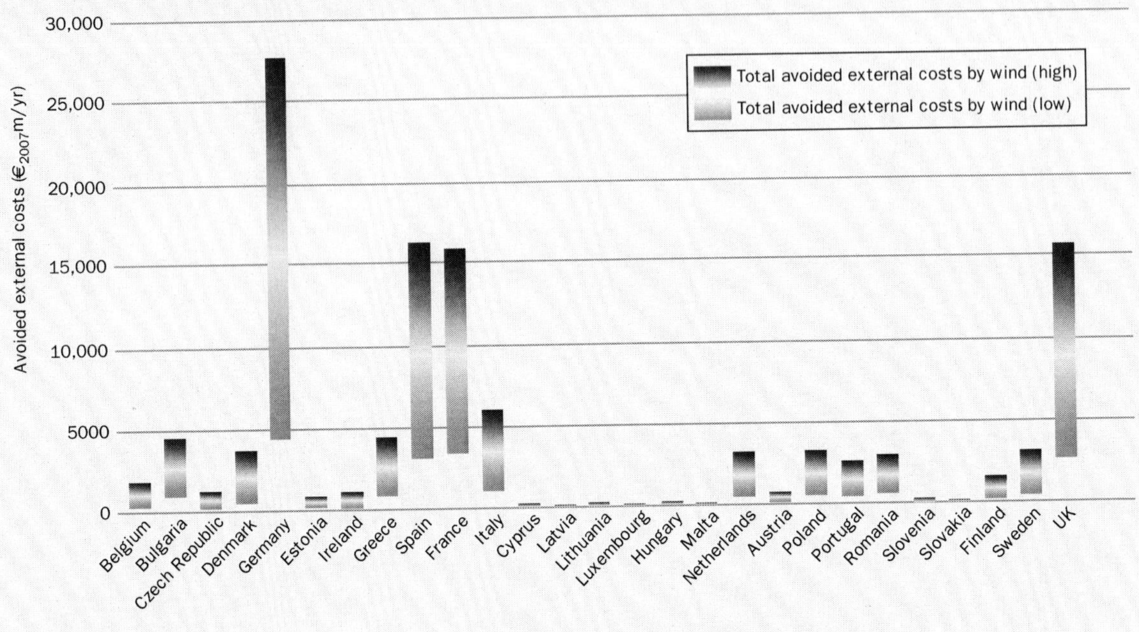

Table V.5.1: Empirical results on avoided emissions and external cost due to wind generation in the EU-27 Member States in 2007

2007	Total emissions from fossil fuel-based electricity generation in 2007			Total emissions from fossil fuel-based electricity generation avoidable by wind (and other renewables) in 2007			Annual wind generation in 2007	Total emissions from fossil fuel-based electricity generation avoided by wind in 2007			Specific external cost from fossil fuel-based electricity generation in 2007	Total external cost from fossil fuel-based electricity generation avoided by wind in 2007
	CO_2 [kt/yr]	SO_2 [kt/yr]	NO_x [kt/yr]	CO_2 [kt/yr]	SO_2 [kt/yr]	NO_x [kt/yr]	[GWh/yr]	CO_2 [kt/yr]	SO_2 [kt/yr]	NO_x [kt/yr]	[€$_{2007}$/kWh]	[€$_{2007}$m]
Belgium	23,424	44	43	20,585	34	38	470	310	0.5	0.6	9.6	51
Bulgaria	26,508	994	72	5044	116	15	17	16	0.4	0.0	20.5	4
Czech Rep.	44,937	183	123	7231	20	21	110	70	0.2	0.2	9.3	12
Denmark	16,642	7	21	8353	2	10	7965	4486	1.3	5.4	5.7	518
Germany	305,795	155	196	97,131	39	64	39,446	24,796	9.9	16.2	6.8	3027
Estonia	15,763	83	15	15,753	83	15	118	184	1.0	0.2	16.1	21
Ireland	14,952	71	39	9345	27	25	2175	1224	3.6	3.2	6.4	157
Greece	63,885	366	334	19,340	62	116	2320	2138	6.9	12.8	15.3	400
Spain	132,708	1081	382	78,259	400	226	38,423	23,545	120.5	68.0	9.1	3968
France	27,447	73	86	16,855	37	52	3763	1502	3.3	4.6	7.4	316
Italy	136,248	333	253	112,504	245	206	4483	2336	5.1	4.3	7.4	377
Cyprus	3573	38	6	3573	38	6	0	0	0.0	0.0	9.0	0
Latvia	456	0	1	456	0	1	57	17	0.0	0.0	3.1	2
Lithuania	1454	5	3	1454	5	3	0	0	0.0	0.0	3.8	0
Luxembourg	949	0	3	949	0	3	64	20	0.0	0.1	3.2	2
Hungary	16,163	256	24	8273	32	13	460	276	1.1	0.4	8.2	43
Malta	1829	20	3	1829	20	3	0	0	0.0	0.0	9.0	0
Netherlands	40,402	12	37	29,262	4	25	3569	1477	0.2	1.3	4.3	172
Austria	14,435	11	11	9553	4	7	1919	1013	0.4	0.7	5.7	124
Poland	124,587	716	221	30,007	120	56	541	401	1.6	0.8	10.6	65
Portugal	23,090	150	74	15,295	95	48	4496	2550	15.9	8.0	7.6	388
Romania	36,059	697	99	17,162	147	46	0	0	0.0	0.0	20.2	0
Slovenia	5790	94	12	1908	29	4	0	0	0.0	0.0	24.8	0
Slovakia	6742	47	24	2707	9	10	8	5	0.0	0.0	9.1	1
Finland	10,217	14	19	5253	3	10	275	95	0.1	0.2	3.4	11
Sweden	1929	2	2	1503	2	2	1743	823	0.8	0.9	4.4	86
UK	172,260	1011	418	100,133	350	234	6278	3128	10.9	7.3	6.7	472
EU-27 TOTAL	1,268,247	6461	2520	619,719	1925	1258	118,700	70,412	183.7	135.3	9.1 (Average)	10,216

Source: Based on Auer et al. (2007)

Table V.5.2: Empirical results on avoided emissions and external cost due to wind generation in the different wind deployment scenarios in the EU-27 Member States in 2020

2020	Annual wind generation scenarios in 2020			Total emissions from fossil fuel-based electricity generation *avoided* by wind in the different scenarios in 2020									Total external cost (average values) from fossil fuel-based electricity generation *avoided* by wind in the different scenarios in 2020		
	Low Scenario	Ref. Scenario	High Scenario	Low Scenario			Reference Scenario			High Scenario			Low Scenario	Ref. Scenario	High Scenario
	[TWh/yr]	[TWh/yr]	[TWh/yr]	CO$_2$ [kt/yr]	SO$_2$ [kt/yr]	NO$_x$ [kt/yr]	CO$_2$ [kt/yr]	SO$_2$ [kt/yr]	NO$_x$ [kt/yr]	CO$_2$ [kt/yr]	SO$_2$ [kt/yr]	NO$_x$ [kt/yr]	[€$_{2007}$m]	[€$_{2007}$m]	[€$_{2007}$m]
Belgium	5.1	6.8	7.9	2386	3.8	3.6	3154	5.0	4.7	3665	5.8	5.5	386	510	593
Bulgaria	7.1	8.7	10.1	6216	43.3	11.0	7593	52.9	13.4	8878	61.8	15.7	1396	1705	1994
Czech Rep.	4.3	5.2	6.1	2640	6.4	7.8	3200	7.7	9.5	3744	9.0	11.1	415	503	588
Denmark	11.3	15.4	17.9	6659	1.8	6.8	9041	2.4	9.2	10,499	2.8	10.7	650	882	1024
Germany	82.1	111.4	129.2	39,181	18.4	24.0	53,158	25.0	32.6	61,645	29.0	37.8	4869	6605	7660
Estonia	1.3	1.6	1.9	2176	11.7	2.0	2718	14.6	2.5	3171	17.0	3.0	254	317	370
Ireland	3.6	5.0	5.7	1715	1.8	3.9	2340	2.4	5.4	2713	2.8	6.2	212	289	335
Greece	9.1	11.5	13.4	6577	10.3	28.1	8265	13.0	35.3	9641	15.1	41.2	1157	1454	1696
Spain	50.3	63.8	73.8	25,164	53.5	28.6	31,617	67.2	35.9	36,884	78.4	41.9	4386	5511	6429
France	45.7	60.0	69.7	12,975	31.4	40.2	17,042	41.2	52.8	19,811	47.9	61.4	3053	4010	4662
Italy	27.8	34.1	39.9	12,386	5.4	9.0	15,178	6.7	11.1	17,742	7.8	12.9	1863	2283	2669
Cyprus	0.7	0.9	1.0	507	4.1	1.0	645	5.3	1.2	752	6.1	1.4	64	82	95
Latvia	1.3	1.6	1.9	474	0.6	1.0	591	0.7	1.3	690	0.8	1.5	55	68	80
Lithuania	1.2	1.5	1.8	646	0.0	1.6	795	0.0	1.9	929	0.0	2.3	72	89	104
Luxembourg	0.2	0.3	0.3	58	0.0	0.2	71	0.0	0.2	82.6	0.0	0.2	7	8	10
Hungary	1.1	1.4	1.6	526	2.0	0.8	638	2.4	0.9	747	2.8	1.1	125	151	177
Malta	0.2	0.2	0.3	100	0.0	0.2	141	0.0	0.2	163	0.0	0.3	13	19	21
Netherlands	10.3	15.6	17.9	3301	0.3	1.6	4981	0.4	2.4	5733	0.5	2.8	389	588	676
Austria	4.7	5.7	6.7	2347	0.9	1.5	2844	1.1	1.8	3328	1.3	2.1	283	344	402
Poland	8.9	11.2	13.0	5878	11.9	11.5	7376	14.9	14.4	8605	17.4	16.8	851	1068	1246
Portugal	8.5	11.4	13.2	3763	3.6	7.8	5011	4.8	10.3	5819	5.5	12.0	507	675	783
Romania	6.4	7.8	9.2	4116	68.0	10.4	5017.9	82.9	12.7	5868	96.9	14.9	948	1155	1351
Slovenia	0.3	0.3	0.4	253	2.2	0.5	307	2.6	0.6	359	3.1	0.7	85	103	121
Slovakia	0.5	0.7	0.8	217	0.1	0.6	263	0.1	0.7	307	0.1	0.8	35	42	50
Finland	8.6	11.1	12.9	3408	1.4	4.6	4381	1.9	5.9	5102	2.2	6.8	379	487	567
Sweden	13.2	18.1	21.0	4629	4.7	7.4	6348.6	6.5	10.1	7356	7.5	11.7	531	729	844
UK	46.1	66.2	76.5	17,066	12.2	25.2	24,519	17.5	36.3	28,315	20.2	41.9	2252	3236	3737
EU-27 TOTAL	360.3	477.4	554.0	165,365	299.7	240.8	217,236	379.1	313.5	252,550	442.0	364.8	25,237	32,913	38,284

Source: Based on Auer et al. (2007)

Table V.5.3: Empirical results on avoided emissions and external cost due to wind generation in the High Scenario in the EU-27 Member States in 2030

2030	Annual wind generation in the *high* scenario in 2030	Total *emissions* from fossil fuel-based electricity generation *avoided* by wind in the *high* scenario in 2030			Total *external cost* (average values) from fossil fuel-based electricity generation *avoided* by wind in the *high* scenario in 2030
	[TWh/yr]	CO_2 [kt/yr]	SO_2 [kt/yr]	NO_x [kt/yr]	[€$_{2007}$m]
Belgium	15.7	6729	11.4	10.1	1123
Bulgaria	13.7	11,526	81.4	20.3	2669
Czech Rep.	7.8	4631	11.4	13.7	740
Denmark	39.4	21,751	6.0	22.2	2120
Germany	276.7	126,329	60.9	77.6	15,828
Estonia	2.9	4653	24.9	4.3	545
Ireland	12.8	5655	6.1	13.0	707
Greece	21.2	14,632	23.6	62.5	2634
Spain	111.9	52,999	116.6	60.4	9506
France	142.8	38,389	96.4	119.3	9389
Italy	55.0	22,807	10.6	16.7	3517
Cyprus	1.7	1220	10.0	2.3	156
Latvia	2.9	970	1.3	2.1	113
Lithuania	2.5	1204	0.0	2.9	135
Luxembourg	0.4	97	0.0	0.3	12
Hungary	2.1	919	3.5	1.3	225
Malta	0.7	388	0.0	0.7	51
Netherlands	53.2	15,796	1.4	7.6	1877
Austria	8.6	4006	1.7	2.5	488
Poland	20.5	12,967	26.8	25.3	1907
Portugal	27.0	11,121	11.2	23.0	1524
Romania	12.2	7305	127.9	18.6	1766
Slovenia	0.5	442	3.8	0.9	153
Slovakia	1.0	371	0.2	1.0	61
Finland	22.8	8525	3.8	11.4	950
Sweden	47.7	15,731	16.9	25.2	1815
UK	200.1	68,257	52.2	101.0	9167
EU-27 TOTAL	1103.8	459,422	709.8	646.4	69,180

Source: Based on Auer et al. (2007)

environmental benefits of wind generation (compared to fossil fuel-based electricity generation) are CO_2: 459,422 kt per year, SO_2: 709.8 kt per year and NO_x: 646.4 kt per year in terms of avoided emissions and €$_{2007}$69,180 million in terms of avoided external cost.

The analyses and results presented above impressively underline the importance of further significantly increasing the share of wind deployment (onshore and offshore) in the EU-27 Member States in the next decades. However, a precondition for the full implementation of the environmental benefits estimated here is both continuous adaption of financial support instruments and the removal of several barriers for market integration of wind energy.

Annex V.5

Table V.5.A briefly summarises the methodology for specific countries or regions of each of the 'early' studies on external costs in the late 1980s and the beginning of the 1990s.

Table V.5.A: Methodologies of external cost studies

Study	Methodology	Location
Hohmeyer (1988a)	Top-down apportioning of total environment damages in Germany to the fossil fuel sector	Germany; existing power plants
Ottinger et al. (1990)	Damage-based approach in which values were taken from a literature review or previous studies	US; existing power plants
Pearce (1992)	Literature survey to identify values used in a damage-based approach to calculate damages; in some respects an update of Ottinger et al. (1990) study from Pace University Center	Estimates for a new and old coal power plant in the UK
US Department of Energy ORNL/RFF (1994)	Damage function, or impact pathway, approach; detailed examination and use of scientific literature; emphasis on developing methodology, rather than on numerical results of specific examples	Estimates for new power plants in rural southwest and southeast US
RCG/Tellus (1994)	Damage function approach; developed EXMOD software	Sterling, rural area in New York state

Source: Lee (1996)

The Social Acceptance of Wind Energy: An Introduction to the Concept

Wind energy, being a clean and renewable energy source in a global context of increasing social concerns about climate change and energy supply, is traditionally linked to very strong and stable levels of public support. The most recent empirical evidence on public opinion towards wind energy at both the EU and the country level fully supports such favourable perception of this energy source among European citizens. Nevertheless, experience in the implementation of wind projects shows that social acceptance is crucial for the successful development of specific wind energy projects. Thus we should look at the main singularities of the social acceptance of wind energy compared to the social acceptance of other energy technologies:

- the (very) high and stable levels of general public support: at an abstract level about 80 per cent of EU citizens support wind energy;
- the higher number of siting decisions to be made due to the current relatively small-scale nature of the energy source;
- the visibility of wind energy devices and the proximity to the everyday life of citizens (if compared with the 'subterranean' and distant character of conventional power generation and fossil fuels extraction); and
- the tensions between support and opposition concerning specific wind power developments at the local level: large majorities of people living near wind farm sites are in favour of their local wind farm (Warren el al. 2005), but wind planning and siting processes are facing significant challenges in some countries across Europe (Wolsink, 2007).

Consequently, the social acceptance of wind power entails both the general positive attitude towards the wind energy technology together with the increasing number of 'visible' siting decisions to be made at the local level. Importantly, it is at the local level where the 'technical' characteristics of wind energy interact with the everyday life of the individual, and the social and institutional environments of the communities hosting such developments. As we will see, the general positive attitudes towards wind power are not necessarily linked to the local acceptance of wind energy projects (Johansson and Laike, 2007).

This is the context in which we find the most recent formulation of the concept of 'social acceptance' linked to renewable energies (Wüstenhagen et al., 2007), the so-called 'triangle model', which distinguishes three key dimensions of social acceptance:

1. socio-political acceptance;
2. community acceptance; and
3. market acceptance.

- **Socio-political acceptance** refers to the acceptance of both technologies and policies at the most general level. Importantly, this general level of socio-political acceptance is not limited to the 'high and

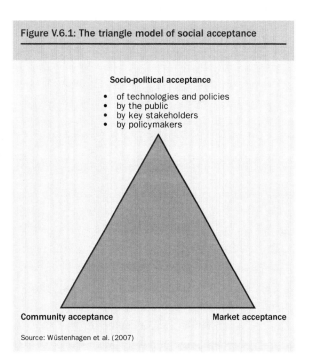

Figure V.6.1: The triangle model of social acceptance

Socio-political acceptance
- of technologies and policies
- by the public
- by key stakeholders
- by policymakers

Community acceptance   Market acceptance

Source: Wüstenhagen et al. (2007)

stable' levels of acceptance by the general public, but includes acceptance by key stakeholders and policymakers. Stakeholders and policymakers involved in discussing 'renewable policies' become crucial when addressing planning issues or promoting local involvement initiatives. Thus the assessment of their levels of acceptance is an area of increasing interest for social researchers.

- **Community acceptance** refers to the acceptance of specific projects at the local level, including potentially affected populations, key local stakeholders and the local authorities. This is the area where social debate around renewables arises and develops, and the one that has attracted most of the social research traditionally carried out in the wind energy field.

- **Market acceptance** refers to the process by which market parties adopt and support (or otherwise) the energy innovation. Here we find processes such as green power marketing and willingness to pay for green power. Market acceptance is proposed in a wider sense, including not only consumers, but also investors and, very significant, intra-firm acceptance.

Interestingly, this 'triangle model' works well with the 'three discourses' scheme suggested by another recent conceptual approach to the social perception of energy technologies (Prades et al., 2008): the 'siting discourse' (where the technology is experienced in terms of a proposed construction of some facility in a given locality); the 'energy-innovation discourse' (where the technology is experienced as an innovation that may or may not fit in with preferred ways of life); and the 'investment discourse' (where the technology is experienced as an investment opportunity that is acceptable, or otherwise, in the light of the possible gains it will produce). Moreover, this 'triangle model' has been proposed as the conceptual framework in a recent task of the International Energy Agency – Wind (Implementing Agreement for Cooperation in the Research, Development and Deployment of Wind Energy Systems) dealing with the social acceptance of wind energy projects: 'Winning hearts and minds' (IEA Wind, 2007a).

The next sections will introduce the main findings of the social research with regards to socio-political acceptance (the acceptance of technologies and policies by both the general public and key stakeholders and policymakers) and community acceptance (the acceptance of specific projects at the local level).

The Social Research on Wind Energy Onshore

Social research on wind energy has primarily focused on three main areas:

1. **public acceptance**: the assessment (and corroboration) of the (high and stable) levels of public support (by means of opinion polls and attitude surveys);
2. **community acceptance**: the identification and understanding of the dimensions underlying social controversy at the local level (by means of single or multiple case studies, including surveys); and
3. **stakeholder acceptance**: social acceptance by key stakeholders and policymakers (by means of interviews and multiple case studies); recent approaches are paying increasing attention to this field.

The following section looks at what social research on wind energy tells us about the social acceptance of wind developments by such a wide range of social actors and levels.

PUBLIC ACCEPTANCE OF WIND ENERGY (SOCIO-POLITICAL ACCEPTANCE)

One of the traditional focuses of social research on wind energy has been the assessment of the levels of public support for wind energy by means of opinion polls and attitude surveys (Walker, 1995). Among opinion polls, the strongest indicator allowing comparisons of the level of support in different countries is the Eurobarometer Standard Survey (EB), carried out twice yearly and covering the population of the EU aged 15 and over. Over the 30 years that these surveys have been conducted, they have proved to be a helpful source of information for EU policymakers on a broad range of economic, social, environmental and other issues of importance to EU citizens. Recent EB data on public opinion (EC, 2006c and 2007c) confirm the strongly positive overall picture for renewable energies in general, and for wind energy in particular, at the EU level, and not only for the present but also for the future (see Figure V.6.2).

When EU citizens are asked about their preferences in terms of the use of different energy sources, renewable energies in general, and wind energy in particular, are rated highly positively (especially when compared with nuclear or fossil fuels). The highest support is for solar energy (80 per cent), closely followed by wind energy, with 71 per cent of EU citizens firmly in favour of the use of wind power in their countries, 21 per cent expressing a balanced view and only 5 per cent are opposed to it. After solar and wind, we find hydroelectric energy (65 per cent support), ocean energy (60 per cent) and biomass (55 per cent). According to this EB survey, only a marginal number of respondents opposed the use of renewable energy sources in their countries. As regards fossil fuels, 42 per cent of the EU citizens favoured the use of natural gas and about one-quarter accepted the use of oil (27 per cent) and coal (26 per cent). Nuclear power seems to divide opinions, with the highest rates of opposition (37 per cent) and balanced opinions (36 per cent) and the lowest rate of support.

Focusing on the use of wind energy, on a scale from 1 (strongly opposed) to 7 (strongly in favour), the EU average is 6.3. Even higher rates of support arose in some countries, for example Denmark (6.7), Greece (6.5), and Poland, Hungary and Malta (6.4). The UK shows the lowest support figure of the EU (5.7), closely followed by Finland and Germany (5.8).

EU citizens also demonstrated a very positive view of renewable energy in general and of wind energy in

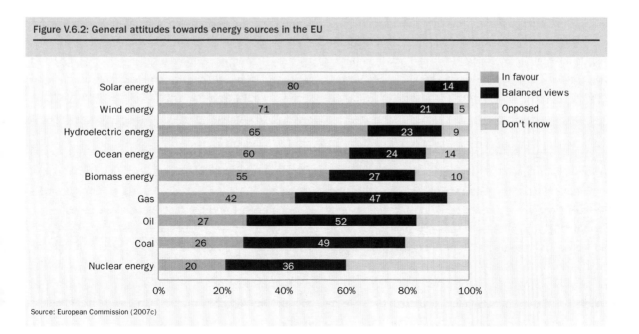

Figure V.6.2: General attitudes towards energy sources in the EU

Source: European Commission (2007c)

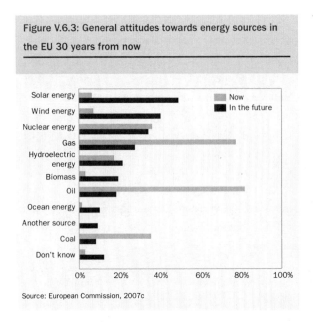

Figure V.6.3: General attitudes towards energy sources in the EU 30 years from now

Source: European Commission, 2007c

Wind Energy Association, Associazione Nazionale Energia Del Vento (Italy) and or the Austrian Wind Energy Association, supports this positive overall scenario with regards to the use of wind energy, both at present and in the future.

One interesting question is the association between these high levels of general public support for wind energy and the actual implementation of wind power in each country. This could be analysed through the correlation of two variables: percentage of people strongly in favour of wind power, from the EB, and wind capacity in kW/1000 inhabitants (Figure V.6.4). The bivariate analysis shows a *low* and *not significant* linear correlation: the highest levels of public enthusiasm about wind power in our sample of countries were not associated with the highest levels of wind capacity per habitant. In line with the most recent formulation of the 'social acceptability' of wind farms, this result may indicate that the generally favourable public support for the technology of wind power does not seem to be directly related to the installed wind capacity (that is linked to positive social and institutional decision-making as will be seen later on). Thus it is very important to properly differentiate between the 'public acceptance' of wind energy and the 'community acceptance' (and stakeholders' acceptance) of specific wind developments.

Finally, from the methodological perspective, it should be noted that, despite a profusion of quantitative surveys, and with a few notable exceptions (Wolsink, 2000) there is still a lack of valid and reliable quantitative methodological tools for understanding general public perceptions of wind farms (Devine-Wright, 2005).

To conclude, the key messages regarding 'public acceptance' are the very high levels of public support for wind energy and the fact that this favourable general condition does not seem to be directly related to the installed wind capacity. Thus there is a need to also look at the perceptions of the other key actors involved in wind development: the local communities hosting the wind farms and the key stakeholders involved in such developments.

particular when asked about their expectations regarding the three most used energy sources 30 years from now. Results showed that wind energy is expected to be a key energy source in the future – just after solar. Respondents in all countries except the Czech Republic, Italy, Slovenia, Slovakia and Finland mentioned wind energy among the three energy sources most likely to be used in their countries 30 years from now. The expected increase in the use of wind energy from 2007 to 2037 is very important in all countries (that mentioned wind), with an average expected increase of 36.35 per cent.

The latest EB on 'Attitudes towards energy' (EC, 2006c) further corroborates this positive picture of wind at the EU level. For EU citizens, the development of the use of wind energy was the third preferred option to reduce our energy dependence on foreign, expensive and highly polluting sources (31 per cent), after the increase in the use of solar energy (48 per cent) and the promotion of advance research on new energy technologies (41 per cent). Importantly, further evidence at the country level gathered by several national wind energy associations, such as the British

Figure V.6.4: Wind acceptability and wind capacity in the EU

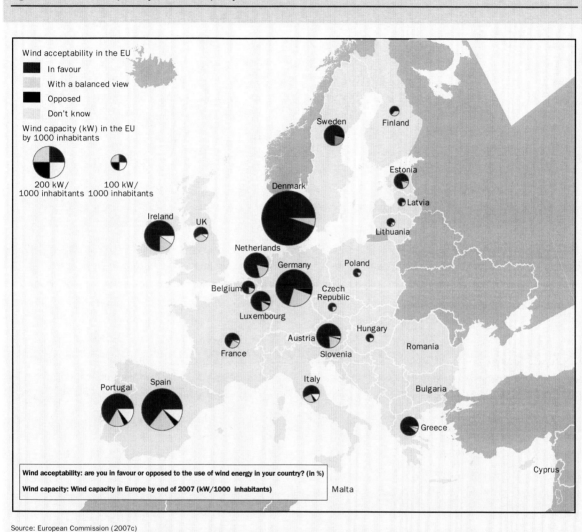

Source: European Commission (2007c)

COMMUNITY ACCEPTANCE OF WIND ENERGY AND WIND FARMS

A wide variety of studies based on different approaches and methodologies have been carried out to identify the key elements involved in the interaction between wind energy developments and the communities hosting them. Importantly, these case studies have allowed a better understanding of the factors explaining success and failure of wind developments, and this may indeed provide useful insights to more evidence-based decision-making in the future.

Recent research into how wind projects interact with the local community questions the traditional explanation of local rejection to technological projects based on the NIMBY acronym (Not In My Back Yard), as this term may give an incorrect or only partial explanation of all the variables involved in the planning process

(Krohn and Damborg, 1999; Wolsink, 2000, 2007). According to the NIMBY idea, resistance is explained in terms of motives of local citizens, but the latest findings suggest that such an interpretation would be too simplistic considering the number of views and circumstances involved in the local planning of wind projects (Warren et al., 2005). The NIMBY label 'leaves the cause of opposition unexplained' (Kempton et al., 2005) and consequently it lacks explanatory value. One of the key messages from social research points out that how wind farms are developed and how people make sense of the impact of wind farms upon the places in which they live may be more important in shaping pubic reactions to new projects than the purely physical or technical factors. As Wolsink (2000) suggests, local opposition is often based on distrust, negative reactions to the actors (developers, authorities and energy companies) trying to build the turbines, and the way projects are planned and managed, and not to wind turbines themselves.

Thus, according to the social sciences literature, when trying to understand the community acceptance of wind farms, there are some potential errors that could lead to misunderstandings. One type of error, as stated in the previous section, is not considering *community acceptance* as a social phenomenon with a *different dynamic* than *public acceptance* of wind power as a reliable source of energy. Another potential error is interpreting public attitudes towards wind farms as merely influenced by the characteristics of the technology, without properly considering how the implementation of the technology is part of a socio-technical system that interacts with the local community, the local environment, the key stakeholders and the project developers.

Three categories have been established that help explain the social response to wind energy (see Prades and González Reyes, 1995). First, we consider those physical, technical and environmental characteristics of the technology that affect how the public perceives wind farms. Second, we analyse the different individual and psycho-social factors of those living in the hosting communities, such as knowledge, general attitudes or familiarity, which might shape views of wind farms. And third, we consider the social and institutional elements governing the interaction between the technology and the hosting community, such as planning characteristics or level of engagement, and how they might influence public attitudes towards, and acceptance of, wind farm projects.

Table V.6.1: Factors affecting public perceptions of wind farms and other energy innovations

Perceptions of physical and environmental factors	Psycho-social factors	Social and institutional factors
Visual impact: • Landscape characteristics • Turbine colour • Turbine and farm size • Unity of the environment (as designed by the authors) • Wind farm design • Turbine noise • Distance to turbines • Ecological site characteristics (birds and other wildlife)	• Familiarity • Knowledge • General attitudes • Perceived benefits and costs • Socio-demographics • Social network influences	• Participatory planning • Public engagement • Justice and fairness issues • Local ownership • Policy frameworks • Centralisation/decentralisation • Campaigns by action groups

Source: CIEMAT

Physical, Environmental and Technical Attributes of Wind Farms

As with any other technological development, the specific physical and technical attributes of the implementation of the technology itself are significant predictors of public attitudes. Consequently, social research on wind farm projects has attempted to identify how such wind power attributes are perceived by the public. One of the most relevant early research findings (Thayer and Freeman, 1987; Wolsink, 1988 and 1989) in this regard was the identification of visual impacts and noise as important issues in the social response to wind energy (Devine-Wright, 2005).

Visual impact has been considered the main influence on public attitudes towards wind farms, as 'aesthetic perceptions, both positive and negative, are the strongest single influence on public attitudes' (Wolsink, 2000). The perceived impact on landscape seems to be the crucial factor in this regard, and opposition to the visual despoliation of valued landscapes has been analysed as the key motivation to opposition to wind farms (Warren et al., 2005). A study on how perceptual factors influence public intention to oppose local wind turbines (Johansson and Laike, 2007) found that 'perceived unity' of (or harmony with), the environment is the most important perceptual dimension. Those who perceived the turbines to have a high degree of unity with the landscape express a low degree of opposition. With regard to colour, a higher level of public support seems to exist for turbines that are painted neutral colours. In relation to size, studies in the UK, Denmark, The Netherlands and Ireland found a systematic preference for smaller groups of turbines over large-scale installations (Devine-Wright, 2005).

Visual intrusion and noise were the key anticipated problems by respondents in a survey carried out in Ireland (Warren et al., 2005). However, the same study found that noise pollution and visual impacts were less important to the public than anticipated before the project construction, concluding that respondents' fears had not been realised. The limited effect of noise disturbance on acceptance levels has also been found in other contexts (Krohn and Damborg, 1999). On a more detailed level, Pedersen and Waye (2007) found, in different areas in Sweden, that the visual factor of the fit of the turbines to the landscape has a stronger impact than the sound levels.

Danger to birds and other wildlife is considered to be one of the more important environmental impacts of wind energy developments. As stated in Chapter V.2 ('Environmental impacts'), bird mortality caused by wind farms seems to be a sporadic event and dependent on different elements such as the season, the specific site, the species and the type of bird activity. Studies on community acceptance (Wolsink, 2000; Simon, 1996) have shown that the concern about hazards to birds, when present, has only a small impact on individuals' perceptions of wind farms. However, in ecological areas with threatened or vulnerable species, impacts from wind farms on birds and habitats might generate opposition from environmental and other public interest groups, media attention and increase local concern.

Another element investigated by the empirical research has been the effect of distance to the wind farms on perceptions. In Denmark, different studies, to some surprise, have found that people living closer to wind farms tend to be more positive about wind turbines than people living farther away (Scottish Executive Central Research Unit, 2000). As we will see, familiarity with wind farms could be one possible explanation of this phenomenon.

To conclude, research has shown that the physical, environmental and technical attributes of wind farms and the selected site are significant predictors of public attitudes and, consequently, issues such as harmony with the landscape and turbine/wind farm size and colour should be carefully considered when planning wind energy developments. Nevertheless, social acceptance of wind farms is not merely influenced by the characteristics of the technology: more important are the implementation of the technology and how it interacts with the local community, the key stakeholders and the project developers.

Psycho-social Factors

Psycho-social factors have become crucial dimensions to explain how local communities interact with, and react to, new wind farm developments. Familiarity with the technology is a significant element widely explored by social research. The familiarity hypothesis refers to the fact that those who experience wind farms generally become more positive towards them (Wolsink, 1994; Krohn and Damborg, 1999). This phenomenon has been represented as the 'U-shape curve' (Wolsink, 1994). Based on empirical data, this model states that public attitudes change from very positive, before the announcement of the project, to negative when the project is announced, to positive again after the construction. This important result shows the dynamic nature of public attitudes. Opinions on technological developments may change as citizens are confronted with specific developments. However, as has been documented (Wolsink, 2007), the improvement of attitudes after a facility has been constructed is not guaranteed.

Separate to the familiarity dimension is the degree of knowledge about wind energy and its effects on individuals' perceptions of wind farms. Although some studies have found a positive relation between knowledge and attitude (Krohn and Damborg, 1999), there is little evidence of a significant correlation between level of knowledge of wind power and its acceptance (Wolsink, 2007; Ellis et al., 2007). This does not mean, however, that providing clear and honest information about the technology and the project does not play an important role in increasing public understanding: it is essential in the process of creating trust between developers, authorities and local communities.

General attitudes towards wind energy are another key element influencing public perceptions of wind farms. As seen in the previous section, general attitudes towards wind power are very positive. A recent study by Johansson and Laike (2007) found that the general attitude towards wind power was one of the most significant predictors in the response to a local project, with those more positive about wind power more in favour of the specific project. Pedersen and Waye (2008) have also revealed that people with anti-wind-energy views perceive wind turbines to be much noisier and more visually intrusive than those who are optimistic about wind power.

The effects of socio-demographic variables on individuals' views of wind farms have also been studied. Age, gender, experience with wind farms, and use of the land and/or beach were found to be slightly correlated with the attitudes towards wind power in a Danish study dealing with public perceptions of on-land or offshore wind turbines (Ladenburg, 2008).

Devine-Wright (2005) has pointed out other psycho-social factors less explored by social research on public reactions to wind farms, such as the role of social networks in 'how people come to hear about proposed wind farm developments and whom they trust, as well as the eventual perceptions that they choose to adopt' (Devine-Wright, 2005, p136). In this framework, social trust, considered as the level of trust individuals have with organizations and authorities managing technological projects, is increasingly regarded as a significant element in social reactions to technological developments (Poortinga and Pidgeon, 2006). In the wind power context, Eltham et al. (2008) have documented, through the study of public opinions of a local population living near a wind farm, how suspicion of the developers' motives by the public, distrust of the developers and disbelief in the planning system may impede the success of wind farm projects. Trust can be created in careful, sophisticated decision-making processes that take time, but it can be destroyed in an instant by processes that are perceived as unfair (Slovic, 1993; Poortinga and Pidgeon, 2004). Trust is an interpersonal and social variable, linking attitudinal processes with institutional practices.

To conclude, psycho-social factors such as familiarity (or otherwise) with wind technology, general attitudes towards the 'energy problem' and/or socio-demographic variables do play a role in the shaping of wind energy acceptance and should properly be considered when planning wind energy developments.

Social and Institutional Factors

The notion of 'citizen engagement' has become a central motif in public policy discourse within many democratic countries, as engagement – 'being responsive to lay views and actively seeking the involvement of the lay public in policymaking and decision-making' (Horlick-Jones et al., 2007) – is acknowledged as an important component of good governance (National Research Council, 1996). Consequently, the analysis of the social acceptance of technologies is increasingly recognising the importance of the 'institutional arrangements', in other words the relationships between the technology, its promoters and the community (Rogers, 1998; Kunreuther et al., 1996). This is precisely the focus of the most recent investigations on the sources of success or failure of wind farms projects: the relationship between local resistance and levels of community engagement, fairness and compensation (Loring 2007; Wolsink, 2007).

One of the most substantive questions in this regard is whether local involvement and participatory planning in wind farms increases local support. Recent studies agree that successful wind farm developments are linked to the nature of the planning and development process, and that public support tends to increase when the process is open and participatory (Warren et al., 2005; Wolsink, 2007; Loring, 2007). It is also suggested that collaborative approaches to decision-making in wind power implementation will be more effective than top-down imposed decision-making (Wolsink, 2007) and that public engagement may serve to reduce opposition and to increase levels of 'conditional supporters' to wind power developments (Eltham et al., 2008). As Wolsink (2007, p1204) states, 'the best way to facilitate the development of wind projects is to build institutional capital (knowledge resources, relational resources and the capacity for mobilisation) through collaborative approaches to planning'.

There is little doubt that fairness issues may shape a local community's reactions to wind developments in siting contexts. Findings from research indicate that perceptions of fairness influence how people perceive the legitimacy of the outcome (Gross, 2007). It is assumed that a fairer process helps the creation of mutual trust, and hence it will increase acceptance of the outcome. As has been stated by other authors, the underlying reason for NIMBY attitudes is not selfishness, but a decision-making process perceived to be unfair (Wolsink, 2007).

In the review of factors shaping public attitudes towards wind farms, it has been emphasised (Devine-Wright, 2005; Krohn and Damborg, 1999) that there exists a significant relationship between share ownership and perceptions. Individuals who own shares in a turbine have a more positive attitude towards wind energy than those with no economic interest. Although limited to the Danish context, it has been found that in some communities, members of wind cooperatives are more willing to accept more turbines in their locality in comparison with non-members.

The influence of policy frameworks on the social acceptance of wind energy has also been analysed through case studies (Jobert et al., 2007). Results from German and French cases underline the relevance of factors directly linked with the implementation of the project: local integration of the project developer, creation of a network of support and access to ownership. According to the authors, the French policy framework makes developers more dependent on community acceptance, and therefore the French case studies show much more conflict resolution and networking among key local actors than in the German one. The planning problem and the role of national and local policies are also being analysed as key dimensions in the social acceptance of wind power in Scotland and Wales (Cowell, 2007).

The role of action groups in a wind farm planning decision (Parkhill 2007) is also receiving quite a lot of attention from social research, as evidence is showing their substantial influence on wind farm planning decisions at the sub-national level (Bell et al., 2005; Toke, 2005; Boström, 2003). The strength of local opposition groups has been considered as an important social and institutional factor causing distrust during the planning and siting stages (Eltham et al., 2008).

To conclude, social research is highlighting the complexity and multidimensionality of the factors underlying community acceptance of wind energy projects. Recent evidence is increasingly demonstrating that 'how' wind farms are developed may be more important in shaping pubic reactions to new projects than the purely physical or technical factors.

STAKEHOLDERS' AND POLICYMAKERS' ACCEPTANCE (SOCIO-POLITICAL ACCEPTANCE)

As most social research on wind energy developments has focused either on 'public acceptance' or 'community acceptance', the first issue to be highlighted is that exploring the acceptance of wind energy by key stakeholders and policymakers (at all levels: EU, national and local) clearly requires further efforts. Nevertheless, the available evidence on stakeholders' and policymakers' acceptance does provide essential insights. The very first investigation on stakeholders' acceptance was carried out in the 1980s, when it was first acknowledged that 'the siting of wind turbines is also a matter of ... political and regulatory acceptance' (Carlman, 1984) and the need to analyse the views of politicians and decision-makers was recognised. The pioneer study on 'institutional frameworks' is from the mid-1990s, when energy policy, policy performance and policy choices related to wind energy in the Dutch context were first analysed (Wolsink, 1996).

The most recent research on stakeholders' acceptance is paying special attention to the so-called 'institutional landscapes' and how diverse types of such landscapes are related to different levels of wind implementation (and ways of achieving it) at the EU level.

With this aim, how key stakeholders in the energy field perceive issues such as political commitment (and the perceived 'urgency' of energy-related matters), financial incentives (models of local financial participation) and planning systems (patterns of early local involvement in the decision-making process) have been analysed in multiple cases studies from several EU countries (The Netherlands, the UK, and the German state of North Rhine Westphalia) from the 1970s to 2004 (Breukers and Wolsink, 2007). A very similar approach was proposed by Toke et al.(2008) to understand the different outcomes of implementation of wind power deployment in five EU countries: Denmark, Spain, Scotland, The Netherlands and the UK). Different national traditions related to four key institutional variables (planning systems; financial support mechanisms; presence and roles of landscape protection organisations; and patterns of local ownership) were examined to identify and understand their inter-relations and how they might be related to the different levels of wind power implementation between countries. This recent research on stakeholders' and policymakers' acceptance has allowed the identification of two crucial factors for the successful implementation of wind energy: the financial incentives and the planning systems. With regard to financial incentives, evidence shows that participation or co-ownership is crucial in successful developments (the feed-in system in combination with support programmes promoting the involvement of a diversity of actors has proved to be the most efficient policy) (Breukers and Wolsink, 2007). As far as the planning system is concerned, evidence shows that planning regimes supporting collaborative practices of decision-making increase the correspondence between policy intentions and the outcome of the process (bottom-up developments have also proved to be the most successful ones) (Toke et al., 2008). Results of an extensive stakeholder consultation carried out on behalf of the European Commission to identify, among other things, the main 'institutional' barriers to exploiting renewable energy sources for electricity production

(Coenraads et al., 2006; see also Chapter IV.5 of this volume) fully supports this picture, as the 'administrative' and 'regulatory' barriers were perceived to be the most severe.

Consequently, and in line with the latest findings of the social research on community acceptance, a key message can be drawn from the most recent analysis on stakeholders' and policymakers' acceptance: facilitating local ownership and institutionalising participation in project planning could allow a better recognition and involvement of the compound interests (environmental, economic and landscape) that are relevant for the implantation of wind energy.

The Social Research on Wind Energy Offshore

In recent years, there has been a growing interest in the analysis of the public reactions to offshore wind power. Although the available empirical evidence is much more limited than that available for onshore wind development, studies from different countries have explored the main factors shaping public attitudes towards offshore projects as well as whether public acceptance differs between offshore and onshore.

As was the case with wind onshore, at the first stages of the technological development the physical and environmental attributes are the ones attracting more attention from social researchers. Offshore projects could also face negative reactions and promoters should be aware that 'coastal communities are just as sensitive to threats to seascapes as rural society is to visual disturbance in highland areas' (Ellis et al., 2007, p536). A study of residents near a proposed development near Cape Cod in the US (Firestone and Kempton, 2007) found that the main factors affecting individuals' reaction to the offshore project were damage to marine life and the environment. The next most frequently mentioned effects were aesthetics, impacts on fishing or boating, and electricity rates. The majority of participants expected negative impacts from the project. However, the project was more supported if turbines were located further offshore. Bishop and Miller (2007), by means of a survey using offshore wind farm simulations in Wales, found less negative response, in terms of perceived visual impact, to moving than to static turbines and to distant than to near turbines.

The available evidence on the psycho-social, social and institutional factors underlying the acceptance of wind offshore in the EU is to be found in Denmark, the country with the longest experience with offshore developments. A longitudinal study using qualitative techniques compared the reactions to Horns Rev and Nysted offshore wind farms (Kuehn, 2005; Ladenburg, in press). The authors found that at both sites, support for the project was linked to environmental attitudes, in a context of climate change and commitments to reduce CO_2 emissions. Another key argument for support was the expected occupational impact at the local level, in other words the expected number of jobs to be generated by the wind development. On the other hand, negative attitudes were based on concerns about visibility and negative impacts on the horizon. At Horns Rev opposition centred on business interests and tourism, while at Nysted the crucial issue was the need to not interfere with nature and preserve it as it was ('intact'). Regarding the planning process, most interviewees showed a feeling of being ignored in the decision-making process, as the decision on the wind park had already been taken by central authorities. Supporters tended to be active in the local debates. A recent study in Denmark (Ladenburg, 2008) has compared local attitudes towards offshore and onshore projects. The study finds that respondents tend to prefer offshore to onshore. Even if onshore wind power is perceived as an acceptable solution to the Danish public (only 25 per cent were opposed to an increase in the number of turbines onshore), respondents were more positive to more offshore wind turbines (only 5 per cent of respondents were opposed).

As far as key stakeholders in the development of renewable energies are concerned, as mentioned in the previous section, the OPTRES study demonstrates that the administrative and regulatory barriers are perceived to be the most severe to the development of wind offshore.

Even though, as the Figure V.6.5 illustrates, the importance attributed to the different barriers varies substantially from onshore to offshore. The accumulated experience in onshore and the lack of it in offshore may be a relevant factor in this regard, together with the management of the interaction between the technology and the community. Consequently, another important message can be drawn from this extensive stakeholder consultation: offshore and onshore wind seem to present relevant differences in the relative perceived importance of the barriers to their development.

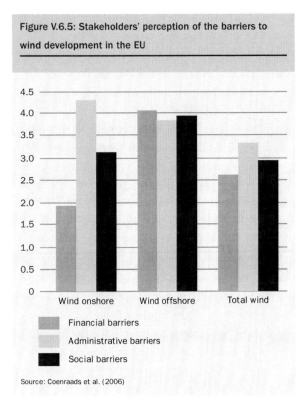

Figure V.6.5: Stakeholders' perception of the barriers to wind development in the EU

Financial barriers
Administrative barriers
Social barriers

Source: Coenraads et al. (2006)

Conclusions

As shown in this chapter, social research linked to wind energy developments has increased in the last few years, and such efforts have allowed a better understanding of the complexity and multidimensionality underlying the social acceptance of wind energy. Thus social research in the wind energy field has allowed the characterisation of factors explaining success or failure of wind developments. A general typology of factors involved has been proposed:

- factors related to the technical characteristics of the technology (physical and environmental characteristics of the site and technical attributes of wind energy);
- factors related to the individual and collective profile of the community hosting such technology (psycho-social factors); and
- factors related to the interaction between technology and society (social and institutional factors).

In this sense, and in order to capture the wide range of factors involved in the development of wind energy, we have provided a more complete formulation of the concept of 'social acceptance'. Three key dimensions have been identified: community acceptance ('the siting discourse'), market acceptance ('the investment discourse') and socio-political acceptance ('the energy-innovation discourse'). Future research needs to focus on these different dimensions as well as how they interact. In terms of socio-political acceptance, special attention should be paid to the development and implementation of suitable financial instruments and fair planning policies. In terms of community acceptance, proper institutional arrangements (including a comprehensive consideration of landscape issues) that could support trust-building processes also require further efforts. Methodological and conceptual improvements, integrated frameworks, and the evaluation of a citizen engagement process will be key elements in the social research on wind energy in the coming years.

A proper consideration of this wide range of issues may provide significant insights to a more evidence-based decision-making process on wind energy developments. There are no recipes to manage social acceptance on technological issues, but more precise knowledge may help promoters and authorities learn from past experiences and find mechanisms to improve citizen engagement with wind energy development.

Part V Notes

[1] For more information on the ECLIPSE project, visit http:// 88.149.192.110/eclipse_eu/index.html

[2] More information on the EC project NEEDS is available at www.needs-project.org/.

[3] For more information on the EC project CASES, see www. feem-project.net/cases/.

[4] The Ecoinvent data v2.0 contains international industrial life-cycle inventory data on energy supply, resource extraction, material supply, chemicals, metals, agriculture, waste management services, and transport services. More information on the Ecoinvent database is accessible at http://www.ecoinvent.org/

[5] Life Cycle Assessments, 2005 Report, Vestas Wind Systems A/S, available at http://www.vestas.com/en/ about-vestas/sustainability/wind-turbines-and-the-environment/life-cycle-assessment-(lca).aspx.

[6] There is often confusion between 'Annex I' (to the UNFCCC) and 'Annex B' (to the Kyoto Protocol). The list of 'industrialised countries' in each is the same, except that Turkey and Belarus are in Annex I but not Annex B. Belarus applied to join Annex B at COP 12 and has been submitted to the Parties for ratification of the amendment to Annex B. Turkey has recently decided to ratify Kyoto, and may apply to join Annex B. Liechtenstein, Slovenia, Slovakia, Croatia and the Czech Republic are in Annex B but not in Annex I.

SCENARIOS AND TARGETS

Acknowledgements

Part VI was compiled by Arthouros Zervos of the National Technical University of Athens, Greece (www. ntua.gr), and Christian Kjaer of EWEA.

PART VI   INTRODUCTION

In December 2008, the EU agreed to a 20 per cent binding target for renewable energy for 2020. The agreement means that more than one third of the EU's electricity will come from renewable energy in 2020, up from 15 per cent in 2005. To achieve this, the European Commission has calculated that 12 per cent of EU electricity should come from wind power.

Part VI takes different scenarios and targets for wind energy development from the industry, the International Energy Agency (IEA) and the European Commission and compares them. It makes sense of what they mean in financial, environmental, industrial and political terms, both for the EU and globally.

It explains how factors such as energy efficiency, offshore development and political decision-making will have a significant effect on whether current scenarios for total installed capacity and the percentage of electricity coming from wind power hold true. Moreover, fluctuating oil prices affect avoided fuel costs, and carbon prices determine how much wind energy saves in avoided CO_2.

These uncertainties have made it necessary for the European Wind Energy Association (EWEA), the Global Wind Energy Council (GWEC), the European Commission and the IEA to develop differing scenarios for wind energy development to 2020 and 2030.

Part VI of this volume uses a wide variety of graphs and charts to depict and compare the various possibilities. It looks at what these translate into in terms of electricity production from wind. It discusses the potential evolution of the cost of installed wind power capacity and of the expenditure avoided thanks to wind's free fuel, again comparing EWEA, European Commission and IEA scenarios.

Overall, the chapters in this final part demonstrate through detailed analysis the relatively indefinite, albeit bright, future of wind energy in Europe and worldwide. Wind energy is set to continue its impressive growth and become an ever more mainstream power source. Yet specific scenarios will remain open to conjecture and modification due to the vast quantity of unknowns to which wind energy development is subject.

Overview and Assessment of Existing Scenarios

The European Commission's 1997 White Paper on renewable sources of energy set the goal of doubling the share of renewable energy in the EU's energy mix from 6 per cent to 12 per cent by 2010. It included a target of 40,000 MW of wind power in the EU by 2010, producing 80 TWh of electricity and saving 72 million tonnes (Mt) of CO_2. The 40,000 MW target was reached in 2005. Another target of the White Paper was to increase the share of electricity from renewable energy sources from 337 TWh in 1995 to 675 TWh in 2010. By the end of 2007, there was 56,535 MW of wind power capacity installed in the EU, producing 119 TWh of electricity and saving approximately 90 Mt of CO_2 annually.

The European Commission's White Paper was followed by Directive 2001/77/EC on the promotion of electricity from renewable energy sources. This important piece of legislation for renewables has led the 27 Member States to develop frameworks for investments in renewable energy. These frameworks had to include financial instruments and reduce both administrative and grid access barriers.

The directive set national indicative targets for the contribution of electricity from renewables as a percentage of gross electricity consumption. The overall goal set out in the directive was to increase the share of electricity coming from renewables from 14 per cent in 1997 to 22 per cent (21 per cent after enlargement) in 2010. With the latest EU directive for the promotion of renewables, more than one third of the EU's electricity will come from renewable energy in 2020.

The 40,000 MW goal from the European Commission's White Paper formed EWEA's target in 1997, but three years later, due to the strong developments in the

German, Spanish and Danish markets for wind turbines, EWEA increased its target by 50 per cent to 60,000 MW by 2010 (and 150,000 MW by 2020). In 2003, EWEA once again increased its target, this time by 25 per cent to 75,000 MW by 2010 (and 180,000 MW by 2020). Due to the expansion of the EU with 12 new Member States, EWEA has now increased its prediction for 2010 to 80,000 MW, while maintaining its 2020 target of 180,000 MW and setting a target of 300,000 MW by 2030.

EWEA/Winter

VI.1 SCENARIOS FOR THE EU-27

While EWEA is confident that its predictions for wind power capacity in the EU to 2010 will be met, there is uncertainty about the projections for 2020 and 2030. The likelihood of a significant market for offshore wind power has been pushed beyond the 2010 timeframe, predominantly as a result of strong onshore wind market growth in the US, China and India in recent years. Much also depends on the future EU regulatory framework for the period after 2010.

In 2008, EWEA published three scenarios – low, reference and high – for the development of wind energy up to 2030.[1]

Much of the development over the coming two decades will depend on the evolution of the offshore market, over which there is currently some uncertainty. In December 2007, the European Commission announced a Communication on Offshore Wind Energy. As mentioned, EWEA's reference scenario assumes 180 GW of installed wind energy capacity in 2020 and 300 GW in

2030. The EU will have 350 GW (including 150 GW off-shore) in the high scenario and 200 GW (including 40 GW offshore) in the low scenario in 2030.

The 56.5 GW of installed capacity in the EU-27 by the end of 2007 produces, in a normal wind year, 119 TWh of electricity, enough to meet 3.7 per cent of EU electricity demand.

In terms of wind power's electricity production and its share of total EU power demand, there are large differences between the three scenarios. Much depends on whether total electricity demand in the EU increases according to the European Commission's business-as-usual (BAU) scenario or stabilises according to its energy efficiency (EFF) scenario.

As can be seen from Table VI.1.1, wind power will produce between 176 TWh (low scenario) and 179 TWh (high scenario) in 2010, between 361 TWh and 556 TWh in 2020, and between 571 TWh and 1104 TWh in 2030.

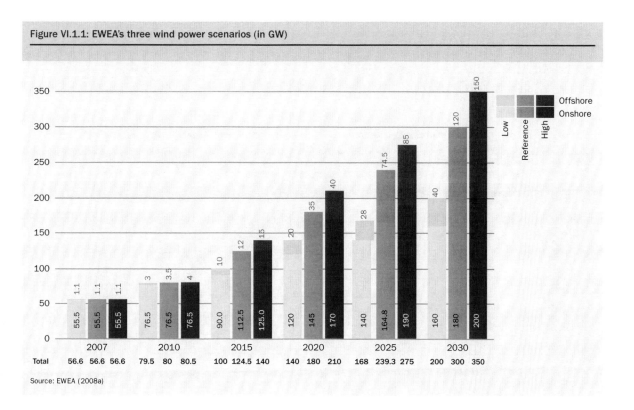

Figure VI.1.1: EWEA's three wind power scenarios (in GW)

	2007			2010			2015			2020			2025			2030		
Total	56.6	56.6	56.6	79.5	80	80.5	100	124.5	140	140	180	210	168	239.3	275	200	300	350

Source: EWEA (2008a)

Table VI.1.2 shows that in EWEA's reference scenario, wind energy meets between 5.0 per cent (BAU) and 5.2 per cent (EFF) of EU electricity demand in 2010, between 11.6 per cent and 14.3 per cent in 2020, and between 20.8 per cent and 28.2 per cent in 2030, depending on how overall electricity consumption develops in the EU between now and 2030.

The calculations in the following sections are based on EWEA's reference scenario and the European Commission's BAU scenario for electricity consumption.

It is assumed that the average capacity factor of all wind turbines in the EU will increase from 24 per cent in 2007 to 25.3 per cent in 2010 and 30.3 per cent in 2020. The increase will be due to better design, exploiting the resources in more windy areas of Europe, technology improvements and a larger share of offshore wind. In Germany, average capacity factors will only start increasing if older turbines start being replaced and offshore wind power takes off. It should be noted that for a technology that makes use of a free resource, a high capacity factor is not a goal in itself. It is not technically problematic to increase capacity factors, but doing so affects grid integration, modelling and generation costs.

Table VI.1.1: Electricity production (in TWh) for EWEA's three scenarios

	Low			Reference			High		
	Onshore	Offshore	Total	Onshore	Offshore	Total	Onshore	Offshore	Total
2007	115	4	119	115	4	119	115	4	119
2010	165	11	176	165	13	177	165	15	179
2015	204	37	241	255	45	299	283	56	339
2020	285	76	361	344	133	477	403	152	556
2025	350	109	459	412	289	701	475	330	805
2030	415	156	571	467	469	935	519	586	1,104

Table VI.1.2: Share of EU electricity demand from wind power, for EWEA's three scenarios and the two EC projections for electricity demand

	Low			Reference			High		
	Onshore	Offshore	Total	Onshore	Offshore	Total	Onshore	Offshore	Total
2007 share EFF				3.5%	0.1%	3.7%			
2007 share BAU				3.5%	0.1%	3.7%			
2010 share EFF	4.9%	0.3%	5.2%	4.9%	0.4%	5.2%	4.9%	0.4%	5.3%
2010 share BAU	4.6%	0.3%	4.9%	4.6%	0.4%	5.0%	4.6%	0.4%	5.0%
2020 share EFF	8.5%	2.3%	10.8%	10.3%	4.0%	14.3%	12.1%	4.6%	16.6%
2020 share BAU	6.9%	1.9%	8.8%	8.4%	3.2%	11.6%	9.8%	3.7%	13.5%
2030 share EFF	12.5%	4.7%	17.2%	14.1%	14.1%	28.2%	15.6%	17.6%	33.2%
2030 share BAU	9.2%	3.5%	12.7%	10.4%	10.4%	20.8%	11.5%	13.0%	24.5%

VI.2 PROJECTING TARGETS FOR THE EU-27 UP TO 2030

Targets for 2010

EWEA's target for 2010 assumes that approximately 23.5 GW of wind energy will be installed in 2008–2010. The Danish wind energy consultancy BTM Consult is more optimistic than EWEA, and foresees a cumulative installed capacity of 91.5 GW by the end of 2010. The main growth markets it highlights are Portugal, France and the UK.

By the end of 2007, 1.9 per cent of wind capacity in the EU was in offshore installations, producing 3.4 per cent of total wind power in Europe. In 2010, EWEA expects 4.4 per cent of total capacity and 16 per cent of the annual market to be covered by offshore wind. Offshore wind power's share of total EU wind energy production will increase to 7 per cent by 2010.

The 56.5 GW of installed capacity in the EU-27 by the end of 2007 will, in a normal wind year, produce 119 TWh of electricity, enough to meet 3.7 per cent of EU electricity demand. The capacity installed by the end of 2010 will produce 177 TWh in a normal wind year, equal to 5 per cent of demand in 2010 (5.7 per cent of 2006 demand). With efficiency measures, wind power's share would cover 5.2 per cent of electricity demand in 2010.

Germany is projected to reach 25 GW and Spain 20 GW of wind capacity in 2010. France, the UK, Italy, Portugal and The Netherlands constitute a second wave of stable markets and will install 42 per cent of new EU capacity over the 2008–2010 period.

For 2008, the annual EU market is expected to fall back to its 2006 level and then increase slightly up to 2010, when it should reach 8200 MW. The forecast assumes that the negotiations on a new EU Renewable Energy Directive and the subsequent development of national action plans in the Member States could cause some legal uncertainty until implemented.

In the three-year period from 2007 to 2010, EWEA forecasts that 23.5 GW of wind energy capacity, including 2.4 GW offshore, will be installed. This will equate to total investments of €31 billion.

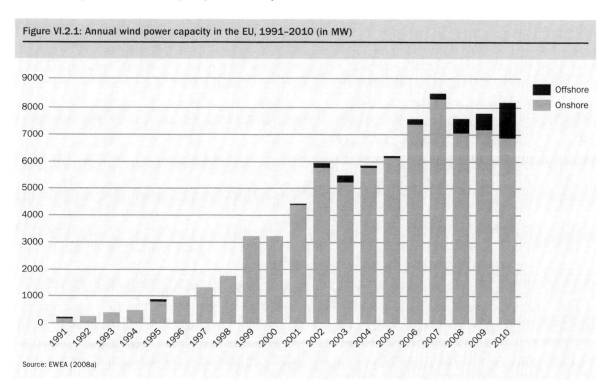

Figure VI.2.1: Annual wind power capacity in the EU, 1991–2010 (in MW)

Source: EWEA (2008a)

Figure VI.2.2: Cumulative capacity in the EU, 1991–2010 (in MW)

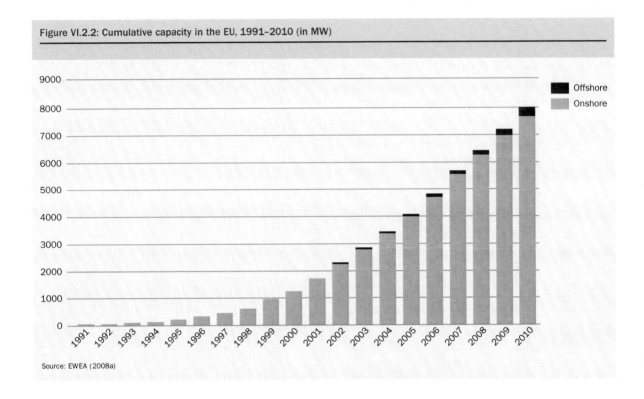

Source: EWEA (2008a)

Over the same three-year period, Germany and Spain's share of the European annual market will be 34 per cent, compared to 60 per cent in 2007 and 80 per cent in 2002, confirming the healthy trend towards less reliance on the first-mover markets. The largest markets in the period are expected to be Spain (20.7 per cent), Germany (14.4 per cent), France (12.1 per cent), the UK (11.6 per cent) and Italy (7.6 per cent). The total includes an additional 102 MW of capacity that should be built to replace turbines installed prior to 1991.

Targets for 2020

On 9 March 2007, the European Heads of State agreed on a binding target of 20 per cent renewable energy by 2020. The 2005 share of renewable energy was approximately 7 per cent of primary energy and 8.5 per cent of final consumption. In January 2008, the European Commission proposed a new legal framework for renewables in the EU, including a distribution of the 20 per cent target between Member States and national action plans containing sectoral targets for electricity, heating and cooling, and transport.

To meet the 20 per cent target for renewable energy, the European Commission expects 34 per cent[2] of electricity to come from renewable energy sources by 2020 (43 per cent of electricity under a 'least cost' scenario[3]) and believes that 'wind could contribute 12 per cent of EU electricity by 2020'.

In 2005 (the reference year of the proposed directive), approximately 15 per cent of EU electricity demand was covered by renewables, including around 10 per cent from large hydro and about 2.1 per cent from wind energy. Excluding large hydropower, for which the realisable European potential has already been reached, and assuming that electricity demand does not increase, the share of renewable electricity in the EU will need to grow fivefold – from approximately 5 per cent to 25 per cent – to reach the electricity target.

Figure VI.2.3: New wind power capacity in the EU, 2008–2010 (total 23,567 MW)

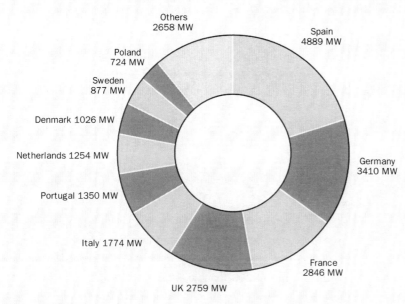

Others	
Greece	629 MW
Ireland	521 MW
Belgium	513 MW
Austria	218 MW
Czech Republic	134 MW
Bulgaria	130 MW
Finland	110 MW
Estonia	92 MW
Hungary	85 MW
Latvia	73 MW
Lithuania	50 MW
Romania	42 MW
Slovenia	25 MW
Slovakia	20 MW
Luxembourg	15 MW
Malta	0 MW
Cyprus	0 MW

Source: EWEA (2008a)

Figure VI.2.4: National overall targets for the share of RES in final energy consumption, 2020

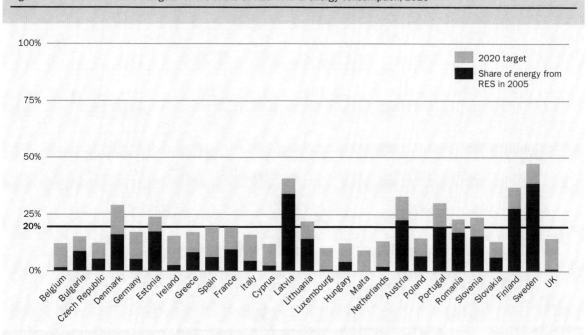

Source: European Commission draft proposal for a Directive on the promotion of the use of energy from renewable sources, EWEA (2008a)

Table VI.2.1: Targets for RES, electricity from RES and wind energy for 2020

	2005	2020
Renewable energy sources (RES)	8.5%	20%
Electricity from RES	15%	34%
Wind energy	2.1%	12–14%
Offshore wind energy	0	3.2–4%

With increased demand, renewable electricity other than large hydropower will need to grow even more.

EWEA maintains the target it set in 2003 of 180 GW by 2020, including 35 GW offshore in its reference scenario. That would require the installation of 123.5 GW of wind power capacity, including 34 GW offshore, in the 13-year period from 2008 to 2020; 16.4 GW of capacity is expected to be replaced in the period.

The 180 GW would produce 477 TWh of electricity in 2020, equal to between 11.6 per cent and 14.3 per cent of EU electricity consumption, depending on the development in demand for power. Twenty-eight per cent of the wind energy would be produced offshore in 2020.

Between 2011 and 2020, the annual onshore market for wind turbines will grow steadily from around 7 GW per year to around 10 GW per year. The offshore market will increase from 1.2 GW in 2011 to reach 6.8 GW in 2020. Throughout the period of the reference scenario, the onshore wind power market exceeds the offshore market in the EU.

A precondition for reaching the EWEA target of 180 GW is that the upcoming Renewable Energy Directive establishes stable and predictable frameworks in the Member States for investors. Much also depends on the European Commission's Communication on Offshore Wind Energy (scheduled for the second half of 2008) and a subsequent adoption of a European policy for offshore wind power in the EU.

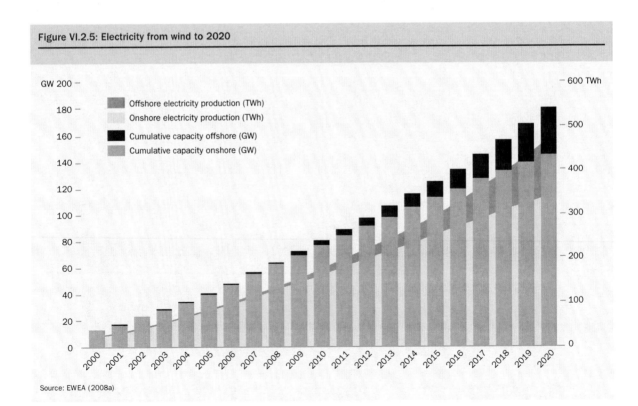

Figure VI.2.5: Electricity from wind to 2020

Offshore electricity production (TWh)
Onshore electricity production (TWh)
Cumulative capacity offshore (GW)
Cumulative capacity onshore (GW)

Source: EWEA (2008a)

Figure VI.2.6: Wind energy annual installations, 2000–2020 (in GW)

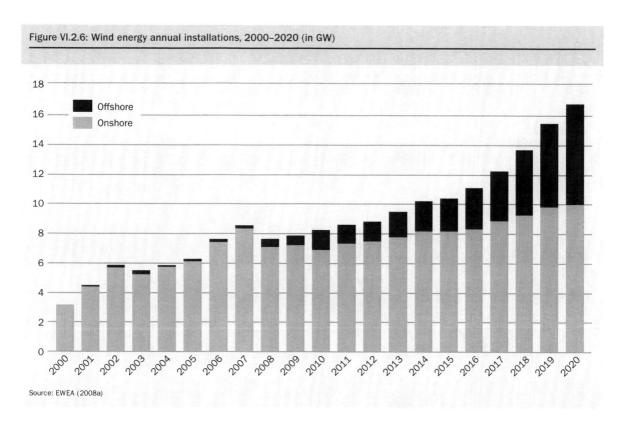

Source: EWEA (2008a)

Targets for 2030

In the EWEA reference scenario, 300 GW of wind power will be operating in the EU in 2030, including 120 GW (40 per cent) of offshore wind power. In the decade from 2021 to 2030, 187 GW will be installed. Of this, 67 GW will be needed to replace decommissioned capacity, predominantly onshore. Onshore will represent 54 per cent (101 GW) of the capacity installed during that decade and the onshore market will remain larger than the offshore market throughout, although the gap narrows towards the end. By 2030, the annual onshore market will be 9.9 GW and the offshore market 9.6 GW, representing investments of €19 billion. In 2025, the offshore market is expected to reach the size of the 2008 onshore market (8.5 GW).

Total installations in the period from 2008 to 2030 will be 327 GW, made up of 207 GW onshore and

120 GW offshore. Of this, 83 GW will come from the replacement of decommissioned onshore capacity. Total investments between 2008 and 2030 will be €339 billion.

By 2030, wind energy will produce 935 TWh of electricity, half of it from offshore wind power, and cover between 21 per cent and 28 per cent of EU electricity demand, depending on future power consumption.

The onshore market will stabilise at approximately 10 GW per year throughout the decade 2020–2030 and 72 per cent of the onshore market will come from the replacement of older wind turbines. The offshore segment increases from an annual installation of 7.3 GW in 2021 to 9.5 GW in 2030.

The wind power production in 2030 will avoid the emission of 575 Mt of CO_2, the equivalent of taking more than 280 million cars off the roads. In 2004 there were 216 million cars in the EU-25.

Figure VI.2.7: Electricity from wind to 2030

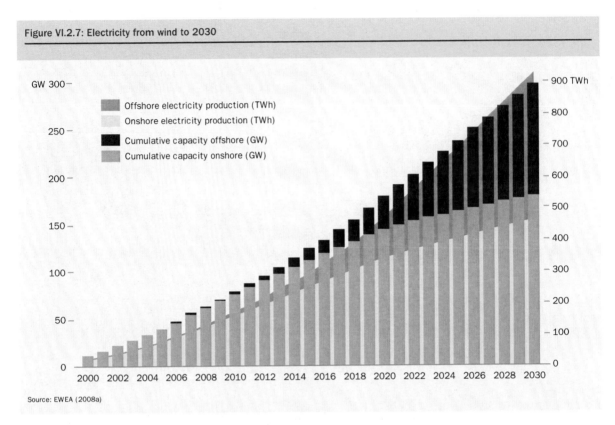

Source: EWEA (2008a)

Figure VI.2.8: Wind energy annual installations, 2000–2030 (in GW)

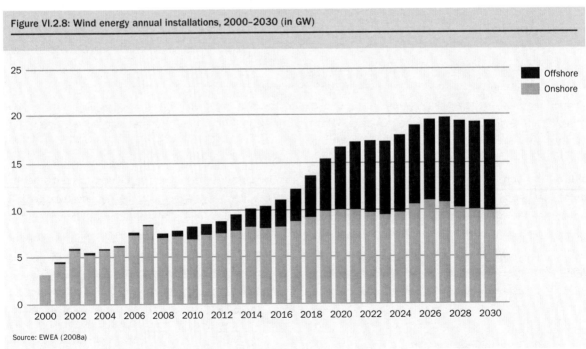

Source: EWEA (2008a)

VI.3 CONTRIBUTION OF WIND POWER TO ELECTRICITY GENERATION AND GENERATION CAPACITY IN THE EU-27

Contribution of Wind Power to Electricity Generation

European electricity generation is projected to increase at an average annual rate of 1.8 per cent between 2000 and 2010, 1.3 per cent in the decade 2010–2020, and 0.8 per cent in the decade up to 2030.

If the reference scenario is reached, wind power production will increase to 177 TWh in 2010, 477 TWh in 2020 and 935 TWh in 2030. The European Commission's baseline scenario assumes an increase in electricity demand of 33 per cent between 2005 and 2030 (4408 TWh). Assuming that EU electricity demand develops as projected by the European Commission, wind power's share of EU electricity consumption will reach 5 per cent in 2010, 11.7 per cent in 2020 and 21.2 per cent in 2030.

If political ambitions to increase energy efficiency are fulfilled, wind power's share of future electricity demand will be greater than the baseline scenario. In 2006, the European Commission released new scenarios to 2030 on energy efficiency and renewables. If EU electricity demand develops as projected in the European Commission's 'combined high renewables and efficiency' (RE & Eff) case, wind energy's share of electricity demand will reach 5.2 per cent in 2010, 14.3 per cent in 2020 and 28.2 per cent in 2030.

Contribution of Wind Power to Generation Capacity

The IEA expects 5087 GW of electricity generating capacity to be installed worldwide in the period 2005–2030, requiring investments of US$5.2 trillion in power generation, $1.8 trillion in transmission grids and $4.2 trillion in distribution grids. The IEA expects 862 GW of this total to be built in the EU, requiring investments of $925 billion in new generation, $137 billion in transmission and $429 billion in distribution grids.

As already mentioned, wind power's contribution to new power capacity in the EU was exceeded only by gas in the last eight years. Thirty per cent of all installed capacity in the period 2000 to 2007 was wind power, 55 per cent was natural gas and 6 per cent was coal-based.

Spare electricity generating capacity is at a historic low and phase-out policies in the EU Member States require 27 GW of nuclear plants to be retired. Europe has to invest in new capacity to replace aging plants and meet future demand. Between 2005 and 2030, a total of 862 GW of new generating capacity needs to be built, according to the IEA – 414 GW to replace aging power plants and an additional 448 GW to meet the growing power demand. The capacity required exceeds the total capacity operating in Europe in 2005 (744 GW).

Table VI.3.1: Wind power's share of EU electricity demand

	2000	2007	2010	2020	2030
Wind power production (TWh)	23	119	177	477	935
Reference electricity demand (TWh)	2577	3243	3568	4078	4408
RE & Eff case electricity demand (TWh)	2577	3243	3383	3345	3322
Wind energy share (reference) (%)	0.9	3.7	5.0	11.7	21.2
Wind energy share (RE & Eff case) (%)	0.9	3.7	5.2	14.3	28.2

Sources: Eurelectric, EWEA and European Commission

Figure VI.3.1: Wind power's share of EU electricity demand

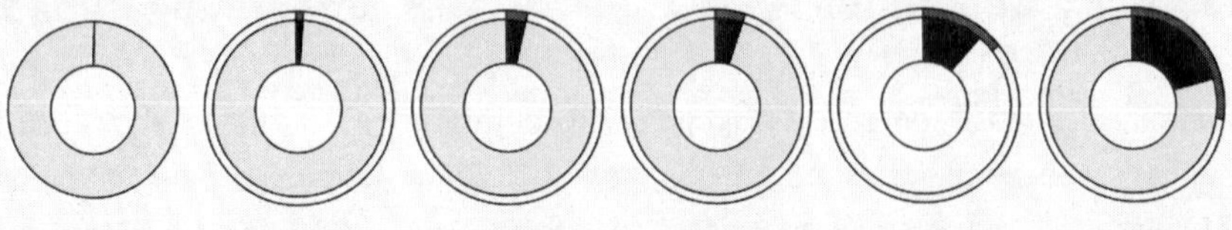

	1995	2000	2007	2010	2020	2030
Wind energy share (reference)	0.2%	0.9%	3.7%	5.0%	11.6%	20.8%
Wind energy (RE & Eff. case)		0.9%	3.7%	5.2%	14.3%	28.2%
Wind power production (TWh)		23	119	176	477	935
Reference electricity demand (TWh)		2577	3243	3554	4107	4503
RE & Eff. case electricity demand (TWh)		2577	3243	3383	3345	3322

Source: EWEA (2008a)

The IEA is less optimistic about the development of wind energy than EWEA. Hence, it is necessary to adjust the IEA figures for total generating capacity and new capacity to take account of the fact that wind energy's capacity factor is lower than that of the average coal, gas or oil plant. Adjusting for the capacity factor adds 18 GW to total generating capacity in 2030 to make a total of 1176 GW, and 26 GW to the figure for new generating capacity between 2005 and 2030 to make a total of 889 GW over the period.

In 2005, 5.4 per cent of all electricity generating capacity in the EU was wind energy. That share is forecast to increase to 9.9 per cent in 2010, 18.1 per cent in 2020 and 25.5 per cent in 2030. Wind power's share of new generating capacity is forecast to be 34 per cent in the period 2005–2020 and 46 per cent in the decade up to 2030. Wind power's share of new capacity in Europe in the 25-year period 2005–2030 should be 39 per cent.

Table VI.3.2: Wind power's share of installed capacity

	2005	2010	2020	2030
Total installed capacity (GW)	744	811	997	1176
Total installed wind capacity (GW)	40	80	180	300
Wind power's share of installed capacity (%)	5.4	9.9	18.1	25.5

Scenarios of the European Commission and the IEA

BASELINE SCENARIOS

Both the European Commission and the International Energy Agency (IEA) publish baseline scenarios for the development of various electricity-generating technologies, including wind energy. In 1996, the European Commission estimated that 8000 MW would be installed by 2010 in the EU. The 8000 MW target was reached in 1999. The Commission's target for 2020 was set at 12,300 MW and reached, two decades ahead of schedule, in 2000.

Since 1996, the European Commission has changed its baseline scenario five times. Over the 12-year period, targets for wind energy in 2010 and 2020 have been increased almost tenfold, from 8 GW to 71 GW (2010) and from 12 GW to 120 GW (2020) in the European Commission's latest baseline scenario from 2008. Surprisingly, the baseline scenario from 2008 gives significantly lower figures for wind energy than the baseline scenario from 2006. The 71 GW projection for 2010 implies that the wind energy market in Europe will decrease by approximately 50 per cent over the next three years with respect to the present market. In the light of the current market achievements, growth

Figure VI.3.2: Wind power's share of installed capacity

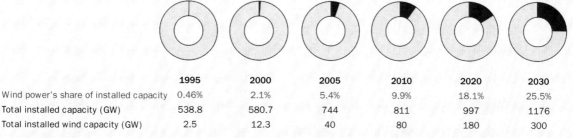

	1995	2000	2005	2010	2020	2030
Wind power's share of installed capacity	0.46%	2.1%	5.4%	9.9%	18.1%	25.5%
Total installed capacity (GW)	538.8	580.7	744	811	997	1176
Total installed wind capacity (GW)	2.5	12.3	40	80	180	300

Source: EWEA (2008a)

Table VI.3.3: Wind power's share of new capacity

	2005–2010	2011–2020	2021–2030
New generating capacity (GW)	117	368	404
New wind generating capacity (GW)	46	117	187
Wind power's share of new capacity (%)	39	32	46

trends and independent market analyses, the European Commission's baseline scenario seems completely out of touch with the market reality, and clearly underestimates the sector's prospects.

Figure VI.3.4 shows the forecast for average annual installations (GW) up to 2030 according to the European Commission's 2008 baseline scenario and to EWEA's baseline, or 'reference', scenario compared with the 2007 market level.

Historically, EWEA's scenarios have been somewhat conservative, and its targets have been revised upwards numerous times. EWEA's 2010 target (based on its 'reference' scenario) was doubled from 40 GW (in 1997) to 80 GW (in 2006). The EWEA reference scenario for 2020 is 60 GW higher than the Commission's baseline scenario. For 2030, the Commission assumes 146 GW while EWEA assumes 300 GW.

Figure VI.3.3: Wind power's share of new capacity

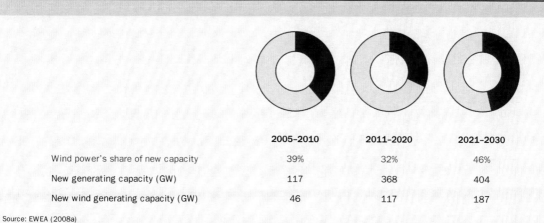

	2005-2010	2011-2020	2021-2030
Wind power's share of new capacity	39%	32%	46%
New generating capacity (GW)	117	368	404
New wind generating capacity (GW)	46	117	187

Source: EWEA (2008a)

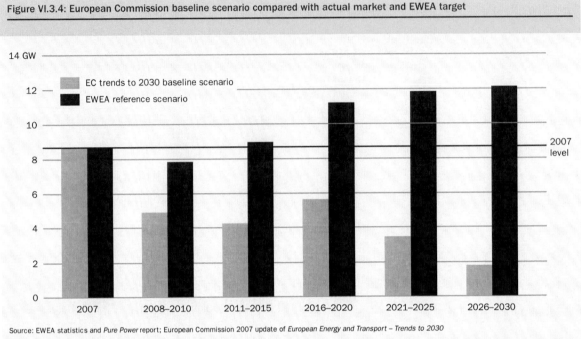

Figure VI.3.4: European Commission baseline scenario compared with actual market and EWEA target

Source: EWEA statistics and *Pure Power* report; European Commission 2007 update of *European Energy and Transport – Trends to 2030*

Table VI.3.4 shows the European Commission's various scenarios for wind energy installations up to 2030, compared with the actual market up to 2008 and EWEA's 2007 scenario up to 2030.

Figure VI.3.5 shows the European Commission's 2008 baseline scenario compared with the EWEA target up to 2030.

The IEA also produces baseline scenarios for the development of wind power. In 2002, the Agency estimated that 33 GW would be installed in Europe in 2010, 57 GW by 2020 and 71 GW by 2030. Two years later, in 2004, it doubled its forecast for wind energy to 66 GW in 2010, and more than doubled its 2020 and 2030 business-as-usual scenarios for wind in the EU to 131 GW in 2020 and 170 GW in 2030. In 2006, the IEA again increased its 2030 target for wind power in the EU to 217 GW (its alternative policy scenario assumes 227 GW). The IEA's reference scenario

Table VI.3.4: European Commission scenarios compared with actual market, EWEA 2008 reference scenario

	1995	2000	2005	2010	2015	2020	2025	2030
EC 1996		4.4	6.1	8.0	10.1	12.3		
EC 1999			15.3	22.6		47.2		
EC 2003				69.9		94.8		120.2
EC 2004	2.5	12.8		72.7		103.5		134.9
EC 2006		12.8	37.7	78.8	104.1	129.0	165.8	184.5
EC 2008 reference scenario			40.8	71.3	92.2	120.4	137.2	145.9
Actual market/EWEA 2007 target	2.497	12.887	40.5	80.0	124.5	180.0	239.3	300.0

Figure VI.3.5: European Commission's 2008 baseline scenario compared with the EWEA target up to 2030 (in GW)

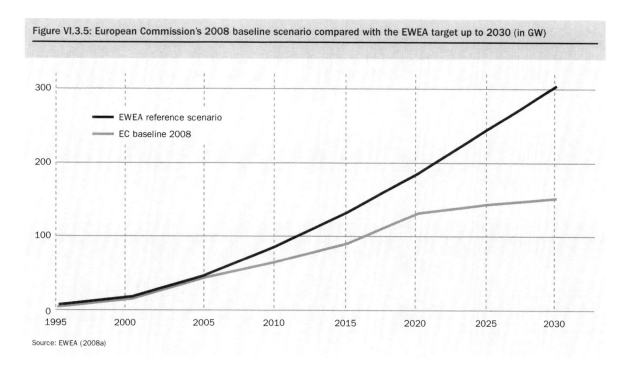

Source: EWEA (2008a)

assumes 68 GW in 2010, 106 GW in 2015, 150 GW in 2020 and 217 GW in 2030. EWEA's reference scenario assumes 80 GW in 2010, 125 GW in 2015, 180 GW in 2020 and 300 GW in 2030.

The European Commission's baseline scenario claims to take 'into account the high energy import price environment', by assuming an oil price of US$55/barrel in 2005, $44.6/barrel in 2010 and $62.8/barrel in 2030. In its 2006 scenario, the IEA assumes an oil price of $47 in 2015, reaching $55 in 2030. In July

2008, the crude oil prices[4] reached an all-time high of $147 a barrel. At the time of writing, there are indications that the IEA will increase its oil price forecast for 2020 to the $100–$120 range.

Table VI.3.5 shows the IEA's various scenarios for wind energy installations in Europe up to 2030, compared with the actual market up to 2007, followed by EWEA's 2008 scenario up to 2030.

Figure VI.3.6 shows the IEA's 2006 reference scenario compared with the EWEA target up to 2030.

Table VI.3.5: IEA's scenarios up to 2030 compared with actual market/EWEA 2007 target

	1995	2000	2005	2010	2015	2020	2025	2030
IEA 2002				33.0		57.0		71.0
IEA 2004				66.0		131.0		170.0
IEA 2006 – reference				68.0	106.0	150.0		217.0
IEA 2006 – APS*				71.0	108.0	151.0		223.0
Actual market/EWEA 2007 target	2.5	12.9	40.5	80.0	124.5	180.0	239.3	300.0

*Alternative policy scenario

Figure VI.3.6: IEA's 2006 baseline scenario compared with the EWEA target up to 2030 (in GW)

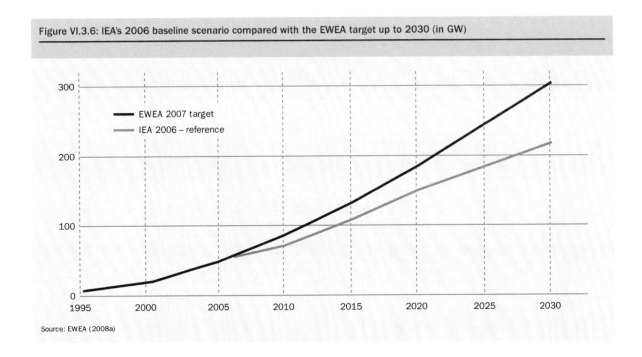

Source: EWEA (2008a)

Table VI.3.6 shows EWEA's various scenarios for wind energy installations up to 2030, compared with the actual market up to 2007.

In its *World Energy Outlook 2006*, the IEA adopts a rather pessimistic view towards future wind energy installations around the globe, particularly as far as the US and the Chinese markets are concerned. Table VI.3.7 shows that a yearly averaging out of the installations required to reach the IEA 2015 cumulative target results in installation figures significantly below current market levels. At the time of writing, the IEA's

World Energy Outlook 2008 has not been published, but there are indications that the Agency's forecast for global wind energy development will be increased to better reflect market expectations.

ADVANCED SCENARIOS

In addition to the baseline/business-as-usual scenarios, the European Commission and the IEA have in recent years published more advanced scenarios with less static assumptions. The European Commission's

Table VI.3.6: EWEA's scenarios up to 2030 compared with the actual market/EWEA 2007 target

	1995	2000	2005	2010	2015	2020	2025	2030
EWEA 1997				40				
EWEA 2000				60		150		
EWEA 2003				75		180		
Actual market/ EWEA 2007 target	2.5	12.9	40.5	80	125	180	165	300

Table VI.3.7: GWEC and IEA 2015 cumulative targets, present market levels and projection of average yearly installations to reach the IEA 2015 target (reference scenario, GW)

	2007 cumulative (GWEC)	2015 cumulative (IEA)	2007 annual (GWEC)	Average/year 2008–2015 (IEA)
World	93.9	168	19.9	9.2
OECD North America	18.7	30	5.6	1.4
European Union	56.5	106	8.5	6.2
China	5.9	7	2.6	0.1

Sources: GWEC (2008) and IEA *World Energy Outlook 2006*

new scenarios on energy efficiency and renewables from 2006 assume that 'agreed policies will be vigorously implemented in the Member States and that certain new targets on the overall share of renewables in 2020 will be broadly achieved'. However, the underlying estimates of fuel and carbon prices are no different from the baseline scenario.

Both the European Commission's and the IEA's advanced scenarios from 2004 are in line with the 80 GW target in 2010 from EWEA. However, the 2020 and 2030 targets from the IEA and the European Commission are significantly below EWEA's targets. The 2006 IEA alternative policy scenario for the EU (151 GW in 2020) is, somewhat surprisingly, only 1 GW higher than its reference scenario. Its 2030 alternative policy scenario is a mere 6 GW higher than its reference scenario (217 GW). The European Commission's advanced 2006 scenarios are more in line with the EWEA targets, and even exceed EWEA's targets for 2020.

Figure VI.3.7: Advanced scenarios for 2010, 2020 and 2030 (in GW)

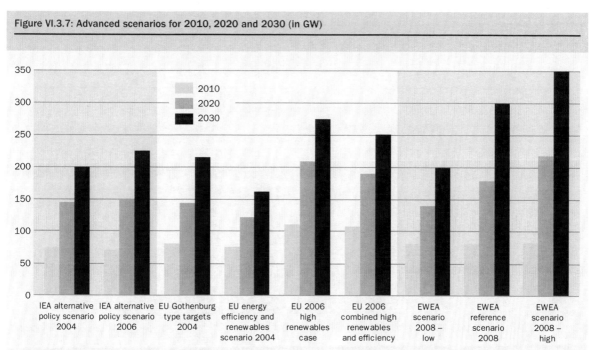

Source: EWEA (2008a)

VI.4 COSTS AND BENEFITS OF WIND DEVELOPMENT IN THE EU-27

Generation Costs and Investments

One of the significant advantages of wind power is that the fuel is free. Therefore, the total cost of producing wind energy throughout the 20- to 25-year lifetime of a wind turbine can be predicted with great certainty. Neither the future prices of coal, oil or gas, nor the price of carbon, will affect the cost of wind power production significantly.

In order to calculate the wind power investments needed to reach EWEA's reference scenario, it is necessary to make assumptions regarding the future cost of installed wind power capacity. For some years, it has been assumed as a rule of thumb that installed wind power capacity costs approximately €1000/kW. That is probably still valid. However, since 2000 there have been quite large variations in the price (not necessarily the cost) of installing wind power capacity; these were described in Part III – The Economics of Wind Power.

In the period 2001 to 2004, the global market for wind power capacity grew less than expected, and

created a surplus in wind turbine production capacity. Consequently, the price of wind power capacity went down dramatically – to €700–800/kW for some projects. In the three years to 2007, the global market for wind energy increased by 30–40 per cent annually, and demand for wind turbines surged, leading to increases in prices.

The European Commission, in its Renewable Energy Roadmap,[5] assumes that onshore wind energy cost €948/kW in 2007 (in €2005). It assumes that costs will drop to €826/kW in 2020 and €788/kW in 2030. That long-term cost curve may still apply for a situation where there is a better balance between demand and supply for wind turbines than at the present time.

Figure VI.4.1 shows the European Commission's assumptions on the development of onshore and offshore wind power capacity costs up to 2030. In addition, there are two curves that reflect the effect of the current demand/supply situation on wind turbine prices in recent years. EWEA assumes onshore wind energy prices of €1300/kW in 2007 (€2005 prices) and offshore prices of €2300/kW. The steep increase in

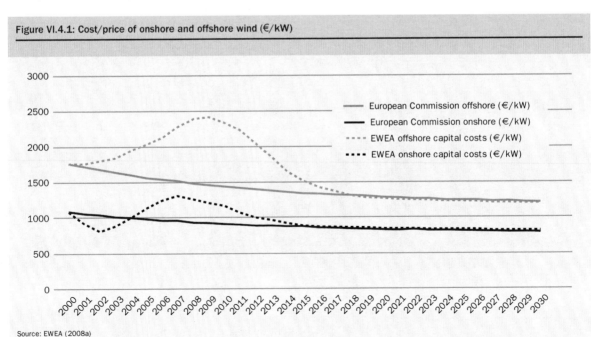

Figure VI.4.1: Cost/price of onshore and offshore wind (€/kW)

— European Commission offshore (€/kW)
— European Commission onshore (€/kW)
···· EWEA offshore capital costs (€/kW)
▪▪▪▪ EWEA onshore capital costs (€/kW)

Source: EWEA (2008a)

offshore prices reflects the limited number of manufacturers in that market, the current absence of economies of scale due to low market deployment and bottlenecks in the supply chain.

Based on the EWEA reference scenario for installed capacity up to 2030 and the wind power capacity prices above, Figure VI.4.2 shows the expected annual wind power investments from 2000 to 2030. The market is expected to stabilise at around €10 billion per year up to 2015, with a gradually increasing share of investments going to offshore. By 2020, the annual market for wind power capacity will have grown to €17 billion annually, with approximately half of investments going to offshore. By 2030, annual wind energy investments in the EU-27 will reach almost €20 billion, with 60 per cent of investments offshore.

Cumulative investments in wind energy over the three decades from 2000 to 2030 will total €390 billion. According to EWEA's reference scenario, approximately €340 billion will be invested in wind energy in the EU-27 between 2008 and 2030. This can be broken down into €31 billion in 2008–2010, €120 billion in 2011–2020 and €188 billion in 2021–2030.

The IEA (2006) expects that €925 billion of investment in electricity generating capacity will be needed for the period 2005 to 2030 in the EU. According to the EWEA reference scenario, €367 billion – or 40 per cent – of that would be investment in wind power.

Avoided Fuel Costs

Fuel is not required to produce wind power. When wind energy is produced, it saves significant amounts of fuel costs in the form of coal, gas and oil that would otherwise have been needed for power production. In addition to these avoided costs, the production of wind energy reduces demand for imported fuel (and thereby the cost of fuel), while reducing the rate of depletion of Europe's remaining fossil fuel reserves.

Naturally, the avoided fuel costs of wind energy depend on the assumptions made about future fuel prices. Oil and gas prices are very closely linked, and coal also follows, to a lesser extent, the price of oil. Both the IEA and the European Commission have for many years made predictions on future coal, gas and oil prices, and most governments base their energy policies on the IEA's fuel price scenarios. Historically, the IEA and European Commission scenarios have been similar, and both institutions have been very consistent in underestimating the future fuel prices.

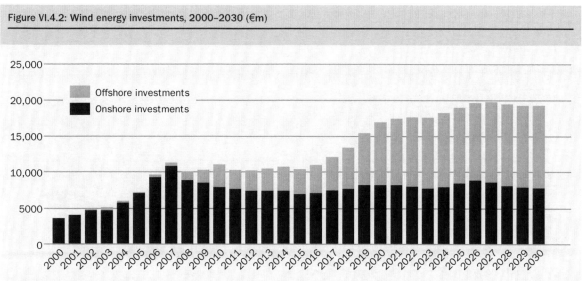

Figure VI.4.2: Wind energy investments, 2000–2030 (€m)

Source: EWEA (2008a)

Table VI.4.1: Oil price assumptions

Oil price assumptions (in US$2005)*	2000	2005	2007	2010	2015	2020	2025	2030
European Commission, 2007	31.3	57.1	68.9	54.5	57.9	61.1	62.3	62.8
International Energy Agency, 2007	31.5	57.1	68.9	57.2	55.5	57.0	58.5	60.1
EWEA, 2008	31.3	57.1	68.9	100.0	105.0	110.0	115.0	120.0

* Adjusted to 2005 prices/actual prices until 2007.

A barrel of oil cost US$100 at the start of 2008, and reached a record $147 in July. The IEA predicts that the oil price will fall to $57 in 2010. In 2004, the IEA predicted that oil would cost $22 a barrel in 2010, $26 in 2020 and $29 in 2030 (in year-2000 dollars).

Table VI.4.1 shows the latest oil price estimates from the European Commission (2007) and the IEA (2007) and an alternative oil price scenario from EWEA. As the table shows, the European Commission believes that the price of oil in 2010 will be approximately 60 per cent lower than today (around $120 in September 2008), while the IEA estimates a drop in the price of oil to circa $57 three years from now. Both institutions believe that the price of oil in 2030 will be approximately $60 a barrel – 50 per cent lower than today.

Nobody can predict oil prices, but it should be a minimum requirement that the European Commission

and the IEA include fuel price sensitivity analysis in their scenarios for the future development of the energy markets.

The fuel costs avoided due to wind energy production can be calculated on the basis of the European Commission's fuel price assumptions for coal, oil and gas up to 2030. As Figure VI.4.3 shows, wind energy avoided €3.9 billion of fuel costs in 2007: €1.7 billion worth of gas, €1.2 billion worth of coal, €0.7 billion worth of oil and €0.3 billion worth of biomass/waste. In EWEA's reference scenario, wind energy will avoid fuel costs of €4.4 billion in 2010, €12 billion in 2020 and €24 billion in 2030, based on the European Commission's fuel price assumptions. Similar results emerge from using the IEA fuel price assumptions.

Assuming fuel prices equivalent to US$90 per barrel of oil, rather than the European Commission's

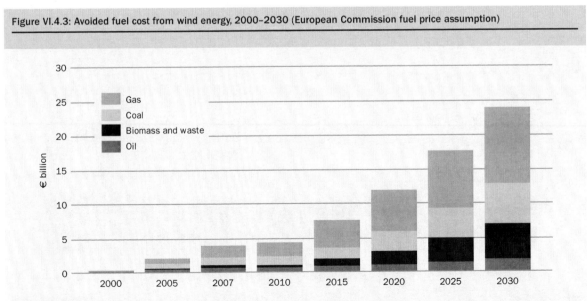

Figure VI.4.3: Avoided fuel cost from wind energy, 2000–2030 (European Commission fuel price assumption)

Source: EWEA (2008a)

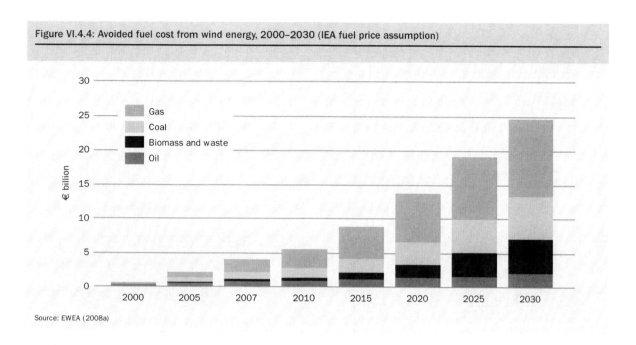

Figure VI.4.4: Avoided fuel cost from wind energy, 2000–2030 (IEA fuel price assumption)

Source: EWEA (2008a)

assumptions, fuel costs avoided due to wind would be €5 billion in 2007, €8.3 billion in 2010, €20.5 billion in 2020 and €34.6 billion in 2030 (see Figure VI.4.5).

The calculations here are based on an €/US$ exchange rate of 0.6838 (February 2008). Fluctuations in exchange rates can have a profound effect on the avoided fuel cost. Had the €/$ exchange rate been 1, wind energy's avoided fuel cost would have been €50.5 billion in 2030 instead of €34.6 billion. However, it could reasonably be argued that the price of oil would be lower if the US dollar were stronger.

In EWEA's fuel price scenario – the oil price increases gradually from $90 to $120 in 2030, and the relationship between oil, gas and coal remains unchanged from the Commission's scenario – wind energy would avoid fuel costs worth €9.2 billion in 2010, €24.6 billion in 2020 and €44.4 billion in 2030 (see Figure VI.4.6).

Investments and Total Avoided Lifetime Cost

So far, Part VI has looked at wind energy's contribution to electricity, CO_2 reductions, avoided fuel cost and so on from a perspective of total installed capacity by the end of each individual year. In this chapter, a lifetime approach is used in order to determine how much CO_2 and fuel cost are avoided from wind power investments made in a given year over the entire lifetime of the capacity. For example, the 300 GW of wind power capacity installed in the EU in 2030 will avoid the emission of 576 Mt of CO_2 in the same year. What has not been taken into account so far in this report is that the wind energy capacity installed – for example, the 19.5 GW that will be installed in 2030 – will continue to produce electricity and avoid CO_2 and fuel costs beyond 2030 – some CO_2 and fuel costs will be avoided right up to 2055.

Figure VI.4.7 (the scenario with oil at $90 and CO_2 at €25) shows the total CO_2 costs and fuel costs avoided during the lifetime of the wind energy capacity installed for each year from 2008 to 2030, assuming a technical lifetime for onshore wind turbines of 20 years and for offshore wind turbines of 25 years. Furthermore, it is assumed that wind energy avoids 690 g of CO_2 per kWh produced, that the average price of a CO_2 allowance is €25/t and that €42 million worth of fuel is avoided for each TWh of wind power produced, equivalent to an oil price throughout the period of $90 per barrel.

Figure VI.4.5: Avoided fuel cost from wind energy, 2000–2030 (fuel price equivalent to January 2008 – US$90/barrel – until 2030)

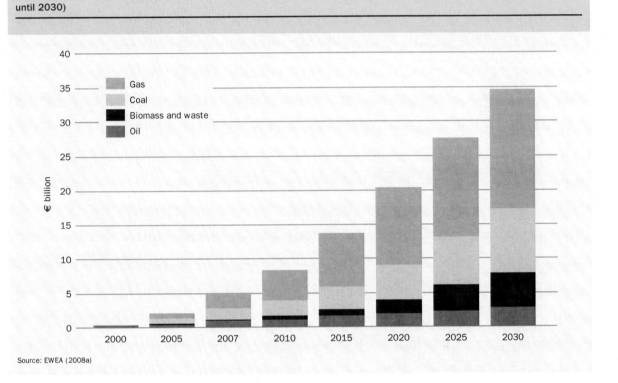

Source: EWEA (2008a)

Figure VI.4.6: Avoided fuel cost from wind energy, 2000–2030 (fuel price increase to US$100 in 2010, $110 in 2020 and $120 in 2030)

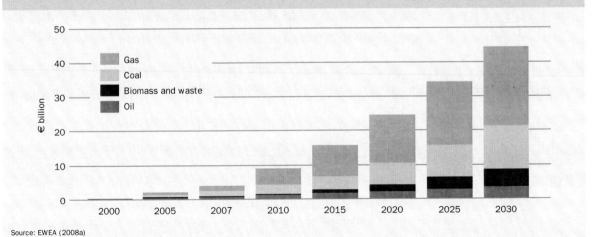

Source: EWEA (2008a)

Figure VI.4.7: Wind investments compared with lifetime avoided fuel and CO_2 costs (oil at US$90/barrel; CO_2 at €25/t)

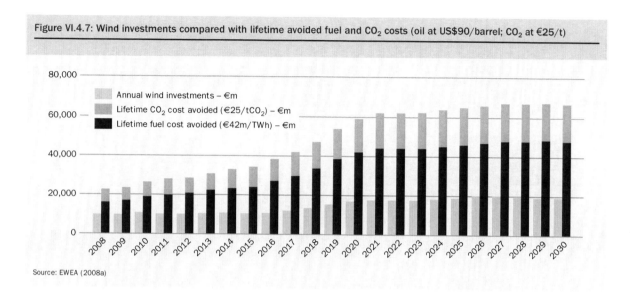

Source: EWEA (2008a)

For example, the 8554 MW of wind power capacity that was installed in the EU in 2007 had an investment value of €11.3 billion and will avoid CO_2 emissions worth €6.6 billion throughout its lifetime and fuel costs of €16 billion throughout its lifetime, assuming an average CO_2 price of €25/t and average fuel prices (gas, coal and oil) based on $90/barrel of oil.

Similarly, the €152 billion of investments in wind power between 2008 and 2020 will avoid €135 billion worth of CO_2 and €328 billion in fuel cost under the same assumptions. For the period up to 2030, wind power investments of €339 billion will avoid €322 billion in CO_2 cost and €783 billion worth of fuel.

It is important to note that these calculations only compare the capital cost of wind energy to avoided CO_2 and fuel cost. The operation and maintenance cost (low because the fuel is free) has not been taken into account. In addition, it would be reasonable to assume that some components of the wind turbine would need replacing during their technical lifetime.

Figure VI.4.8: Wind investments compared with lifetime avoided fuel and CO_2 costs (oil at US$50/barrel; CO_2 at €10/t)

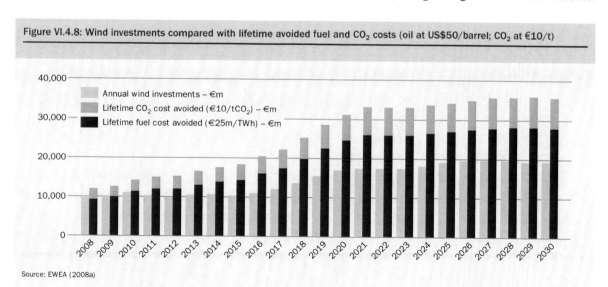

Source: EWEA (2008a)

Figure VI.4.9: Wind investments compared with lifetime avoided fuel and CO_2 costs (oil at US$120/barrel; CO_2 at €40/t)

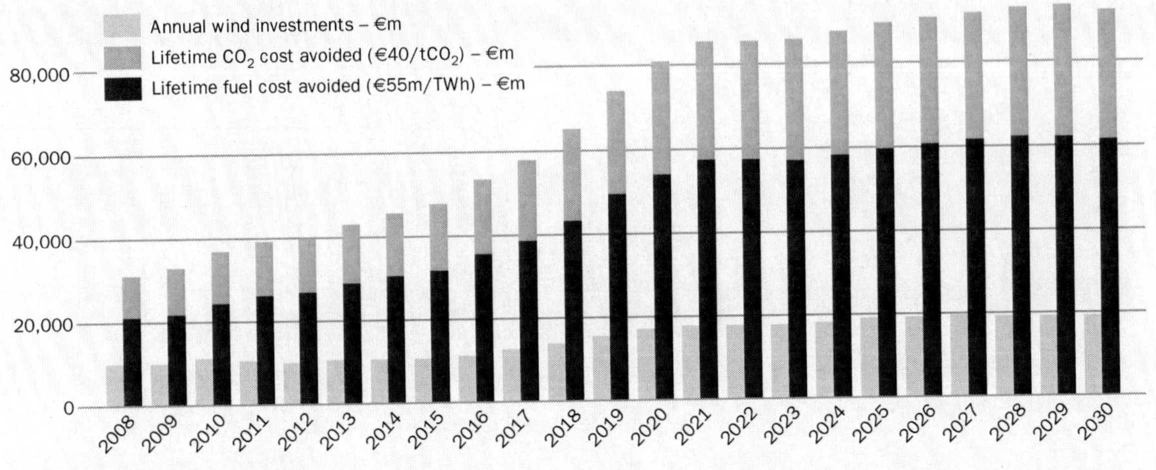

Source: EWEA (2008a)

Table VI.4.2: The different savings made depending on the price of oil (per barrel) and CO_2 (per tonne)

Totals (oil US$90: CO_2 €25)	2008–2010	2011–2020	2021–2030	2008–2020	2008–2030
Investment	31,062	120,529	187,308	151,591	338,899
Avoided CO_2 cost	21,014	113,890	186,882	134,904	321,786
Avoided fuel cost	51,165	277,296	455,017	328,462	783,479

Totals (oil US$50: CO_2 €10)	2008–2010	2011–2020	2021–2030	2008–2020	2008–2030
Investment	31,062	120,529	187,308	151,591	338,899
Avoided CO_2 cost	8,406	45,556	74,753	53,962	128,714
Avoided fuel cost	30,456	165,057	270,843	195,513	466,356

Totals (oil US$120; CO_2 €40)	2008–2010	2011–2020	2021–2030	2008–2020	2008–2030
Investment	31,062	120,529	187,308	151,591	338,899
Avoided CO_2 cost	33,623	182,223	299,011	215,846	514,857
Avoided fuel cost	67,002	363,126	595,856	430,128	1,025,984

Source: EWEA

This has not been taken into account either. The purpose is simply to compare the investment value in an individual year with the avoided fuel and CO_2 cost over the lifetime of the wind turbines.

As can be seen from Table VI.4.2, changing the CO_2 and fuel price assumptions has a dramatic impact on the result. With low CO_2 prices (€10/tonne) and fuel prices (equivalent to $50/barrel of oil) throughout the period, wind power investments over the next 23 years avoid €466 billion instead of €783 billion. With high prices for CO_2 (€40/tonne) and fuel (equivalent to $120/barrel of oil), wind power would avoid fuel and CO_2 costs equal to more than €1 trillion over the three decades from 2000 to 2030.

VI.5 GLOBAL SCENARIOS

Global Market Forecast for 2008–2012

The Global Wind Energy Council (GWEC) predicts that the global wind market will grow by over 155 per cent from 2007 to reach 240.3 GW of total installed capacity by 2012 (GWEC, 2008). This would represent an addition of 146.2 GW in five years, attracting investment of over €180 billion (US$277 billion, both in 2007 values). The electricity produced by wind energy will reach over 500 TWh in 2012 (up from 200 TWh in 2007), accounting for around 3 per cent of global electricity production (up from just over 1 per cent in 2007).

The main areas of growth during this period will be North America and Asia, more specifically the US and China. The emergence of significant manufacturing capacity in China by foreign and domestic companies will also have an important impact on the growth of the global markets. While tight production capacity is going to remain the main factor limiting further market growth, Chinese production may help take some of the strain out of the current supply situation.

The average growth rates during this five-year period in terms of total installed capacity are expected to be 20.7 per cent, compared with 23.4 per cent during 2003–2007. In 2012, Europe will continue to house the largest wind energy capacity, with a total of 102 GW, followed by Asia with 66 GW and North America with 61.3 GW.

The yearly additions in installed capacity are predicted to grow from 19.9 GW in 2007 to 36.1 GW in 2012, with an average growth rate of 12.7 per cent. Considering that annual markets have been increasing by an average of 24.7 per cent over the last five years, growth could be much stronger in the future were it not for continuing supply chain difficulties which will considerably limit the growth of annual markets for the next two years. This problem should be overcome by 2010, and along with the development of the offshore market, growth rates are expected to recover in the next decade.

GWEC predicts that Asia will install 12.5 GW of new wind generating capacity in 2012, up from 5.2 GW in 2007. This growth will be mainly led by China, which since 2004 has doubled its total capacity every year, thereby consistently exceeding even the most optimistic predictions. By 2010, China could be the biggest national market globally. This development is underpinned by a rapidly growing number of domestic and foreign manufacturers operating in the Chinese market.

While China will emerge as the continental leader in Asia, sustained growth is also foreseen in India, while other markets such as Japan, South Korea and Taiwan will also contribute to the development of wind energy on the continent.

By 2012, the European market should stand at 10.3 GW – the same size as the North American market (10.5 GW). Overall, this means that over 29 per cent of global new installations will take place in Europe in 2012. In terms of total installed capacity, Europe will continue to be the biggest regional market, with 42.4 per cent of all wind power capacity installed in the world by the end of 2012.

The large-scale development of offshore wind energy will only start to have a significant impact on European market growth towards the end of the time period under consideration. However, it is expected that offshore development will lend momentum to growth in Europe during the next decade.

In Europe, Germany and Spain will remain the leading markets, but their relative weight will decrease as other national markets emerge on the scene. While the spectacular growth of the Spanish market in 2007, with over 3.5 GW of new installations, will not be sustained, a stable pace of 2–2.5 GW per year on average can be expected, enabling Spain to reach the government's 2010 target of 20 GW. The size of the German annual market will decrease, but it will remain the second strongest European market for the 2008–2012 period and the biggest in terms of total installed capacity. By 2010, offshore developments will give new impetus to the German market, resulting in stronger

Figure VI.5.1: Offshore wind in the EU

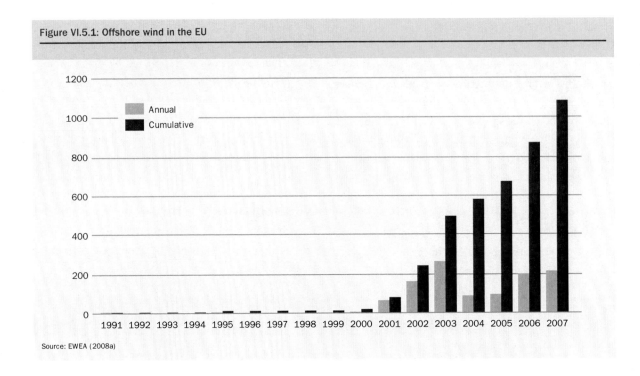

Source: EWEA (2008a)

growth. Other important markets in Europe will be France and the UK, each increasing by an average of 1 GW per year.

The North American market will see strong growth, led by the US, with the Canadian market maintaining its development. In total, North America will see an addition of 42.6 GW in the next five years, reaching 61.3 GW of total capacity in 2012. This represents an average of 8.5 GW of new capacity added every year (the bulk of which is in the US).

Figure VI.5.2: Germany, Spain and Denmark's share of EU market, 2000–2007

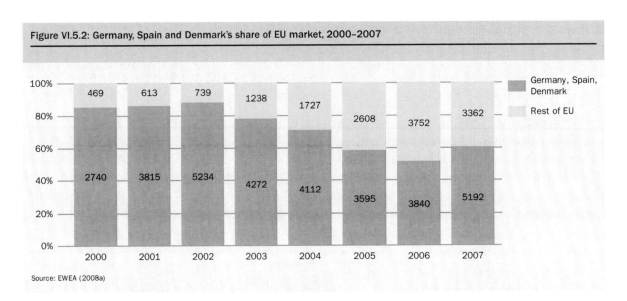

Source: EWEA (2008a)

Figure VI.5.3: Annual global installed capacity, 2007–2012

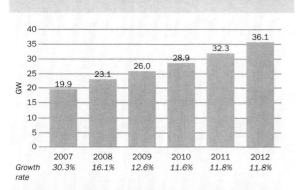

Source: GWEC

Figure VI.5.4: Cumulative global installed capacity, 2007–2012

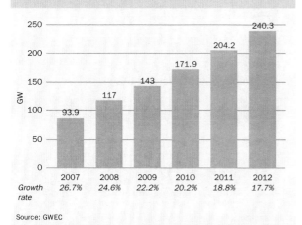

Source: GWEC

These figures assume that the US Production Tax Credit (PTC) will be renewed in time for the current strong growth to continue. If it is not, the 2009 market could suffer. However, the high-level engagement of an increasing number of US states, 27 of which have already introduced Renewable Portfolio Standards, will also assure sustained growth. A change in the US administration may further underpin this development.

Figure VI.5.5: New global installed capacity, 2008–2012

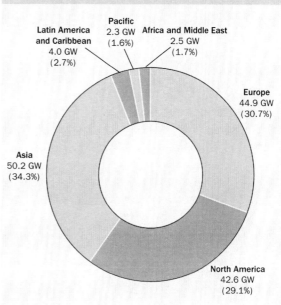

Source: GWEC

Figure VI.5.6: Cumulative global installed capacity, end 2007

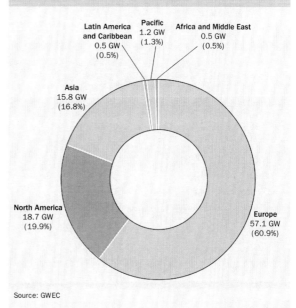

Source: GWEC

Figure VI.5.7: Annual capacity in 2007

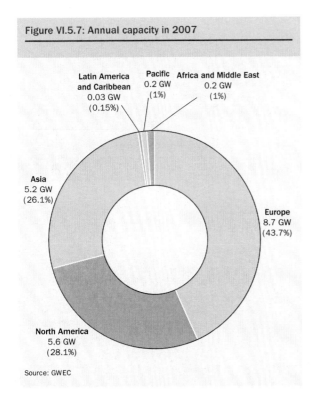

Source: GWEC

Figure VI.5.9: Annual capacity in 2012

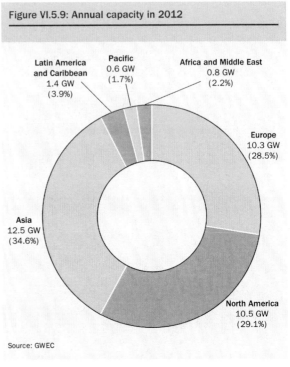

Source: GWEC

Figure VI.5.8: Cumulative capacity, end 2012

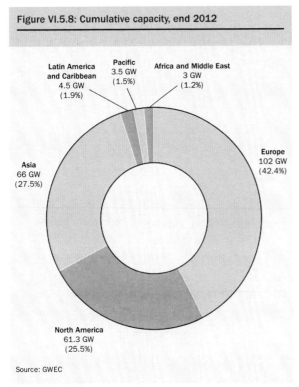

Source: GWEC

Latin America is expected to contribute more substantially to the global total in the future, mainly driven by Brazil, Mexico and Chile. By 2012, the total installed capacity in Latin America and the Caribbean will increase eightfold to reach 4.5 GW, and an annual market of 1.4 GW. However, despite its tremendous potential, Latin America is likely to remain a small market until the end of the period under consideration, progressing towards more significant development in the next decade.

The Pacific region will see around 2.3 GW of new installations in 2008–2012, bringing the total up to 3.5 GW. While in Australia, wind energy development slowed down considerably in 2006 and 2007, the outlook for the future is more optimistic, mainly thanks to the change in federal government at the end of 2007, the ratification of the Kyoto Protocol and the pledge to implement a new target for 20 per cent of electricity from renewables by 2020. New Zealand, however, got new impetus with 151 MW of new installations in

2007, and many more projects are at various stages of development.

Africa and the Middle East will remain the region with the smallest wind energy development, with a total installed capacity of 3 GW by 2012, up from 500 MW in 2012. However, it is expected that market growth will pick up in the coming five years, with annual additions reaching around 800 MW by 2012. This development will be driven by Egypt and Morocco, with some development also predicted in other North African and Middle Eastern countries.

VI.6 THE 'GLOBAL WIND ENERGY OUTLOOK' SCENARIOS

The Global Wind Energy Outlook scenarios as presented by GWEC and Greenpeace (GWEC/Greenpeace, 2008) examine the future potential of wind power up to 2030, starting from a range of assumptions which will influence the development of the wind industry.

This exercise has been carried out jointly by GWEC, Greenpeace International and the German Aerospace Centre (DLR). Projections on the future of wind energy development have been extrapolated from a larger study of global sustainable energy pathways up to 2030, conducted by DLR for Greenpeace and the European Renewable Energy Council (EREC).

Scenario Methodology

REFERENCE SCENARIO

There are three different Global Wind Energy Outlook scenarios looking at the future growth of wind energy around the world. The most conservative 'reference' scenario is based on the projections in the *World Energy Outlook 2007* report from the IEA. This only takes existing energy policies into account, though including assumptions such as continuing electricity and gas market reform, the liberalisation of cross-border energy trade, and recent policies aimed at combating pollution. Based on the IEA's figures, the scenario then projects the growth of wind power up to 2030.

MODERATE SCENARIO

The 'moderate' scenario takes into account all existing or planned policy measures from around the world that support renewable energy. It also assumes that the targets set by many countries for either renewables or wind energy are successfully implemented. Moreover, it assumes renewed investor confidence in the sector established by a successful outcome from the current round of climate change negotiations, which are set to culminate at the UNFCCC COP 15 in Copenhagen in December 2009.

ADVANCED SCENARIO

The most ambitious scenario, the 'advanced' version examines the extent to which this industry could grow in a best-case 'wind energy vision'. The assumption here is that all policy options in favour of renewable energy, following the industry's recommendations, have been selected, and that the political will is there to carry them out.

Up to 2012, the figures for installed capacity are closer to being forecasts than scenarios. This is because the data available from the wind energy industry shows the expected growth of worldwide markets over the next five years based on orders for wind turbines that have already been received. After 2012, the pattern of development is clearly much more difficult to predict. Nonetheless, the scenario still shows what could be achieved if the wind energy market is given the encouragement it deserves.

Energy Efficiency Projections

These three scenarios for the global wind energy market are then set against two projections for the future growth of electricity demand. Most importantly, these projections do not just assume that growing demand by consumers will inevitably need to be matched by supply options. On the basis that demand will have to be reduced if the threat of climate change is to be seriously tackled, they take into account an increasing element of energy efficiency.

The more conservative of the two global electricity demand projections is again based on data from the IEA's *World Energy Outlook 2007*, extrapolated forwards to 2050. This is the 'reference' projection. It does not take into account any possible or likely future policy initiatives and assumes, for instance, that there will be no change in national policies on nuclear power. The IEA's assumption is that 'in the absence of new government policies, the world's energy needs will rise inexorably'. Global demand would therefore almost double from the baseline 12,904 TWh in 2002 to reach 29,254 TWh by 2030 and continue to grow to 42,938 TWh by 2050.

The IEA's expectations on rising energy demand are then set against the outcome of a study on the potential effect of energy-efficiency savings developed by DLR and the Ecofys consultancy. The study describes an ambitious development path for the exploitation of energy-efficiency measures. It focuses on current best practice and available technologies in the future, and assumes that continuous innovation takes place. The most important sources of energy saving are in efficient passenger and freight transport and in better insulated and designed buildings: together these account for 46 per cent of worldwide energy savings.

Under the 'high energy efficiency' projection, input from the DLR/Ecofys models shows the effect of energy-efficiency savings on the global electricity demand profile. Although this assumes that a wide range of technologies and initiatives have been introduced, their extent is limited by the potential barriers of cost and other likely roadblocks. This still results in global demand increasing by much less than under the reference projection, to reach 21,095 TWh in 2030. By the end of the scenario period in 2050, demand is 35 per cent lower than under the reference scenario.

Main Assumptions and Parameters

GROWTH RATES

Market growth rates in this scenario are based on a mixture of historical figures and information obtained from analysts of the wind turbine market. Annual growth rates of more than 20 per cent per annum, as envisaged in the advanced version of the scenario, are high for an industry which manufactures heavy equipment. The wind industry has experienced much higher growth rates in recent years, however. In the five years up to 2007 the average annual increase in global cumulative installed capacity was 25 per cent.

It should also be borne in mind that, whilst growth rates eventually decline to single figures across the range of scenarios, the level of wind power capacity envisaged in 40 years' time means that even small percentage growth rates will by then translate into large figures in terms of annually installed megawatts.

TURBINE CAPACITY

Individual wind turbines have been steadily growing in terms of their nameplate capacity – the maximum electricity output they can achieve when operating at full power. The average nameplate capacity of wind turbines installed globally in 2007 was 1.49 MW. The largest turbines on the market are now 6 MW in capacity.

GWEC's scenarios make the conservative assumption that the average size will gradually increase from today's figure to 2 MW in 2013 and then level out. It is possible, however, that this figure will turn out to be greater in practice, requiring fewer turbines to achieve the same installed capacity. It is also assumed that each turbine will have an operational lifetime of 20–25 years, after which it will need to be replaced. This 'repowering' or replacement of older turbines has been taken into account in the scenarios.

CAPACITY FACTORS

'Capacity factor' refers to the percentage of its nameplate capacity that a turbine installed in a particular location will deliver over the course of a year. This is primarily an assessment of the wind resource at a given site, but capacity factors are also affected by the efficiency of the turbine and its suitability for the particular location. As an example, a 1 MW turbine operating at a 25 per cent capacity factor will deliver 2190 MWh of electricity in a year.

From an estimated average capacity factor today of 25 per cent, the scenario assumes that improvements in both wind turbine technology and the siting of wind farms will result in a steady increase. Capacity factors are also much higher out to sea, where winds are stronger and more constant. The growing size of the offshore wind market, especially in Europe, will therefore contribute to an increase in the average.

The scenario foresees the average global capacity factor increasing to 28 per cent by 2012.

CAPITAL COSTS AND PROGRESS RATIOS

The capital cost of producing wind turbines has fallen steadily over the past 20 years, as manufacturing techniques have been optimised, turbine design has been largely concentrated on the three-bladed upwind model with variable speed and pitch regulation, and mass production and automation have resulted in economies of scale.

The general conclusion from industrial learning curve theory is that costs decrease by some 20 per cent each time the number of units produced doubles. A 20 per cent decline is equivalent to a progress ratio of 0.80.

In the calculation of cost reductions in this report, experience has been related to numbers of units, i.e. turbines, and not megawatt capacity. The increase in average unit size is therefore also taken into account.

The progress ratio assumed here is at 0.90 up until 2009. After that it goes down to 0.80 before steadily rising again from 2016 onwards.

The reason for this graduated assumption, particularly in the early years, is that the manufacturing industry has not so far gained the full benefits of series production, especially due to the rapid upscaling of products. Neither has the full potential of future design optimisations been realised.

Contrary to this theory, the past few years, particularly since 2006, have seen a marked increase in the price of new wind turbines. This has been triggered by a mixture of rising raw material prices and shortages in the supply chain for turbine components. Examples of raw materials whose price has increased substantially are steel (used in towers, gearboxes and rotors), copper (used in generators) and concrete (used in foundations and towers). Global steel prices have almost doubled in the current year up to August 2008, while copper prices have quadrupled in the last five years. In addition, rising energy prices have also driven up the cost of manufacturing and transporting wind turbines. Supply chain pressures have included in particular a shortage of gearboxes and of the range of bearings used throughout the manufacturing of turbines. These shortages are being addressed by the component manufacturers, who are building new production capacity and opening up new manufacturing bases, for example in China. Some observers predict that component supply may catch up with demand by 2010.

Even so, the cost of wind turbine generators has still fallen significantly overall, and the industry is recognised as having entered the 'commercialisation phase', as understood in learning curve theories.

Capital costs per kilowatt of installed capacity are taken as an average of €1300 in 2007, rising to €1450 in 2009. They are then assumed to fall steadily from 2010 onwards to about €1050. From 2020 the scenario assumes a levelling out of costs. All figures are given at 2007 prices.

Scenario Results

An analysis of the Global Wind Energy Outlook scenarios shows that a range of outcomes is possible for the global wind energy market. The outcomes differ according to the choice of demand-side options and the assumptions for growth rates on the wind power supply side.

REFERENCE SCENARIO

The reference scenario, which is derived from the IEA's *World Energy Outlook 2007*, starts off with an assumed growth rate of 27 per cent for 2008, decreasing to 10 per cent by 2010, then falling to 4 per cent by 2030.

As a result, the scenario foresees cumulative global capacity reaching 139 GW, producing 304 TWh per year and covering 1.7 per cent of the world's electricity demand by the end of this decade. By 2020, global capacity would stand at 352 GW, growing to almost 500 GW by 2030, with an annual capacity increase of around 30 GW.

The relative penetration of wind energy into the global electricity supply system varies according to which demand projection is considered. Around 864 TWh produced in 2020 would account for between 3.6 per cent and 4.1 per cent of the world's electricity production, depending on the extent of the energy-efficiency measures introduced. By 2030, production of 1218 TWh would only meet 4.2–5.1 per cent of global demand.

MODERATE SCENARIO

In the moderate wind energy scenario, growth rates are expected to be substantially higher than in the reference version. The assumed cumulative annual growth rate starts at 27 per cent for 2008, decreases

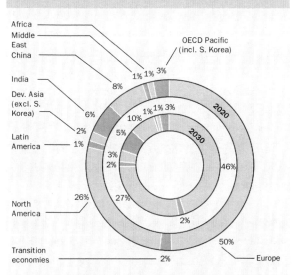

Figure VI.6.1: Regional distribution – Reference scenario 2020 and 2030

2020		2030	
Europe	176 GW	Europe	227 GW
Transition economies	7 GW	Transition economies	11 GW
North America	92 GW	North America	132 GW
Latin America	5 GW	Latin America	8 GW
Dev. Asia (excl. S. Korea)	7 GW	Dev. Asia (excl. S. Korea)	16 GW
India	20 GW	India	27 GW
China	27 GW	China	49 GW
Middle East	2 GW	Middle East	4 GW
Africa	4 WG	Africa	7 GW
OECD Pacific (incl. S. Korea)	12 GW	OECD Pacific (incl. S. Korea)	16 GW

Source: GWEC

Figure VI.6.2: Regional distribution – Moderate scenario 2020 and 2030

2020		2030	
Europe	182 GW	Europe	306 GW
Transition economies	9 GW	Transition economies	34 GW
North America	214 GW	North America	366 GW
Latin America	50 GW	Latin America	103 GW
Dev. Asia (excl. S. Korea)	40 GW	Dev. Asia (excl. S. Korea)	140 GW
India	69 GW	India	142 GW
China	101 GW	China	201 GW
Middle East	8 GW	Middle East	20 GW
Africa	10 GW	Africa	21 GW
OECD Pacific (incl. S. Korea)	30 GW	OECD Pacific (incl. S. Korea)	70 GW

Source: GWEC

to 19 per cent by 2010, and continues to fall gradually to 11 per cent by 2020 and 3 per cent by 2030.

The result is that by the end of this decade, the global wind power capacity is expected to reach 172 GW, with annual additions of 28.9 GW. By 2020, the annual market grows to 81.5 GW, and the cumulative global wind power capacity reaches a level of over 700 GW. By 2030, a total of over 1420 MW would be installed, with annual installations in the region of 84 GW.

In terms of generated electricity, this would translate into over 1700 TWh produced by wind energy in 2020 and 3500 TWh in 2030. Depending on demand-side development, this would supply 7.3–8.2 per cent of global electricity demand in 2020 and 11.9–14.6 per cent in 2030.

ADVANCED SCENARIO

In the advanced wind energy scenario, an even more rapid expansion of the global wind power market is envisaged. The assumed growth rate starts at 27 per cent in 2008, falls to 22 per cent by 2010, then to 12 per cent by 2020 and 5 per cent by 2030.

The result is that by the end of this decade, global capacity reaches 186 GW, with annual additions of around 36.5 GW. By 2020, global capacity is over 1000 GW, with annual additions of around 142 GW, and by 2030, the total wind generation capacity reaches almost 2400 GW. The annual market then stabilises at around 165 GW.

In terms of generated electricity, this translates into 2600 TWh produced by wind energy in 2020 and 5700 TWh in 2030. Again depending on the increase in demand by that time, wind power would cover 11.2–12.6 per cent of global electricity demand in 2020 and as much as 19.7–24.0 per cent in 2030 – in other words meeting between a fifth and a quarter of the world's electricity needs.

REGIONAL BREAKDOWN

All three scenarios for wind power are broken down into geographical regions based on the methodology used by the IEA. For the purposes of this analysis, the regions are defined as Europe, the transition economies, North America, Latin America, China, India, the Pacific (including Australia, South Korea and Japan), developing Asia (the rest of Asia), and the Middle East and Africa.

Figure VI.6.3: Regional distribution – Advanced scenario 2020 and 2030

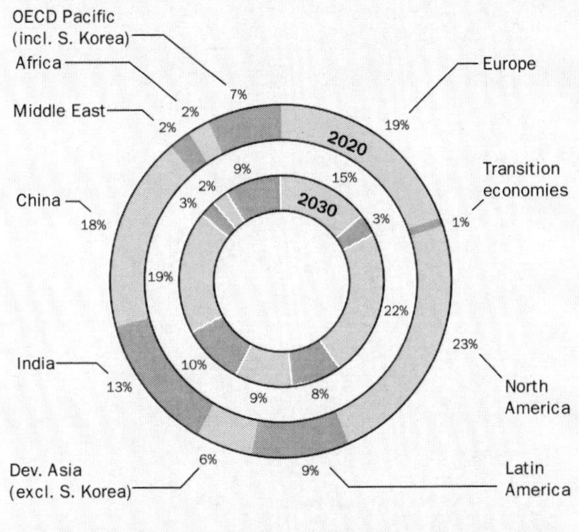

OECD Pacific (incl. S. Korea)
Africa
Middle East
Europe
China
Transition economies
India
North America
Dev. Asia (excl. S. Korea)
Latin America

2020		2030	
Europe	213 GW	Europe	353 GW
Transition economies	10 GW	Transition economies	75 GW
North America	243 GW	North America	520 GW
Latin America	100 GW	Latin America	201 GW
Dev. Asia (excl. S. Korea)	61 GW	Dev. Asia (excl. S. Korea)	211 GW
India	138 GW	India	235 GW
China	201 GW	China	451 GW
Middle East	25 GW	Middle East	63 GW
Africa	17 GW	Africa	52 GW
OECD Pacific (incl. S. Korea)	75 GW	OECD Pacific (incl. S. Korea)	215 GW

Source: GWEC

This breakdown of world regions has been used by the IEA in the ongoing series of World Energy Outlook publications. We chose to use it here in order to facilitate comparison with the IEA projections and because the IEA provides the most comprehensive global energy statistics.

The level of wind power capacity expected to be installed in each region of the world by 2020 and 2030 is shown in Figures VI.6.1 to VI.6.3. These show that Europe would continue to dominate the world market under the least ambitious reference scenario. By 2030, Europe would still have 46 per cent of the global wind power market, followed by North America with 27 per cent. The next largest region would be China with 10 per cent.

The two more ambitious scenarios envisage much stronger growth in regions outside Europe. Under the moderate scenario, Europe's share will be 23 per cent by 2030, with North America dominating the global market at 27 per cent and major contributions coming from China (14 per cent), India (10 per cent) and developing Asia (10 per cent). Latin America (7 per cent) and the Pacific region (5 per cent) will play a smaller role than previously estimated.

The advanced scenario predicts an even stronger growth for China, which would see its share of the world market increasing to 19 per cent by 2030. The North American market accounts for 22 per cent of global wind power capacity, whilst Europe's share is 15 per cent, followed by India (10 per cent), developing Asia (9 per cent), the Pacific region (9 per cent) and Latin America (8 per cent). In both scenarios, Africa and the Middle East would play only a minor role in the timeframe discussed (1 per cent of global capacity in the moderate and 2 per cent in the advanced scenario).

In all three scenarios it is assumed that an increasing share of new capacity is accounted for by the replacement of old power plants. This is based on a wind turbine average lifetime of 20 years. Turbines replaced within the timescale of the scenarios are assumed to be of the same cumulative installed capacity as the original smaller models. The result is that an increasing proportion of the annual level of installed capacity will come from repowered turbines. These new machines will contribute to the overall level of investment, manufacturing output and employment. As replacement turbines, their introduction will not, however, increase the total figure for global cumulative capacity.

The German Aerospace Centre

The German Aerospace Centre (DLR) is the largest engineering research organisation in Germany. Among its specialities are the development of solar thermal power station technologies, the utilisation of low- and high-temperature fuel cells, particularly for electricity generation, and research into the development of high-efficiency gas and steam turbine power plants.

The Institute of Technical Thermodynamics at the DLR (DLR-ITT) is active in the field of renewable energy research and technology development for efficient and low-emission energy conversion and utilisation. Working in cooperation with other DLR institutes, industry and universities, research is focused on solving key problems in electrochemical energy technology and solar energy conversion. This encompasses application-orientated research, the development of laboratory and prototype models, and the design and operation of demonstration plants. System analysis and technology assessment are used to help prepare strategic decisions in the field of research and energy policy.

Within the DLR-ITT, the System Analysis and Technology Assessment Division has long-term experience in the assessment of renewable energy technologies. Its main research activities are in the field of techno-economic utilisation and system analysis, leading to the development of strategies for the market introduction and dissemination of new technologies, mainly in the energy and transport sectors.

Scenario Background

The DLR was commissioned by Greenpeace International and EREC to conduct a study on global sustainable energy pathways up to 2050. This so-called 'Energy revolution' scenario published in early 2007 is a blueprint on how to cut global CO_2 emissions by 50 per cent by 2050, while maintaining global economic growth. Part of the study examines the future potential for renewable energy sources; together with input from the wind energy industry and analysis of regional projections for wind power around the world, it forms the basis of the Global Wind Energy Outlook scenario.

Part VI Notes

1 See EWEA report, 'Pure power: Wind energy scenarios up to 2030', EWEA, March 2008.

2 Renewable Energy Roadmap, COM(2006)848 final, European Commission.

3 Renewable Energy Roadmap – Impact Assessment, SEC(2006)1720, European Commission.

4 West Texas Intermediate.

5 See: http://ec.europa.eu/energy/energy_policy/doc/03_renewable_energy_roadmap_en.pdf.

APPENDICES

APPENDIX A: ONSHORE WIND MAPS

EUROPE

Figure A.1: European wind atlas

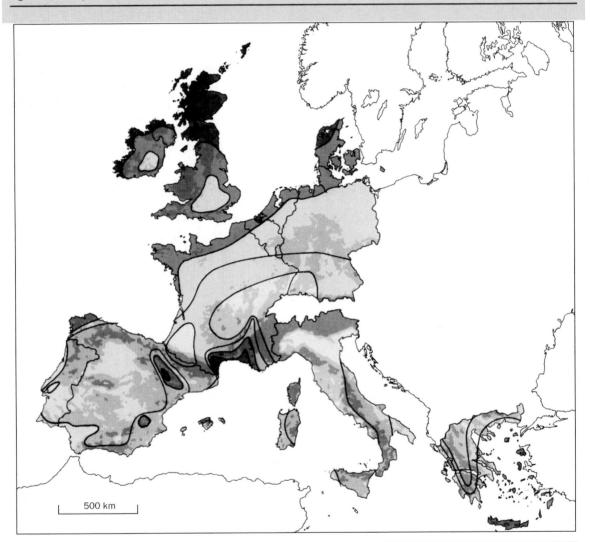

	Wind resources at 50 metres above ground level for five different topographic conditions									
	Sheltered terrain		Open terrain		At a sea coast		Open sea		Hills and ridges	
	m/s	W/m²	m/s	W/m²	m/s	W/m²	m/s	W/m²	m/s	W/m²
	>6.0	>250	>7.5	>500	>8.5	>700	>9.0	>800	>11.5	>1800
	5.0–6.0	150–250	6.5–7.5	300–500	7.0–8.5	400–700	8.0–9.0	600–800	10.0–11.5	1200–1800
	4.5–5.0	100–150	5.5–6.5	200–300	6.0–7.0	250–400	7.0–8.0	400–600	8.5–10.0	700–1200
	3.5–4.5	50–100	4.5–5.5	100–200	5.0–6.0	150–250	5.5–7.0	200–400	7.0–8.5	400–700
	<3.5	<50	<4.5	<100	<5.0	<150	<5.5	<200	<7.0	<400

Source: Risø DTU

DENMARK

Figure A.2: Denmark wind atlas

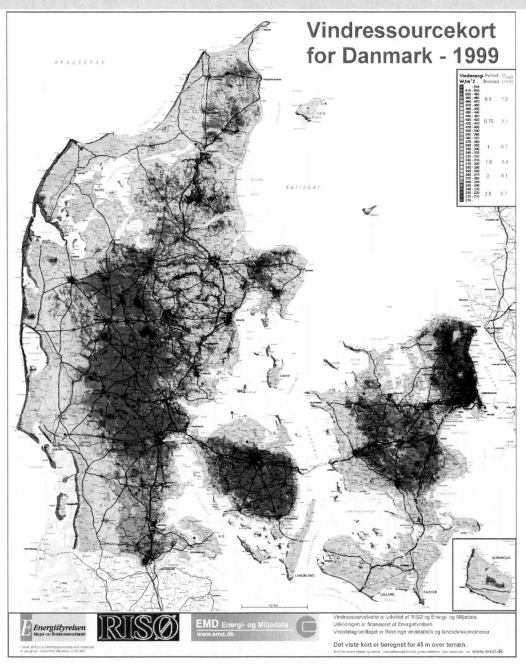

Source: [2] in the list of references

FINLAND

Figure A.3: Finland wind atlas

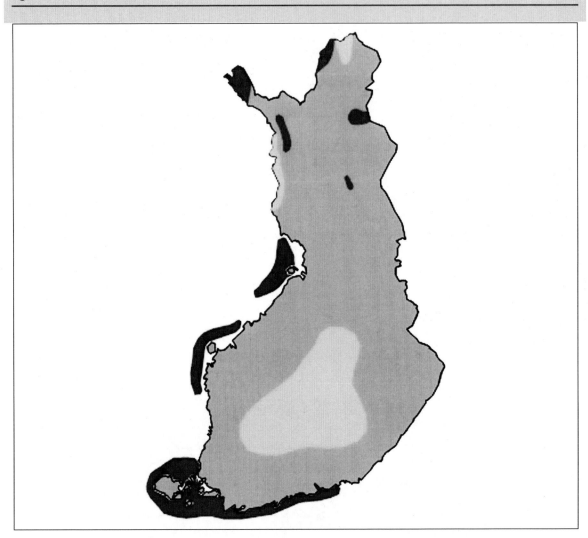

Wind resources at 50 metres above ground level for five different topographic conditions										
	Sheltered terrain		Open terrain		At a sea coast		Open sea		Hills and ridges	
	m/s	W/m²	m/s	W/m²	m/s	W/m²	m/s	W/m²	m/s	W/m²
	>6.0	>250	>7.5	>500	>8.5	>700	>9.0	>800	>11.5	>1800
	5.0–6.0	150–250	6.5–7.5	300–500	7.0–8.5	400–700	8.0–9.0	600–800	10.0–11.5	1200–1800
	4.5–5.0	100–150	5.5–6.5	200–300	6.0–7.0	250–400	7.0–8.0	400–600	8.5–10.0	700–1200
	3.5–4.5	50–100	4.5–5.5	100–200	5.0–6.0	150–250	5.5–7.0	200–400	7.0–8.5	400–700
	<3.5	<50	<4.5	<100	<5.0	<150	<5.5	<200	<7.0	<400

Source: [4] in the list of references

GREECE

Figure A.4: Greece wind atlas

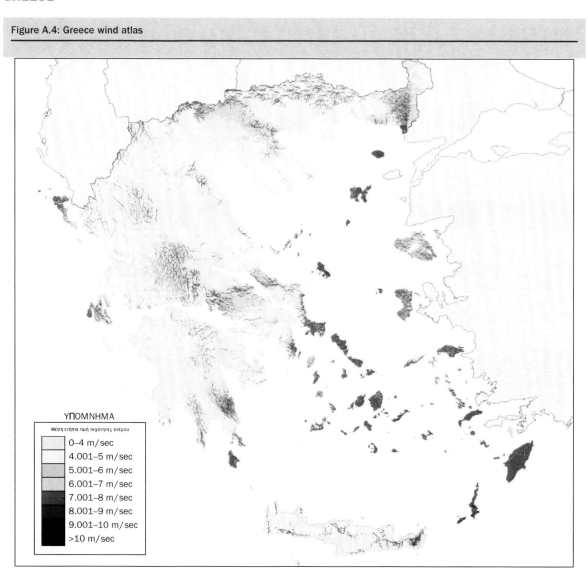

Source: [6] in the list of references

IRELAND

Figure A.5: Ireland wind atlas

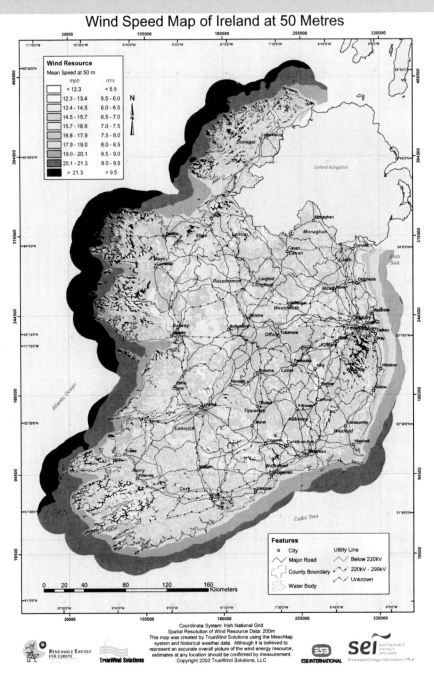

Source: [8] in the list of references

UK

CENTRAL EUROPEAN COUNTRIES

Figure A.6: UK wind atlas

Figure A.7: Central European wind atlas

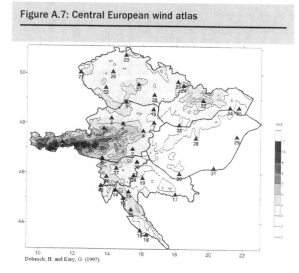

Source: [1] in the list of references

NOABL-Wind resources at 45 m above ground level		
<5.0 m/s	5.0–6.0	6.0–7.0	7.0–8.0
100.0 m/s	8.0–9.0	9.0–10.0

Source: [12] in the list of references

ARMENIA

Figure A.8: Armenia wind atlas

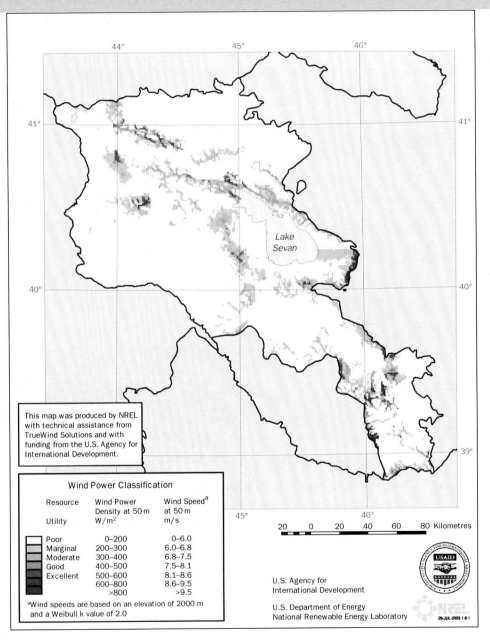

This map was produced by NREL with technical assistance from TrueWind Solutions and with funding from the U.S. Agency for International Development.

Wind Power Classification

Resource Utility	Wind Power Density at 50 m W/m²	Wind Speed[a] at 50 m m/s
Poor	0–200	0–6.0
Marginal	200–300	6.0–6.8
Moderate	300–400	6.8–7.5
Good	400–500	7.5–8.1
Excellent	500–600	8.1–8.6
	600–800	8.6–9.5
	>800	>9.5

[a]Wind speeds are based on an elevation of 2000 m and a Weibull k value of 2.0

U.S. Agency for
International Development

U.S. Department of Energy
National Renewable Energy Laboratory

29-JUL-2003 1.6.1

Source: [22] in the list of references

BULGARIA

Figure A.9: Bulgaria wind atlas

Source: [14] in the list of references

ESTONIA

Figure A.10: Estonia wind atlas

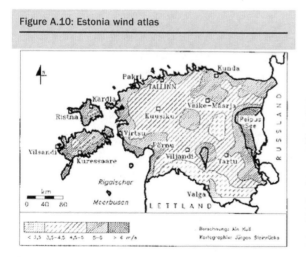

Source: [15] in the list of references

Figure A.11: Estonia wind atlas

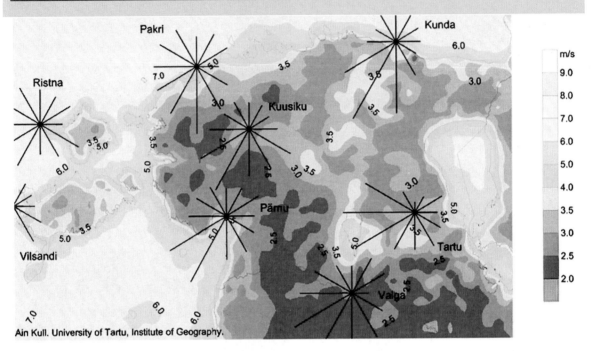

Source: [16] in the list of references

RUSSIA

Figure A.12: Russian wind atlas

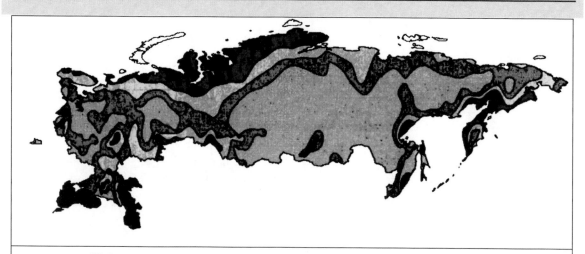

	Wind resources at the height of 50 metres above ground level for five different topographic conditions									
	Sheltered terrain		Open terrain		Sea coast		Open sea		Hills and ridges	
	m/s	W/m^2	m/s	W/m^2	m/s	W/m^2	m/s	W/m^2	m/s	W/m^2
	>6.0	>250	>7.5	>500	>8.5	>700	>9.0	>800	>11.5	>1800
	5.0–6.0	150–250	6.5–7.5	300–500	7.0–8.5	400–700	8.0–9.0	600–800	10.0–11.5	1200–1800
	4.5–5.0	100–150	5.5–6.5	200–300	6.0–7.0	250–400	7.0–8.0	400–600	8.5–10.0	700–1200
	3.5–4.5	50–100	4.5–5.5	100–200	5.0–6.0	150–250	5.5–7.0	200–400	7.0–8.5	400–700
	<3.5	<50	<4.5	<100	<5.0	<150	<5.5	<200	<7.0	<400

Source: [24] in the list of references

APPENDIX B: OFFSHORE WIND SPEEDS MODELLED IN 'STUDY OF OFFSHORE WIND ENERGY IN THE EC'

Figure B.1: Denmark and Germany

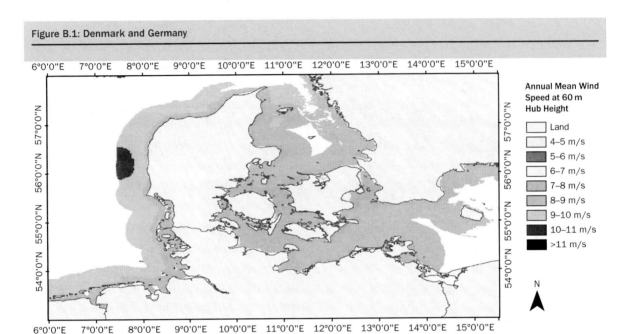

Source: Matthies and Garrad (1993)

Figure B.2: France – Atlantic

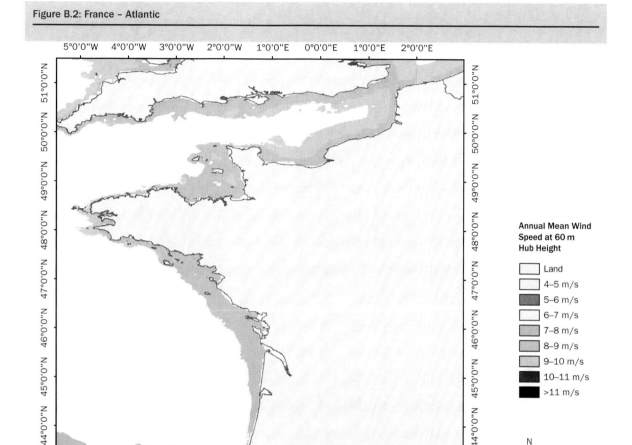

Annual Mean Wind
Speed at 60 m
Hub Height

- Land
- 4–5 m/s
- 5–6 m/s
- 6–7 m/s
- 7–8 m/s
- 8–9 m/s
- 9–10 m/s
- 10–11 m/s
- >11 m/s

N

Source: Matthies and Garrad (1993)

Figure B.3: France – Mediterranean

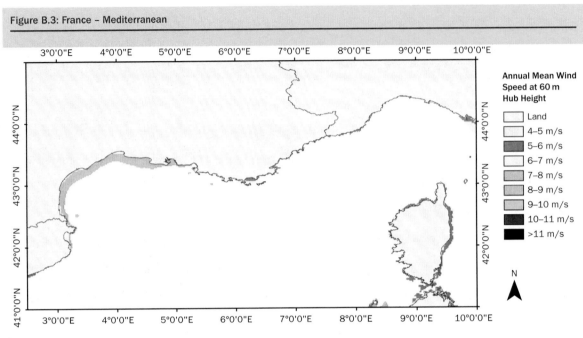

Source: Matthies and Garrad (1993)

Figure B.4: Great Britain – North

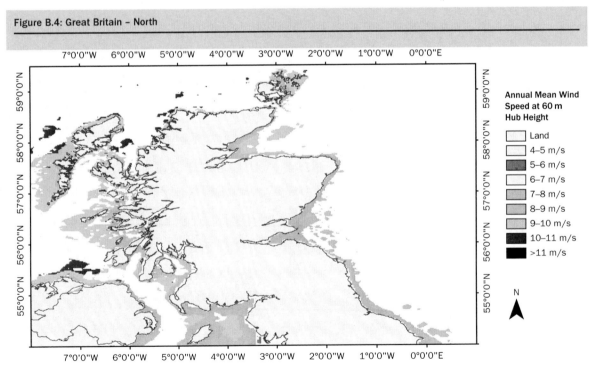

Source: Matthies and Garrad (1993)

Figure B.5: Great Britain – South

Source: Matthies and Garrad (1993)

Figure B.6: Greece

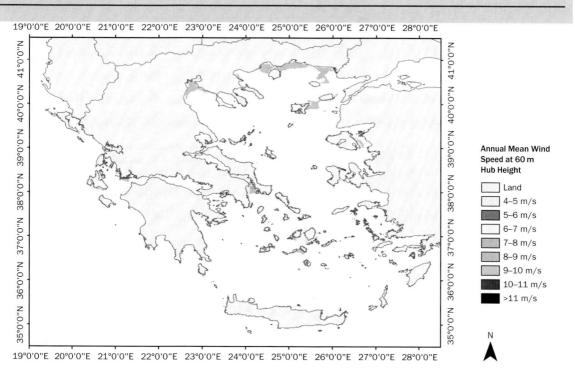

Source: Matthies and Garrad (1993)

Figure B.7: Ireland

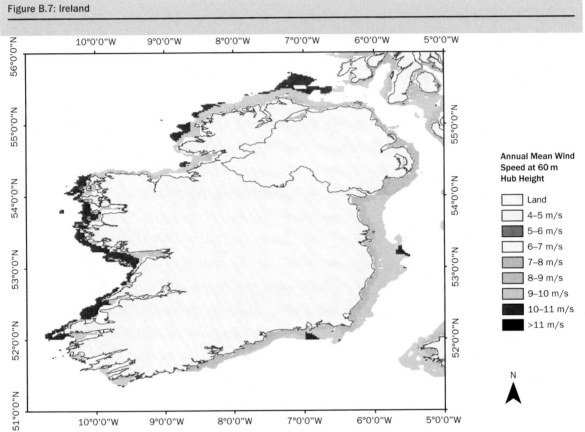

Annual Mean Wind
Speed at 60 m
Hub Height

	Land
	4–5 m/s
	5–6 m/s
	6–7 m/s
	7–8 m/s
	8–9 m/s
	9–10 m/s
	10–11 m/s
	>11 m/s

Source: Matthies and Garrad (1993)

Figure B.8: Italy

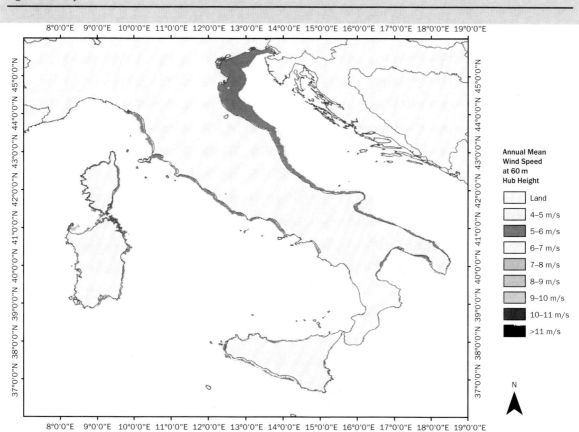

Source: Matthies and Garrad (1993)

Figure B.9: The Netherlands and Belgium

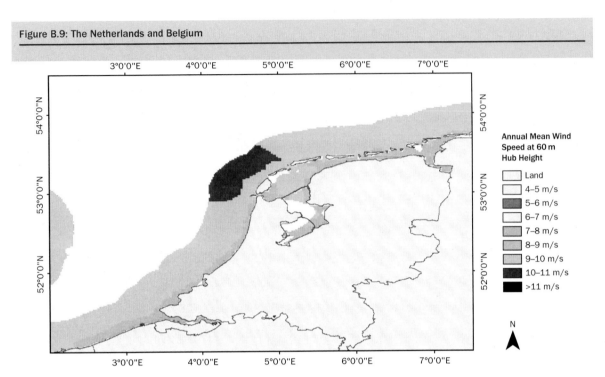

Source: Matthies and Garrad (1993)

Figure B.10: Spain and Portugal

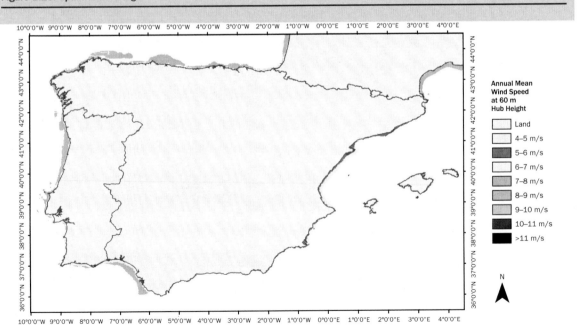

Source: Matthies and Garrad (1993)

APPENDIX C: WORKED EXAMPLE FOR CULLIAGH WIND FARM, IRELAND

Introduction

The main text has provided a general discussion of the assessment of the wind resource and energy production. This appendix is included in order to provide a 'worked example'. It demonstrates all the different aspects of the process outlined in the main text. The project considered is the Culliagh Wind Farm in Ireland, which consists of 18 Vestas V47 wind turbines and was constructed in 2000. The following specific analyses are presented:

1. the results of the pre-construction projection of the expected energy production of the wind farm, including uncertainty analysis;
2. the review of the actual production of the wind farm over a 17-month period; and
3. the results of a 'wind in–energy out' validation test of the predictive methodologies employed in (1).

Airtricity, a leading international wind farm developer, owns the Culliagh Wind Farm and thanks are to be extended to them for allowing their proprietary data to be used for this case study. A photograph of the wind farm is presented in Figure C.1

Description of the Site and Monitoring Equipment

The location of the site is shown in Figure C.2. The site lies in central County Donegal approximately 14 km southwest of Letterkenny. The location of Malin Head Meteorological Station is also marked on the figure. The wind farm site lies on Culliagh Mountain with maximum elevation of approximately 360 m, as shown in Figure C.3.

The site at Culliagh Mountain has had one 30 m and two 10 m temporary meteorological masts installed since mid-1997. The 10 m data is not considered further in this report.

The wind data from the 30 m site mast have been recorded using NRG sensors with a maximum of 40 anemometer and wind vane at 10 m and 30 m. An NRG 9210 logger was programmed to record hourly mean wind speed, wind speed standard deviation, 3-second gust and direction.

Malin Head Meteorological Station

The assessment of the wind climate at the site uses data recorded at a nearby meteorological station, Malin Head, which is situated on the coast approximately 65 km north-northeast of the Culliagh site. From discussions with Met Éireann (the Irish meteorological service) staff and consideration of other meteorological stations in the region, it was concluded that Malin Head was the most appropriate reference meteorological station for this analysis. Data from 1979 to 2000 have been used in the analysis reported here. Discussions with Met Éireann staff indicate that there has been no change during this period which will have a significant effect on the consistency of the measurements. This is important since the analysis method used here relies on long-term consistency of the

Figure C.1: The Culliagh Wind Farm

Figure C.2: Location of the proposed Culliagh Mountain Wind Farm and Malin Head Meteorological Station

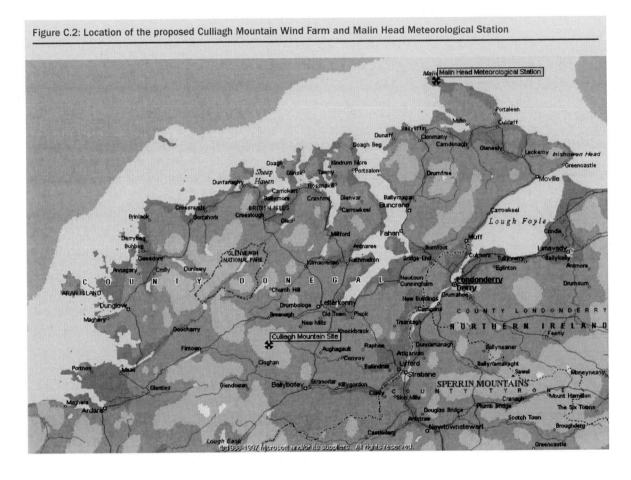

measurements at the meteorological station. The location of the Malin Head Meteorological Station is presented in Figure C.4 and a photograph of the meteorological mast is presented in Figure C.5.

Wind Data

The data sets from Malin Head and the Culliagh site, as used in the analyses described in the following sections, are summarised in Table C.1.

Description of the Proposed Wind Farm

The wind turbine model selected for the proposed Culliagh Mountain Wind Farm is the Vestas V47 660 kW model with a hub height of 45 m. The basic parameters of the turbine are presented in Table C.2.

The power curve used in the analysis has been supplied for an air density of 1.225 kg/m^3 and is presented in Table C.3.

From data recorded at local meteorological stations and with standard lapse rate assumptions, the Culliagh Mountain site is predicted to have an air density of 1.205 kg/m^3. Since the predicted mean air density at the site differs from the air density for which the power curves were supplied, a small air density adjustment following IEC 61400-12:1998 was made to the power curves used in the analysis.

The power curve for the Vestas V47 660 kW turbine has been compared to a reference curve from an independent test of the performance of the turbine. It was found that the reference curve outperformed the supplied curve by 2 per cent for the wind regime at the Culliagh site. This result indicates that the supplied

Figure C.3: The Culliagh Mountain site

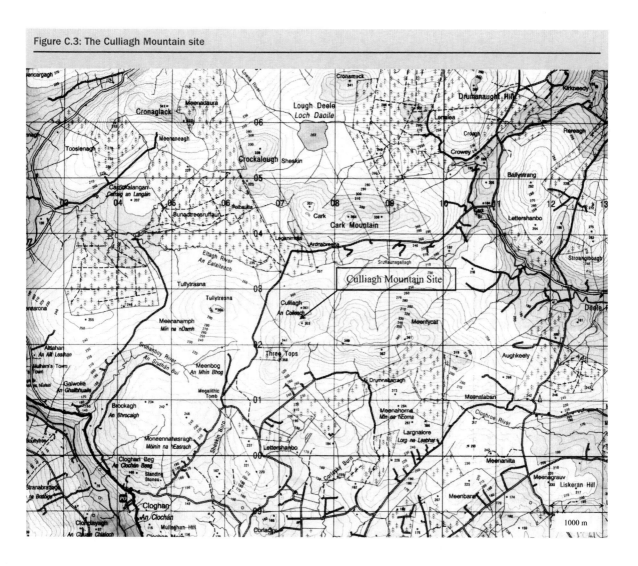

curve is broadly in line with the performance that may be expected.

The proposed Culliagh Wind Farm is designed to have a total nameplate capacity of just under 12 MW. The wind farm layout has been supplied by the client and the layout is presented in Figure C.6. Also shown in Figure C.6 are the locations of the meteorological masts.

The Culliagh Mountain Wind Farm is located approximately 1.5 km south of the existing Cark Wind Farm. The effect of these turbines on the predicted energy production of the Culliagh development was also estimated.

Results of the Analysis

The analysis to determine the wind regime and expected energy production of the proposed Culliagh Wind Farm involved several steps:

- the directional correlations between wind speeds recorded at Culliagh Mast 05 at 30 m and at Malin Head were established;
- the correlation relationships were applied to historical wind data recorded at Malin Head to produce a description of the long-term wind regime at Culliagh Mast 05;

Figure C.4: Area surrounding the Malin Head Meteorological Station

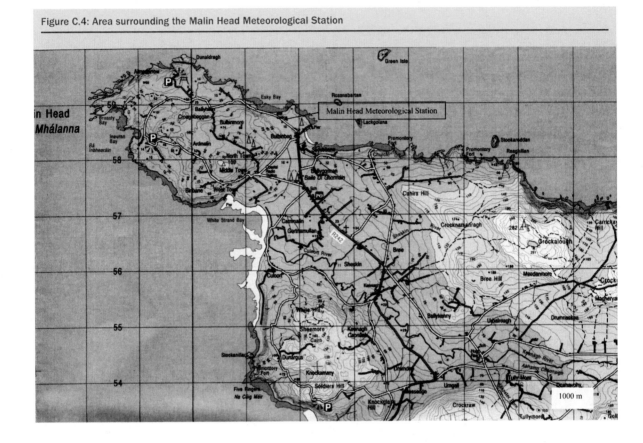

- wind flow modelling was carried out to determine the hub height wind speed variations over the site relative to the 30 m anemometry mast;
- the energy production of the wind farm was calculated, taking account of array losses and topographic effects;
- the seasonal variation in the energy production of the wind farm was calculated; and
- sources of uncertainty in the wind speed and energy production estimates were identified and quantified.

Correlation of Wind Regime at Culliagh Mountain and Malin Head

The measured wind direction at Culliagh Mast 05 at 30 m is compared to the concurrent wind direction

measured at Malin Head in Figure C.7. The directions recorded between the two locations show some scatter but are generally well correlated for the most frequent sectors.

The monitored wind speeds at 30 m height in each of twelve 30-degree direction sectors are compared to the concurrent wind speed at Malin Head in Figure C.8. The quality of the correlation is considered to be reasonable for all direction sectors. The wind speed ratios for each direction sector are presented in Table C.4.

Long-Term Mean Wind Speed at Culliagh Mountain

The wind speed ratios listed in Table C.4 were used to factor the long-term wind speeds at Malin Head for the period 1979 to 1998. By this method, the long-term

Figure C.5: The Malin Head anemometry tower

mean wind speed at Culliagh Mast 05 at 30 m was calculated to be 7.2 m/s.

The corresponding joint wind speed and direction frequency distribution for Culliagh Mast 05 over the historical period 1979 to 1998 is presented in Figure C.9 in the form of a wind rose.

Site Wind Speed Variations at Culliagh Mountain

The variation in wind speed over the Culliagh Mountain site has been predicted using the WAsP computational flow model, details of which are given in the appendix to the study. WAsP was used to model the wind flow over the site, being initiated from the long-term wind speed and direction frequency distribution derived for Mast 05 at 30 m.

Table C.5 shows the predicted long-term mean wind speed at each wind turbine location at hub height. The average long-term mean wind speed at a hub height of 45 m for the whole wind farm was found to be 8.1 m/s.

Projected Energy Production

The energy production for each of the wind farm layouts is detailed in Table C.6 (the energy capture of individual turbines is given in Table C.5).

The energy production predictions include calculation of the array and topographic effects, an estimate of availability and electrical loss and factors to account for wind turbine icing, high wind hysteresis and the wake effect of existing turbines. Other potential

Table C.1: Data available from Culliagh and from Malin Head		
Culliagh Mountain Mast 05 (206940, 402500)	Hourly mean wind speed, standard deviation, gust and direction at 30 m.	5 July 1997–24 January 1999
	Hourly mean wind speed, standard deviation and direction at 10 m.	
Malin Head Meteorological Station (241950, 458550)	Hourly record of ten-minute mean wind speed and direction (time series data).	May 1997–January 2000
	Hourly record of ten-minute mean wind speed and direction (frequency table).	1979–1998

Table C.2: Main parameters of the Vestas V47 660 kW wind turbine

Diameter	47.0 m
Hub height	45.0 m
Rotor speed	28.5 rpm
No of blades	3
Nominal rated power	660 kW

sources of energy loss are also listed. It is recommended to carefully reconsider these issues, since at the time of this energy assessment there was insufficient information to estimate the effect on the predicted energy production.

Table C.3: Performance data for the Vestas V47 660 kW wind turbine

Wind speed (m/s at hub height)	V47 power output (kW)
4	2.9
5	43.8
6	96.7
7	166
8	252
9	350
10	450
11	538
12	600
13	635
14	651
15	657
16	659
17	660
18	660
19	660
20	660
21	660
22	660
23	660
24	660
25	660

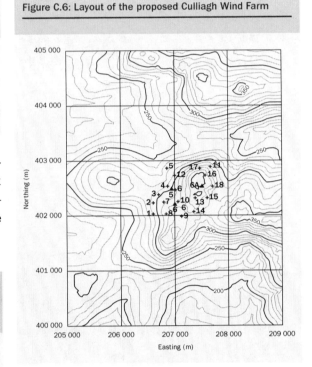

Figure C.6: Layout of the proposed Culliagh Wind Farm

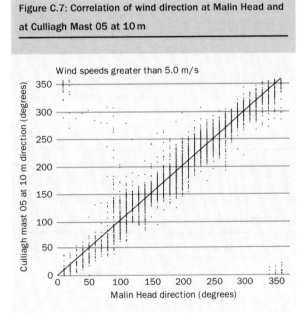

Figure C.7: Correlation of wind direction at Malin Head and at Culliagh Mast 05 at 10 m

Figure C.8: Correlation of wind speed at Malin Head and at Culliagh Mast 05 at 30 m

(a)

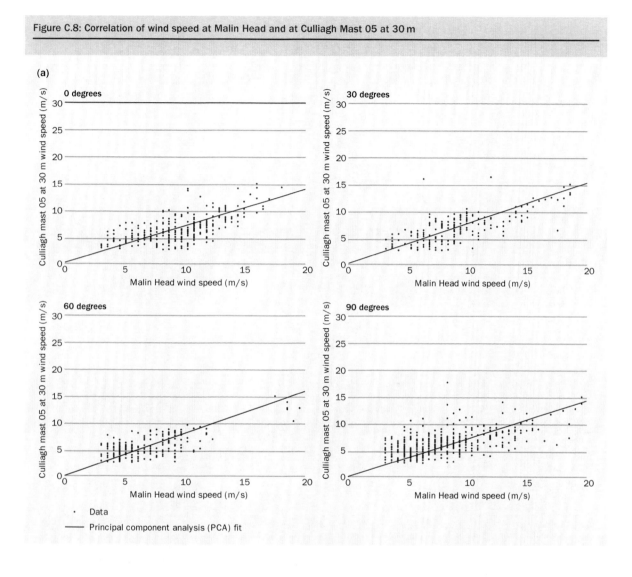

- Data

—— Principal component analysis (PCA) fit

Seasonal Variations

The monthly energy production of the wind farm is presented in Table C.7. There is a large seasonal variation of the predicted long-term monthly energy production, with winter and summer months producing approximately 140 per cent and 60 per cent, respectively, of the long-term mean monthly energy production.

Uncertainty Analysis

The main sources of deviation from the central estimate have been quantified and are shown in Tables C.8a and Table C.8b, which consider future periods of ten years and one year, respectively.

The figures in these tables, when added as independent errors, give the following uncertainties in net energy production: 4.5 GWh/annum for a future

Figure C.8: Correlation of wind speed at Malin Head and at Culliagh Mast 05 at 30 m – Continued

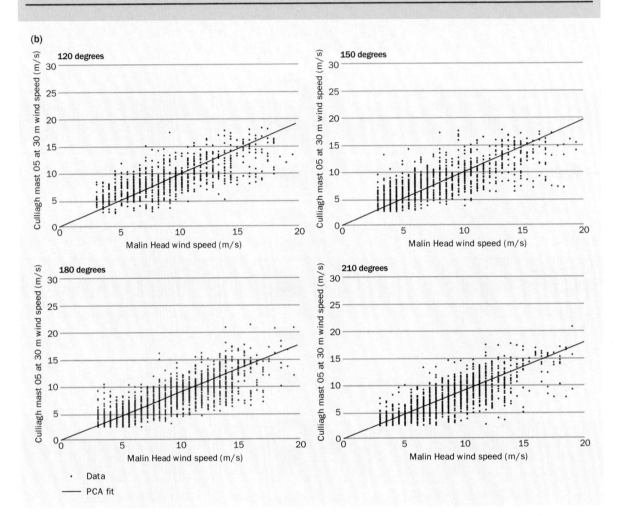

- Data
— PCA fit

one-year period and 2.7 GWh/annum for a future ten-year period. The detailed derivation of the above uncertainties is presented below.

There are four main categories of uncertainty associated with the site wind speed prediction at Culliagh Mountain:

1. There is an uncertainty associated with the measurement accuracy of the site anemometers. The instruments used on this site have not been individually calibrated to MEASNET standards and a consensus calibration has been applied. Batch

calibration of NRG Maximum 40 anemometers have shown them to conform to the consensus calibration to within 1.5 per cent. Therefore a figure of 2 per cent is assumed here so as to account for other second-order effects such as over-speeding, degradation, air density variations and sensor mounting. No allowance has been made for uncertainty in the Malin Head anemometer, as consistency and not absolute accuracy is important.

2. An error analysis was carried out on the correlation for each direction sector and from this the standard

Figure C.8: Correlation of wind speed at Malin Head and at Culliagh Mast 05 at 30 m – Continued

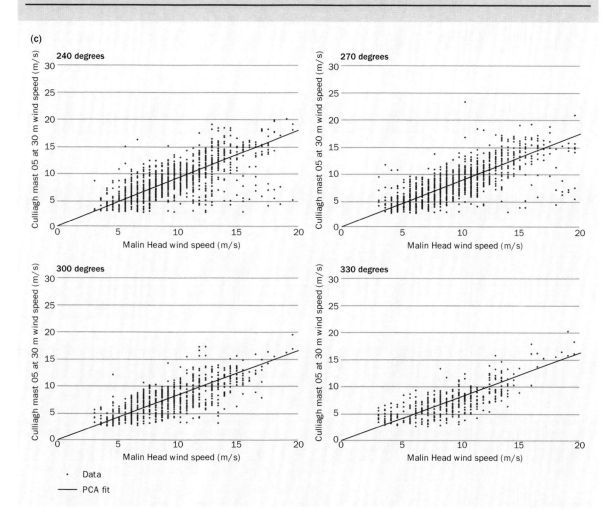

(c)

240 degrees

270 degrees

300 degrees

330 degrees

· Data
—— PCA fit

error for the long-term mean wind speed was determined. This was carried out for the correlation between Malin Head and Culliagh Mountain.

3. There is an uncertainty associated with the assumption made here that the historical period at the meteorological site is representative of the climate over longer periods. A study of historical wind records from a number of reference stations indicates an average variability of 6 per cent in the annual mean wind speed. This figure is used to define the uncertainty in assuming the long-term

mean wind speed is defined by a period 20 years in length.

4. For a finite number of future years, the mean wind speed may differ from the long-term mean due to the natural variability of a random process. Account is taken of the future variability of wind speed in the energy confidence analysis but not the wind speed confidence analysis.

It is assumed that the time series of wind speed is random with no systematic trends. Care was taken to

Table C.4: Wind speed ratios between Culliagh Mast 05 at 30 m and Malin Head

Direction sector	Number of hours analysed	Wind speed ratio
345–15	278	0.701
15–45	194	0.767
45–75	229	0.800
75–105	461	0.718
105–135	795	0.957
135–165	1098	0.976
165–195	1622	0.879
195–225	1208	0.897
225–255	1210	0.894
255–285	1230	0.868
285–315	708	0.834
315–345	421	0.819
All	9454	0.861

ensure that consistency of the Malin Head measurement system and exposure has been maintained over the historical period and no allowance is made for uncertainties arising due to changes in either.

Uncertainties type (1), (2) and (3) above are added as independent errors on a root-sum-square basis to give the total uncertainty in the site wind speed prediction for the historical period considered.

It is considered here that there are four categories of uncertainty in the energy output projection:

1. Long-term mean wind speed-dependent uncertainty is derived from the total wind speed uncertainty (types (1), (2), (3) and (4) above), using a factor for the sensitivity of the annual energy output to

Table C.5: Mean wind speed and projected energy output of individual wind turbines

Turbine number	Mean hub height wind speed[1] (m/s)	Energy output[2] (GWh/annum)
1	7.7	2.2
2	7.8	2.1
3	7.8	2.1
4	7.6	1.9
5	7.4	2.0
6	7.8	2.0
7	8.0	2.1
8	8.1	2.3
9	8.4	2.5
10	8.0	2.2
11	8.2	2.3
12	7.6	2.0
13	8.6	2.4
14	8.2	2.3
15	8.2	2.3
16	8.8	2.5
17	8.5	2.4
18	8.3	2.3
Overall	8.1	

Notes: [1]Wind speed at location of turbines at 45 m height, not including wake effects; [2]Individual turbine output includes topographic and array effects only.

Figure C.9: Annual wind rose for Culliagh Mast 05 at 30 m

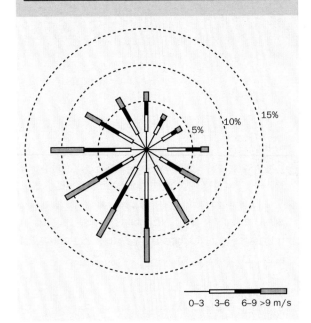

0–3   3–6   6–9 >9 m/s

Table C.6: Predicted energy production of Culliagh Mountain Wind Farm

Ideal energy production	40.2	GWh/annum
Topographic effect	107.0%	Calculated
Array effect	92.7%	Calculated
Electrical transmission efficiency	97.0%	Estimate
Availability	97.0%	Estimate
Icing and blade fouling	99.0%	Estimate
High wind hysteresis	99.6%	Estimate
Substation maintenance	100.0%	Not considered
Utility downtime	100.0%	Not considered
Power curve adjustment	100.0%	Not considered
Columnar control losses	100.0%	Not considered
Wake effect of existing wind farms	99.8%	Estimate
Net energy production	36.9	GWh/annum

Table C.8a: Uncertainty in projected energy output of the proposed wind farm – Ten-year future period

	Wind speed		Energy output[1]	
Source of uncertainty	**(%)**	**(m/s)**	**(%)**	**(GWh/annum)**
Anemometer accuracy	2.0	0.14		=
Correlation accuracy		0.19		
Period representative of long-term	1.3	0.10		
Total wind		**0.26**		**2.22**
Wake and topographic calculation	n/a	n/a	3.0	1.11
Wind variability (10 years)	1.9	0.14		1.19
Overall (10 years)				**2.75**

Note: [1]Sensitivity of net production to wind speed is calculated to be 8.68 GWh/annum/(m/s).

changes in annual mean wind speed. This sensitivity is derived by a perturbation analysis about the central estimate.

2. Wake and topographic modelling uncertainties. Validation tests of the methods used here, based on full-scale wind farm measurements made at small wind farms, have shown that the methods are accurate to 2 per cent in most cases. For this development, an uncertainty in the wake and topographic modelling of 3 per cent is assumed.

Table C.7: Monthly variation of the projected energy output of the wind farm

Month	Energy output[1] (GWh)
January	4.27
February	3.87
March	3.84
April	2.53
May	2.16
June	1.86
July	2.05
August	2.21
September	2.85
October	3.60
November	3.67
December	3.99

Note: [1]Energy output includes all losses.

Table C.8b: Uncertainty in projected energy output of the proposed wind farm – One-year future period

	Wind speed		Energy output[1]	
Source of uncertainty	**(%)**	**(m/s)**	**(%)**	**(GWh/annum)**
Anemometer accuracy	2.0	0.14		=
Correlation accuracy		0.19		
Period representative of long-term	1.3	0.10		
Total wind		**0.26**		**2.22**
Wake and topographic calculation	n/a	n/a	3.0	1.11
Wind variability (1 year)	6.0	0.43		3.75
Overall (1 year)				**4.49**

Note: [1]Sensitivity of net production to wind speed is calculated to be 8.68 GWh/annum/(m/s).

3. Future wind speed-dependent uncertainties described in (4) above have been derived using a factor for the sensitivity of the annual energy output to changes in annual mean wind speed. This sensitivity is derived by a perturbation analysis about the central estimate.

4. Turbine uncertainties are generally the subject of contract between the developer and turbine supplier and we have therefore made no allowance for them in this work.

Again those uncertainties which are considered are added as independent errors on a root-sum-square basis to give the total uncertainty in the projected energy output.

Summary of the Results of the Analysis

Wind data have been recorded at the Culliagh Mountain site for a period of 18 months. Based on the results from the analysis of these data, in combination with concurrent data and historical wind data recorded at Malin Head Meteorological Station, the following conclusions are made concerning the wind regime at the Culliagh Mountain site:

- The long-term mean wind speed is estimated to be 7.2 m/s at a height of 30 m above ground level.
- The standard error associated with the predicted long-term mean wind speed at 30 m is 0.26 m/s. If a normal distribution is assumed, the confidence limits for the prediction are as given in Table C.9.

Site wind flow and array loss calculations have been carried out, and from these we draw the following conclusions:

- The long-term mean wind speed averaged over all turbine locations at 45 m is estimated to be 8.1 m/s.
- The projected net energy capture of the proposed Culliagh Mountain Wind Farm is predicted to be 36.9 GWh/annum.

These predictions of net energy include topographic effects, array losses, availability, electrical transmission losses, air density adjustments, and factors to account for turbine icing, high wind hysteresis and the wake effect of existing turbines.

The net energy predictions presented above represent the long-term mean, 50 per cent exceedence levels, for the annual energy production of the wind farm. These values are the best estimate of the long-term mean value to be expected from the project. There is therefore a 50 per cent chance that, even when taken over very long periods, the mean energy production will be less than the value given in Table C.8. Estimates of long-term mean values with different levels of exceedence are set out in Table C.9.

- The standard error associated with the prediction of energy capture has been calculated and the confidence limits for the prediction are given in Table C.10.

Table C.9: Confidence limits – Wind speed

Probability of exceedence (%)	Long-term mean wind speed at 30 m (m/s)
90	6.9
75	7.0
50	7.2

Table C.10: Confidence limits – Energy

Probability of exceedence (%)	Net energy output (GWh/annum) 1-year average	Net energy output (GWh/annum) 10-year average
90	31.1	33.4
75	33.9	35.1
50	36.9	36.9
75	39.9	38.7
90	42.7	40.4

Actual Production of the Wind Farm

The Commissioning of the Culliagh wind farm took place in late 2000, and by November 2000 the wind farm was in full commercial operation. A review of the performance of the wind farm was undertaken in early 2002.

Table C.11 presents the expected long-term monthly energy production of the wind farm along with the actual energy production of the wind farm over the period from November 2000 to March 2002. It can be seen that individual months can deviate substantially from long-term expectations, for example February 2001 experienced production which was only 74 per cent of the long term expectations for this month, while in June 2001, 140 per cent of the long-term expectations for energy production in this month was

produced. Over the 17 month period for which data are available, the actual production of the wind farm has been 1.6 per cent below long-term expectations. This figure is well within the 75 and 90 per cent exceedence levels for the prediction presented above. A detailed assessment of the availability of the wind farm over the above operational period has not been undertaken, but it is understood that high availability levels have been achieved.

The data recorded at Malin Head indicate that the windiness of the period from November 2000 to March 2002 was some 4.9 per cent down on long-term expectations, making suitable assumptions about the seasonal variation of wind speed. This implies that over the longer term it is likely that the energy production of the wind farm will in fact exceed the central estimate value of 36.9 GWh/annum and may settle at a level which is close to the 25 per cent exceedence level presented in Table C.10. A more detailed assessment which includes issues such as wind direction, air density and availability would be required to provide a revised central estimate of wind farm production.

A separate validation of the accuracy of the modelling techniques employed to predict the long-term energy production of the Culliagh Mountain Wind Farm was undertaken. A comparison was made between the expected energy production of the wind farm, based on the actual mean wind speed recorded at Malin Head Meteorological Station, and the actual wind farm energy production. This was undertaken on an hourly basis. Thus the accuracy of the correlation relationships between Malin Head and the site and of the site flow model and turbine wake models was assessed using a 'wind in–energy out' test. Suitable adjustments were made to reflect the actual air density at the site. The comparison was undertaken for the operational period described above and data were only compared where all turbines were available and when wind farm SCADA data and data from Malin Head Meteorological Station were also available. Using these criteria, a comparison was made over a total of approximately

Table C.11: Expected and actual production of Culliagh Mountain Wind Farm

Month	Year	Expected production (GWh)	Actual production (GWh)
Nov	2000	3.670	3.703
Dec	2000	3.990	3.530
Jan	2001	4.270	3.546
Feb	2001	3.870	2.876
Mar	2001	3.840	3.410
Apr	2001	2.530	2.850
May	2001	2.160	1.699
Jun	2001	1.860	2.608
Jul	2001	2.050	1.813
Aug	2001	2.210	1.538
Sep	2001	2.850	2.941
Oct	2001	3.600	4.369
Nov	2001	3.670	3.645
Dec	2001	3.990	3.679
Jan	2002	4.270	4.801
Feb	2002	3.870	4.604
Mar	2002	3.840	4.037
	Total	56.540	55.649

Figure C.10: Cumulative plot showing measured energy against concurrent expected energy for the operating period

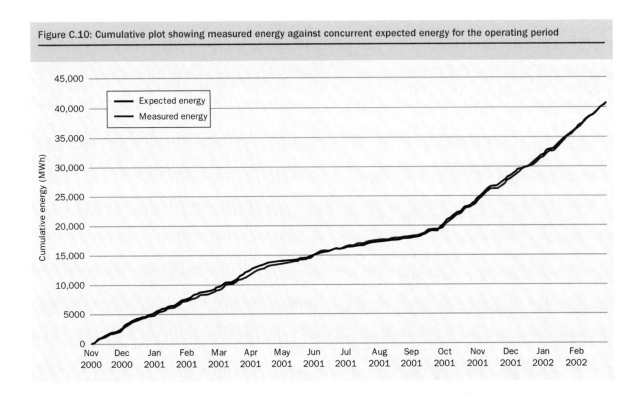

8300 hours. The results of the comparison of the expected and actual energy production of the wind farm are presented in Figure C.10 as a cumulative plot. Over the full period considered, the actual production was 99.7 per cent of the expected energy production of the wind farm, which provides confidence in the accuracy of the methods employed. It is noted that for individual months and for individual turbines larger discrepancies between the expected and actual energy production are observed.

Concluding Remarks

Appendix C has shown that the techniques outlined in the main text can be used to predict the behaviour of a wind farm with a good level of agreement. It has also demonstrated that the methods can be used to determine both mean values and associated uncertainties. It is hoped that it has proved a useful illustration of the techniques which are presently used by the industry.

APPENDIX D: DETAILED DESCRIPTION OF CORRELATION TECHNIQUES

Over the past decade, there has been an ongoing industry debate over different correlation methodologies which can be used for the prediction of the long-term mean wind speed at a site. All correlation methods have a common feature in that they:

1. establish a relationship between the concurrent data recorded at the site and reference station; and
2. apply the relationship to the historical data recorded at the reference station to predict the long-term wind regime at the site.

Such methodologies are commonly called measure correlate predict (MCP) analyses. Variables in such correlation analyses mooted over the past decade include those defined in Tables D.1 and D.2.

The tables present a bewildering array of options. While the technical merit of some methods over other methods can be argued, experience has shown that where the wind regimes at the site and reference meteorological station are well correlated, the results obtained tend to be relatively insensitive to the specific correlation methodology adopted. For cases where the correlation between the site and reference station is less good, then significant divergence is sometimes seen between the results obtained with different methods. In such circumstances, careful checks are required to ensure that

Table D.1: Prediction methodologies based on ten-minute or hourly data

Technique	Option 1	Option 2	Others
Directional bin size	30 degrees	Other	
Regression analysis technique	Principal component analysis	Least squares fit	
Fitting method	One parameter fit	Two parameter fit	Non-linear
Low wind speed cut-off	Exclude lowest wind speed data	Include lowest wind speed data	

Table D.2: Prediction methodologies based on longer-term data

Technique	Option 1	Option 2	Others
Averaging period	Monthly	Daily	
Fitting method	One parameter fit	Two parameter fit	Non-linear
Threshold for data coverage	Varies		

the correlation is sufficiently good to justify the use of the reference meteorological station at all. Due consideration also needs to be given to the interpretation of the uncertainty associated with a specific correlation methodology.

The methods based on ten-minute data or hourly data typically use the long-term wind rose recorded at the reference meteorological station. Those based on daily or monthly correlations are dependent on the site wind rose. It is often pragmatically observed that where hourly or ten-minute correlations between a site and reference station are poor, a reasonable correlation is observed over longer time periods such as a month.

Detailed Description of a Measure Correlate Predict Analysis

A detailed description of the steps within a measure correlate predict analysis is described below based on hourly data from the site and reference station. As indicated in the previous section, different approaches may be used. In the following discussion the proposed wind farm site is referred to as the 'target site' and the meteorological station is referred to as the 'reference site'.

The first stage in the approach is to record, over a period of about one year, concurrent wind data from both the target site and the nearby reference site for which well-established long-term wind records are available. The short-term measured wind data are then used to establish the correlation between the winds at the two locations. Finally, the correlation is used to adjust the long-term historical data recorded at the reference site to calculate the long-term mean wind speed at the site.

The concurrent data are correlated by comparing wind speeds at the two locations for each of twelve 30 degree direction sectors, based on the wind direction recorded at the reference site. This correlation involves two steps:

1. Wind directions recorded at the two locations are compared to determine whether there are any local features influencing the directional results. Only those records with speeds in excess of, say, 5 m/s at both locations are used.

2. Wind speed ratios are determined for each of the direction sectors using a principal component analysis.

In order to minimise the influence of localised winds on the wind speed ratio, the data are screened to reject records where the speed recorded at the reference site falls below 3 m/s (or a slightly different level) at the target site. The average wind speed ratio

is used to adjust the 3 m/s wind speed level for the reference site to obtain the different level for the target site, which ensures an unbiased exclusion of data. The wind speed at which this level is set is a balance between excluding low winds from the analysis and still having sufficient data for the analysis. The level used only excludes wind speeds below the cut-in wind speed of a wind turbine, which do not contribute to the energy production.

The result of the analysis described above is a table of wind speed ratios, each corresponding to one of 12 direction sectors. These ratios are used to factor the wind data measured at the reference site over the historical reference period to obtain the long-term mean wind speed at the target site. This estimate therefore includes the following influences:

- 'speed-up' between the target site and the reference site on a directional basis; this can be a very important characteristic, and sometimes speed-ups differ by a factor of as much as two; and
- the wind patterns at the reference site have been translated through the correlation process, so the long-term pattern at the target site has also been established.

APPENDIX E: SWT MANUFACTURERS AND THEIR MODELS

Less than 1 kW

The existing models and main features in this range are shown in Table E.1.

Table E.1: Existing wind turbine models, less than 1 kW

Wind turbine	Rated power (kW)	Rotor diameter (m)	Rotor type/no of blades	Generator type	Manufacturer/country
WG 503	0.025	0.51	HAWT (6)	PMG	Rutland (UK)
WG 910-3	0.090	0.91	HAWT (6)	PMG	Rutland (UK)
VT-60	0.12	0.9	HAWT (6)	PMG	Technoelektro (KRO)
VT-120	0.12	1.2	HAWT (5)	PMG	Technoelektro (KRO)
WS-0,15B/0,15C	0.12	0.30 (× 0.5)	VAWT	PMG	Windside (FIN)
WS-0,30 A	0.12	0.30 (× 1)	VAWT	PMG	Windside (FIN)
Pacific 100	01	0.928	HAWT (6)	PMG	Ampair (UK)
Flip 100	0.1	1.2	HAWT (3)	PMG	S&W Team (GER)
Inclin 250	0.25	1.35	HAWT (2)	PMG	Bornay (SP)
Twister 300 T	0.25	1.0 (× 1)	VAWT	PMG	Marc (GER)
Pacific 300	0.3	1.2	HAWT (3)	PMG	Ampair (UK)
Velter B	0.3	1.7	HAWT (3)	PMG	Solenersa (SP)
Speedy Vertical	0.3	1.2 (× 0.8)	VAWT (3)	PMG	Ropatec (IT)
FM 1803	0.34	1.8	HAWT (2)	PMG	Rutland (UK)
Superwind 350	0.35	1.12	HAWT (3)	PMG	Superwind (GER)
Air-X	0.4	1.14	HAWT (3)	PMG	Southwest (US)
StealthGen D-400	0.4	1.10	HAWT (5)	PMG	Eclectric (UK)
Aerocraft 502	0.5	2.4	HAWT (3)	PMG	Aerocraft (GER)
Enflo Windtec	0.5	0.71	HAWT (5)	PMG	Enflo Windtec (SWI)
Ampair Pacific	0.6	1.7	HAWT (3)	PMG	Ampair (UK)
Inclin 600	0.6	2.0	HAWT (2)	PMG	Bornay (SP)
Proven WT 600	0.6	2.55	HAWT (3)	PMG	Proven (UK)
Velter D	0.7	2.2	HAWT (3)	PMG	Solenersa (SP)
Aerocraft 752	0.75	2.4	HAWT (3)	PMG	Aerocraft (GER)
Espada	0.8	2.2	HAWT (2)	PMG	Fortis (NED)
Aerocraft 1002 H	1.0	2.4	HAWT (3)	PMG	Aerocraft (GER)
BWC Excell XL1	1.0	2.5	HAWT (3)	PMG	Bergey (US)
Lakota	1.0	2.1	HAWT (3)	PMG	Aeromax (US)
Whisper 100/200	0.9/1.0	2.1/3	HAWT (3)	PMG	Southwest (US)
Airdolphin Z-1000	1.0	1.8	HAWT (3)	PMG	Zephyr (JAP)
WS-1000	1.0	1.75	HAWT (3)	PMG	Windsave (UK)
Twister 300 T	1.0	1.9 (× 1.9)	VAWT (3)	PMG	Marc (GER)
WS-2AK/WS-2B	1.0	1.0 (× 2)	VAWT	PMG	Windside (FIN)
Easy Vertical	1.0	1.8 (× 1.15)	VAWT (3)	PMG	Ropatec (IT)

1 kW < SWT < 7 kW

The existing models and main features in this range are shown in Table E.2.

Table E.2: Existing wind turbine models, 1–7 kW

Wind turbine	Rated power (kW)	Rotor diameter (m)	Rotor type/no of blades	Generator type	Manufacturer/country
WS-4A/4AK/4C	1.2	1.0 (× 4)	VAWT	PMG	Windside (FIN)
Passaat	1.4	3.12	HAWT (3)	PMG	Fortis (NED)
Butterfly I	1.5	3.0	HAWT (3)	PMG	Energotech (GER)
SG 280	1.5	2.88	HAWT (3)	PMG	Geiger (GER)
Inclin 1500	1.5	2.86	HAWT (2)	PMG	Bornay (SP)
Velter I	1.5	3.1	HAWT (3)	PMG	Solenersa (SP)
Butterfly 1K	1.5	3.0	HAWT (3)	PMG	Energotech (GER)
Skystream 3,7	1.8	3.72	HAWT (3)	PMG	Southwest (US)
Antaris 2,5 KS	2.5	3/3.5	HAWT (3)	PMG	Heyde Windtechniks (GER)
Pawicon-2500	2.5	3.5	HAWT (3)	PMG	Pawicon (GER)
WT 2500	2.5	3.5	HAWT (3)	PMG	Proven (UK)
Tulipo	2.5	5.0	HAWT (3)	Asynchro + convert	WES (NED)
ARE 110	2.5	3.6	HAWT (3)	PMG	Abundant RE (US)
Turby 2,5	2.5	2 (× 2.65)	VAWT (3)	PMG	Turby (NED)
Inclin 3000	3.0	4.0	HAWT (2)	PMG	Bornay (SP)
Westwind 3	3.0	3.7	HAWT (3)	PMG	GP & GF Hill (AUS)
Simply Vertical	3.0	3.0	VAWT (3)	PMG	Ropatec (IT)
Whisper H175	3.2	4.5	HAWT (2)	PMG	Southwest (US)
Butterfly 3K	3.5	4.3	HAWT (3)	PMG	Energotech (GER)
Vento 5	5.0	5.0	HAWT (3)	PMG	Windeco (US)
ATS-1	5.0	5.4	HAWT (3)	PMG	Iskra (UK)
Aerosmart 5	5.0	5.1	HAWT (3)	Asynchro + gear	SMA (GER)
Montana	5.0	5.0	HAWT (3)	PMG	Fortis (NED)
Westwind 5	5.5	5.10	HAWT (3)	PMG	GP & GF Hill (AUS)
SWT 6000 AC	6.0	6.0	HAWT (4)	Asynchro + gear	Conergy (GER)
Inclin 6000	6.0	4.0	HAWT (3)	PMG	Bornay (SP)
WT 6000	6.0	5.5	HAWT (3)	PMG	Proven (UK)
Siroco	6.0	5.6	HAWT (2)	PMG	Eoltec (FRA)
QR 5	6.0	3.1 (× 5)	VAWT	PMG	QR (UK)
Maxi Vertical	6.0	4.7 (× 2.5)	VAWT (3)	PMG	Ropatec (IT)
AV-7	6.5	12.8	HAWT (3)	PMG	Aventa (GER)
Butterfly 6K	7.0	4.6	HAWT (3)	PMG	Energotech (GER)

7 kW < SWT < 50 kW

The existing models and main features in this range are shown in Table E.3.

Table E.3: Existing wind turbine models, 7–50 kW

Wind turbine	Rated power (kW)	Rotor diameter (m)	Rotor type/no of blades	Generator type	Manufacturer/country
SWT-7500	7.5	6.0	HAWT (4)	Asynchro + gear	Conergy (GER)
BWC EXCEL-R	7.5/10	6.7	HAWT (3)	PMG	Bergey (US)
WT 8000	8.0	5.4	HAWT (3)	PMG + gear	Webs (GER)
Aeroturbine	9.0	8.0	HAWT (3)	Synchron + gear	Aeroturbine (GER)
Aircon 10 S	9.8	7.1	HAWT (3)	PMG	Aircon (GER)
Alize	10.0	7.0	HAWT (3)	PMG	Fortis (NED)
Enwia E0	10.0	9.0	HAWT (3)	Synchron + gear	Alex Giersh (POL)
ARE 442	10.0	7.2	HAWT (3)	PMG	Abundant RE (US)
Westwind 10	10.0	6.20	HAWT (3)	PMG	GP & GF Hill (AUS)
Gaia Wind	11.0	13.0	HAWT (2)	Asynchro + gear	Gaia (DK)
WT 15000	15.0	9.0	HAWT (3)	PMG	Proven (UK)
Velter XV	15.0	7.2	HAWT (3)	PMG	Solenersa (SP)
GEV 10/20	15/20	10	HAWT (2)	Asynchro + gear	Vergnet (FRA)
Westwind 20	20.0	10.4	HAWT (3)	PMG	GP & GF Hill (AUS)
Gazelle 20	20.0	11.0	HAWT (3)	Asynchro + gear	Gazelle (UK)
Jacobs 20	20.0	9.5	HAWT (3)	PMG	Jacobs (US)
JIMP 20	20.0	8–10	HAWT (3)	PMG	Jonica Impiati (IT)
Big Star Vertical	20.0	8.5 (× 4.3)	VAWT (5)	PMG	Ropatec (IT)
WS-12	25.0	2.0 (× 6)	VAWT	PMG	Windside (FIN)
Wind Runner	25.0	11.0	HAWT (3)	PMG	Eoltec (FRA)
P14-30	30.0	14.0	HAWT (2)	PMG	Pitchwind (SWE)
Enwia E40	30.0	10.0	HAWT (3)	Synchron + gear	A Giersch (POL)
FL30	30.0	13.0	HAWT (3)	Asynchro + gear	Furlaender (GER)
Subaru 15/40	40.0	15.0	HAWT (3)	PMG	Subaru (JAP)
WT 50	50.0	11.5	HAWT (3)	PMG + gear	Webs (GER)
Vertikon H 50	50.0	12.0 (× 12.5)	VAWT (3)	PMG	MARC (GER)
EW15	50.0	15	HAWT (3)	Asynchro + gear	Entegrity Wind (US)

SWT greater than 50 kW

The existing models and main features in this range are shown in Table E.4.

Table E.4: Existing wind turbine models, 50–100 kW

Wind turbine	Rated power (kW)	Rotor diameter (m)	Rotor type/ no of blades	Generator type	Manufacturer/country
WT 50 SC	55.0	13.5	HAWT (3)	Asynchro + gear	Windtower (CZECH)
WES 18	80.0	18.0	HAWT (2)	Asynchro + convert	WES (NED)
E-20	100.0	20.0	HAWT (3)	Synchronous multipole	Enercon (GER)
V20	100.0	20.0	HAWT (2)	Asynchro + gear	Ventis (GER)
FL 100	100.0	21.0	HAWT (3)	Asynchro + gear	Furlaender (GER)
Enertech 100	100.0	*	HAWT (3)	*	Enertech (GER)
Subaru 22/100	100.0	22.0	HAWT (3)	PMG	Fuji Heavy Industries (JAP)
Northwind 100	100.0	19/20/21	HAWT (3)	PMG	DES (US)

* Data not yet available

APPENDIX F: CURRENT NATIONAL AND EUROPEAN R&D

The Main EU-Funded Projects

In 2006 more than 20 R&D projects were running with the support of FP6 and FP7 (the Framework Programmes are the main EU-wide tool for supporting strategic research areas).

The management and monitoring of projects is divided between two European Commission Directorate-Generals: the Directorate-General for Research (DG Research), for projects with medium- to long-term impact, and the Directorate-General for Transport and Energy (DG TREN), for demonstration projects with short- to medium-term impact on the market.

Private Initiatives

EOLIA (SPAIN)

The government-funded programme CENIT (the National Strategic Consortia for Technological Research) is focused on research activities. The target of the CENIT funding programme is to support large public–private consortiums and is aimed at overcoming strategic issues. In this framework, the private initiative EOLIA was launched in 2007.

The purpose of EOLIA is to carry out the research needed for the new technologies for offshore wind in deep waters. This covers a broad range of topics, from support structures, cables and moorings to project development (environmental impact assessment, wind resource and planning). It includes future applications and synergies (desalination and aquiculture).

The project's total budget of €34 million is being supported with €17 million from the Centre for the Development of Industrial Technology (CDTI). It started in 2007 and will be completed in 2010.

Several companies from the Acciona Group are participating in EOLIA (energy, wind turbines, desalination, infrastructure, engineering), together with major partners such as ABB, Construcciones Navales del Norte

Shipyard, General Cable, Ingeteam Group, Ormazabal and Vicinay, smaller partners such as Tinamenor and IMATIA, and the participation of research centres such as the National Renewable Energy Centre (CENER).

International Networks

THE EUROPEAN ACADEMY OF WIND ENERGY (EAWE)

The EAWE is a cooperation initiative on wind energy R&D made up of research institutes and universities in seven countries: Germany, Denmark, Greece, The Netherlands, Spain, the UK and Norway. The Academy was founded to formulate and execute shared R&D projects and to coordinate high-quality scientific research and education on wind energy at a European level. The core group is made up of 25 bodies, representing seven EU countries and more than 80 per cent of long-term research activity in the field of wind energy.

The activities of the EAWE are split into:

- integration activities such as PhD exchanges, exchange of scientists and the exploitation of existing research infrastructures;
- activities for the spreading of excellence, through the development of international training courses, dissemination of knowledge, support to SMEs and standardisation; and
- long-term research activities (see below).

Table F.1 lists thematic areas and topics that have been identified as first priority long-term R&D issues for EAWE's joint programme of activities.

THE EUROPEAN RENEWABLE ENERGY CENTRES AGENCY

The European Renewable Energy Centres Agency (EUREC) was established as a European economic interest grouping in 1991 to strengthen and rationalise

Table F.1: Priority long-term R&D issues for EAWE's joint programme of activities

Long-term wind forecasting	• Wind resources
	• Micro-siting in complex terrain
	• Annual energy yield
	• Design wind conditions (turbulence, shear, gusts, extreme winds) offshore, onshore and in complex terrain
Wind turbine external conditions	• Characteristics of wind regime and waves
	• Atmospheric flow and turbulence
	• Interaction of boundary layer and large wind farms
	• Prediction of exceptional events
Wind turbine technology	• Aerodynamics, aeroelasticity and aeroacoustics
	• Electrical generators, power electronics and control
	• Loads, safety and reliability
	• Materials, structural design and composite structures
	• Fracture mechanisms
	• Material characterisation and life-cycle analysis
	• New wind turbine concepts
Systems integration	• Grid connection and power quality issues
	• Short-term power prediction
	• Wind farm and cluster management and control
	• Condition monitoring and maintenance on demand
	• New storage, transmission and power compensation systems
Integration into the energy economy	• Integration of wind power into power plant scheduling and electricity trading
	• Profile-based power output and virtual power plants
	• Transnational and transcontinental supply structures
	• Control of distributed energy systems

the European research, demonstration and development efforts in all renewable energy technologies. As an independent, member-based association, it incorporates 48 prominent research groups from all over Europe.

EUREC members' research fields include solar buildings, wind, photovoltaics, biomass, small hydro, solar thermal power stations, ocean energy, solar chemistry and solar materials, hybrid systems, developing countries, and the integration of renewable energy into the energy infrastructure.

THE EUROPEAN WIND ENERGY TECHNOLOGY PLATFORM (TPWind)

TPWind's task is to identify and prioritise areas for increased innovation and new and existing R&D tasks. Its primary objective is to make overall reductions in the social, environmental and technological costs of wind energy. This is reflected in TPWind's structure, where the issues raised by the working groups (see below) are focused on areas where technological improvement leads to significant cost reductions.

This helps to achieve the EU's renewable electricity production targets. The Platform develops coherent recommendations, with specific tasks, approaches, participants and the necessary infrastructure. These are given in the context of private R&D and EU and Member State programmes, such as the EU's FP7.

TPWind is a network of more than 150 members, representing the whole industry from all over the EU. It is split into seven technical working groups, covering the issues of:

1. wind conditions;
2. wind power systems;
3. wind energy integration;
4. offshore deployment and operations;
5. market and economics;
6. policy and environment; and
7. R&D financing.

It comprises a Mirror Group, which includes representatives of the Member States, and a Steering Committee representing the whole industry. Detailed information is available at www.windplatform.eu.

GEO – THE WIND ENERGY COMMUNITY OF PRACTICE

The Wind Energy Working Group is part of the Energy Community of Practice, which is a section of the Group on Earth Observation (GEO). Under the auspices of the G8, GEO is an international initiative, aiming to establish the Global Earth Observation System of Systems (GEOSS) within the next ten years.

The Wind Energy Working Group directly contributes to the goals of one of the nine societal benefit areas of GEOSS, the energy area, for the improved management of energy resources. Specifically:

GEOSS outcomes in the energy area will support environmentally responsible and equitable ener-

gy management; better matching of supply and demand of energy; reduction of risks to energy infrastructure; more accurate inventories of greenhouse gases and pollutants; and a better understanding of renewable energy potential. (GEOSS 10-Year Implementation Plan, Section 4.1.3)

THE INTERNATIONAL ELECTROTECHNICAL COMMISSION (IEC)

The IEC, through its Technical Committee 88, is responsible for the development of standards relevant to wind turbine generator systems. It has produced standards for design requirements, power curve measurement, power quality control, rotor blade testing, lightning protection, acoustic noise measurement techniques, measurement of mechanical loads, and communications for monitoring and control of wind power plants.

Its current work programme includes both standards and design requirements for offshore wind turbines, for gearboxes and for wind farm power performance testing.

THE INTERNATIONAL MEASURING NETWORK OF WIND ENERGY INSTITUTES (MEASNET)

MEASNET is a cooperation of institutes that are engaged in the field of wind energy and want to ensure high-quality measurements and the uniform interpretation of standards and recommendations and obtain interchangeable results. The members have established an organisational structure for MEASNET, and they periodically perform mutual quality assessments of their harmonised measurements and evaluations.

This network was founded in 1997. It now has ten full members and five associate members.

THE EUROPEAN COMMITTEE FOR ELECTROTECHNICAL STANDARDIZATION (CENELEC)

CENELEC was created in 1973 as a result of the merging of two previous European organisations: CENELCOM and CENEL. Nowadays, CENELEC is composed of the National Electrotechnical Committees of 30 European countries. In addition, eight National Committees from neighbouring countries participate in CENELEC's work with affiliate status.

CENELEC's mission is to prepare voluntary electrotechnical standards that will help develop the Single European Market/European Economic Area for electrical and electronic goods and services, removing barriers to trade, creating new markets and cutting compliance costs.

THE INTERNATIONAL ENERGY AGENCY (IEA)

In its report 'Long-term research and development needs for wind energy for the time frame 2000 to 2020' (IEA, 2001), the Executive Committee of the IEA's Implementing Agreement for Wind Energy stated that continued R&D is essential for providing the reductions in cost and uncertainty that are necessary for reaching the anticipated deployment levels of wind energy.

In the mid-term, the report suggests the following R&D areas of major importance for the future deployment of wind energy: forecasting techniques, grid integration, public attitudes and visual impact.

In the long term, the Implementing Agreement sees R&D focusing on closer interaction of wind turbines and their infrastructure as a priority.

Since its inception, the Executive Committee of the Implementing Agreement has been involved in a wide range of R&D activities. The current research and development activities are organised into seven tasks (referred to as 'annexes'), giving an insight into its perception of current R&D priorities:

- **Annex XI: Base technology information exchange**. This refers to coordinated activities and informa-

tion exchange in two areas: i) the development of recommended practices for wind turbine testing and evaluation, including noise emissions and cup anemometry, and ii) joint actions in specific research areas such as turbine aerodynamics, turbine fatigue, wind characteristics, offshore wind systems and forecasting techniques.

- **Annex XIX: Wind energy in cold climates**. The objectives here include i) gathering and sharing information on wind turbines operating in cold climates, ii) establishing a formula for site classification, aligning meteorological conditions with local needs, and iii) monitoring the reliability and availability of standard and adapted turbine technology, as well as the development of guidelines.
- **Annex XX: HAWT aerodynamics and models from wind tunnel tests and measurements**. The main objective is to gather high-quality data on aerodynamic and structural loads for HAWTs, to model their causes and to predict their occurrence in full-scale machines.
- **Annex XXI: Building dynamic models of wind farms for power system studies that aim to assist in the planning and design of wind farms**. These studies develop models for use in combination with software packages for the simulation and analysis of power system stability.
- **Annex XXIII: Offshore wind energy technology development**. The aim is to identify and conduct R&D activities towards the reduction of costs and uncertainties and to identify and organise joint research tasks between interested countries.
- **Annex XXIV: Integration of wind and hydropower systems into the electricity grid**. The goal is to identify feasible wind/hydro system configurations, limitations and opportunities, involving an analysis of the integration of wind energy into grids fed by a significant proportion of hydropower, and opportunities for pumped hydro storage.
- **Annex XXV**: The 'design and operation of power systems with large amounts of wind power production' has recently been added as an additional task.

THE OFFSHORE WIND ENERGY NETWORK (OWE)

OWE is an independent source of information for professionals working in the field of offshore wind energy. It is also a gateway to several research projects on offshore wind energy. It provides a survey of the existing offshore wind farms, and information on existing offshore-related research projects and networks (for example CA-OWEE, COD and WE@SEA).

National Networks

DENMARK

Megavind

The Megavind partnership is the result of a government initiative for the development of environmentally effective wind technology. It addresses the challenges Denmark is facing in order to maintain its position as an internationally leading centre of competence within the field of wind power.

The partnership is the catalyst and initiator of a strengthened testing, demonstration and research strategy within the field of wind power in Denmark. It aims to think innovatively in regard to validation, testing and demonstration within wind power technology and the integration of wind power into the entire energy system. It therefore recommends creating an accumulated strategy for testing and demonstrating:

- components and turbine parts;
- wind turbines and wind farms; and
- wind power plants in the energy system.

Long-term university research and education in general should make a priority of the fundamental or generic technologies that are part of the development of wind turbines and wind power plants. These include:

- turbine design and construction;
- blades – aerodynamics, structural design and materials;
- wind loads and siting;
- the integration of wind power into the energy system; and
- offshore technology.

Megavind's recommendations will function as a reference for strategic research within wind power in the coming years, thus becoming the valid research strategy for wind power in Denmark.

GERMANY

The Centre of Excellence for Wind Energy (CE Wind)

The research network CE Wind, founded in 2005, includes the universities of Schleswig-Holstein. Through scientific research, CE Wind deals with fundamental questions relating to the wind turbines of the future, wind parks and the corresponding infrastructure.

CE Wind looks at the main issues regarding grid connection and integration, the design of rotor blades, generators, towers and foundations, operation monitoring and maintenance, impact on the environment of turbines in the multi-megawatt class, and operation in extreme local conditions.

ForWind

ForWind was founded in August 2003. It combines the interdisciplinary competencies of the universities of Oldenburg and Hanover and of its associated members, the universities of Stuttgart and Essen, in the field of wind power utilisation.

ForWind bridges basic research at the universities with demands from the industry for applied innovative wind energy conversion techniques. The research performed ranges from estimation of the wind resource to the grid integration of wind power. The research priorities are:

- wind resources and offshore meteorology;
- aerodynamics of rotor blades;
- turbulence and gusts;
- wave and wake loads;

- analysis of Scour Automatic System and load identification;
- material fatigue and lifetime analysis;
- material models for composite rotor blades;
- structural health monitoring for blades, tower and the converting system;
- hydro-noise reduction;
- interaction of ground and foundation structure;
- grouted joints for offshore constructions;
- electrical generator power system simulation and analysis of power quality; and
- grid connection of large-scale wind farms.

Research at Alpha Ventus (RAVE)

To help launch the deployment of offshore wind in German waters, the German Federal Ministry for the Environment (BMU) will support the offshore test wind farm Alpha Ventus in the North Sea with a research budget of about €50 million over the next few years.

This research initiative was named RAVE – Research at Alpha Ventus – and consists of a variety of projects connected with the installation and operation of Alpha Ventus. The different project consortia in RAVE are made up of most of the offshore research groups in Germany. RAVE is represented and coordinated by the ISET institute in Kassel.

In order to provide all participating research projects with detailed data, the test site will be equipped with extensive measurement instrumentation. The overall objective of the research initiative is to reduce the costs of offshore wind energy deployment in deep water. The institutes and companies participating in the RAVE initiative have prepared projects on the following topics so far:

- Obtaining joint measurements and data management;
- Analysis of loads and modelling, and further development of the different components of offshore wind turbines;

- loads at offshore foundations and structures;
- monitoring of the offshore wind energy deployment in Germany – 'Offshore WMEP';
- grid integration of offshore wind energy;
- further development of Lidar wind measuring techniques, analysis of external conditions and wakes;
- measurement of the operating noises and modelling of the sound propagation between tower and water; and
- environmental research.

SPAIN

The Spanish Wind Energy Technology Platform (REOLTEC)

REOLTEC (Techno-Scientific Wind Energy Network) was created in July 2005 with the aim of integrating and coordinating actions focused on research, development and innovation activities in the field of wind energy in Spain. In the last two years, the network has created working groups focused on different topics related to wind energy: wind turbines, applications, resource and siting, offshore, grid integration, certification, and social impact.

REOLTEC has the full support of AEE (the Spanish Wind Energy association). It is made up of the main players in the wind energy companies, research centres, universities and government agencies in Spain. This gives the network a wide-ranging point of view on the best path to follow in the coming years.

THE NETHERLANDS

INNWIND

The long-term R&D programme of the INNWIND consortium is funded by the government of The

Netherlands. The budget is €1.5 million per year. The consortium partners are:

- the Energy Research Centre of The Netherlands;
- Delft University of Technology; and
- the Knowledge Centre WMC (Wind Turbine Materials and Constructions).

The aim of the programme is to develop expertise, concepts, computer models and material databases that will be made available and applicable through a new generation of software tools. This is to enable the construction of large, robust, reliable, low-maintenance and cost-effective offshore wind turbines that are readily available for developers.

The INNWIND R&D priority areas are:

- concepts and components;
- aeroelasticity;
- materials and constructions;
- model development and realisation of an integrated modular design tool; and
- design guidelines.

We@Sea

We@Sea is a body funded by the Government of The Netherlands. It focuses on the national target of 6 GW offshore for 2020. The total budget is €26 million for five years. The We@Sea research priorities are:

- integration of wind power, and scenarios for its development;
- offshore wind power generation;
- spatial planning and environment;
- energy transportation and distribution;
- the energy market and financing;
- installation, exploitation, maintenance and dismantling; and
- training, education and dissemination of knowledge.

THE UK

Collaborative Offshore Wind Farm Research into the Environment (COWRIE)

COWRIE is an independent company set up to raise awareness and understanding of the potential environmental impacts of the UK's offshore wind farm programme. Identified research areas are:

- birds and benthos;
- electromagnetic fields;
- marine bird survey methodology;
- remote techniques; and
- underwater noise and vibration.

The Offshore Wind Energy Network (OWEN)

OWEN is a joint collaboration between industry and researchers. It promotes research on all issues connected with the development of the UK offshore wind energy resource (for example shallow water foundation design, submarine cabling, power systems, product reliability and impacts on the coastal zone).

The main aims of OWEN are:

- to identify, in detail, the research required by the UK wind energy industry so that the offshore wind energy resource can be developed quickly, effectively and efficiently;
- to provide a forum where specific research or development issues can be discussed;
- to ensure that regular reports of ongoing research projects are disseminated to relevant academic and industrial partners; and
- to ensure that the final results of any research project are widely publicised through tools such as conferences, newsletters and journals, whilst remaining aware of the need to preserve commercial confidentiality in the relevant cases.

The UK Energy Research Centre (UKERC)

The UK Energy Research Centre's mission is to be the UK's pre-eminent centre of research and source of authoritative information and leadership on sustainable energy systems.

UKERC undertakes world-class research addressing whole-system aspects of energy supply and use, while developing and maintaining ways of enabling cohesive research on energy. Research themes include:

- demand reduction;
- future sources of energy;
- energy infrastructure and supply;
- energy systems and modelling;
- environmental sustainability; and
- materials for advanced energy systems.

ITI Energy

ITI Energy is a private company, part of ITI Scotland Ltd. Its aims are the funding and managing of early stage technology development. It benefits from a long-term direct funding commitment from the Scottish Government through Scottish Enterprise. The available budget is £150 million over ten years. The ITI Energy programme includes:

- battery management systems;
- composite pipeline structure;
- hydrogen handling materials;
- interior surface coating;
- large-scale power storage;
- rechargeable batteries;
- wind turbine access systems;
- active power networks; and
- offshore renewables programmes.

The Energy Technologies Institute (ETI)

The ETI is an energy, research and development institute that is planned to begin operating in the UK in 2008. It is being set up by the UK government to 'accelerate the development of secure, reliable and cost-effective low-carbon energy technologies towards commercial deployment'. This new institute is supported by a number of companies as a 50:50 public–private partnership. The institute is expected to work with a range of academic and commercial bodies.

Conclusion

This large number of networks shows the willingness of the research sector to coordinate its efforts. It demonstrates the need for research, and the quest for improved efficiency through knowledge-sharing.

Building a research network is a way to strengthen the whole wind energy community, and to improve its attractiveness for the private sector, which can take advantage of a high level of expertise and information.

The European Wind Energy Technology Platform is the instrument that brings together institutes, research networks and private companies in order to set the research and market development priorities for the wind energy sector up to 2030.

Special Focus: Design Software for Wind Turbines

Currently used design tools are only partially suitable for the reliable design of very large wind turbines, and have only been validated and verified by means of measurements on what are now 'medium size' machines. Some physical properties that are irrelevant in small and medium-sized turbines cannot be neglected in the design of large, multi-megawatt turbines.

It is difficult to define for these machines a clear upper limit to which existing design tools can be applied. However, it is generally acknowledged by experts that the design risks increase considerably for machines with rotors of over around 125 m in diameter.

For this reason, new design tools are needed, supplemented with new features that take into account such issues as extreme blade deflections and wave loading of support structures in the case of offshore turbines. Such new tools will be essential if a new generation of wind turbines is to be designed and manufactured in a cost-effective way.

Moreover, in the case of offshore and complex or forested terrains, large uncertainties remain on the evaluation of local wind resources and loads. At this stage, advanced flow modelling for wind loads and resources has still not been verified and validated at a satisfactory level. These uncertainties on loadings should be taken into account in the design process.

High-quality design tools reduce the need for elaborating and performing expensive testing of prototypes, reduce the time required to market innovative concepts, and provide manufacturers with a competitive advantage. The probability of failure of wind turbines newly introduced to the market will also reduce, providing financiers and end users with a lower risk profile, less uncertainty and consequently lower electricity costs to the consumer.

One of the key challenges in developing design tools suitable for very large wind turbines is to understand and model aeroelastic phenomena. Figure F.1

gives an impression of the many dynamic external forces that act on wind turbines and the many ways the wind turbine structure may be distorted and may vibrate.

On the left, all the external dynamic forces that expose a turbine to extreme fatigue loading are indicated. On the right the various vibration and deflection modes of a wind turbine can be seen.

From the dynamic point of view, a wind turbine is a complex structure to design reliably for a given service lifetime.

In fact, the fatigue loading of a wind turbine is more severe than that experienced by helicopters, aircraft wings and car engines. The reason is not only the magnitude of the forces but also the number of load cycles that the structure has to withstand during its lifetime of 20 or more years (see Figure F.2).

The larger the wind turbine becomes, the more extreme the fatigue loading becomes. Thus, system identification and inverse methods for providing the loads under real conditions are required.

Computational fluid dynamics (CFD) tools are currently being developed into the design codes of the future. Large-scale wind turbines can be equipped with

---

Figure F.1: Modelling of the complete aeroelastic system of a wind turbine using symbolic programming

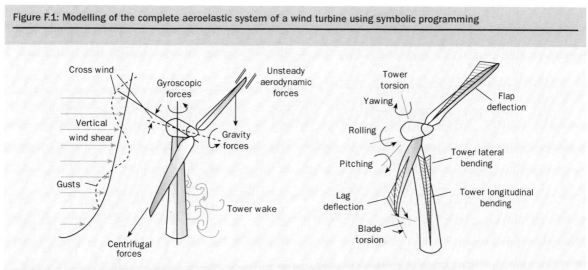

Source: Kießling F, Modellierung des aeroelastischen Gesamtsystems einer Windturbine mit Hilfe symbolischer Programmierung. DFVLR-Report, DFVFLR-FB 84-10, 1984.

Figure F.2: The fatigue loading of a wind turbine during its lifetime is large compared to, for example, bridges, helicopters, aeroplanes and bicycles

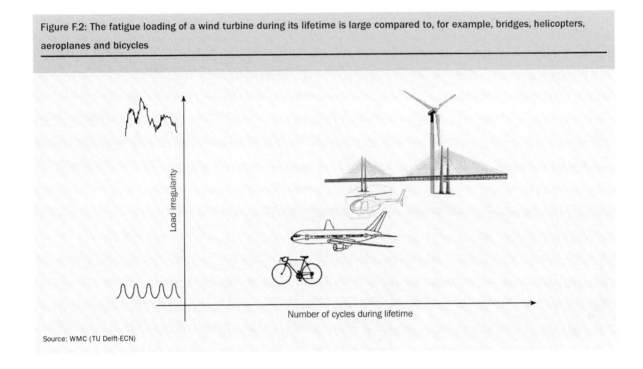

Source: WMC (TU Delft-ECN)

sensors that record dynamic behaviour. Once developed, experimental verification for virtually all new design tools needs to be carried out, taking into account external flow conditions.

Figure F.3 illustrates the complexity of rotor flow. A number of numerical codes exist to analyse and design components and subsystems such as drive trains, rotor blades, drive train dynamics and tower dynamics.

Currently, different design packages are not fully compatible with each other. It is therefore not possible to consider the system as a whole in the design phase. In terms of system optimisation, this implies that partial optimisation is performed on subsystems. This local optimisation is unlikely to be equivalent to the result of a global approach to optimisation.

Integral design methods include sub-design routines such as those for blades, power electronic systems,

mechanical transmission, support structures and transport, and installation loads. These methods should be thoroughly verified during their development and introduced into the standard design and certification processes. Through its dedicated work package, 'Integral design approaches and standards', the current UpWind project will bring solutions to this specific issue.

Many of the elements necessary for an integral design base are available. However, existing knowledge is not fully applied. Future research should therefore focus not only on improving the methodology, but also on improving wind turbine manufacturers', component manufacturers' and certification bodies' access to the know-how.

The interaction between flow and blade deformation is very complex. Three-dimensional aspects (tip vortices), axial flow, flow detachment (stall) and flow-induced vibrations all have to be taken into account in

Figure F.3: Sketch of three-dimensional flow, stall-induced vibrations and centrifugal effects on flow

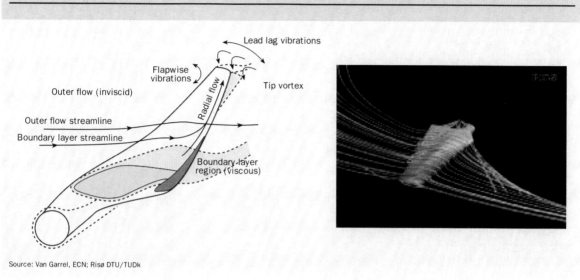

Source: Van Garrel, ECN; Risø DTU/TUDk

order to guarantee stable operation of the blade and accurate calculation of its lifetime. CFD is likely to be used in the future for detailed flow calculations as the computing time is reduced and the non-linear effects are better understood and modelled. The picture on the right in Figure F.3 shows the result of a CFD calculation of the flow around a rotor blade. The future vision is that integral design of a wind turbine will be able to be carried out so reliably that no extensive field tests will be needed before market introduction.

APPENDIX G: TRANSMISSION SYSTEMS IN EUROPE

Power system operators have been cooperating for decades, mainly to maximise system reliability and quality of power supply, while optimising the use of primary energy and capacity resources. As a result, five regional zones have emerged in Europe:

1. the synchronous zone of the Nordic countries;
2. the synchronous zone of the UCTE countries;
3. the synchronous zone of Great Britain;
4. the synchronous zone of the island of Ireland (Republic of Ireland and Northern Ireland); and
5. the Baltic Interconnected Power System.

The Synchronous Zone of the Nordic Countries

This synchronous zone comprises the power systems of Finland, Sweden, Norway and Eastern Denmark. The capacity of these power plants is around 90 GW and the annual electricity production is nearly 400 TWh, serving around 25 million people. The total primary control reserve is 1600 MW (operating reserve 600 MW and disturbance reserve 1000 MW). The transmission system operators (TSOs) of these countries have organised a cooperative body, NORDEL, whose primary objective is to create the conditions for, and to develop further, an efficient and harmonised Nordic electricity market.

This synchronous zone is interconnected by DC lines to Poland, Germany and Russia.

The Synchronous Zone of the UCTE Countries

The Union for the Coordination of Transmission of Electricity (UCTE) is the association of transmission system operators in continental Europe for 23 countries. The UCTE network ensures electricity supply for some 500 million people in one of the biggest electrical synchronous interconnections in the world. The estimated plant capacity is 603 GW (end 2004) and the total primary control reserve is 3 GW.

This synchronous area is interconnected both internally and across borders.

The Synchronous Zone of Great Britain

The National Grid Company (NGC) is now the system operator of the electricity transmission system on the islands of Great Britain, including England, Wales and Scotland. In April 2005, the Scottish system came under NGC control, although ownership is still separate.

Distribution is handled by several separate companies and the capacity of power plants is about 81 GW. The system is interconnected by DC lines with France (2000 MW) and Northern Ireland (450 MW) and is able to sustain a loss of 1320 MW.

The Synchronous Zone of (the Island of) Ireland

This smallest synchronous zone is operated by two TSOs: ESB and SONI. Their power system has a total installed capacity of power plants of about 7.6 GW and is connected to the Great Britain synchronous zone by a DC cable of 450 MW. The system reserve is 400 MW.

The Baltic Interconnected Power System

The interconnected grid of the Baltic States, Lithuania, Latvia and Estonia, is not synchronously linked to the power grids of other EU countries. There is a link with Finland, however, and links are also planned with Poland and Sweden. In 2006, the TSOs of these three countries established a cooperative organisation, BALTSO.

APPENDIX H: BASICS CONCERNING THE OPERATION AND BALANCING OF POWER SYSTEMS

In power systems, the balance between generation and consumption must be maintained continuously. The essential parameter in controlling the energy balance in the system is system frequency. If generation exceeds consumption, the frequency rises and if consumption exceeds generation, the frequency falls. As shown in Figure H.1, power system operation covers several timescales, ranging from seconds to days. Ultimately, it is the responsibility of the system operator to ensure that the power balance is maintained at all times.

Primary and Secondary Control

To start with, the *primary reserve* is activated automatically by frequency fluctuations. Generators on primary control respond rapidly, typically within 30–60 seconds. Imbalances may occur due to the tripping of a thermal unit or the sudden disconnection of a significant load. An immediate response from primary control is required to reinstate the power balance, so that the system frequency becomes stable again. For this near-immediate response to power imbalances, adequate generation reserves must be available from generation units in operation.

The *secondary reserve* is activated either manually or automatically, within 10 to 15 minutes following frequency deviation from nominal frequency. It backs up the primary reserve and stays operational until long-term reserves take over, as illustrated in Figure H.2. The secondary reserve consists of a spinning reserve (hydro or thermal plants in part-load operation) and a

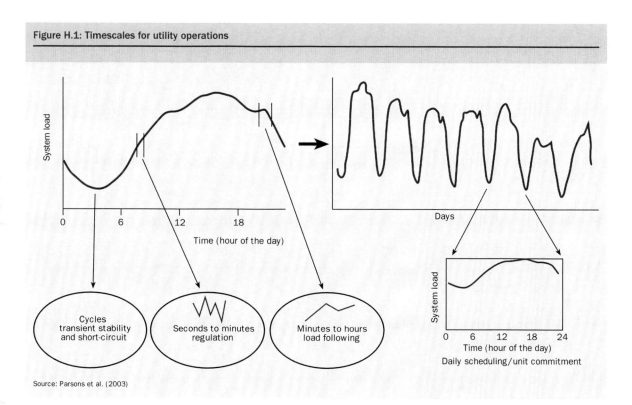

Figure H.1: Timescales for utility operations

System load

Time (hour of the day)

0   6   12   18

Days

Cycles
transient stability
and short-circuit

Seconds to minutes
regulation

Minutes to hours
load following

System load

0   6   12   18   24
Time (hour of the day)

Daily scheduling/unit commitment

Source: Parsons et al. (2003)

standing reserve (rapidly starting gas turbine power plants and load shedding).

Since changes in loads and generation, which result in a power imbalance, are not predictable or scheduled in advance, primary and secondary controls operate continuously, to ensure that system frequency remains close to its nominal value.

Tertiary Control, Unit Commitment

Electrical power consumption varies by the minute, hour and day. As the power balance must be continuously maintained, generation is scheduled to match the longer-term variations. Such economic dispatch decisions are made in response to anticipated trends in demand (while primary and secondary control continues to respond to unexpected imbalances). For example, an increase in load usually occurs from around 7.00 am to midday, or early afternoon. After the daily peak is reached, the load typically falls over the next few hours, finally reaching a daily minimum late at night.

Some generators require several hours to get started and synchronised to the grid, which means that the generation available for the midday peak must have been initiated hours in advance, in anticipation of this peak. In many cases, the shutdown process is also lengthy, and units may require several hours of cooling prior to restarting. The decision to use this type of unit often means that it must run for several days, prior to shutting down, in order to be economically viable. This timescale is called 'unit commitment', and it can range from several hours to several days, depending on specific generator characteristics and operational practice.

Figure H.2: Activation of power reserves and frequency of power system as a function of time when a large power plant is disconnected from the power system

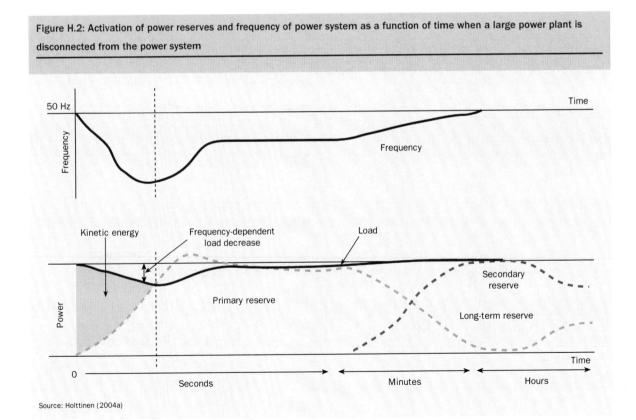

Source: Holttinen (2004a)

Task of the System Operator

During operational hours, the balancing task is usually taken over from individual power producers by the system operator. This is cost-effective, as the deviations of individual producers and loads equal out when aggregated, and only the net imbalances in the system area need to be balanced to control the frequency.

System operators have the information on schedules for production, consumption and interconnector usage. They either draw up these schedules themselves or obtain them from the electricity market or other parties involved (producers, balance responsible players or programme responsible parties) and they may also use online data and forecasts of load and wind power to assist in their operational duties. During operational hours, they follow the power system operation and call producers that have generators or loads as reserves, which can be activated depending on the need to balance power system net imbalances.

APPENDIX I: DETAILED COUNTRY REPORTS

Austria

MARKET STRUCTURE

With a share of 70 per cent RES-E (electricity from renewable energy sources) of gross electricity consumption in 1997, Austria was the leading EU Member State for many years. Large hydropower is the main source of RES-E in Austria. More recently, a steady rise in the total energy demand has taken place, and a decrease in the share of RES-E has been noted.

MAIN SUPPORTING POLICIES

Austrian policy supports RES-E through feed-in tariffs (FITs) that are annually adjusted by law. The responsible authority is obliged to buy the electricity and pay a FIT. The total available budget for RES-E support was decreased in May 2006, and tariff adjustments that are adjusted annually have been implemented. Within the new legislation, the annual allocated budget for RES support has been set at €17 million for 'new RES-E' up to 2011. This yearly budget is pre-allocated among different types of RES (30 per cent to biomass, 30 per cent to biogas, 30 per cent to wind, and 10 per cent to PV and the other remaining RES). Within these categories, funds will be given on a 'first come, first served' basis.

At present, a new amendment is being verified, suggesting an increase in the annual budget for support of new RES-E from €17 to €21 million. Consequently, the duration of FIT fuel-independent technologies might be extended to 13 years (now 10 years) and fuel-dependent technologies to 15 years (now 10 years), on behalf of the Minister of Economics. Moreover, investment subsidies of up to 15 per cent are in place for small hydro plants (> 1 MW). Emphasis is placed on 700 MW wind power, 700 MW small hydro power and 100 MW biomass.

FUTURE TARGETS

The RES-E target to be achieved in Austria by 2010 is 78.1 per cent of gross electricity consumption. In 2004, the share of renewable energy in gross electricity consumption reached 62.14 per cent, compared to 70 per cent in 1997.

Belgium

MARKET STRUCTURE

With a production of 1.1 per cent RES-E of gross electricity consumption in 1997, Belgium was at the bottom of the EU-15. National energy policies are implemented separately among the three regions of

Table I.1: Feed-in tariffs (valid for new RES-E plants permitted in 2006 and/or 2007) in Austria		
Technology	**Duration fixed years**	**2006–2007 fixed €/MWh**
Small hydro		31.5–62.5
PV systems		300–490
Wind systems		76.5 (2006); 75.5 (2007)
Geothermal energy	Year 10 and 11 at 75% and year 12 at 50%	74 (2006); 73 (2007)
Solid biomass and waste with large biogenic fraction Note: Expressed values refer to 'green' solid biomass (such as wood chips or straw). Lower tariffs in case of sawmill, bark (−25% of default) or other biogenic waste streams (−40 to −50%)		113–157 (2006); 111–156.5 (2007) 64 (2006); 63 (2007) – max 50% for hybrid plants
Biogas		115–170 (2006); 113–169.5 (2007)
Sewage and landfill gas		59.5–60 (2006); 40.5–41 (2007)
Mid-scale hydro power plants (10–20 MW) and CHP plants receive investment support of up to 10% of the total investment costs		

the country, leading to different supporting conditions and separate, regional markets for green certificates. Policy measures in Belgium contain incentives to use the most cost-effective technologies. Biomass is traditionally strong in Belgium, but both hydro power and onshore wind generation have shown strong growth in recent years.

KEY SUPPORT SCHEMES

Two sets of measures are the key to the Belgian approach to RES-E:

1. Obligatory targets have been set (obligation for all electricity suppliers to supply a specific proportion of RES-E) and guaranteed minimum prices or 'fallback prices' have been foreseen. In the Walloon region, the CWaPE (Commission Wallonne pour l'Energie) has registered an average price of €92/MWh per certificate during the first three months of 2006. In Flanders, the average price during the first half of 2006 has been around €110/MWh (VREG – Regulator in Flanders). In all three of the regions, a separate market for green certificates has been created. Due to the low penalty rates,

which will increase over time, it is currently more favourable to pay penalties than to use the certificates. Little trading has taken place so far.

2. Investment support schemes for RES-E investments are available. Among them is an investment subsidy for PV.

FUTURE TARGETS

For Belgium, the target for RES-E has been set at 6 per cent of gross electricity consumption by 2010. Nationally, the target for renewable electricity is 7 per cent by 2007 in the Walloon region, 6 per cent by 2010 in Flanders and 2.5 per cent by 2006 in Brussels.

Bulgaria

MARKET STRUCTURE

Bulgaria is approaching its RES-E target for 2010. Large-scale hydro power is currently the main source of RES-E, but its technical and economic potential is already fully exploited. Good opportunities exist for biomass, since 60 per cent of land consists of agricultural

Table I.2: Implementation of RES-E in Belgium

			Flanders	Walloon	Brussels	Federal
Target	%		2010: 6%	2007: 7% RES-E & CHP	2004: 2.00% 2005: 2.25% 2006: 2.50%	
Duration	years		10	10		
Min price[1] (fixed)	€/MWh	Wind offshore	n/a	n/a	n/a	90[2]
	€/MWh	Wind onshore	80			50
	€/MWh	Solar	450	65 all RES-E		150
	€/MWh	Biomass and other	80			20
	€/MWh	Hydro	95			50
Penalty	€/MWh		€125 (2005–2010)	€100 (2005–2007)	€75 (2005–2006) €100 (2007–2010)	

Notes: [1]Min prices: for the Federal State the obligation to purchase at a minimum price is on the TSO; for the regions the obligation is on the distribution system operator (DSO).
[2]Wind, first 216 MW installed capacity: €107/MWh

land, and about 30 per cent is forest cover. Bulgaria's RES-E share of gross electricity consumption increased from 7.2 per cent in 1997 to 9.28 per cent in 2004.

KEY SUPPORT SCHEMES

RES-E policy in Bulgaria is based on the following key mechanisms:
- Mandatory purchase of electricity at preferential prices will be applied until the planned system of issuing and trading green certificates comes into force (expected by 2012).
- A Green Certificate Market is planned to be put in place from 2012. A regulation will determine the minimum mandatory quotas of renewable electricity that generation companies must supply as a percentage of their total annual electricity production. Highly efficient CHP will also be included under the tradable green certificate scheme. Under the green certificate scheme there will still be a mandatory purchase of electricity produced for production up to 50 MW.

FUTURE TARGETS

The RES-E target to be achieved in 2010 is about 11 per cent for electric energy consumption. The goal of Bulgaria's National Programme on Renewable Energy Sources is to significantly increase the share of non-hydroelectric RES in the energy mix. A total wind power capacity of around 2200–3400 MW could be installed. Solar potential exists in the East and South of Bulgaria, and 200 MW could be generated from geothermal sources.

Cyprus

MARKET STRUCTURE

In Cyprus, an issue regarding policy integration has been observed, since investments in a new fossil fuel power plant creating excess capacity are underway. Until 2005, measures that proactively supported renewable energy production, such as the New Grant Scheme, were not very ambitious. In Cyprus, targets

Table I.3: Actual mandatory purchase prices, determined by the State Energy Regulation Commission, in Bulgaria

Technology	Duration	Preferential price 2008[1]
Wind Plants with capacity up to 10 MW for all installation committed before 1 January 2006	12 years	€61.4 / MWh
Wind new installations produced after 1 January 2006 effective operation >2250 h/a	12 years	€79.8 / MWh
Wind new installations produced after 1 January 2006 effective operation <2250 h/a	12 years	€89.5 / MWh
Hydro with top equaliser	12 years	€40.9 / MWh
Hydro <10 MW	12 years	€43.6 / MWh
Solar PV <5 kW	12 years	€400 / MWh
Solar PV >5 kW	12 years	€367 / MWh
Other RES	12 years	€40.6 / MWh

Note: [1]VAT not included. Currently, the Bulgarian Government is considering whether to keep such differentiated levels of support for the different renewable resources, or to set a uniform preferential price for all types of RES.

Table I.4: FITs in Cyprus

Technology	Capacity restrictions	Duration fixed years	2005 fixed €/MWh	2006 fixed €/MWh	Note
Wind	No limit	First 5 yrs	92	92	Based on mean annual wind speed
		Next 10 yrs	48–92	48–92	Varies according to annual operation hours: < 1750–2000 h: €85–92/MWh 2000–2550 h: €63–85/MWh 2550–3300 h: €48–63/MWh
Biomass, landfill and sewage gas	No limit	15	63	63	A more generous scheme is currently being developed for biomass electricity; up to €128/MWh is expected, depending on the category of investment
Small hydro	No limit	15	63	63	
PV	Up to 5 kW	15	204	204	
	Without investment subsidy	15	x	337–386	Households receive higher tariff than companies

Note: Exchange rate €1 = CYP0.58.

are not being met. In 2006, a New Enhanced Grant Scheme was agreed upon. The leading RES in Cyprus is PV; wind power has a high potential.

KEY SUPPORT SCHEMES

RES-E policy in Cyprus is made up of the following components:
- The New Grant Scheme, valid from 2004 until 2006. A tax of 0.22c€/kWh on every category of electricity consumption is in place. The income generated by this tax is used for the promotion of RES.
- The New Enhanced Grant Scheme was installed in January 2006. Financial incentives (30–55 per cent of investments) in the form of government grants and FITs are part of this scheme.
- Operation state aid for supporting electricity produced by biomass has been suggested, and forwarded to the Commission for approval.

FUTURE TARGETS

The Action Plan for the Promotion of RES determines that the contribution of RES to the total energy consumption of Cyprus should rise from 4.5 per cent in 1995

to 9 per cent in 2010. The RES-E target to be achieved in 2010 from the EU Directive is 6 per cent. In Cyprus, the RES share of total energy consumption decreased from 4.5 per cent in 1995 to 4 per cent in 2002.

Czech Republic

MARKET STRUCTURE

The Czech Republic's legislative framework in relation to renewable energy sources has been strengthened by a new RES Act, adopted in 2005, and a Government Order regulating the minimum amount of biofuels or other RES fuels that must be available for motor fuel purposes. Targets for increasing RES in total primary energy consumption have been set at national level. The use of biomass in particular is likely to increase as a result of the new legislation.

KEY SUPPORT SCHEMES

In order to stimulate the growth of RES-E, the Czech Republic has decided on the following measures:
- A feed-in system for RES-E and cogeneration, which was established in 2000.

Table I.5: Key support schemes in the Czech Republic

Technology	Duration fixed years	premium years	2005 fixed €/MWh	2006 fixed €/MWh	premium €/MWh	2007 fixed €/MWh	premium €/MWh
Wind energy			87	85	70	88–114	70–96
Small hydro (up to 10 MW)			68	81	49	60–85	23–48
Biomass combustion			84	79–101	46–68	84–121	44–81
Biomass co-firing with fossil fuels	Equals the lifetime	Set annually	17	*	19–41		9–55
Biogas			81	77–103	44–69	81–108	41–69
Geothermal electricity			117	156	126	161	125
PV			201	456	435	229–481	204–456

Note: * The Energy Regulatory Office (ERO) cannot reduce this by more than 5 per cent each year. Exchange rate €1 = CZK27.97.

- A new RES Act, adopted in 2005, extending this system by offering a choice between a FIT (a guaranteed price) or a 'green bonus' (an amount paid on top of the market price). Moreover, the FIT is index-linked whereby an annual increase of at least 2 per cent is guaranteed.

FUTURE TARGETS

A 15–16 per cent share of RES in total primary energy consumption by 2030 has been set as a target at national level. For RES-E, the target to be achieved is 8 per cent in 2010. The Czech Republic's RES percentage of total primary energy consumption is currently approximately 3 per cent. A very gradual increase can be observed in the RES-E share of gross electricity consumption (3.8 per cent in 1997, 4.1 per cent in 2004).

Denmark

MARKET STRUCTURE

Due to an average growth of 71 per cent per year, Danish offshore wind capacity remains the highest per capita in Europe (409 MW in total in 2007). Denmark is at present close to reaching its RES-E target for 2010. Two new offshore installations, each of 200 MW, are planned. RES other than offshore wind are slowly but steadily penetrating the market supported by a wide array of measures such as a new re-powering scheme for onshore wind.

KEY SUPPORT SCHEMES

In order to increase the share of RES-E in the overall electricity consumption, Denmark has implemented the following measures:

- A tendering procedure has been used for two new large offshore installations. Operators will receive a spot price and initially a settling price as well. Subsequent offshore wind farms are to be developed on market conditions.
- A spot price, an environmental premium (€13/MWh) and an additional compensation for balancing costs (€3/MWh) for 20 years is available for new onshore wind farms.
- Fixed FITs exist for solid biomass and biogas under certain conditions.
- Subsidies are available for CHP plants based on natural gas and waste.

Table I.6: Key support schemes in Denmark

Technology	Duration	Tariff	Note
Wind onshore	20 years	Market price plus premium of €13/MWh	Additionally balancing costs are refunded at €3/MWh, leading to a total tariff of approx. €57/MWh
Wind offshore	50.000 h operation	€66–70/MWh spot market price plus a €13/MWh premium	A tendering system was applied for the last two offshore wind parks; balancing costs are by the owners
Solid biomass and biogas	10 years following 10 years	€80/MWh €54/MWh	New biogas plants are only eligible for the tariff if they are grid connected before end of 2008
Natural gas and waste CHP plants	20 years 20 years	Individual grant, depending on previous grants Three-time tariff	Above 10 MW only; annual, non-production-related grant 5–10 MW can choose the support scheme; below 5 MW only three-time tariff (receive a subsidy depending on when electricity production takes place, and this combined with the electricity market price, provides a three-tier tariff)
PV	Not determined	€200–250/MWh	'Meter running backwards' principle applied in private houses

FUTURE TARGETS

In Denmark, the RES-E target from the EU Directive is 29 per cent of gross electricity consumption by 2010. With an increase from 8.7 per cent RES-E in 1997 to 26.30 per cent in 2004, Denmark is nearing its target of 29 per cent RES-E of gross electricity consumption in 2010.

Estonia

MARKET STRUCTURE

Estonia has extensive fossil fuel reserves, including a large oil shale industry. However, the average annual growth rate for RES-E stands at 27 per cent. Estonia's largest RES potential is to be found in the biomass sector, but possibilities also exist in the areas of wind power, biogas electricity and small hydropower.

KEY SUPPORT SCHEMES

Estonian legislation relevant to RES-E includes:
- An obligation on the grid operator to buy RES-E providing that the amount 'does not exceed the network losses during the trading period' which came into force in 2005.
- A voluntary mechanism involving green energy certificates was also created by the grid operator (the state-owned Eesti Energia Ltd) in 2001.

Renewable electricity is purchased for a guaranteed fixed price of 81 EEKcents/kWh (5.2c€/kWh). Before, the Electricity Market Act (EMA) prices were linked to the sales prices of the two major oil-shale-based power plants.

The EMA states that the preferential purchase price for wind electricity is guaranteed for 12 years, but all current support mechanisms will be terminated in 2015. There is no information on legislation planned to replace this after 2015.

FUTURE TARGETS

In Estonia, the share of electricity produced from renewable energy sources is projected to reach 5.1 per cent in 2010. For RES-E, an average annual growth rate of 27 per cent has been registered between 1997 and 2004. Estonia's share of RES-E stood at 0.7 per cent in 2004, compared to 0.2 per cent in 1997. Dominant sources of RES-E in Estonia are solid biomass and small-scale hydropower.

Table I.7: Key support schemes in Estonia

Technology	Duration fixed years	2003–present fixed €/MWh
All RES	Wind: 12 Current support mechanisms will be terminated in 2015	52

Finland

MARKET STRUCTURE

Finland is nearing its RES-E target for 2010, and continues to adjust and refine its energy policies in order to further enhance the competitiveness of RES. Through subsidies and energy tax exemptions, Finland encourages investment in RES. Solid biomass and large-scale hydropower plants dominate the market, and biowaste is also increasing its share. Additional support in the form of FITs based on purchase obligations or green certificates is being considered for onshore wind power.

KEY SUPPORT SCHEMES

Finland has taken the following measures to encourage the use of RES-E:

- Tax subsidies: RES-E has been made exempt from the energy tax paid by end users.
- Discretionary investment subsidies: new investments are eligible for subsidies up to 30 per cent (40 per cent for wind).
- Guaranteed access to the grid for all electricity users and electricity-producing plants, including RES-E generators (Electricity Market Act – 386/1995).

Table I.8: Key support schemes in Finland

Technology	2003–present Tax reimbursement €/MWh
Wind and forest chip	6.9
Recycled fuels	2.5
Other renewables	4.2

FUTURE TARGETS

By 2025, Finland wants to register an increase in its use of renewable energy by 260 PJ. With regard to RES-E, the target to be met is 31.5 per cent of gross electricity consumption in 2010. With figures of 24.7 per cent in 1997 and 28.16 per cent in 2004, Finland is progressing towards its RES-E target of 31.5 per cent in 2010.

France

MARKET STRUCTURE

France has centred its RES approach around FITs on the one hand and a tendering procedure on the other. Hydropower has traditionally been important for electricity generation, and the country ranks second when it comes to biofuel production, although the biofuels target for 2005 was not met.

KEY SUPPORT SCHEMES

The French policy for the promotion of RES-E includes the following mechanisms:

- FITs (introduced in 2001 and 2002, and modified in 2005) for PV, hydro, biomass, sewage and landfill gas, municipal solid waste, geothermal, offshore wind, onshore wind and CHP; and
- a tender system for large renewable projects.

FUTURE TARGETS

The RES-E target from the EU Directive for France is 21 per cent RES-E share of gross electricity consumption in 2010. France's share of RES-E decreased from 15 per cent in 1997 to 12.64 per cent in 2004. France has vast resources of wind, geothermal energy and biomass, and wind power and geothermal electricity have shown growth. In addition, there is potential in the area of solid biomass.

Table I.9: Key support schemes in France

Technology	Duration	Tariff	Note
Wind onshore	10 years following 5 years	€82/MWh €28–82/MWh	Depending on the local wind conditions
Wind offshore	10 years following 10 years	€130/MWh €30–130/MWh	Depending on the local wind conditions
Solid biomass	15 years	€49/MWh	Standard rate, including premium up to €12/MWh
Biogas	15 years	€45–57.2/MWh	Standard rate, including premium up to €3/MWh
Hydropower	20 years	€54.9–61/MWh	Standard rate, including premium up to €15.2/MWh
Municipal solid waste	15 years	€45–50/MWh	Standard rate, including premium up to €3/MWh
CHP plants		€61–95/MWh	
Geothermal	15 years 15 years	€120/MWh €100/MWh	Standard rate In metropolis only Plus an efficiency bonus of up to €30/MWh
PV	20 years 20 years	€300/MWh €400/MWh	In metropolis In Corsica, DOM and Mayotte Plus €250/MWh and €150/MWh respectively if roof-integrated

Germany

KEY ISSUES

Germany is an EU leader in wind utilisation, PV, solar thermal installations and biofuel production. Its onshore wind capacity covers approximately 50 per cent of the total installed capacity in the EU. A stable and predictable policy framework has created conditions favourable to RES penetration and growth. FITs for RES-E have proven a successful policy, leading to a very dynamic market for RES.

KEY SUPPORT SCHEMES

With the aim of promoting RES-E, Germany has introduced the following schemes through its Renewable Energy Act of 2004:
- FITs for onshore wind, offshore wind, PV, biomass, hydro, landfill gas, sewage gas and geothermal; and

Table I.10: Key support schemes in Germany

Technology	Duration	Tariff	Note
Wind onshore	20 years	€83.6/MWh €52.8/MWh	For at least 5 years Further 15 years, annual reduction of 2% is taken into account
Wind offshore	20 years	€91/MWh €61.9/MWh €30–130/MWh	For at least 12 years Further 8 years, annual reduction of 2% is taken into account
Solid biomass and biogas	20 years 20 years	€81.5–111.6/MWh €64.5–74.4/MWh additional €20/MWh	Annual reduction of 1.5% Annual reduction of 1.5% In CHP applications only
Hydropower up to 5 MW	30 years	€66.5–96.7/MWh	Lower FITs also for hydro plants up to 150 MW
Geothermal	20 years	€71.6–150/MWh	Annual reduction of 1% from 2010 on
PV	20 years	€406–568/MWh	Annual reduction of 6.5%; prices vary depending on the location

Table I.11: Key support schemes in Greece

RES-E Technology	Mainland €/MWh	Autonomous islands €/MWh
Wind onshore	73	84.6
Wind offshore	90	90
Small hydro (<20 MW)	73	84.6
PV system (≤100 kWp)	450	500
PV system (>100 kWp)	400	450
Solar thermal power plants (≤5 MWp)	250	270
Solar thermal power plants (>5 MWp)	230	250
Geothermal	73	84.6
Biomass and biogas	73	84.6
Others	73	84.6

- large subsidised loans available through the DtA (Deutsche Ausgleichsbank) Environment and Energy Efficiency Programme.

FUTURE TARGETS

Overall, Germany would like to register a 10 per cent RES share of total energy consumption in 2020. The RES-E targets set for Germany are 12.5 per cent of gross electricity consumption in 2010 and 20 per cent in 2020. Substantial progress has already been made towards the 2010 RES-E target. Germany's RES-E share in 1997 was 4.5 per cent, which more than doubled to 9.46 per cent by 2004.

Greece

MARKET STRUCTURE

Hydropower has traditionally been important in Greece, and the markets for wind energy and active solar thermal systems have grown in recent years. Geothermal heat is also a popular source of energy. The Greek Parliament has recently revised the RES policy framework, partly to reduce administrative burdens on the renewable energy sector.

KEY SUPPORT SCHEMES

General policies relevant to RES include a measure related to investment support, a 20 per cent reduction of taxable income on expenses for domestic appliances or systems using RES, and a concrete bidding procedure to ensure the rational use of geothermal energy. In addition, an inter-ministerial decision was taken in order to reduce the administrative burden associated with RES installations.

Greece has introduced the following mechanisms to stimulate the growth of RES-E:

- FITs were introduced in 1994 and amended by the recently approved Feed-in Law. Tariffs are now technology-specific, instead of uniform, and a guarantee of 12 years is given, with a possibility of extension to up to 20 years.
- Liberalisation of RES-E development is the subject of Law 2773/1999.

FUTURE TARGETS

According to the EU Directive, the RES-E target to be achieved by Greece is 20.1 per cent of gross electricity consumption by 2010. In terms of RES-E share of gross electricity consumption, the 1997 figure of 8.6 per cent increased to 9.56 per cent in 2004.

Hungary

KEY ISSUES

After a few years of little progress, major developments in 2004 brought the Hungarian RES-E target within reach. Geographical conditions in Hungary are favourable for RES development, especially biomass.

Between 1997 and 2004, the average annual growth of biomass was 116 per cent. Whilst environmental conditions are the main barrier to further hydropower development, other RES such as solar, geothermal and wind energy are hampered by administrative constraints (for example the permit process).

KEY SUPPORT SCHEMES

The following measures exist for the promotion of RES-E:

- A feed-in system is in place. It has been using technology-specific tariffs since 2005, when Decree 78/2005 was adopted. These tariffs are guaranteed for the lifetime of the installation.
- A green certificate scheme was introduced with the Electricity Act (2001, as amended in 2005). This act gives the government the right to define the start date of implementation. At that time, FITs will cease to exist.

Nevertheless, from 2007, subsidies for cogeneration power and RES will be decreased, since national goals of production from RES were already achieved in 2005.

FUTURE TARGETS

The Hungarian Energy Saving and Energy Efficiency Improvement Action Programme expresses the country's determination to reach a share of renewable energy consumption of at least 6 per cent by 2010. The target set for Hungary in the EU Directive is a RES-E share of 3.6 per cent of gross electricity consumption. Progress is being made towards the 3.6 per cent RES-E target. Hungary's RES-E share amounted to 0.7 per cent in 1997 and 2.24 per cent in 2004.

Ireland

MARKET STRUCTURE

Hydro and wind power make up most of Ireland's RES-E production. Despite an increase in the RES-E share over the past decade, there is still some way to go before the target is reached. Important changes have occurred at a policy level. Ireland has selected the renewable energy feed-in tariff (REFIT) as its main instrument. From 2006 onwards, this new scheme is expected to provide some investor certainty, due to a

Table I.12: Key support schemes in Hungary

Technology		Duration fixed years	2005 fixed Ft./kWh	2005 fixed €/MWh	2006 fixed Ft./kWh	2006 fixed €/MWh
Geothermal, biomass, biogas, small hydro (<5 MW) and waste	Peak		28.74	117	27.06	108
	Off-peak		16.51	67	23.83	95
	Deep off-peak		9.38	38	9.72	39
Solar, wind	Peak	According to the lifetime of the technology	n/a	n/a	23.83	95
	Off-peak		n/a	n/a	23.83	95
	Deep off-peak		n/a	n/a	23.83	95
Hydro (>5 MW), cogeneration	Peak		18.76	76	17.42	69
	Off-peak		9.38	38	8.71	35
	Deep off-peak		9.38	38	8.71	35

Note: Exchange rate 1 Ft. = 0.004075 euros (1 February 2005) and 1 Ft. = 0.003975 euros (1 February 2006).

Table I.13: Key RES-E support schemes in Ireland

Technology	Tariff duration fixed years	2006 fixed €/MWh
Wind >5 MW plants		57
Wind <5 MW plants		59
Biomass (landfill gas)	15	70
Other biomass		72
Hydro		72

15-year FIT guarantee. No real voluntary market for renewable electricity exists.

KEY SUPPORT SCHEMES

Between 1995 and 2003, a tender scheme (the Alternative Energy Requirement – AER) was used to support RES-E. Since early 2006, the REFIT has become the main tool for promoting RES-E. €119 million will be used over 15 years from 2006 to support 55 new renewable electricity plants with a combined capacity of 600 MW. FITs are guaranteed for up to 15 years, but may not extend beyond 2024. During its first year, 98 per cent of all the REFIT support has been allocated to wind farms.

FUTURE TARGETS

The RES-E target for Ireland, set by the EU Directive, to be met by 2010, is 13.2 per cent of gross electricity consumption. The country itself would like to reach an RES-E share of 15 per cent by that time. The European Energy Green Paper, published in October 2006, sets targets over longer periods. In relation to Ireland, it calls for 30 per cent RES-E by 2020. Ireland is making some modest progress in relation to its RES-E target, with 3.6 per cent in 1997 and 5.23 per cent in 2004.

Italy

KEY ISSUES

Despite strong growth in sectors such as onshore wind, biogas and biodiesel, Italy is still a long way from the targets set at both national and European level. Several factors contribute to this situation. First, there is a large element of uncertainty, due to recent political changes and ambiguities in the current policy design. Second, there are administrative constraints, such as complex authorisation procedures at local level. And third, there are financial barriers, such as high grid connection costs.

In Italy, there is an obligation on electricity generators to produce a certain amount of RES-E. At present, the Italian Government is working out the details of more ambitious support mechanisms for the development and use of RES.

KEY SUPPORT SCHEMES

In order to promote RES-E, Italy has adopted the following schemes:
- Priority access to the grid system is guaranteed to electricity from RES and CHP plants.

Table I.14: Key support schemes in Italy

Technology	Capacity	Duration fixed years	2006 fixed €/MWh
Solar PV	<20 kW		44.5*
	≤50 kW		46
	50 < P < 1000 kW		49
Building-integrated PV	<20 kW	20	48.9*
	≤50 kW		50.6
	>50 kW		max 49 + 10%

Note: *From February 2006, these tariffs are also valid for PV with net metering ≤20 kW.

- An obligation for electricity generators to feed a given proportion of RES-E into the power system. In 2006, the target was 3.05 per cent. In cases of non-compliance, sanctions are foreseen, but enforcement in practice is considered difficult because of ambiguities in the legislation.
- Tradable green certificates (which are tradable commodities proving that certain electricity is generated using renewable energy sources) are used to fulfil the RES-E obligation. The price of such a certificate stood at €109/MWh in 2005.
- A FIT for PV exists. This is a fixed tariff, guaranteed for 20 years and adjusted annually for inflation.

FUTURE TARGETS

According to the EU Directive, Italy aims for a RES-E share of 25 per cent of gross electricity consumption by 2010. Nationally, producers and importers of electricity are obliged to deliver a certain percentage of renewable electricity to the market every year. No progress has been made towards reaching the RES-E target. While Italy's RES-E share amounted to 16 per cent in 1997, it decreased slightly to 15.43 per cent in 2004.

Latvia

MARKET STRUCTURE

In Latvia, almost half the electricity consumption is provided by RES (47.1 per cent in 2004), with hydropower being the key resource. The growth observed between 1996 and 2002 can be ascribed to the so-called 'double tariff', which was phased out in 2003. This scheme was replaced by quotas that are adjusted annually. A body of RES-E legislation is currently under development in Latvia. Wind and biomass would benefit from clear support, since the potential in these areas is considerable.

KEY SUPPORT SCHEMES

The two main RES-E policies that have been followed in Latvia are:
1. fixed FITs, which were phased out in 2003; and
2. a quota system, which has been in force since 2002, with authorised capacity levels of installations determined by the Cabinet of Ministers on an annual basis.

The main body of RES-E policy in Latvia is currently under development. Based on the Electricity Market Law of 2005, the Cabinet of Ministers must now develop and adopt regulations in 2006 to deal with the following areas:
- pricing for renewable electricity;
- eligibility criteria to determine which renewable energy sources qualify for mandatory procurement of electricity; and
- the procedure for receiving guarantees of origin for renewable electricity generated.

FUTURE TARGETS

According to the EU Directive, the RES-E share that Latvia is required to reach is 49.3 per cent of gross electricity consumption by 2010. Between 1997 and 2004, the Latvian RES-E share of gross electricity consumption increased from 42.4 per cent to 47.1 per cent.

Lithuania

MARKET STRUCTURE

Lithuania depends, to a large extent, on the Ignalina nuclear power plant, which has been generating 75–88 per cent of the total electricity since 1993. In 2004, Unit 1 was closed, and the shutdown of Unit 2 is planned before 2010. In order to provide alternative sources of energy, in particular electricity, Lithuania has set a national target of 12 per cent RES by 2010

Table I.15: Key support schemes in Lithuania

Technology	Duration fixed years	2002–present fixed €/MWh
Hydro	10	57.9
Wind	10	63.7
Biomass	10	57.9

(8 per cent in 2003). The implementation of a green certificate scheme was, however, postponed for 11 years. The biggest renewables potential in Lithuania can be found in the field of biomass.

KEY SUPPORT SCHEMES

The core mechanisms used in Lithuania to support RES-E are the following:
- FITs: in 2002, the National Control Commission for Prices and Energy approved the average purchase prices of green electricity. The tariffs are guaranteed for a fixed period of 10 years.

- After 2010, a green certificate scheme should be in place. The implementation of this mechanism has been postponed until 2021.

FUTURE TARGETS

At national level, it has been decided that the RES share of Lithuania's total energy consumption should reach 12 per cent by 2010. The RES-E EU Directive has fixed a RES-E target of 7 per cent of gross electricity consumption by 2010. In 2003, RES accounted for about 8 per cent of the country's energy supply. Between 1997 and 2004, an increase of 0.41 per cent in the RES-E share of consumption was noted (3.71 per cent in 2004 compared to 3.3 per cent in 1997).

Luxembourg

MARKET STRUCTURE

Despite a wide variety of support measures for RES and a stable investment climate, Luxembourg has not

Table I.16: Key support schemes in Luxembourg

Technology	Tariff duration fixed years	2001 to September 2005		From October 2005	
		Capacity	Tariff fixed €/MWh	Capacity	Tariff fixed €/MWh
Wind	10	Up to 3000 kW	25	<501 kW	77.6
Hydro					
Biomass				<501 kW	102.6 (77.6 + 25 for biomass)
Biogas (including landfill and sewage)					
Wind	10	x	x	500–10,001 kW	max 77.6 Lower for higher capacities
Hydro					
Biomass				500–10,001 kW	max 102.6 Lower for higher capacities
Biogas (including landfill and sewage)					
PV – municipalities	20	Up to 50 kW	250	No capacity restriction	280
PV – non-municipalities			450–550		560

made significant progress towards its targets in recent years. In some cases, this has been caused by limitations on eligibility and budget. While the electricity production from small-scale hydropower has stabilised in recent years, the contribution from onshore wind, PV and biogas has started to increase.

KEY SUPPORT SCHEMES

The 1993 Framework Law (amended in 2005) determines the fundamentals of Luxembourgian RES-E policy:

- Preferential tariffs are given to the different types of RES-E for fixed periods of 10 or 20 years. The feed-in system might be subject to change, due to further liberalisation of the sector.
- Subsidies are available to private companies that invest in RES-E technologies, including solar, wind, biomass and geothermal technologies.

FUTURE TARGETS

The RES-E target to be achieved in 2010, as set by the EU Directive, is 5.7 per cent of gross electricity consumption. A slight increase in Luxembourg's RES-E share can be noted. In 2004, the RES-E share amounted to 2.8 per cent of gross electricity consumption, compared to 2.1 per cent in 1997.

Malta

MARKET STRUCTURE

The market for RES in Malta is still in its infancy, and at present, penetration is minimal. RES has not been adopted commercially, and only solar energy and biofuels are used. Nevertheless, the potential of solar and wind is substantial. In order to promote the uptake of RES, the Maltese Government is currently creating a framework for support measures. In the meantime, it has set national indicative targets for RES-E lower than those agreed in its Accession Treaty (between 0.31 per cent and 1.31 per cent, instead of 5 per cent).

KEY SUPPORT SCHEMES

In Malta, RES-E is supported by a FIT system and reduced value-added tax systems.

FUTURE TARGETS

The RES-E target set by the EU Directive for Malta is 5 per cent of gross electricity consumption in 2010. However, at national level, it has been decided to aim for 0.31 per cent, excluding large wind farms and waste combustion plants; or for 1.31 per cent in the event that the plans for a land-based wind farm are implemented. The total RES-E production in 2004 was 0.01 GWh and, therefore, the RES-E share of gross electricity consumption was effectively zero.

The Netherlands

MARKET STRUCTURE

After a period during which support was high but markets quite open, a system was introduced (in 2003) that established sufficient incentives for domestic RES-E production. Although successful in encouraging investments, this system, based on premium tariffs, was abandoned in August 2006 due to budgetary constraints. Political uncertainty concerning renewable energy support in The Netherlands is compounded by an increase in the overall energy demand. Progress towards RES-E targets is slow, even though growth in absolute figures is still significant.

Table I.17: Key support schemes (FIT) in Malta

Technology	Support system	Comments
PV < 3.7 kW	€46.6 / MWh	Feed-in
Solar	5–15%	VAT reduction

Note: A framework for measures to further support RES-E is currently being examined.

MAIN SUPPORTING POLICIES

RES-E policy in The Netherlands is based on the 2003 MEP policy programme (Environmental Quality of Power Generation) and is composed of the following strands:

- Source-specific premium tariffs, paid for ten years on top of the market price. These tariffs were introduced in 2003 and are adjusted annually. Tradable certificates are used to claim the FITs. The value of these certificates equals the level of the FIT. Due to budgetary reasons, most of the FITs were set at zero in August 2006.
- An energy tax exemption for RES-E was in place until 1 January 2005.
- A guarantee of origin system was introduced simply by renaming the former certificate system.

The premium tariffs are given in Table I.18.

FUTURE TARGETS

In its climate policy, The Netherlands set a global target of 5 per cent renewable energy by 2010 and 10 per cent by 2020. According to the EU Directive, the RES-E share of The Netherlands should reach 9 per cent of gross electricity consumption in 2010. Between 1997 and 2004, progress was made towards the RES-E target. In 1997, the RES-E share was 3.5 per cent and by 2004 it had risen to 4.60 per cent.

Poland

MARKET STRUCTURE

Progress towards the RES-E target in Poland is slow and the penalties designed to ensure an increased supply of green electricity have not been adequately used. Despite the high potential of hydropower plants, they have not been fully used to date; biomass resources (in the form of forestry residues, agricultural residues and energy crops) are plentiful in Poland, and landfill gas is also promising.

MAIN SUPPORTING POLICIES

The Polish RES-E policy includes the following mechanisms:

- Tradable Certificates of Origin were introduced by the April 2005 amendment of the Law on Energy (1997).
- The Obligation for Power Purchase from Renewable Sources (2000, amended in 2003) involves a requirement on energy suppliers to provide a certain minimum share of RES-E (3.1 per cent in 2005, 3.6 per cent in 2006, 4.8 per cent in 2007 and 7.5 per cent in 2010). Failure to comply with this

Table I.18: Key support systems – premium tariffs in the Netherlands				
Technology	**Duration years**	**1 July–31 December 2004 premium, €/MWh**	**1 January 2005–30 June 2006 premium, €/MWh**	**Since August 2006 premium, €/MWh**
Mixed biomass and waste		29	29	0
Wind onshore		63	77	0
Wind offshore	10	82	97	0
Pure biomass large scale >50 MW		55	70	0
Pure biomass small scale <50 MW		82	97	97*
PV, tidal and wave, hydro		82	97	0

Note: *Only for installations using biogas from manure digestion and having a capacity below 2 MW. Total premium is limited to €270 million for the complete duration period.

legislation leads, in theory, to the enforcement of a penalty; in 2005, this was not adequately enforced.

- An excise tax exemption on RES-E was introduced in 2002.

FUTURE TARGETS

Poland has a RES-E and primary energy target of 7.5 per cent by 2010. Steady but modest progress is being made with regard to the RES-E target, since the RES-E share of gross energy consumption was about 2.6 per cent in 2005, compared to 2.2 per cent in 2004 and 1.6 per cent in 1997. The potential of hydropower, biomass and landfill gas is high in Poland.

Portugal

MARKET STRUCTURE

The measures adopted so far in Portugal in relation to renewable energy constitute a comprehensive policy mix, complete with monitoring system. Between 1997 and 2004, Portugal has moved further away from its RES-E target. Due to the fact that this target is not entirely realistic, since it was based on the exceptional hydropower performance of 1997, Portugal is not expected to reach its target, even if measures are successful.

KEY SUPPORT SCHEMES

In Portugal, the following measures have been taken to stimulate the uptake of RES-E:

- Fixed FITs per kWh exist for PV, wave energy, small hydro, wind power, forest biomass, urban waste and biogas.
- Tendering procedures were used in 2005 and 2006 in connection with wind and biomass installations.
- Investment subsidies up to 40 per cent can be obtained.
- Tax reductions are available.

Table I.19: Key support schemes for RES-E in Portugal

Technology		Duration fixed years	2004 fixed €/MWh	2006[1] fixed €/MWh
Photovoltaics	<5 kW		450	450
Photovoltaics	>5 kW		245	310
Wave			247	n/a
Small hydro	<10 MW	15	78	75
Wind			90[2]	74
Forest biomass			78	110
Urban waste			70	75
Biogas			n/a	102

Notes: [1]Stated 2006 tariffs are average tariffs. Exact tariff depends on a monthly correction of the inflation, the time of feed-in (peak/off peak) and the technology used.

[2]Tariff only up to 2000 full load hours; 2006 tariff for all full load hours.

The Decreto Lei 33-A/2005 has introduced new FITs as listed in Table I.19.

FUTURE TARGETS

The RES-E target to be achieved by Portugal in 2010 is 39 per cent of gross electricity consumption. Portugal, which nearly met its RES-E target for 2010 in 1997, has now moved further away from this target. A sharp decline between 38.5 per cent in 1997 to only 23.84 per cent in 2004 was observed.

Romania

MARKET STRUCTURE

In terms of RES-E objective of gross electricity consumption, Romania is on target. In 2004, the majority of all RES-E was generated through large-scale hydropower. To a large extent, the high potential of small-scale hydropower has remained untouched. Between 1997 and 2004, both the level of production and the growth rate of most RES has been stable. Provisions

Table I.20: Key support schemes for RES-E in Romania

Period	Penalties for non-compliance
2005–2007	€63/GC
2008–2012	€84/GC (or CV – Certificat Verde)

The quota system is imposed on power suppliers, trading electricity between the producers and consumers.

for public support are in place, but renewable energy projects have so far not been financed.

KEY SUPPORT SCHEMES

Romania introduced the following measures to promote RES-E:

- A quota system, with tradable green certificates (TGCs) for new RES-E, has been in place since 2004, with a mandatory quota increase from 0.7 per cent in 2005 to 8.3 per cent in 2010–2012. TGCs are issued to electricity production from wind, solar, biomass or hydropower generated in plants with less than 10 MW capacity.
- Mandatory dispatching and priority trading of electricity produced from RES has existed since 2004.

FUTURE TARGETS

In Romania, the RES target to be achieved is 11 per cent of gross energy in 2010. The RES-E target was set at 33 per cent of gross electricity consumption in 2010; however, the share decreased from 31.3 per cent in 1997 to 29.87 per cent in 2004.

Slovakia

MARKET STRUCTURE

In the Slovak Republic, large-scale hydro energy is the only renewable energy source with a notable share in total electricity consumption. Between 1997 and 2004, this market share stabilised. The share taken up by small-scale hydro energy has decreased by an average of 15 per cent per year over the same period. An extended development programme, with 250 selected sites for building small hydro plants has been adopted. The government has decided to use only biomass in remote, mountainous, rural areas, where natural gas is unavailable. Between 1997 and 2004, the Slovak Republic moved further away from its RES target.

KEY SUPPORT SCHEME

RES-E policy in the Slovak Republic includes the following measures:

- A measure that gives priority regarding transmission, distribution and supply was included in the 2004 Act on Energy.
- Guarantees of origin are being issued.
- Tax exemption is granted for RES-E. This regulation is valid for the calendar year in which the facility commenced operation and then for five consecutive years.
- A system of fixed FITs has been in place since 2005.
- Subsidies up to €100.000 are available for the (re)construction of RES-E facilities.

Table I.21: Key support schemes (FITs) in the Slovak Republic

Technology	2006 fixed SKK/MWh	2006 fixed €/MWh	2007* fixed SKK/MWh	2007* fixed €/MWh
Wind	2800	75.1	1950–2565	55–72
Hydro <5 MW	2300	61.7	1950–2750	55–78
Solar	8000	214.6	8200	231
Geothermal	3500	93.9	3590	101
Biogas	x	x	2560–4200	72–118
Biomass combustion	2700	72.4	2050–3075	58–87

Note: *Exact level of FIT depends on the exchange rate (here €1 = SKK35.458). The prices have been set so that a rate of return on the investment is 12 years when drawing a commercial loan. These fixed tariffs will be inflation adjusted the following year.

Decree No 2/2005 of the Regulatory Office for Network Industries (2005) set out the fixed FITs available for RES-E.

FUTURE TARGETS

In terms of its primary energy consumption, the Slovak Republic has fixed the target of 6 per cent renewable energy consumption by 2010. The target set by the EU Directive for RES-E is 31 per cent in 2010. Currently, renewable energy represents about 3.5 per cent of the total primary energy consumption in the Slovak Republic. Between 1997 and 2004, the share of RES-E decreased from 17.9 per cent to 14.53 per cent of gross energy consumption. In the Slovak Republic, the highest additional mid-term potential of all RES lies with biomass.

Slovenia

MARKET STRUCTURE

Slovenia is currently far from meeting its RES targets. Solid biomass has recently started to penetrate the market. Hydropower, at this time the principal source of RES-E, relies on a large amount of very old, small hydro plants; and the Slovenian Government has made the refurbishment of these plants part of the renewable energy strategy. An increase in capacity of the larger-scale units is also foreseen. In Slovenia, a varied set of policy measures has been accompanied by administrative taxes and complicated procedures.

KEY SUPPORT SCHEMES

In Slovenia, the RES-E policy includes the following measures:

- RES-E producers can choose to receive either fixed FITs or premium FITs from the network operators. A Purchase Agreement is concluded, valid for 10 years. According to the Law on Energy, the uniform annual prices and premiums are set at least once a year. Between 2004 and 2006, these prices stayed the same.
- Subsidies or loans with interest-rate subsidies are available. Most of the subsidies cover up to 40 per cent of the investment cost. Investments in rural

Table I.22: Key support schemes in Slovenia

| Technology | Capacity | Duration | | 2004–present | | | |
		fixed years	premium years	fixed SIT/MWh	premium SIT/MWh	fixed €/MWh	premium €/MWh
Hydro	<1 MW			14.75	6.75	62	28
	1–10 MW			14.23	6.23	59	26
Biomass	<1 MW			16.69	8.69	70	36
	>1 MW			16.17	8.17	68	34
Biogas (landfill and sewage gas)	<1 MW	After 5 years tariff reduced by 5%. After 10 years tariff reduced by 10%		12.67	–	53	–
	>1 MW			11.71	–	49	–
Biogas (animal waste)	–			28.92	–	121	–
Wind	<1 MW			14.55	6.55	61	27
	>1 MW			14.05	6.05	59	25
Geothermal	–			14.05	6.05	59	25
Solar	<36 kW			89.67	81.67	374	341
	>36 kW			15.46	7.46	65	31

areas with no possibility of connection to the electricity network are eligible to apply for an additional 20 per cent subsidy.

FUTURE TARGETS

At national level, a target to increase the share of RES in total primary energy consumption from 8.8 per cent in 2001 to 12 per cent by 2010 has been set. The RES-E target to be achieved in 2010, as a result of the EU Directive, is 33.6 per cent in Slovenia. At present, the contribution of RES to the national energy balance is about 9 per cent. In 2004, the Slovenian RES-E share of gross electricity consumption was 29.9 per cent. The potential of solid biomass is high, with over 54 per cent of land covered by forests.

Spain

MARKET STRUCTURE

Spain is currently far from its RES-E target. In 1997, a strong support programme in favour of RES was introduced. In 2004, hydropower still provided 50 per cent of all green electricity, while onshore wind and biomass had started penetrating the market. PV energy is also promising, with an average growth rate of 54 per cent per year. Proposed changes to the FITs and the adoption of a new Technical Buildings Code (2006) show increased support for biomass, biogas, solar electricity and solar thermal.

KEY SUPPORT SCHEMES

RES-E in Spain benefits from the following support mechanisms:

- A FIT or a premium price is paid on top of the market price. The possibility of a cap and floor mechanism for the premium is being considered. In the draft law published on 29 November 2006, reduced support for new wind and hydro plants and increased support for biomass, biogas and solar thermal electricity were proposed.
- Low-interest loans that cover up to 80 per cent of the reference costs are available.

Fixed and premium FITs for 2004, 2005 and 2006 are shown in Table I.23.

Table I.23: Fixed and premium FITs in Spain							
		2004		2005		2006	
Technology	**Duration 2004–2006**	**fixed €/MWh**	**premium €/MWh**	**fixed €/MWh**	**premium €/MWh**	**fixed €/MWh**	**premium €/MWh**
PV <100 kWp		414.4	x	421.5	x	440.4	x
PV >100 kWp		216.2	187.4	219.9	190.6	229.8	199.1
Solar thermal electricity		216.2	187.4	219.9	190.6	229.8	199.1
Wind <5 MW		64.9	36.0	66.0	36.7	68.9	38.3
Wind >5 MW	No limit, but fixed tariffs are reduced after either 15, 20 or 25 years depending on technology	64.9	36.0	66.0	36.7	68.9	38.3
Geothermal <50 MW		64.9	36.0	66.0	36.7	68.9	38.3
Mini hydro <10 MW		64.9	36.0	66.0	36.7	68.9	38.3
Hydro 10–25 MW		64.9	36.0	66.0	36.7	68.9	38.3
Hydro 25–50 MW		57.7	28.8	58.6	29.3	61.3	30.6
Biomass (biocrops, biogas)		64.9	36.0	66.0	36.7	68.9	38.3
Agriculture + forest residues		57.7	28.8	58.6	29.3	61.3	30.6
Municipal solid waste		50.5	21.6	51.3	22.0	53.6	23.0

FUTURE TARGETS

The Spanish *Plano de Energías Renovables 2005–2010* sets the goal of meeting 12 per cent of total energy consumption from RES in 2010. The target to be achieved in 2010, under the RES-E Directive, is 29.4 per cent of gross electricity consumption. The revised *Plano* of 2005 sets capacity targets for 2010, which include wind (20,155 MW), PV (400 MW), solar thermal (4.9 million m²), solar thermal electric (500 MW) and biomass (1695 MW). In Spain, the RES-E share of gross electricity consumption was 19.6 per cent in 2004, compared to 19.9 per cent in 1997.

Sweden

MARKET STRUCTURE

Sweden is moving away from its RES-E target. In absolute figures, RES-E production decreased between 1997 and 2004, mainly due to a lower level of large-scale hydro production. However, other RES, such as biowaste, solid biomass, offshore wind and PV have shown significant growth. In Sweden, a comprehensive policy mix exists, with tradable green certificates as the key mechanism. This system creates both an incentive to invest in the most cost-effective solutions and uncertainty for investment decisions due to variable prices.

KEY SUPPORT SCHEMES

Swedish RES-E policy is composed of the following mechanisms:

- Tradable Green Certificates were introduced in 2003. The Renewable Energy with Green Certificates Bill that came into force on 1 January 2007 shifts the quota obligation from electricity users to electricity suppliers.
- The environmental premium tariff for wind power is a transitory measure and will be progressively phased out by 2009 for onshore wind.

FUTURE TARGETS

The RES-E target from the EU Directive for Sweden is 60 per cent of gross electricity consumption by 2010. The Swedish Parliament decided to aim for an increase in RES of 10 TWh between 2002 and 2010, which corresponds to a RES-E share of around 51 per cent in 2010. This deviates from the target originally set by the Directive. In June 2006, the Swedish target was amended to increase the production of RES-E by 17 TWh from 2002 to 2016. The Swedish share of RES-E for gross electricity consumption has decreased from 49.1 per cent in 1997 to 45.56 per cent in 2004 and approximately 38 per cent at the present time.

The UK

MARKET STRUCTURE

In the UK, renewable energies are an important part of the climate change strategy and are strongly supported by a green certificate system (with an obligation on suppliers to purchase a certain percentage of electricity from renewable energy sources) and several grants programmes. Progress towards meeting the target has been significant (electricity generation from renewable energies increased by around 70 per cent between 2000 and 2005), although there is still some way to go to meet the 2010 target. Growth has been mainly driven by the development of significant wind energy capacity, including offshore wind farms.

KEY SUPPORT SCHEMES

The UK's policy regarding renewable energy sources consists of four key strands:

1. Obligatory targets with tradable green certificate (ROC) system (a renewables obligation on all electricity suppliers in Great Britain). The non-compliance 'buy-out' price for 2006/2007 was set

Table I.24: Key support schemes in the UK

Year	Targets % supply of consumption target	Non-compliance buy-out price		Amount recycled (England and Wales)	Total 'worth' of ROC (England and Wales) (buy-out + recycle)	
		£/MWh	€/MWh*	£/MWh	£/MWh	€/MWh
2002–2003	3	X	x	X	X	X
2003–2004	4.3	30.51	44.24	22.92	53.43	77.47
2004–2005	4.9	31.39	45.52	13.66	45.05	65.32
2005–2006	5.5	32.33	46.88			
2006–2007	6.7	33.24	48.20			
2007–2008	7.9					
2008–2009	9.1					
2009–2010	9.7					
2010–2011	10.4				Not yet known	
2011–2012	11.4	Increases in line with retail price index				
2012–2013	12.4					
2013–2014	13.4					
2014–2015	14.4					
2015–2016	15.4					
Duration	One ROC is issued to the operator of an accredited generating station for every MWh of eligible renewable electricity generated with no time limitations					
Guaranteed duration of obligation	The Renewables Obligation has been guaranteed to run until at least 2027. Supply targets increase to 15.4% in 2015, and are guaranteed to remain at least at this level until 2027					

at £33.24/MWh (approx. €48.20/MWh), which will be annually adjusted in line with the retail price index.

2. The climate change levy: RES-E is exempt from the climate change levy on electricity of £4.3/MWh (approx. €6.3/MWh)

3. Grants schemes: funds are reserved from the New Opportunities Fund for new capital grants for investments in energy crops/biomass power generation (at least £33 million or €53 million over three years), for small-scale biomass/CHP heating (£3 million or €5 million), and planting grants for energy crops (£29 million or €46 million for a period of seven years). A £50 million (€72.5 million) fund, the Marine Renewables Deployment Fund, is available for the development of wave and tidal power.

4. Development of a regional strategic approach for planning/targets for renewable energies.

Annual compliance periods run from 1 April one year to 31 March the following year. ROC auctions are held quarterly. In the April 2006 auction over 261,000 ROCs were purchased at an average price of £40.65 (the lowest price for any lot was £40.60).

The following limits have been placed on biomass co-firing within the ROC:

- In the 2009–2010 compliance period, a minimum of 25 per cent of co-fired biomass must be energy crops;
- In 2010–2011, a minimum of 50 per cent of co-fired biomass must be energy crops;
- In 2011–2016, a minimum of 75 per cent of co-fired biomass must be energy crops; and
- After 2016, co-firing will not be eligible for ROCs.

FUTURE TARGETS

The RES-E target to be achieved by the UK in 2010 is 10 per cent of gross electricity consumption. An indicative target of 20 per cent for RES-E for 2020 has been set. After a relatively stable share in the early 2000s, growth over the past couple of years has been significant. In 2005, the share of renewable sources in electricity generation reached 4.1 per cent.

APPENDIX J: STUDIES ON EMPLOYMENT CREATION IN THE WIND ENERGY SECTOR

Table J.1: Summary of the main recent studies dealing with employment creation in the wind energy sector

Source	Title	Geographical coverage	Methodology	Main results
ADEME, 2008	'ADEME&Vous, Stratégie & Études. Maîtrise de l'énergie et développement des énergies renouvelables'	France	Net production/employment ratios (imports have been disregarded).	7000 jobs in the manufacturing of wind turbines and major sub-components; 500 in companies operating wind energy farms.
AEE, 2007	'Eólica 07. Todos los datos, análisis y estadísticas del sector eólico'	Spain	Questionnaires to Spanish wind energy companies, complemented by information from official tax-related registries.	There are more than 300 wind energy companies in Spain, creating 15,450 direct jobs and another 19,560 indirect jobs. This figure may go up to 58,800 if government objectives (20,000 MW in 2012) are achieved. 29.97% of the jobs are in the O&M sub-sector, 22.72% in the manufacturing of the machines, 19.42% in technical and engineering services, 9.12% in manufacturing, 3.24% in R&D, and 4.53% in 'others'.
Algoso and Rusch, 2004	'Job growth from renewable energy development in the Mid-Atlantic'	Mid-Atlantic States of the US: Maryland, Delaware, New Jersey and Pennsylvania	The number of jobs was calculated with the I-O Renewable Energy Policy Project. The technical coefficients were estimated by means of a survey sent to 19 wind energy companies in 2001. Indirect employment figures derive from the Texas Comptroller's office.	A capacity installed of 10,200 MW in 2015 would entail 11,100 year-long jobs in wind turbine manufacturing and installation; 740 permanent jobs in O&M and supporting areas; and around 12,700 indirect jobs. The jobs/MW ratio is 2.48. Choosing wind energy over a comparable amount of natural gas installations would create more than twice as many jobs.
DWIA, 2008	Sector statistics	Denmark	Questionnaires to Danish wind energy companies.	In 2006, 23,500 people worked for the wind energy sector (direct). 13,000 of these constitute direct employment in wind turbine and blade manufacturing companies.
European Commission, 2006a	European Commission's 2007 Impact Assessment on the Renewable Energy Roadmap	EU (27 Member States)	I-O tables, based on Green-X, PRIMES and ASTRA models.	Meeting the 20% RE target in 2020 will entail a net increase of 650,000 jobs in the EU, half of which may come from the biomass sector. The increase of RE will favour changes in the composition of the labour market, rather than its size.
European Parliament, 2007	'Employment potential of renewable forms of energy and increased efficiency of energy use'	EU	Meta-analysis of past employment studies.	A more rapid switch to renewables appears to have an unambiguous benefit in terms of overall employment. The growth of a particular segment of the clean energy business (renewables, energy efficiency or sustainable transport) is often partially dependent on growth in other areas, since the markets for products and technologies are linked. Workers that lose their jobs in the fossil fuel industry should have the opportunity to retrain for employment in the clean energy industry.

Source	Title	Region	Method	Results
EREC, 2007	'New renewable energy target for 2020 – A renewable energy roadmap for the EU'	EU (15 Member States)	I-O tables, based on Saphire model.	The wind energy sector will account for around 184,000 jobs in 2010 (direct, indirect and induced effects) and 318,000 in 2020 (if the 20% RE target is reached).
EWEA, 2003	Survey for *Wind energy – The facts*	EU (15 Member States)	Survey of wind energy manufacturers, supplemented by the use of technical coefficients.	Direct employment in wind turbine manufacturing in Europe for 2002 accounted for 30,946 jobs; turbine installation for another 14,649; O&M for 2768.
EWEA and Greenpeace, 2005	'Wind Force 12: A blueprint to achieve 12% of the world's electricity from wind power by 2020'	World	Meta-analysis of past employment studies.	2.3 million jobs will be linked to the wind energy sector worldwide in 2020 if the 12% target is reached. 444,000 jobs in North America; 222,000 in Europe; 251,800 in Latin America; 44,400 in Africa, 44,400 in the Middle East; 325,600 in Eastern European and transition economies; 444,000 in China; 148,000 in East Asia; 266,400 in South Asia and 148,000 in OECD Pacific.
Federal Ministry for the Environment, Nature Conservation and Nuclear Safety, 2006	'Renewable energy: Employment effects. Impact of the expansion of renewable energy on the German labour market'	Germany	Comprehensive study, using a questionnaire and extensive theoretical models (I-O table). The study presents net results on the overall economy – direct, indirect and induced impacts.	The wind energy sector is responsible for around 64,000 jobs in Germany (2004 data). Half of them are direct jobs. By 2030, around 300,000 new jobs will be created in the renewable energy sources sector. The net impact can be situated between 80,000 and 130,000, depending on the future energy prices.
Federal Ministry for the Environment, Nature Conservation and Nuclear Safety, 2008	'Kurz – Und langfristige Auswirkungen des Ausbaus der erneuerbaren Energien auf den deutschen Arbeitsmarkt', interim report	Germany	Update of the 2006 report (questionnaire + I-O table).	84,300 employees in the wind energy sector by the end of 2007 (direct + indirect).
Kammen et al., 2004	'Putting renewables to work: How many jobs can the clean industry generate?'	EU and US	Meta-analysis of past employment studies.	The renewable energy sector generates more jobs per MW of power installed, per unit of energy produced and per dollar of investment than the fossil fuel-based energy sector. The distribution of employment benefits across regions can vary considerably. In the US, a 20% RE share by 2020 can create between 176,440 and 240,850 new jobs, as compared with a figure of 86,369 in the business-as-usual scenario. The jobs/MW ratio for wind power ranges between 0.71 and 2.79.

continued

Table J.1: (continued)

Source	Title	Geographical coverage	Methodology	Main results
Lehr et al., 2008	'Renewable energy and employment in Germany'	Germany	Comprehensive study, using a questionnaire and extensive I-O tables, INFORGE and PANTA RHEI models.	Gross employment figures in 2004: 63,944 workers. The wind sector lacks skilled personnel, but the situation is expected to improve in 2010. Global market share of wind energy products coming from Germany was 40% in 2004, and is expected to decrease to between 15 and 20% in 2020.
Pedden, 2005	'Analysis: Economic impacts of wind applications in rural communities'	US	Meta-analysis of 13 studies.	Wind installations create significant direct impact on the economies of the local communities, especially those with few supporting industries. The number of local and construction and operation jobs created by a wind energy installation depends upon the skills available in the local community. The jobs/MW ratio is highly variable: from 0.36 to 21.37.
Pfaffenberger et al., 2006	'Renewable energies – Environmental benefits, economic growth and job creation'	EU, with emphasis on Germany	Meta-analysis of previous studies.	All studies predict a growth in gross employment. The net employment impacts are substantially less, and can even be negative. None of the studies has taken into account the recent increase of energy prices, which will tend to increase the positive effect of RES on employment.
UNEP, ILO and ITUC, 2007	'Green jobs: Towards sustainable work in a low-carbon world', preliminary report	World	Meta-analysis of previous studies.	The wind energy sector created 300,000 jobs in 2006 worldwide. The jobs/MW ratio in manufacturing, construction and installation can be estimated between 0.43 and 2.51; 0.27 for O&M and 0.70 to 2.78 in total.
Whiteley et al., 2004	MITRE project 'Meeting the targets and putting RE to work' overview report	EU (15 Member States)	I-O tables, based on Saphire, model.	The wind energy sector will create between 162,000 (with current policies) and 368,000 (with advanced renewable strategies) new jobs in the EU (net effect; direct, indirect and induced jobs) by 2020. After 2010, the employment levels will be maintained only if the sector is capable of keeping its leading role and finding new markets outside the EU.

GLOSSARY

A

Active power is a real component of the apparent power, usually expressed in kilowatts (kW) or megawatts (MW), in contrast to **reactive power**.

Adequacy: a measure of the ability of the power system to supply the aggregate electric power and energy requirements of the customers within component ratings and voltage limits, taking into account planned and unplanned outages of system components. Adequacy measures the capability of the power system to supply the load in all the steady states in which the power system may exist, considering standard conditions.

Ancillary services are interconnected operations services identified as necessary to effect a transfer of electricity between purchasing and selling entities and which a provider of transmission services must include in an open access transmission tariff.

Annualised net metering is the same as **net metering**, but in this case the regulator averages a user's net electricity consumption or production over the span of one full year, rather than a shorter period.

ASACS: UK Air Surveillance and Control Systems.

Auxiliary costs are other than those of the turbine itself, in other words foundation, grid connection, electrical installation, road construction, financial charges and so on.

Availability describes the amount of the time that the wind turbine is actually functional, not out of order or being serviced.

B

Balance of Plant (BOP): the infrastructure of a wind farm project, in other words all elements of the wind farm, excluding the turbines. Includes civil works, **SCADA** and internal electrical system. It may also include elements of the grid connection.

Black start capability: some power stations have the ability to start up independently of a power grid. This is an essential prerequisite for system security, as these plants can be called on during a blackout to re-power the grid.

Boundary layer profile: see **wind shear profile**.

C

Capacity is the rated continuous load-carrying ability of generation, transmission or other electrical equipment, expressed in megawatts (MW) for **active power** or megavolt-amperes (MVA) for apparent power.

Capacity credit: a wind turbine can only produce when the wind blows and therefore is not directly comparable to a conventional power plant. The capacity credit is the percentage of conventional capacity that a given turbine can replace. A typical value of the capacity credit is 25 per cent (see **capacity factor**).

Capacity factor (load factor) is the ratio between the average generated power in a given period and the installed (rated) power.

Capital costs are the total investment costs of the turbine, including auxiliary costs.

Carbon dioxide (CO_2) is a naturally occurring gas, and also a by-product of burning fossil fuels and biomass, as well as land-use changes and other industrial processes. It is the principal anthropogenic greenhouse gas that affects the Earth's radiative balance.

Citizen engagement can be defined as being responsive to lay views and actively seeking the involvement of the lay public in policymaking and decision-making. Considered a central motif in public policy discourse within many democratic countries, it is acknowledged as an important component of good governance.

Climate change is a change of climate attributed directly or indirectly to human activity which alters the composition of the global atmosphere and which is in addition to natural climate variability observed over comparable time periods.

Cogging: variation in speed of a generator due to variations in magnetic flux as rotor poles pass stator poles. Cogging in permanent magnet generators can hinder the start-up of small wind turbines at low wind speeds.

Community acceptance refers to the acceptance of specific projects at the local level, including affected populations, key local stakeholders and local authorities.

Contingency is the unexpected failure or outage of a system component, such as a generator, transmission line, circuit breaker, switch or other electrical element. A contingency may also include multiple components which are related by situations leading to simultaneous component outages.

Control area is a coherent part of the UCTE **Interconnected System** (usually concurrent with the territory of a company, a country or a geographical area, and physically demarcated by the position of points for measurement of the interchanged power and energy to the remaining interconnected network), operated by a single **transmission system operator (TSO)**, with physical loads and controllable generation units connected within the control area. A control area may be a coherent part of a **control block** that has its own subordinate control in the hierarchy of **secondary control**.

Control block comprises one or more **control areas**, working together in the **secondary control** function, with respect to the other control blocks of the **synchronous area** to which it belongs.

Costs of generated wind power: see **levelised costs**.

Curtailment means a reduction in the scheduled capacity or energy delivery.

D

D is the wind turbine rotor diameter (measured in metres).

Darrieus rotor is a sleek vertical axis wind turbine developed by French invertor G. J. M. Darrieus in 1929 based on aerodynamic profiles.

dB(A): The human ear is more sensitive to sound in the frequency range 1 kHz to 4 kHz than to sound at very low or high frequencies. Therefore, sound meters are normally fitted with filters adapting the measured sound response to the human ear.

Decibel (dB) is a unit of measurement that is used to indicate the relative amplitude of a sound or the ratio of the signal level such as sound pressure. Sound levels in decibels are calculated on a logarithmic scale.

Diffuser is a downwind device that diffuses the wind stream through a rotor.

Direct drive is a drive-train concept for wind turbines in which there is no gearbox and the wind turbine rotor is connected directly to a low-speed electrical generator.

Direct employment is the total number of people employed in companies belonging to a specific sector.

Discount rate is the interest rate used to calculate the present-day costs of turbine installations.

Distributed generation means single or small clusters of wind turbines spread across the landscape, in contrast to the concentration of wind turbines in large arrays or wind power plants.

Doppler shift principle: when a source generating waves moves relative to an observer, or when an observer moves relative to a source, there is an apparent shift in frequency. If the distance between the observer and the source is increasing, the frequency apparently decreases, while the frequency apparently increases if the distance between the observer and the source is decreasing. This relationship is called the Doppler effect (or Doppler shift) after Austrian physicist Christian Johann Doppler (1803–1853).

Doubly fed induction generator (DFIG) is an electrical machine concept in which variable-speed operation is provided by using a relatively small power electronic converter to control currents in the rotor, such that the rotor does not necessarily rotate at the synchronous speed of the magnetic field set up in the stator.

DTI: Department of Trade and Industry of the UK Government.

E

Efficiency for a turbine describes the amount of active electrical power generated as a percentage of the wind power incident on the rotor area.

Electricity demand is the total electricity consumption in GWh consumed by a nation annually.

Emissions are the discharges of pollutants into the atmosphere from stationary sources such as smokestacks, other vents, surface areas of commercial or industrial facilities, and mobile sources such as motor vehicles, locomotives and aircraft. With respect to climate change, emissions refer to the release of greenhouse gases into the atmosphere over a specified area and period of time.

Energy payback is the time period it takes for a wind turbine to generate as much energy as is required to produce the turbine in the first place, install it, maintain it throughout its lifetime and, finally, scrap it. Typically, this takes 2–3 months at a site with reasonable exposure.

Equivalent sound level (dB$_{Leq}$) quantifies the environmental noise as a single value of sound level for any desired duration. The environmental sounds are usually described in terms of an average level that has the same acoustical energy as the summation of all the time-varying events.

ETSU: Energy Technology Support Unit of the UK Government.

EWEA: European Wind Energy Association.

Experience curve relates the cumulative quantitative development of a product with the development of the specific costs. The more this product is produced, the more efficient the production process and the cheaper it becomes.

External costs are those costs incurred in activities which may cause damage to a wide range of receptors, including human health, natural ecosystems and the built environment, and yet are not reflected in the price paid by consumers.

F

Fault ride-through (FRT) is a requirement of many network operators, such that the wind turbine remains connected during severe disturbances on the electricity system, and returns to normal operation very quickly after the disturbance ends.

FINO 1 is an offshore research platform in the North Sea, off Germany.

Fuel cycle: the impacts of power production are not exclusively generated during the operation of the power plant, but also in the entire chain of activities needed for the electricity production and distribution, such as fuel extraction, processing and transformation, construction and installation of the equipment, and the disposal of waste. These stages, which constitute the chain of electricity production and distribution, are known as the fuel cycle.

Full load hours is the turbine's average annual production divided by its rated power. The higher the number of full load hours, the higher the tubine's production at the chosen site.

Furling is a passive overspeed control mechanism which functions by reducing the projected swept area, by turning the rotor out of the incident wind direction.

G

Gate closure is the point in time when generation and demand schedules are notified to the system operator.

Generation mix is the percentage distribution by technology (nuclear, thermal, large hydro, renewables) of MWs from operational generation plants.

Geographical information system (GIS) is a software system which stores and processes data on a geographical or spatial basis.

Giromill (or cycloturbine) is a vertical axis H-configuration wind turbine with articulating straight blades.

Greenhouse gases are those gaseous constituents of the atmosphere, both natural and anthropogenic, that absorb and emit radiation at specific wavelengths within the spectrum of infrared radiation emitted by the Earth's surface, the atmosphere and clouds. Human-made greenhouse gases in the atmosphere such as halocarbons and other chlorine- and bromine-containing substances are dealt with under the Montreal Protocol. Beside carbon dioxide, nitrous oxide and methane, the Kyoto Protocol deals with the greenhouse gases sulphur hexafluoride, hydrofluorocarbons and perfluorocarbons.

Grid-connected: a wind turbine is grid-connected when its output is channelled directly into a national grid (see also **stand-alone system**).

Grid reinforcement: a weak grid can be reinforced by up-rating its connection to the rest of the grid. The cost of doing this may fall to the wind farm developer.

H

High voltage (HV): typically 100 to 150 kV.

Horizontal axis wind turbine (HAWT): a wind turbine whose rotor axis is substantially parallel to the wind flow.

Hub: the rotating component of the wind turbine to which the rotor blades are fixed.

Hub height is the height of the rotor axis above the ground.

Hybrid power systems (HPS) are combinations of renewable technologies (such as wind turbines or solar photovoltaics) and conventional technologies (such as diesel generators) that are used to provide power to remote areas.

I

IEC: International Electrotechnical Commission.

Impact pathway approach is developed by ExternE to establish the effects and spatial distribution of the burdens from the fuel cycle to find out their final impact on health and the environment. Subsequently, the economic valuation assigns the respective costs of the damages induced by a given activity.

Independent power producer (IPP): a privately owned and operated electricity production company not associated with national utility firms.

Indirect employment refers to those employed in sectors and activities supplying intermediate products/components to, for example, wind turbine manufacturers. Indirect employment includes employment throughout the production chain.

Input-output: the national accounts of a country's or region's economic transactions keep track of all the inputs and outputs between economic sectors.

Installed capacity is the total MW of operational generation plant of a given technology.

Institutional capacity building refers to the process of creating more effective institutions through the increase of shared knowledge resources, relational resources and the capacity for mobilisation. It is usually related to the capacity to facilitate open policy- and decision-making processes (at national and local levels) that provide access to relevant stakeholders and room for various types of knowledge resources.

Institutional framework is a concept used to refer to the policy and regulatory elements affecting energy developments. In the wind energy context, this would include issues such as political commitment, financial incentives, planning systems, presence and roles of landscape protection organisations, and patterns of local ownership.

Interconnected system: two or more individual electric systems that normally operate synchronously and are physically connected via tie-lines (see also: **synchronous area**).

Interconnection is a transmission link (such as a tie-line or transformer), which connects two **control areas**.

Intermedial load refers to those electricity-generation technologies contributing to satisfy the demand in a range between the base load and peak load of the electricity system. A generating unit that normally operates at a constant output (amount of electricity delivered) takes all or part of the base load of a system. In contrast, a peak load unit is only used to reach specific peak periods of a few hours when the demand is high.

Investment costs are the costs of the turbine itself, including transport from the factory to the place where the turbine is erected.

ISO: International Organization for Standardization.

K

K-factor is a weighting of the harmonic load currents according to their effects on transformer heating. A K-factor of 1.0 indicates a linear load (no harmonics). The higher the K-factor, the greater the harmonic heating effects. The K-Factor is used by transformer manufacturers and their customers to adjust the load rating as a function of the harmonic currents caused by the load(s). Generally, only substation transformer manufacturers specify K-factor load de-rating for their products. So, for K-factors higher than 1, the maximum transformer load is de-rated.

Kilohertz (kHz) is a unit of measurement of frequency. It is a unit of alternating current (AC) or electromagnetic (EM) wave frequency equal to one thousand hertz (1000 Hz).

L

Learning rate is a learning curve parameter. It is estimated on available data for wind turbines and tells you the achieved reduction in specific production costs.

Levelised costs: the present-day average cost per kWh produced by the turbine over its entire lifetime, including all costs – (re-)investments, operation and maintenance. Levelised costs are calculated using the discount rate and the turbine lifetime.

Load means an end-use device or customer that receives power from the electricity system. Load should not be confused with demand, which is the measure of power that a load receives or requires.

Load-frequency control (LFC): see **secondary control**.

Local ownership is a way of community involvement based on the fact that local residents can own shares in and obtain personal benefits from local developments. There is a significant relationship between share ownership and positive attitudes towards wind farms, and local ownership and levels of wind implementation.

Low voltage (LV): below 1000 V.

Low-voltage ride-through (LVRT): see **fault ride-through**.

M

Market acceptance refers to the process by which market parties adopt and support (or otherwise) the energy innovation. Market acceptance is proposed in a wider sense, including not only consumers but also investors and, very significantly, intra-firm acceptance.

Medium voltage (MV): typically 10 to 35 kV.

Met mast: a mast or tower which carries meteorological instrumentation (typically wind speed transducers at several heights and wind direction, air temperature and pressure transducers).

Microvolts/cm (μVcm^{-1}) is a unit of measurement of electrical fields.

Millitesla (mT) is a unit of measurement of static magnetic fields.

Minigrid is a distribution network usually operating only at low voltage and providing electricity supply to a community. It is supplied by one or more diesel generators, wind turbines, mini-hydro generators or solar photovoltaics.

Minute reserve (15-minute reserve): see **tertiary control**.

Multiplier/multiplicator: for employment, this measures the direct and indirect employment effect of producing €1 million worth of output from the wind turbine manufacturing sector. Basically, it assumes that it is valid to multiply total wind turbine manufacturing in euros with a factor giving the necessary employment to produce this output. Series of multipliers for historical national account statistics exist.

N

N-1 criterion is a rule that requires elements remaining in operation after the failure of a single network element (such as a transmission line/transformer or generating unit, or in certain instances a busbar) to be capable of accommodating the change of flows in the network caused by that single failure.

(N-1)-safety means that any single element in the power system may fail without causing a succession of other failures, leading to a total system collapse. Together with avoiding constant overloading of grid elements, (N-1)-safety is a main concern for the grid operator.

Net metering is a policy implemented by some states and utilities to ensure that any extra electricity produced by an on-site generator, such a small wind system, can be sent back into the utility system, and where the owner is charged for energy on the basis of the net import.

Net transfer capacity is the maximum value of generation that can be wheeled through the interface between the two systems without leading to network constraints in either system, taking into account technical uncertainties about future network conditions.

Network power frequency characteristic defines the sensitivity, given in megawatts per hertz (MW/Hz), usually associated with a (single**) control area/block** or the entire **synchronous area**, which relates the difference between scheduled and actual **system frequency** to the amount of generation required to correct the power imbalance for that **control area/block** (or, vice versa, the stationary change of the **system frequency** in the case of a disturbance of the generation/load equilibrium in the **control area** without being connected to other **control areas**). It should not be confused with the K-factor (K). The network power frequency characteristic includes all active **primary control** and self-regulation of load and changes, due to modifications in the generation pattern and demand.

NIMBY is the acronym for 'not in my back yard' and refers to an explanation of the local rejection to technological projects. Recent research is questioning the traditional explanation of local rejection to technological projects based on the NIMBY concept, as this may be giving an incorrect or partial explanation of all the variables at stake.

Nitrogen oxide (NOx) is a product of combustion from transportation and stationary sources. It is a major contributor to acid depositions and the formation of ground-level ozone in the troposphere. It is formed by combustion under high pressure and high temperature in an internal combustion engine. It changes into nitrogen dioxide in the ambient air and contributes to photochemical smog.

Numerical weather prediction (NWP) means weather forecasting by computational simulation of the atmosphere.

O

Offshore: wind generation plant installed in a marine environment.

Offshore wind developments are wind projects installed in shallow waters off the coast.

Onshore wind developments are wind farms installed on land.

P

Participatory planning is a planning process open to higher levels of public engagement. Successful wind farm developments are linked to the nature of the planning and development process, and public support tends to increase when the process is open and participatory. Thus, collaborative approaches to decision-making in wind power implementation are suggested to be more effective than top-down imposed decision-making.

Permanent magnet generator (PMG) is a synchronous electrical generator design based on the use of permanent magnets on the rotor.

Photovoltaic generation (PV) is the generation of electricity from sunlight or ambient light, using the photovoltaic effect.

Point of common coupling (PCC) is the point on the public electricity network at which other customers are, or could be, connected. Not necessarily the same location as **point of connection**.

Point of connection (POC) is the point at which the wind farm electrical system is connected to the public electricity system.

Pollutant: a substance that is present in concentrations that may harm organisms (humans, plants and animals) or exceed an environmental quality standard. The term is frequently used synonymously with contaminant.

Power curve depicts the relationship between net electric output of a wind turbine and the wind speed measured at hub height on a 10-minute average basis.

Primary control (frequency control, primary frequency control) maintains the balance between generation and demand in the network, using turbine speed governors. Primary control is an automatic decentralised function of the turbine governor to adjust the generator output of a unit as a consequence of a frequency deviation/offset in the **synchronous area.** Primary control should be distributed as evenly as possible over units in operation in the synchronous area. The global primary control behaviour of an interconnection partner (**control area/block**) may be assessed by the calculation of the equivalent fallout of the area (basically resulting from the fallout of all generators and the self-regulation of the total demand). By the joint action of all interconnected undertakings, primary control ensures the operational reliability for the power system of the synchronous area.

Primary controller: decentralised/locally installed control equipment for a generation set to control the valves of the turbine, based on the speed of the generator (see **primary control**). The insensitivity of the primary controller is defined by the limit frequencies between which the controller does not respond. This concept applies to the complete primary controller-generator unit. A distinction is drawn between unintentional insensitivity, associated with structural inaccuracies in the unit, and a dead band set intentionally on the controller of a generator.

Primary control power is the power output of a generation set due **primary control**.

Primary control range is the range of adjustment of **primary control power**, within which **primary controllers** can provide automatic control, in both directions, in response to a frequency deviation. The concept of the primary control range applies to each generator, each **control area/block** and the entire **synchronous area**.

Primary control reserve: the (positive/negative) part of the **primary control range** measured from the working point prior to the disturbance up to the maximum **primary control power** (taking account of a limiter). The concept of the primary control reserve applies to each generator, each **control area/block** and the entire **synchronous area**.

Productivity is used here as employees per output unit in fixed prices. The 2 per cent increase in productivity used as a basic assumption implies that 2 per cent less people are needed to produce the same output every year. If additional cost reductions of turbines are assumed, this must partly be attributed to additional productivity increases further reducing the need for employees.

Progress ratio is related to the learning rate (see **learning rate**) – if the learning rate is 15 per cent, then the progress ratio is 85 per cent.

PX is a power exchange scheduling coordinator and is independent of system operators and all other market participants.

R

Rated wind speed is the lowest steady wind speed at which a wind turbine can produce its rated output power.

Reactive power is an imaginary component of the apparent power. It is usually expressed in kilo-vars (kVAr) or mega-vars (MVAr). Reactive power is the portion of electricity that establishes and sustains the electric and magnetic fields of alternating-current equipment. Reactive power must be supplied to most types of magnetic equipment, such as motors and transformers, and causes reactive losses on transmission facilities. Reactive power is provided by generators, synchronous condensers or electrostatic equipment such as capacitors and directly influences the electric system voltage. The reactive power is the imaginary part of the complex product of voltage and current.

Reinvestments are the costs of replacing a larger and more costly part of a turbine.

Reliability describes the degree of performance of the elements of the bulk electric system that results in electricity being delivered to customers within accepted standards and in the amount desired. Reliability at the transmission level may be measured by the frequency, duration and magnitude (or the probability) of adverse effects on the electric supply/ transport/generation. Electric system reliability can be addressed by considering two basic and functional aspects of the electric system:

1. *adequacy*: the ability of the electric system to supply the aggregate electrical demand and energy requirements of customers at all times, taking into account scheduled and reasonably expected unscheduled outages of system elements; and
2. *security*: the ability of the electric system to withstand sudden disturbances such as electric short circuits or unanticipated loss of system elements.

Reynolds number: a dimensionless number describing the aerodynamic state of an operating aerofoil. The number is used along with the angle of attack to describe the limits of a particular aerofoil's lift-to-drag ratio and the conditions at which stall occurs. Small wind turbine aerofoils typically operate in a low Reynolds number range, from 0.150 to 0.5 million.

Rural electrification provides a regular supply of electricity to rural residents. It implies the extension of power lines to rural areas, or the use of stand-alone or isolated power systems.

S

Savonius rotor (S-rotor): a simple drag device producing high starting torque developed by the Finnish inventor Sigurd J. Savonius.

SCADA: see **supervisory control and data acquisition system**.

Secondary control is a centralised automatic function to regulate the generation in a **control area**, based on secondary control reserves in order to maintain its interchange power flow at the control programme with all other control areas (and to correct the loss of capacity in a control area affected by a loss of production) and, at the same time, in the case of a major frequency deviation originating from the control area, particularly after the loss of a large generation unit, to restore the frequency to its set value in order to free the capacity engaged by the **primary control** (and to restore the **primary control reserves**). In order to fulfil these functions, secondary control operates by the Network Characteristic Method. Secondary control applies to selected generator sets in the power plants comprising this control loop. Secondary control operates for periods of several minutes, and is therefore dissociated from primary control. This behaviour over time is associated with the PI (proportional-integral) characteristic of the secondary controller.

Secondary control range: the range of adjustment of the secondary control power, within which the secondary controller can operate automatically, in both directions at the time concerned, from the working point of the secondary control power. The positive/negative secondary control reserve is the part of the secondary control range between the working point and the maximum/minimum value. The portion of the secondary control range already activated at the working point is the secondary control power.

Security limits define the acceptable operating boundaries (thermal, voltage and stability limits). The **TSO** must have defined security limits for his own network and must ensure adherence to these security limits. Violation of these limits for a prolonged period of time could cause damage and/or an outage of another element that could cause further deterioration of system operating conditions.

Small-signal stability is the ability of the electric system to withstand small changes or disturbances without the loss of synchronism among the synchronous machines in the system, while having an adequate damping of system oscillations (sufficient margin to the border of stability).

Small wind turbine (SWT): a system with $300\,m^2$ rotor swept area or less that converts kinetic energy in the wind into electrical energy.

Social acceptance: in the energy and technology policy context, this concept refers to the responses of the lay public (including the hosting communities), and of stakeholders, such as industry and non-governmental, governmental and research organisations, to a specific energy innovation. The most recent and comprehensive approach to the social acceptance of renewable energies proposes the 'triangle model', integrating three key dimensions: socio-political acceptance, community acceptance and market acceptance.

Social trust: in technological and risk contexts, this refers to the level of trust individuals have towards organisations and authorities managing technological projects. It is increasingly regarded as a significant element in social reactions to technological developments. Trust can be created in careful, sophisticated decision-making processes that take time, but it can be destroyed in an instant.

Socio-political acceptance refers to the acceptance of both technologies and policies at the most general level. This general level of socio-political acceptance is not limited to the 'high and stable' levels of acceptance by the general public, but includes acceptance by key stakeholders and policymakers.

Stability is the ability of an electric system to maintain a state of equilibrium during normal and abnormal system conditions or disturbances.

Stand-alone systems are electric power systems independent of the network or grid, often used in remote locations where the cost of providing lines from large central power plants is prohibitive.

Static load flow calculations investigate the risk of system overload, voltage instability and **(N-1)-safety** problems. System overload occurs when the transmitted power through certain lines or transformers is above the capacity of these lines or transformers. System static voltage instability may be caused by a high reactive power demand from wind turbines. Generally speaking, a high reactive power demand causes the system voltage to drop.

Sulphur dioxide (SO_2) is a heavy, pungent, colourless gas formed primarily by the combustion of fossil fuels. It is harmful to human beings and vegetation, and contributes to the acidity in precipitation.

Supervisory control and data acquisition system (SCADA) is the wind farm monitoring system which allows the owner and the turbine manufacturer to be notified of faults or alarms, remotely start and stop turbines, and review operating statistics.

Surface roughness (Z_0) is a parameter used to describe the roughness of the surface of the ground.

Synchronous area: an area covered by **interconnected systems**. These systems' **control areas** are synchronously interconnected with the control areas of members of the association. Within a synchronous area the **system frequency** is commonly steady. A certain number of synchronous areas may exist in parallel on a temporary or permanent basis. A synchronous area is a set of synchronously interconnected systems that has no synchronous interconnections to any other interconnected systems.

System frequency is the electric frequency of the system that can be measured in all network areas of the **synchronous area** under the assumption of a coherent value for the system in a timeframe of seconds (with minor differences between different measurement locations only).

T

Tertiary control is any automatic or manual change in the working points of generators (mainly by rescheduling) in order to restore an adequate secondary control reserve at the right time. The power that can be connected automatically or manually under tertiary control in order to provide an adequate secondary control reserve is known as the tertiary control reserve or minute reserve. This reserve must be used in such a way that it contributes to the restoration of the **secondary control range** when required.

Thrust curve: a graph which shows the force applied by the wind at the top of the tower as a function of wind speed.

Tip speed: speed (in m/s) of the blade tip through the air.

Transformer: a piece of electrical equipment used to step up or down the voltage of an electrical signal. Most turbines have a dedicated transformer to step up their voltage output to the grid voltage.

Transient stability is the ability of an electric system to maintain synchronism between its parts when subjected to a disturbance of specified severity and to regain a state of equilibrium following that disturbance.

Transmission is the transport of electricity on the extra-high or high-voltage network (transmission system) for delivery to final customers or distributors. Operation of transmission includes as well the tasks of system operation concerning the management of energy flows, reliability of the system and availability of all necessary system services / **ancillary services**.

Transmission system operator (TSO): a company that is responsible for operating, maintaining and developing the transmission system for a **control area** and its **interconnections**.

Turbine lifetime is the expected total lifetime of the turbine (normally 20 years).

Turbulence intensity measures the 'roughness' of the wind, calculated for a time series of wind speed data, as the standard deviation divided by the mean wind speed.

U

UNEP-GEF: United Nations Environment Programme, Division of Global Environment Facility Coordination.

Unity (or harmony) with the landscape is the degree to which individuals perceive wind turbines to be integrated with the landscape. The perceived impact on landscape seems to be the crucial factor in public attitudes towards wind farms, and opposition to the visual despoliation of valued landscapes has been analysed as the key motivation for opposition to wind farms.

U-shape curve is a model stating that public attitudes towards wind farms change from being very positive, before the announcement of the project, to negative, when the project is announced, to positive again, after the construction. It is related to the familiarity factor, considered a key element in individuals' perception of technological developments.

Utility is the incumbent electricity supplier to end users (usually state-owned at some period), which may own and operate other electricity supply assets, including transmission networks and usually generation plant.

V

Value chain is the set of interconnected activities, consisting of discrete value-adding market segments, that comprise an industry. In the case of the wind energy industry, this may include (but is not restricted to) wind turbine manufacturing, project development, financing, asset ownership, operations and maintenance, and electricity distribution.

Value of statistical life (VSL) is an approach measuring a society's willingness to pay to avoid additional cases of death. This can be seen in spending for improved safety in the aircraft or car industry. In the EU and the US, values of between 1 and 10 million US$ or € per life saved have been found in different studies. Earlier versions of the ExternE project adopted a figure of US$3 million per life saved for VSL calculations. In these calculations the age of a person saved does not matter.

Vertical axis wind turbine (VAWT): a wind turbine with a vertical rotor axis.

W

Wind Atlas Analysis and Application Program (WAsP): a program for predicting wind climate and energy production from wind farms.

Wind farm design tool (WFDT): software to aid in the design and optimisation of a wind farm.

Wind home system (WHS): a wind-based system to provide basic lighting and entertainment needs to an individual home, with a capacity typically in the range of hundreds of watts.

Wind rose: a circular diagram giving a visual summary of the relative amounts of wind available in each of a number of direction sectors (often 12) at a given location, and the speed content of that wind.

Wind shear profile (α): the increase in wind speed with height above ground or sea level.

Wound rotor: a type of synchronous electrical machine in which the magnetic field on the rotor is established by passing a current through coils on the rotor. The alternative is to establish the magnetic field using permanent magnets (see **PMG**).

Wound rotor induction generator (WRIG): see **doubly fed induction generator**.

Y

Years of life lost (YOLL): the YOLL approach takes into account that due to different causes people in very different age groups may be at risk. In the case of a chronic disease leading to the death of very old people, only the years of life lost due to the disease, as compared to the average life expectancy, are taken into account. For each year of life lost, approximately 1/20th of the value of statistical life is used.

Z

Zone of visual influence (ZVI): the land area around a wind farm from which a specified number of wind turbines can be seen. Often presented as a map, with areas coloured depending on the number of turbines which can be seen.

REFERENCES

Ackermann, T. (2005) *Wind Power in Power Systems*, Wiley, United Kingdom

Ackermann, T. (2007) 'Annex to Report of the Grid Connection Inquiry, "Grid Issues for Electricity Production Based on Renewable Energy Sources in Spain, Portugal, Germany and United Kingdom"', Stockholm, November

ADEME (2008) 'ADEME&Vous, Stratégie & Études. Maîtrise de l'énergie et développement des énergies renouvelables', available at www.ademe.fr

AEE (2007) 'Todos los datos, análisis y estadísticas del sector eólico', *Proceedings Eólica 07*, Asociación Empresarial Eólica, available at www.aeeolica.org/doc/AnuarioAEE_Eolica2007esp_n.pdf

Algoso, D. and Rusch, E. (2004) *Renewables Work. Job Growth from Renewable Energy Development in the Mid-Atlantic*, NJPIRG Law and Policy Centre, Trenton, NJ

American Wind Energy Association (AWEA) (2008a) www.awea.org/

American Wind Energy Association (AWEA) (2008b) 'Small wind turbines market survey 2008', available at www.awea.org/smallwind/pdf/2008_AWEA_Small_Wind_Turbine_Global_Market_Study.pdf

Auer, H., Obersteiner, C., Prüggler, W., Weissensteiner, L., Faber, T. and Resch, G. (2007) 'Guiding a least cost grid integration of RES-electricity in an extended Europe', Action Plan, Project GreenNet-Europe, Vienna, Austria

Auer, H. (2008) 'Overview of the main RES-E support schemes for wind energy in the EU-27 Member States', Vienna, Austria

Awerbuch, S. (2003a) 'Determining the real cost – Why renewable power is more cost competitive than previously believed', *Renewable Energy World*, March/April

Awerbuch, S. (2003b) 'The true cost of fossil-fired electricity', *Power Economics*, May

Azar, C. and Sterner, T. (1996) 'Discounting and distributional considerations in the context of global warming', *Ecological Economics*, vol 19, no 11, pp169–184

Baring-Gould, E. I. (2002) 'Worldwide status of wind-diesel applications', Wind-Diesel Workshop report, United States

Barrios, L. and Rodríguez, A. (2004) 'Behavioural and environmental correlates of soaring – Bird mortality at on-shore wind turbines', *Journal of Applied Ecology*, vol 41, pp72–81

Bell, D., Gray, T. and Haggett, C. (2005) 'The "social gap" in wind farm siting decisions: Explanations and policy responses', *Environmental Politics*, vol 14, no 4, pp460–477

Birdlife (2005) 'Birds and Habitats Directive Task Force. Position statement on wind farms and birds', available at www.birdlife.org/eu/pdfs/Windfarm_position08.pdf

Bishop, I. and Miller, D. (2007) 'Visual assessment of offshore wind turbines: The influence of distance, contrast, movement and social variables', *Renewable Energy*, vol 32, pp814–831

Blanco, M. I. and Rodrigues, G. (2008) 'Can the future EU ETS support wind energy investments', *Energy Policy*, vol 36, issue 4, pp1509–1520

BMU (2002) *Vergleich externer Kosten der Stromerzeugung in Bezug auf das Erneuerbare Energien Gesetz*, Endbericht, Berlin, Germany

BMU (2006) 'Renewable energy: Employment effects: Impact of the expansion of renewable energy on the German labour market', Federal Ministry for the Environment, Nature Conservation and Nuclear Safety, available at www.bmu.de/english

BMU (2008) 'Kurz- und langfristige Auswirkungen des Ausbaus der erneuerbaren Energien auf den deutschen Arbeitsmarkt', interim report, Federal Ministry for the Environment, Nature Conservation and Nuclear Safety

Boesen, C. and Kjaer, J. (2005) 'Review report: The Danish offshore wind farm. Demonstration projects: Horns Rev and Nysted offshore wind farms', environmental impact assessment and monitoring, prepared for The Environmental Group by Elsam Engineering and ENERGI E2, 2004

Boston Consulting Group (2004) 'Donner un nouveau souffle à l'éolien terrestre, développement de l'éolien terrestre en France', June 2004, available at www.ventdecolere.org/archives/doc-references/BCG.pdf

Böstrom, M. (2003) 'Environmental organisations in new forms of political participation: Ecological modernisation and the making of voluntary rules', *Environmental Values*, vol 12, no 2, pp175–193

Breukers, S. and Wolsink, M. (2007) 'Wind power implementation in changing institutional landscapes: An international comparison', *Energy Policy*, vol 35, no 5, pp2737–2750

Brusa, A. and Lanfranconi, C. (2006) 'Guidelines for realization of wind plants and their integration in the territory', Italian Association of Renewable Energy Producers, Milano, Paper presented at EWEC 2006.

BTM-Consult (2008) 'World market update 2007', March

BWEA (2005) *Guidelines for Health and Safety in the Wind Energy Industry*, ISBN 978-1-870064-42-2

Capros, P., Antzos, L., Papandreou, V. and Tasios, N. (2008) 'European energy and transport: Trends to 2030 – Update 2007', Report to the European Commission, ISBN 978-92-79-07620-6, Belgium

Carbon Trust/DTI (2004) 'Renewables network impacts study', available at http://www.carbontrust.org.uk/carbontrust/about/Publications/Renewables%20Network%20Study%20Final.pdf

Carlman, I. (1982) 'Wind energy potential in Sweden: The importance of non-technical factors', *Fourth International Symposium on Wind Energy Systems*, 21–24 September, Stockholm, Sweden, pp335–348

Carlman, I. (1984) 'The views of politicians and decision-makers on planning for the use of wind power in Sweden', *European Wind Energy Conference*, 22–36 October, Hamburg, Germany, pp339–343

CIEMAT, Small Wind Turbine Database, available at www.energiasrenovables.ciemat.es/?pid=17000 (in Spanish), accessed June 2008

Coenraads, R., Voogt, M. and Morotz, A. (2006) 'Analysis of barriers for the development of electricity generation from renewable energy sources in the EU-25', OPTRES Report to the European Commission, ECOFYS, Utrecht, Netherlands, May

Coenraads, R., Reece, G., Voogt, M., Ragwitz, M., Held, A., Resch, G., Faber, T., Haas, R., Konstantinaviciute, I., Krivošík, J. and Chadim, T. (2008) *PROGRESS, Promotion and Growth of Renewable Energy Sources and Systems*, Utrecht, Netherlands, 5 March 2008

Commission of the European Communities (2001) 'Directive 2001/77/EC of the European Parliament and of the Council of 27 September 2001 on the promotion of electricity produced from renewable energy sources in the internal electricity market', *Official Journal of the European Communities*, L 283/33, 27 October, http://eur-lex.europa.eu/pri/en/oj/dat/2001/l_283/l_28320011027en00330040.pdf

Commission of the European Communities (2004) 'Communication from the Commission to the Council and the European Parliament: The share of renewable energy in the EU', Commission Report in accordance with Article 3 of Directive 2001/77/EC, evaluation of the effect of legislative instruments and other Community policies on the development of the contribution of renewable energy sources in the EU and proposals for concrete actions, {SEC(2004) 547}, COM(2004) 366 final, Brussels, Belgium, 26 May, http://ec.europa.eu/energy/res/legislation/country_profiles/com_2004_366_en.pdf

Commission of the European Communities (2005) 'Communication from the Commission: The support of electricity from renewable energy sources', {SEC (2005) 1571}, COM(2005) 627 final, Brussels, 7 December, http://ec.europa.eu/energy/res/biomass_action_plan/doc/2005_12_07_comm_biomass_electricity_en.pdf

Commission of the European Communities (2008) 'Proposal for a Directive of the European Parliament and of the Council on the promotion of the use of energy from renewable sources', presented by the Commission, {COM(2008) 30 final}, {SEC(2008) 57}, {SEC(2008) 85}, COM(2008) 19 final, 2008/0016 (COD), Brussels, 23 January 2008, http://ec.europa.eu/energy/res/legislation/doc/strategy/res_directive.pdf

CORDIS, Community Research and Development Information Center, 'Seventh Framework Programme (FP7)', available at http://cordis.europa.eu/fp7/home_en.html

Council Directive 79/409/EEC of 2 April 1979 on the conservation of wild birds, Official OJ L103/1, 25 April 1979

Council Directive 92/43/EEC of 21 May 1992 on the conservation of natural habitats and of wild fauna and flora, OJ L 206, 22 July 1992

Cowell, R. (2007) 'Wind power and the planning problem: The experience of Wales', *European Environment*, vol 17, no 5, pp291–306

Czisch, G. (2001) 'Global renewable energy potential – Approaches to its use', *Regenerative Energiequellen: Schulung für polnische Führungskräfte*, Magdeburg, April/September, www.iset.uni-kassel.de/abt/w3-w/folien/magdeb030901/

Danish Energy Authority, 'Wind turbines – Introduction and basic facts', www.ens.dk/sw14294.asp, accessed August 2008

DEA (2006) 'Offshore wind farms and the environment: Danish experiences from Horns Rev and Nysted', Danish Energy Authority

De Lucas, M., Janss, G. F. E. and Ferrer, M. (eds) (2007) *Birds and Wind Farms: Risk Assessment and Mitigation*, Quercus, Madrid, Spain

Del Río Gonzalez, P. (2006) 'Harmonisation versus decentralization in the EU ETS. An economic analysis', *Climate Policy*, vol 6, no 4, pp457–475

DENA (2005) 'Planning of the grid integration of wind energy in Germany, onshore and offshore up to the year 2020', DENA grid study, Deutsche Energie-Agentur, March, available at www.dena.de/themen/thema-reg/projektarchiv/

Department for Trade and Industry, Marine Consents and Environment Unit (2004) 'Guidance notes: Offshore wind farm contents process', London, August, www.berr.gov.uk/files/file22990.pdf

Desholm, N., Fox, A. D., Beasley, P. D. L. and Kahlert, J. (2006) 'Remote techniques for counting and estimating the number of bird-wind turbine collisions at sea: A review', *Ibis*, vol 148, pp76–89

Devine-Wright, P. (2005) 'Beyond NIMBYism: Towards an integrated framework for understanding public perceptions of wind energy', *Wind Energy*, vol 8, pp125–139

Devine-Wright, P. and Devine-Wright, H. (2006) 'Social representations of intermittency and the shaping of public support for wind energy in the UK', *International Journal of Global Energy Issues*, vol 25, nos 3–4, pp243–256

Dewi (2002) 'Studie zur aktuellen Kostensituation 2002 der Windenergienutzung in Deutschland' (draft), Deutsches Windenergie-Institut Gmbh, July

Dirksen, S., Spaans, A. L. and Widen, J. v. D. (2007) 'Collision risks for diving ducks at semi-offshore wind farms in freshwater lakes: A case study', in M. de Lucas, G. Janss and M. Ferrer (eds), *Birds and Wind Farms. Risk Assessment and Mitigation*, Quercus, Madrid, Spain

DOE/GO-102008-2567 (2008) '20% wind energy by 2030: Increasing wind energy's contribution to US electricity supply', available at www.nrel.gov/docs/fy08osti/41869.pdf, accessed June 2008

Dowling, P. and Hurley, B. (2004) 'A strategy for locating the least cost wind energy sites within an EU electrical load and grid infrastructure perspective', EWEC, London, UK

Drewitt, A. L. and Langston, R. H. W. (2006) 'Assessing the impacts of wind farms on birds', *Ibis*, vol 148, pp29–42

DTI (2002) 'Wind energy and aviation interest interim guidelines', *Wind Energy*, Defence and Civil Aviation Interest Working Group, DTI, Ministry of Defence, Civil Aviation Authority and BWEA.

DTI (2006) 'The measurement of low frequency noise at three UK wind farms', available at www.berr.gov.uk/files/file31270.pdf

DWIA (2008) 'Environmental and employment benefits of wind, 2008', Danish Wind Industry Association, available at www.windpower.org

ECON Pöyry – MSC WIND (2008) 'A multi-client study – Implications of large-scale wind power in Northern Europe', February 2008

Edge, G. and Blanchard, L. (2007) 'Delivering offshore wind power in Europe', European Wind Energy Association (EWEA), available at www.ewea.org/fileadmin/ewea_documents/images/publications/offshore_report/ewea-offshore_report.pdf

EEA (2008) 'Energy and environment report 2008', available at http://reports.eea.europa.eu/eea_report_2008_6/en/

EEG (2007) 'Green-X data base', Energy Economics Group

Ekraft, Eltra (2004) 'Wind turbines connected to grids with voltages above 100kV: Technical regulation for the properties and the regulation of wind turbines', Denmark, November, www.energinet.dk/NR/rdonlyres/E4E7A0BA-884F-4E63-A2F0-98EB5BD8D4B4/0/WindTurbinesConnectedtoGridswithVoltageabove100kV.pdf

ELINFRASTRUKTURUDVALGET (2008) 'Technical report on the future expansion and undergrounding of the electricity transmission grid: Summary', Denmark

Ellerman, A. D., Buchner, B. and Carraro, C. (2007) *Allocations in the European Emissions Trading Scheme. Rights, Rents and Fairness*, Cambridge Ed. Cambridge

Ellis, G., Barry, J. and Robinson, C. (2007) 'Many ways to say "no", different ways to say "yes": Applying Q-Methodology to understand public acceptance of wind farm proposals', *Journal of Environmental Planning and Management*, vol 50, no 4, pp517–551

Eltham, D. C., Harrison, G. P. and Allena, S. J. (2008) 'Change in public attitudes towards a Cornish wind farm: Implications for planning', *Energy Policy*, vol 36, pp23–33

EREC (2007) *New Renewable Energy Target for 2020 – A Renewable Energy Roadmap for the EU*, European Renewable Energy Council, Brussels, Belgium, available at www.erec.org

Erickson, W., Johnson, G. and Young, D. (2002) 'Summary of anthropogenic causes of bird mortality', Third International Partners in Flight Conference, 20–24 March, Asilomar, CA, United States, available at www.dialight.com/FAQs/pdf/Bird%20Strike%20Study.pdf

Erickson, W. P., Johnson, G. D. and Young, D. P. (2005) 'A summary and comparison of bird mortality from anthropogenic causes with an emphasis on collisions', USDA Forest Gen. Tech. Rep. PSW-GTR-191. 2005, available at http://www.fs.fed.us/psw/publications/documents/psw_gtr191/Asilomar/pdfs/1029-1042.pdf

Erlich, I., Winter, W. and Dittrich, A. (2006) 'Advanced grid requirements for the integration of wind turbines into the German transmission system', IEEE/IEE, Montreal/Canada, available at http://ieeexplore.ieee.org

Estanqueiro, A., Rui, C., Flores, P., Ricardo, J., Pinto, M., Rodrigues, R. and Peças Lopes, J. (2008) 'How to prepare a power system for 15% wind energy penetration: The Portuguese case study', *Wind Energy*, vol 11, no 1, pp75–84

Eurelectric (2006) 'Statistics and prospects for the European electricity sector (1980–1990, 2000–2030) – EURPROG', Brussels, Belgium, December

European Commission DG XII, Science Research and Development (1994) 'Externalities of fuel cycles – ExternE Project, summary report', JOULE, EUR 16521 EN, Brussels, Belgium, and Luxembourg

European Commission DG XII, Science Research and Development (1995a) 'Externalities of fuel cycles – ExternE Project, Volume 2: Methodology', JOULE, EUR 16521 EN, Brussels, Belgium and Luxembourg

European Commission DG XII, Science Research and Development (1995b) 'Externalities of fuel cycles – ExternE Project, Volume 3: Coal and lignite', JOULE, EUR 16521 EN, Brussels, Belgium and Luxembourg

European Commission (2001) 'Directive 2001/77/EC of the European Parliament and of the Council on the promotion of electricity produced from renewable energy sources in the internal electricity market', 27 September

European Commission (2003) 'External costs research results on socio-environmental damages due to electricity and transport external cost', ExternE Project, Brussels, Belgium

European Commission (2006a) 'Accompanying document to the Communication from the Commission to the Council and the European Parliament: Renewable Energy Road Map: Renewable energies in the 21st century: Building a more sustainable future. Impact assessment', Commission Staff Working Document, COM (2006) 848 final

European Commission (2006b) 'European energy and transport: Scenarios on energy efficiency and renewables'

European Commission (2006c) 'Attitudes towards energy', Special Eurobarometer 247/Wave 64.2

European Commission (2007a) 'Communication from the Commission to the Council, the European Parliament, the European Economic and Social Committee, and the Committee of the Regions – A European Strategic Energy Technology Plan (SET-Plan)' COM (2007) 723 final

European Commission (2007b) 'Monitoring industrial research: The 2007 EU industrial R&D investment scoreboard'

European Commission (2007c) 'Energy technologies: Knowledge, perception, measures', Special Eurobarometer 262, Wave 65.3 – TNS Opinion & Social

European Commission DG TREN (2007) 'Renewables make the difference', available at http:// ec.europa.eu/energy/climate_actions/doc/brochure/2008_res_brochure_en.pdf

European Commission DG TREN (2008) 'European energy and transport: Trends to 2030 – update 2007',

available at: http://ec.europa.eu/dgs/energy_transport/figures/trends_2030_update_2007/energy_transport_trends_2030_update_2007_en.pdf

European Commission, DG RTD (2006) 'The state and prospects of European energy research: Comparison of Commission, Member and Non-Member States' R&D Portfolios', EUR 22397

European Commission, DG TREN (2007) 'Intelligent energy Europe, renewable energy projects – Renewable electricity', http://ec.europa.eu/energy/intelligent/projects/elec_en.htm

European Commission, DG TREN (2007) 'Intelligent Energy Europe, project fact sheet: Renewable electricity supply interactions with conventional power generation, networks and demand (RESPOND)', http://ec.europa.eu/energy/intelligent/projects/doc/factsheets/respond.pdf

European Commission, DG RTD (2005) 'Key tasks for future European energy R&D: A first set of recommendations for research and development by the Advisory Group on Energy', EUR 21352

European Commission (2008) 'ExternE: Externalities of energy – A research project of the European Commission', Project website at www.externe.info

European Parliament DG IPOL: Economic and Scientific Policy (2007) 'Employment potential of renewable forms of energy and increased efficiency of energy use', briefing note IP/A/EMPL/FWC/2006-03/SC3

European Wind Energy Technology Platform (2008) 'Strategic research agenda, market deployment strategy from 2008 to 2030', European Wind Energy Technology Platform, Brussels, available at www.windplatform.eu

Eurostat, 'Official 2006 Eurostat statistics of the European Commission, EU Energy in Figures: Pocket Book 2007/2008', available online at DG TREN website of the European Commission, http://ec.europa.eu/dgs/energy_transport/figures/pocketbook/2007_en.htm

Eurostat (2007a) 'Statistical Office of the European Communities – Labour market: Structural indicators', available at http://epp.eurostat.ec.europa.eu/cache/ITY_SDDS/EN/eb031_base.htm

Eurostat (2007b) 'The life of women and men in Europe: A statistical portrait', Statistical Office of the European Communities, available at http://epp.eurostat.ec.europa.eu/portal/page?_pageid=1073,46587259&_dad=portal&_schema=PORTAL&p_product_code=KS-80-07-135

Eurostat (2008a) 'Job vacancy statistics, annual data', Statistical Office of the European Communities

Eurostat (2008b) 'The life of men and women in Europe: Statistical portrait', Statistical Office of the European Communities

EWEA (2003) 'Survey for Wind Energy – The Facts', European Wind Energy Association, Brussels, Belgium

EWEA (2005a) 'Prioritising wind energy research – Strategic research agenda of the wind energy sector', European Wind Energy Association, Brussels, Belgium

EWEA (2005b) 'Large-scale integration of wind energy in the European power supply: Analysis, issues and recommendations', European Wind Energy Association, Brussels, Belgium.

EWEA (2006a) 'Focus on 2030: EWEA aims for 22% of Europe's electricity by 2030', Wind Directions, European Wind Energy Association, Brussels, Belgium, November/December

EWEA (2006b) 'No fuel: Wind power without fuel', European Wind Energy Association briefing, available at www.ewea.org/

EWEA (2007) 'Internal paper: Administrative and grid barriers', European Wind Energy Association, July

EWEA (2008a) 'Pure power – Wind energy scenarios up to 2030, final report', European Wind Energy Association, Brussels, March, available at www.ewea.org/fileadmin/ewea_documents/documents/publications/reports/purepower.pdf

EWEA (2008b) 'The employment report', internal document, European Wind Energy Association

EWEA (2008c) 'Wind at work: wind energy and job creation in the EU', European Wind Energy Association, available at www.ewea.org

EWEA and Greenpeace (2005) 'Wind Force 12: A blueprint to achieve 12% of the world's electricity from wind power by 2020', available at www.ewea.org/fileadmin/ewea_documents/documents/publications/WF12/wf12-2005.pdf

Exo, K. M., Hüppop, O. and Garthe, S. (2003) 'Birds and offshore wind farms: A hot topic in marine ecology', Wader Study Group Bulletin, vol 100, pp50–53

ExternE-Pol (2005) 'Externalities of energy – Extension of accounting framework and policy applications', report to the European Commission (DG RTD), Contract No ENG1-CT2002-00609, Coordination: ARMINES/Ecole des Mines de Paris

Firestone, J. and Kempton, W. (2007) 'Public opinion about large offshore wind power: Underlying factors', Energy Policy, vol 35, pp1584–1598

Forsyth, T. and Baring-Gould, I. (2007) 'Distributed wind market applications', National Renewable Energy Laboratory, United States

Fox, A. D., Desholm, M., Kahler, J., Christensen, T. K. and Petersen, I. K. (2006) 'Information needs to support environmental impact assessment of the effects of European marine offshore wind farms on birds', Ibis, vol 148, pp129–144

Friedrich, R. et al. (1989) 'Externen Kosten der Stromerzeugung', Universität Flensburg, Flensburg/Germany

Friedrich, R. et al. (2004) 'Final report of New-Ext: New elements for the assessment of external costs from energy technologies', European Commission FP5 Project No ENG1-CT2000-00129, Germany

Frischknecht R., Jungbluth N., Althaus H.-J., Doka, G., Dones, R., Hellweg S., Hischier R., Humbert S., Margni M., Nemecek, T. and Spielmann M. (2007) 'Implementation of life cycle impact assessment methods', final report Ecoinvent v2.0 No. 3, Swiss Centre for Life Cycle Inventories, Duebendorf.

García, L. (2007) 'Onshore wind and environmental impact requirements: The experience of ACCIONA', European Wind Energy Conference EWEC 2007, Milan, Italy

Giebel, G. (2005) 'Wind power has a capacity credit – A catalogue of 50+ supporting studies', *WindEng EJournal*, available at www.windeng.net

Gill, A. B. (2005) 'Offshore renewable energy: Ecological implications of generating electricity in the coastal zone', *Journal of Applied Ecology*, vol 42, pp605–615

Gipe, P. (2004) *Wind Power*, James and James, London

Global Wind Energy Council (2008) 'Global wind 2007 report', Brussels, Belgium, pp11–14

Global Wind Energy Council and Greenpeace (2008) 'Global wind energy outlook 2008', Brussels, Belgium

Greenpeace (2005) 'Offshore wind: Implementing a new powerhouse for Europe, grid connection, environmental impact and political framework', Brussels, Belgium

Gross, C. (2007) 'Community perspectives of wind energy in Australia: The application of a justice and community fairness framework to increase social acceptance', *Energy Policy*, vol 35, pp2727–2736

Hecklau, J. (2005) 'Visual characteristics of wind turbines', *Proceedings of NWCC Technical Considerations in Siting Wind Developments*, available at www. nationalwind.org/events/siting/proceedings.pdf

Held, A. (2008) 'PROGRESS: Identification of administrative and grid barriers to the promotion of electricity from renewable energy sources (RES)', Fraunhofer Institute for Systems and Innovation Research, Karlsruhe, Germany

Hepburn, H. and Edworthy, J. (2005) 'Infrasound from wind turbine: Observations from Castle River wind farm', Canadian Wind Energy Conference, Toronto/Canada

Hohmeyer, O. (1988a) *Social costs of energy consumption. External effects of electricity generation in the Federal Republic of Germany*, Springer, Berlin/West

Hohmeyer, O. (1988b) 'Systematic presentation of the clean air policy in individual OECD member states', Country reports, Karlsruhe, Germany

Hohmeyer, O. (1996) 'Social costs of climate change, strong sustainability and social costs', in O.

Hohmeyer et al. (eds), *Social Costs and Sustainability: Valuation and Implementation in the Energy and Transport Sector*, Springer-Verlag, Berlin, Germany

Hohmeyer et al. (eds) (1996) *Social Costs and Sustainability: Valuation and Implementation in the Energy and Transport Sector*, Springer-Verlag, Berlin, Germany

Hohmeyer et al. (eds) (2000) 'Chance Automausstieg – Perspektiven für neue Arbeitsplätze an Atomstandorten', Greenpeace report, Hamburg, Germany

Holttinen, H. (2004a) 'Impacts of hourly wind variations on the system operation in Nordic Countries', European Wind Energy Conference 2004, London, UK

Holttinen, H. (2004b) 'The impact of large scale wind power production on the Nordic electricity system', VTT Publications 554, VTT Processes, Espoo, Finland, www.vtt.fi/inf/pdf/publications/2004/P554.pdf

Holttinen, H. et al. (2007) 'State of the art IEA Task 25 report: Design and operation of power systems with large amounts of wind power', VTT Working Papers 82, VTT Technical Research Centre of Finland, Espoo

Horlick-Jones, T., Walls, J., Rowe, G., Pidgeon, N., Poortinga, W., Murdock, G. and O'Riordan, T. (2007) *The GM Debate: Risk, Politics and Public Engagement*, Routledge, London, UK

Howard, M. and Brown, C. (2004) 'Results of the electromagnetic investigations and assessments of marine radar, communications and positioning systems undertaken at the North Hoyle wind farm', QinetiQ and the Maritime and Coastguard Agency

Hunt, G. (1998) 'Raptor floaters at Moffat's equilibrium', *Oikos*, vol 82, no 1, pp191–197

Hüppop, O., Dierschke, J., Exo, K. M., Fredrich, E. and Hill, R. (2006) 'Bird migration studies and potential collision risk with offshore wind turbines', *Ibis*, vol 148, pp90–109

IDEA (2005) 'Plan de energías renovables en España 2005–2010', www.mityc.es/NR/rdonlyres/C1594B7B-DED3-4105-96BC-9704420F5E9F/0/ResumenPlanEnergiasRenov.pdf

IEA, Energy R&D Statistics Database, available at www.iea.org/Textbase/stats/rd.asp

IEA (1999) Annex XI, IEC 6-1400 Part 12, available at www.measnet.com

IEA (1999, second print 2003) 'Expert study group on recommended practices for wind turbine testing and evaluation, Part XI: Wind speed measurement and use of cup anemometry'

IEA (2001) 'Long-term research and development needs for the time frame 2000 to 2020', IEA R&D Wind Executive Committee, PWT Communications, Boulder US

IEA (2005a) 'Offshore wind experiences', available at www.iea.org/Textbase/Papers/2005/offshore.pdf

IEA (2005b) Annex XI, IEC 6-1400 Part 12, available at http://www.ieawind.org/AnnexXXV.html

IEA (2006) Annual Report 2006, International Energy Agency, Paris, France

IEA Wind (2007a) 'Implementing agreement for cooperation in the research, development, and deployment of wind energy systems: Proposal for a new task. Social acceptance of wind energy projects: Winning hearts and minds', October, www.ieawind.org/iea_wind_pdf/New_Task_Social_Acceptance_29_10_07.pdf

IEA (2007b) Wind Energy, Annual Report, International Energy Agency, Paris, France

IEA (2007c) World Energy Outlook, International Energy Agency, Paris

IEA (2008) 'Recabs-model, developed in the IEA implementing agreement on renewable energy technology deployment', http://recabs.iea-retd.org/energy_calculator

IEC (2003) 'Wind turbine generator systems. Part 11: Acoustic noise measurement techniques', International Electrotechnical Commission 61400-11, document No. 88/166/FDIS, Switzerland

IEC (2005) 'Wind turbines – Part 12-1: Power performance measurements of electricity-producing wind turbines', International Electrotechnical Commission 61400-12:1

IEC (2006) 'Design requirements for small wind turbines' International Electrotechnical Commission 61400-2

IEC (2008) 'Wind turbine generator systems – Part 21: Measurement and assessment of power quality characteristics of grid connected wind turbines, Ed. 2.0', International Electrotechnical Commission 61400-21, FDIS

IFC (2007) 'Environmental, health and safety guidelines for wind energy', International Finance Corporation

ISET (2004) 'Wind energy report Germany 2004', Institut für Solare Energieversorgungstechnik, Universität Kassel, Germany

ISO (1996) 'Acoustics – Attenuation of sound during propagation outdoors – Part 2: General method of calculation', International Organization for Standardization 9613 Part 2

Jacobsen, J. (2005) 'Infrasound emission from wind turbines', Journal of Low Frequency Noise, Vibration and Active Control, vol 24, no 3, pp145–155

Jensen, P. H., Morthorst, P. E., Skriver, S., Rasmussen, M. et al. (2002) 'Economics of wind turbines in Denmark – investment and operation and maintenance costs for selected generations of turbines', Risø, Denmark

Jobert, A., Laborgne, P. and Mimler, S. (2007) 'Local acceptance of wind energy: Factors of success identified in French and German case studies', Energy Policy, vol 35, pp2751–2760

Johansson, M. and Laike, T. (2007) 'Intention to respond to local wind turbines: The role of attitudes and visual perception', Wind Energy, vol 10, no 5, pp435–451

Jones, C. (2005) Applied Welfare Economics, ISBN 978-0-19-928197-8, The Australian National University

Junfeng, L. and Hu, G. (2007) 'China Wind Power Report', China Environmental Science Press, Beijing, available at http://www.gwec.net/uploads/media/wind-power-report.pdf

Kammen, D., Kapadia, K. and Fripp, M. (2004) 'Putting renewables to work: How many jobs can the clean industry generate?', Renewable and Appropriate Energy Laboratory report, University of California, Berkeley CA/United States, available at http://socrates.berkeley.edu/~rael/papers.html

Kempton, W., Firestone, J., Lilley, J., Rouleau, T. and Whitaker, P. (2005) 'The off-shore wind power debate. Views from Cape Cod', *Coastal Management*, vol 33, pp119–149

Köller, J., Köppel, J. and Peters, W. (eds) (2006) *Offshore Wind Energy. Research on Environmental Impacts*, Berlin, Germany

Krohn, S. and Damborg, S. (1999) 'On public attitudes towards wind power', *Renewable Energy*, vol 16, pp954–960

Kuehn, S. (2005) 'Sociological investigation of the reception of Horns Rev and Nysted Offshore wind farms in the local communities', ECON Analyse, March, www.hornsrev.dk/Engelsk/Miljoeforhold/uk-rapporter.htm

Kulisik, B., Loizou, E., Rozakis, S. and Segon, V. (2007) 'Impacts of biodiesel production on Croatian economy', *Energy Policy*, vol 35, pp6036–6045

Kunreuther, H., Slovic, P. and MacGregor, D. (1996) 'Risk perception and trust: Challenges for facility siting', *Risk: Health, Safety and the Environment*, vol 7, pp109–118

Ladenburg, J. (2008) 'Attitudes towards on-land and offshore wind power development in Denmark: Choice of development strategy', *Renewable Energy*, vol 33, pp111–118

Ladenburg, J. (in press) 'Visual impact assessment of offshore wind farms and prior experience', *Applied Energy*

Landberg, L., Giebel, G., Nielsen, H. A., Nielsen, T. and Madsen, H. (2003) 'Short-term prediction – An overview', *Wind Energy*, vol 6, pp273–280

Lee, R. (1996) 'Externalities studies: Why are the numbers different', in O. Hohmeyer et al. (eds) (1996) *Social Costs and Sustainability*, Springer-Verlag, Berlin

Lehr, U., Nitsch, J., Kratzat, M., Lutz, C. and Edler, D. (2008) 'Renewable energy and employment in Germany', *Energy Policy*, vol 36, pp108–117

Lekuona, J. M. and Ursúa, C. (2007) 'Avian mortality in wind power plants of Navarra (Northern Spain)' in M. de Lucas, G. Janss and M. Ferrer (eds) *Birds and Wind Farms. Risk Assessment and Mitigation*, Quercus, Madrid, Spain

Leontief, W. (1986) *Input-Output Economics* (second edition), Oxford University Press, New York, United States

Leventhall, G. (2003) 'A review of published research on low frequency noise and its effects', report for the Department for Environment, Food and Rural Affairs, London, UK

Lindfors, L.-G. et al. (1995) 'Nordic guidelines on life-cycle assessment', Nord 1995:20, Nordic Council of Ministers, Copenhagen

Loring, J. (2007) 'Wind energy planning in England, Wales and Denmark: Factors influencing project success', *Energy Policy*, vol 35, pp2648–2660

Madsen, H., Pinson, P., Kariniotakis, G., Nielsen, H. and Nielsen, T. (2005) 'Standardising the performance evaluation of short term wind power prediction models', *Wind Engineering*, vol 29, no 6, pp475–489

Marbek Resource Consultants Ltd and GPCo Inc. (2005) 'Survey of the small (300W to 300kW) wind turbine market in Canada', submitted to Natural Resources Canada

Marco, J. M., Circe, T. and Guillermo, G. (2007) 'Towards determination of the wind farm portfolio effect based on wind regimes dependency analysis', World Wind Energy Conference, October 2007, Mar del Plata, Argentina

Martf, I. et al. (2000) 'First results of the application of a wind energy prediction model in complex terrain', EWEA Special Topic Conference, Kassel, Germany

Matthies, H.G. and Garrad, A. (1993) 'Study of offshore wind energy in the EC' JOULE 1 (JOUR 0072), Report to the European Commission, Germanischer Lloyd, Hamburg, Germany

MEASNET (2006) 'Measurement procedures', International Measuring Network of Wind Energy

Institutes, available at www.measnet.com/document. html, accessed September 2008

Milborrow, D. J. (2003) *Wind Power Monthly Magazine*, April

Milligan, M. and Kirby, B. (2008) 'Analysis of sub-hourly ramping impacts of wind energy and balancing area size', presented at WindPower 2008, Houston, TX, United States

Moehrlen, C. S. et al. (2000) 'Wind power prediction and power plant scheduling in Ireland', EWEA Special Topic Conference, Kassel, Germany

Moorhouse, A., Hayes, M., von Hünerbein, S., Piper, B. and Adams, M. (2007) 'Research into aerodynamic modulation of wind turbine noise: Final report', University of Salford, UK, available at http://usir. salford.ac.uk/1554/

Mora, D. and Hohmeyer, O. (2005) 'External cost of electricity generation systems: Final report', Re-Xpansion project funded by the European Commission (ALTENER-2002-054), Brussels, Belgium

Mortensen, N. G., Heathfield, D. N., Myllerup, L., Landberg, L. and Rathmann, O. (2007) *Wind Atlas Analysis and Application Program, WAsP 9 Help Facility,* ISBN 978-87-550-3607-9

Murley, A. (2008) 'BWEA annual SWT market report', presentation at All Energy 2008, available at www. all-energy.co.uk/userfiles/file/alew-murley210208.pdf

National Research Council (1996), in P. Stern and H. Fineberg (eds) *Understanding Risk: Informing Decisions in Democratic Society*, National Academies Press, Washington, DC

National Research Council (2007) *Environmental Impacts of Wind-Energy Projects*, National Academies Press, Washington, DC

Nayak, D. R., Miller, D., Nolan, A., Smith, P. and Smith, J. (2008) 'Calculating carbon energy savings from wind farms on Scottish peat lands. A new approach', available at www.scotland.gov.uk/Publications/ 2008/06/25114657/0

Neij, L. (1997) 'Use of experience curves to analyse the prospects for diffusion and adaptation of renew-

able energy technologies', *Energy Policy*, vol 25, no 13, pp1099–1107

Neij, L. et al. (2003) 'Final report of EXTOOL: Experience curves, a tool for energy policy programmes assessment', European Commission FP5 project No. ENG1-CT2000-00116, Lund, Sweden

Nordel (2002) 'Nordel common balance management in the Nordic Countries', Nordic Transmission System Operators

OECD/IEA (2004) *Renewable Energy, Market and Policy Trends in IEA Countries*, OECD and International Energy Agency, Paris, France

OECD/IEA (2007) *World Energy Outlook 2007*, OECD and International Energy Agency, Paris

Ottinger, R. L. et al. (1990) *Environmental Cost of Electricity*, Pace University Centre of Environmental Legal Studies, Oceana Publications Inc, New York, United States

Parkhill, T. (2007) 'Tensions between Scottish national policies for onshore wind energy and local dissatisfaction – Insights from regulation theory', *European Environment*, vol 17, no 5, pp307–320

Parsons, B. et al. (2003) 'Grid impacts of wind power: A summary of recent studies in the United States', National Renewable Energy Laboratory, Conference document of the European Wind Energy Conference 2003, Madrid, Spain

Pedden, M. (2005) 'Analysis: Economic impacts of wind applications in rural communities', National Renewable Energy Laboratory, Subcontract No NREL-SE-500-3909

Pedersen, E. and Person Waye, K. (2004) 'Perception and annoyance due to wind turbine noise – A dose–response relationship', *Journal of Acoustical Society of America*, vol 116, no 6, pp3460–3470

Pedersen, E. and Person Waye, K. (2007) 'Wind turbine noise, annoyance and self-reported health and well-being in different living environments', *Occupational and Environmental Medicine*, vol 64, no 7, pp480–486, available at www.websciences.org/cftemplate/ NAPS/archives/indiv.cfm?ID=20066545

Pedersen, E. and Person Waye, K. (2008) 'Wind turbines – Low level noise sources interfering with restoration?', *Environmental Research Letters*, vol 3

Percival, S. M. (2003) 'Birds and wind farms in Ireland: A review of potential issues and impact assessment', Ecology Consulting

Percival, S. M. (2007) 'Predicting the effects of wind farms on birds in the UK: The development of an objective assessment method', in M. de Lucas, G. Janss and M. Ferrer (eds) *Birds and Wind Farms. Risk Assessment and Mitigation*, Quercus, Madrid

Pfaffenberger, W., Jahn, K. and Djourdjin, M. (2006) 'Renewable energies – Environmental benefits, economic growth and job creation', case study paper, Bremer Energie Institut, Germany

Platts (2008) UDI World Electric Power Plants Database Europe, available at www.platts.com

Poortinga, W. and Pidgeon, N. F. (2004) 'Trust, the asymmetry principle, and the role of prior beliefs', *Risk Analysis*, vol 24, no 6, pp1475–1486

Poortinga, W. and Pidgeon, N. F. (2006) 'Prior attitudes, salient value similarity, and dimensionality: Toward an integrative model of trust in risk regulation', *Journal of Applied Social Psychology*, vol 36, no 7, pp1674–1700

Prades López, A., Horlick-Jones, T., Oltra, C. and Solá, R. (2008) 'Lay perceptions of nuclear fusion: Multiple modes of understanding', *Science and Public Policy*, vol 35, no 2, pp95–105

Prades, A. and González Reyes, F. (1995) 'La percepción Social del Riesgo: Algo más que discrepancia expertos/público', *Revista de la Sociedad Española de Protección Radiológica – Radioprotección*, vol 3, no 10

Raftery, P., Tindal, A. J. and Garrad, A. D. (1997) 'Understanding the risks of financing wind farms', *Proceedings of European Wind Energy Conference*, Dublin, Ireland

Raftery, P., Tindal, A. J., Wallenstein, M., Johns, J., Warren, B. and Dias Vaz, F. (1999) 'Understanding the risks of financing wind farms', *Proceedings of European Wind Energy Conference*, Nice, France

Ragwitz, M. et al. (2007) *Assessment and Optimisation of Renewable Energy Support Schemes in the European Electricity Market*, Fraunhofer IRB Verlag, Germany

RECS International (2005) 'The use of the guarantee of origin', evaluation report, Renewable Energy Certificate System, available at www.recs.org

Redlinger, R. Y., Dannemand Andersen, P. and Morthorst, P. E. (2002) *Wind Energy in the 21st Century*, Palgrave Macmillan, United Kingdom

REN21 (2008) 'Renewables 2007 global status report', Energy Policy Network for the 21st Century, REN21 Secretariat, Paris, and Worldwatch Institute, Washington, DC

Rennings, K. (1996) 'Economic and ecological concepts of sustainable development: External costs and sustainability indicators', in O. Hohmeyer et al. (eds), *Social Costs and Sustainability*, Springer-Verlag, Berlin, Germany

Resch, G., Faber, T., Ragwitz, M., Held, A., Panzer, C. and Haas, R. (2008) 'Futures-e Recommendation report: 20% RES by 2020 – A balanced scenario to meet Europe's renewable energy target', European Commission Intelligent Energy – Europe Project Futures-e EIE/06/143/SI2.444285, Vienna, Austria

Rogers, G. (1998) 'Siting potentially hazardous facilities: What factors impact perceived and acceptable risk?', *Landscape and Urban Planning*, vol 39, pp265–281

Rohrig, K. et al. (2004) 'New concepts to integrate German offshore wind potential into electrical energy supply', ISET, Universität Kassel, given at the European Wind Energy Conference (EWEC 2004), London, UK

Rubio, M. J. and Varas, J. (1999) *'El análisis de la realidad en la intervención social. Métodos y técnicas de investigación social'*, Editorial CCS, Madrid, Spain

Schuman, H. and Stanley, P. (1996) *Questions and Answers in Attitude Surveys: Experiments on Question Form, Wording, and Context* (reprint edition), SAGE Publications, Thousand Oaks, CA, United States

Scott, K. E., Anderson, C., Dunsford, H., Benson, J. F. and MacFarlane, R. (2005) 'An assessment of the sensitivity and capacity of the Scottish seascape in relation to offshore windfarms', Scottish Natural Heritage Commissioned Report No 103 (ROAME No 03AA06)

Scottish Executive Central Research Unit (2000) 'Public Attitudes towards Wind Farms in Scotland', Scottish Executive, Edinburgh, United Kingdom

Scottish Executive (2002) 'Planning Advice Note 45: Renewable Energy Technologies', Edinburgh, United Kingdom, available at http://www.scotland.gov.uk/Publications/2002/02/pan45/pan-45

Scottish Government (2008) 'The economic impacts of wind farms on Scottish tourism. A report for the Scottish Government', Glasgow Caledonian University, Moffatcentre and Cogentsi, Glasgow, United Kingdom, available at www.scotland.gov.uk/Resource/Doc/214910/0057316.pdf

SDC (2005) 'Wind power in the UK. A guide to the key issues surrounding onshore wind power development in the UK', Sustainable Development Commission, London, UK

SEA (2004) 'The electromagnetic compatibility and electromagnetic field implications for wind farming in Australia', Sustainable Energy Australia

Sempreviva, A. M., Barthelmie, R., Giebel, G., Lange, B. and Sood, A. (2003) '2FP6 "POW'WOW" coordination action project', available at http://powwow.risoe.dk/publ/SemprevivaBarthelmieGiebelLangeSood-OffshoreWindResourceAssessment_444_Ewec2007fullpaper.pdf

Sengupta, D. and Senior, T. (1983) 'Large wind turbine sitting handbook: Television interference assessment', final subcontract report

Simon, A. M. (1996) 'A summary of research conducted into attitudes to wind power from 1990 to 1996', British Wind Energy Association, available at www.bwea.com/ref/surveys-90-96.html

Skytte, K. (2008) 'Implication of large-scale wind power in northern Europe', Presentation at EWEC 08, Brussels, available at http://ewecproceedings.info/

Slovic, P. (1993) 'Perceived risk, trust, and democracy', *Risk Analysis*, vol 13, pp675–682

Smith, J. C., Parsons, B., Acker, T., Milligan, M., Zavadil, R., Schuerger, M. and DeMeo, E. (2007) 'Best practices in grid integration of variable wind power: Summary of recent US case study results and mitigation measures', *Proceedings of the 2007 European Wind Energy Conference*, Milan, Italy, May

Soder, L. et al. (2007) 'Experience from wind integration in some high penetration areas', *IEEE Transactions on Energy Conversion*, vol 22, no 1, http://ieeexplore.ieee.org/xpl/freeabs_all.jsp?isnumber=4105991&arnumber=4106017&count=29&index=1

Stanton, C. (2005) 'Visual impacts. UK and European perspectives', *Proceedings of NWCC Technical Considerations in Siting Wind Developments*, available at www.nationalwind.org/events/siting/proceedings.pdf

Strbac, G. and Bopp, T. (2004) 'Value of fault ride through capability for wind farms', report to Ofgem, July, accessible at www.sedg.ac.uk

Szymczak, G. (2007) 'How should the problem of wind-farm connection be resolved in Poland? Examples of procedures for obtaining connection offers in EU countries', *Wokół Energetyki Journal*, Polish Wind Energy Association, April

Tande, J. O., Muljadi, E., Carlson, O., Pierik, J., Estanqueiro, A., Sørensen, P., O'Malley, M., Mullane, A., Anaya-Lara, O. and Lemstrom, B. (2004) 'IEA wind annex XXI: Dynamic models of wind farms for power system studies', European Wind Energy Conference (EWEC 2004), London, UK, available at www.energy.sintef.no/wind/iea_dynamic_models_EWEC'04_paper.pdf

Tande, J. O. G., Korpås, M., Warland, L., Uhlen, K. and Van Hulle, F. (2008) 'Impact of TradeWind offshore wind power capacity scenarios on power flows in the European HV network', Seventh International Workshop on Large-Scale Integration of Wind Power and on Transmission Networks for Offshore Wind Farms, Madrid, Spain, May

Thayer, R. and Freeman, C. M. (1987) 'Altamont: Public perceptions of a wind energy landscape', *Landscape and Urban Planning*, vol 14, pp379–398

Thomsen, F., Lüdemann, K., Kafemann, R. and Piper, W. (2006) *Effects of offshore wind farm noise on marine mammals and fish biola*, Hamburg, Germany, on behalf of COWRIE Ltd., available at www.offshorewind.co.uk

Toke, D. (2005) 'Explaining wind power planning outcomes: Some findings from a study in England and Wales', *Energy Policy*, vol 33, no 12, pp1527–1539

Toke, D., Breukers, S. and Wolsink, M. (2008) 'Wind power deployment outcomes: How can we account for the differences?', *Renewable and Sustainable Energy Reviews*, vol 12, no 4, pp1129–1147

Troen, I. and Lundtang Petersen, E. (1989) *European Wind Atlas*, Risø National Laboratory, ISBN 978-87-550-1482-4

Tsoutsos, T., Gouskos, Z., Karterakis, S. and Peroulaki, E. (2006), European Wind Energy Conference, Athens, Greece

UCTE, 'System adequacy forecast 2007–2020', Union for the Co-ordination of Transmission of Electricity, available at www.ucte.org/_library/systemadequacy/saf/UCTE_SAF_2007-2020.pdf

UCTE, 'Transmission development plan 2008', Union for the Co-ordination of Transmission of Electricity, available at www.ucte.org/_library/otherreports/tdp08_report_ucte.pdf

Ummels, B. C. et al. (2008) 'Energy Storage Options for System Integration of Offshore Wind Power in the Netherlands', *Proceedings of European Wind Energy Conference, Brussels, Belgium, 3–7 April*

UNEP, ILO and ITUC (2007) 'Green jobs: Towards sustainable work in a low-carbon world', preliminary report, Green Jobs Initiative, WorldWatch Institute, UNEP

UNEP Risø CDM/JI (2008) 'Pipeline analysis and database', available at http://cdmpipeline.org/cdm-projects-type.htm

United Nations Framework Convention on Climate Change (UNFCCC) (2006)

University of Newcastle (2002) 'Visual assessment of windfarms: Best practice', Scottish Natural Heritage Commissioned Report F01AA303A

Vereinigung Deutscher Elektrizitätswerke (2000) '*VDEW-Statistik 1998 – Leistung und Arbeit*', VWEW Energieverlag, Frankfurt, Germany (CD-ROM)

Wahlberg, M. and Westerberg, H. (2005) 'Hearing in fish and their reactions to sounds from offshore wind farms', *Marine Ecology Progress Series*, vol. 288, pp295–309

Walker, G. (1995) 'Renewable energy and the public', *Land Use Policy*, vol 12, no 1, pp49–59

Warren, C. R., Lumsden, C., O'Dowd, S. and Birnie, R. V. (2005) 'Green on green: Public perceptions of wind power in Scotland and Ireland', *Journal of Environmental Planning and Management*, vol 48, no 6, pp853–875

Watkiss, P., Downing, T., Handley, C. and Butterfield, R. (2005) 'The impacts and costs of climate change', final report to DG Environment, Brussels, Belgium, September

Weisberg, H. F., Krosnick, J. A. and Bowen, B. D. (1996) *An Introduction to Survey Research, Polling, and Data Analysis* (third edition), Sage, Thousand Oaks, CA, United States

Whiteley, O. et al. (2004) 'MITRE Project overview report: Meeting the targets and putting renewable energies to work', European Commission DG TREN Altener project, available at http://mitre.energyprojects.net/

Wolsink, M. (1988) 'The social impact of a large wind turbine', *Environmental Impact Assessment Review*, vol 32, no 8, pp324–325

Wolsink, M. (1989) 'Attitudes and expectancies about wind turbines and wind farms', *Wind Engineering*, vol 13, no 4, pp196–206

Wolsink, M. (1994) 'Entanglement of interests and motives: Assumptions behind the NIMBY-theory on facility siting', *Urban Studies*, vol 31, no 6, pp851–866

Wolsink, M. (1996) 'Dutch wind power policy – Stagnating implementation of renewables', *Energy Policy*, vol 24, no 12, pp1079–1088

Wolsink, M. (2000) 'Wind power and the NIMBY-myth: Institutional capacity and the limited significance of public support', *Renewable Energy*, vol 21, pp49–64

Wolsink, M. (2007) 'Wind power implementation: The nature of public attitudes: Equity and fairness instead of "backyard motives"', *Renewable and Sustainable Energy Reviews*, vol 11, pp1188–1207

Woyte, A., Gardner, P. and Snodin, H. (2005) 'COD work package 8: Grid issues', European Commission FP5 Project COD NNE5-2001-00633, Ireland, available at www.offshorewindenergy.org/cod/CODReport_Grid.pdf

Woyte, A. et al. (2008) *A North Sea Electricity Grid (r) Evolution: Electricity Output of Interconnected Offshore Wind Power Generation in the North Sea. A Vision on Offshore Wind Power Integration*, Greenpeace, Brussels, Belgium

Wratten, A., Martin, S., Welstead, J., Martin, J., Myers, S., Davies, H. and Hobson, G. (2005) *The Seascape and Visual Impact Assessment Guidance for Offshore Wind Farm Developers*, Enviros Consulting and Department of Trade and Industry

Wüstenhagen, R., Wolsink, M. and Bürer, M. J. (2007) 'Social acceptance of renewable energy innovation: An introduction to the concept', *Energy Policy*, vol 35, pp2683–2691

REFERENCES FOR TABLE I.2.1

[1]   Dobesch, H. and Kury, G. (1997) 'Wind atlas for the Central European Countries Austria, Croatia, Czech Repulic, Hungary, Slovak Republic and Slovenia', *Österreichische Beiträge zu Meteorologie und Geophysik*, Heft 16, Zentralanstalt für Meteorologie und Geodynamik, Vienna, Austria

[2]   Risø EMD (1999) 'Wind resource atlas for Denmark', available at www.windatlas.dk/World/DenmarkWRA.html, accessed September 2008

[3]   Petersen, E. L., Troen, I., Frandsen, S. and Hedegaard, K. (1981) *Wind Atlas for Denmark. A Rational Method for Wind Energy Siting*, Risø-R-428, Risø National Laboratory, Roskilde, Denmark

[4]   Tammelin, B. (1991) *Suomen Tuuliatlas. Vind Atlas för Finland* (Wind Atlas for Finland) (in Finnish and Swedish), Finnish Meteorological Institute, Helsinki, Finland

[5]   Traup, S. and Kruse, B. (1996) *Wind und Windenergiepotentiale in Deutschland. Winddaten für Windenergienutzer* (in German), Selbstverlag des Deutschen Wetterdienstes, Offenbach am Main, Germany

[6]   Foussekis, D., Chaviaropoulos, P., Vionis, P., Karga, I., Papadopoulos, P. and Kokkalidis, F. (2006) 'Assessment of the long-term Greek wind atlas', Centre for Renewable Energy Sources, *Proceedings of the European Wind Energy Conference 2006, Athens, Greece*

[7]   Watson, R. and Landberg, L. (2001) 'The Irish wind atlas', in P. Helm and A. Zervos (eds) *Wind Energy for the New Millennium, Proceedings of 2001 European Wind Energy Conference and Exhibition (EWEC), Copenhagen*, WIP Renewable Energies, Munich, Germany, pp894–897

[8]   ESBI Consultants and TrueWind Solutions (2003) 'Project report, Republic of Ireland wind atlas 2003', Sustainable Energy Ireland, Dublin, June, Report No 4Y103A-1-R1, available at www.sei.ie/uploadedfiles/RenewableEnergy/IrelandWindAtlas2003.pdf (accessed September 2008)

[9]   Podesta, A. et al. (2002) 'The wind map of Italy', presented at Eurosun 2002, Bologna, Italy

[10] Botta, G., Casale, C., Lembo, E., Maran, S., Serri, L., Stella, G., Viani, S., Burlando, M., Cassola, F., Villa, L. and Ratto, C. F. (2007) 'The Italian wind atlas – status and progress', in *Proceedings of European Wind Energy Conference 2007*, EWEA, Brussels, Belgium, available at www.ewec2007proceedings.info/index.php (accessed September 2008)

[11] Krieg, R. (1992) *'Vindatlas för Sverige'* (Wind Atlas for Sweden) (in Swedish), Slutrapport på projekt 506 269-2 på uppdrag av NUTEK, Norrköping; see also Krieg, R. (1999) *'Verifiering af beräknad vind-energiproduktion (Verification of estimated wind power productions)'* (in Swedish), SMHI rapport No 28, Norrköping, Sweden

[12] Burch, S. F. and Ravenscroft, F. (1992) 'Computer modelling of the UK wind energy resource: Overview report', Energy Technology Support Unit Report WN7055, UK Department for Business Enterprise and Regulatory Reform

[13] Vector, A. S. (2003) 'Norwegian wind atlas', available at www.windsim.com/wind_energy/wind_atlas/index.html, accessed September 2008

[14] Ivanov, P., Sabeva, M. and Stanev, S. (1982) *Reference Book for PR of Bulgaria, Volume IV: Wind*, Nauka I Izkustvo Publishing House, Sofia, Bulgaria, available at http://ebrdrenewables.com/sites/renew/countries/Bulgaria/profile.aspx, accessed September 2008

[15] Rathmann, O. (2003) *The UNDP/GEF Baltic Wind Atlas*, Risø-R-1402(EN), Risø National Laboratory, Roskilde, Denmark

[16] Kull, A. and Steinrücke, G. (1996) *Estonia Wind Atlas*, University of Tartu, Institute of Geography, Tartu, Estonia

[17] 'Estonia Wind Map' (no date), available at http://130.226.17.201/extra/web_docs/wind maps/estland.jpg, accessed September 2008

[18] 'Latvian Wind Map', Latvian Association of Wind Energy, available at www.windenergy.lv/en/karte.html, accessed September 2008

[19] Sander and Partner GmbH (2004), 'Wind atlas Poland', available at www.sander-partner.ch/be/en/Polen/index.html, accessed September 2008

[20] ICEMENERG (1993) 'Wind atlas of Romania', available at www.ebrdrenewables.com/sites/renew/countries/Romania/profile.aspx, accessed September 2008

[21] Dündar, C., Canbaz, M., Akgün, N. and Ural, G. (2002) *'Türkiye Rüzgar Atlasi'* (Turkish Wind Atlas), Turkish State Meteorological Service and General Directorate of Electrical Power Resources Survey and Development Administration, available at www.meteoroloji.gov.tr/2006/arastirma/arastirma-arastirma.aspx?subPg=107&Ext=htm

[22] Elliott, D., Schwartz, M., Scott, G., Haymes, S., Heimiller, D. and George, R. (2003) *Wind Energy Resource Atlas of Armenia*, NREL/TP-500-33544, available at www.nrel.gov/docs/fy03osti/33544.pdf, accessed September 2008

[23] Gelovani, M., Chikvaidze, G., Eristavi, V., Lobdjanidze, N., Rogava, S., Rishkov, M., Sukhishvili, E., Tusishvili, O., Zedginidze, A. and Zedginidze, I. (2004) *Wind Energy Atlas of Georgia. Volume I: Regional Estimations*, edited by A. Zedginidze, advisor L. Horowicz, Karenergo Scientific Wind Energy Center, Tbilisi, Georgia, ISBN 978-99928-0-910-5

[24] Starkov, A. N., Landberg, L., Bezroukikh, P. P. and Borisenko, M. M. (2000) *Russian Wind Atlas*, ISBN 978-5-7542-0067-8, Russian-Danish Institute for Energy Efficiency, Moscow, and Risø National Laboratory, Roskilde, Denmark

[25] Suisse Éole, *Wind Energie Karte der Schweiz*, available at http://wind-data.ch/windkarte, accessed September 2008

INDEX

Page numbers in *italics* refer to figures.